# Insect Pathogens

## Molecular Approaches and Techniques

# Insect Pathogens
## Molecular Approaches and Techniques

Edited by

**S. Patricia Stock**

*Department of Entomology*
*University of Arizona*
*USA*

**John Vanderberg**

*USDA-ARS*
*US Plant, Soil and Nutrition Laboratory*
*USA*

**Noël Boemare**

*Institut National de la Recherche Agronomique (INRA)*
*Université Montpellier*
*France*

*and*

**Itamar Glazer**

*Agricultural Research Organization*
*The Volcani Centre*
*Israel*

www.cabi.org

**CABI is a trading name of CAB International**

CABI Head Office
Nosworthy Way
Wallingford
Oxfordshire OX10 8DE
UK

Tel: +44 (0)1491 832111
Fax: +44 (0)1491 833508
E-mail: cabi@cabi.org
Website: www.cabi.org

CABI North American Office
875 Massachusetts Avenue
7th Floor
Cambridge, MA 02139
USA

Tel: +1 617 395 4056
Fax: +1 617 354 6875
E-mail: cabi-nao@cabi.org



A catalogue record for this book is available from the British Library, London, UK.

**Library of Congress Cataloging-in-Publication Data**

Insect pathogens : molecular approaches and techniques/edited by S. Patricia Stock... [et al.].
p. cm.
Includes bibliographical references and index.
ISBN 978-1-84593-478-1 (alk. paper)
1. Insects--Pathogens. 2. Insects--Molecular aspects. I. Stock, S. Patricia. II. Title.

SB942.I57 2009
632'.7--dc22

2008031290

ISBN-13: 978 1 84593 478 1

Typeset by SPi, Pondicherry, India.
Printed and bound in the UK by MPG Books Group.

The paper used for the text pages in this book is FSC certified. The FSC (Forest Stewardship Council) is an international network to promote responsible management of the world's forests.

# Contents

# Contributors

**Byron J. Adams**, *Department of Microbiology and Molecular Biology, Brigham Young University, Provo, UT 84602-5253, USA.*

**Serap Aksoy**, *Department of Epidemiology and Public Health, Section of Vector Biology, Yale University School of Medicine, LEPH 606, 60 College Street, New Haven, CT 06510, USA.*

**Basil Arif**, *Great Lakes Forestry Centre, Sault Ste Marie, Ontario, Canada.*

**Geoffrey M. Attardo**, *Department of Epidemiology and Public Health, Section of Vector Biology, Yale University School of Medicine, LEPH 606, 60 College Street, New Haven, CT 06510, USA.*

**Noël Boemare**, *Institut National de la Recherche Agronomique (INRA), UMR1133 Laboratoire EMIP, F-34095 Montpellier, France; Université Montpellier II, UMR 1133 Laboratoire EMIP, F-34095 Montpellier, France.*

**John P. Burand**, *Division of Entomology, Department of Plant Soil and Insect Science, University of Massachusetts, Amherst, Massachusetts, USA.*

**Louela A. Castrillo**, *Department of Entomology, Cornell University, Ithaca, NY 14853, USA.*

**Todd A. Ciche**, *Michigan State University, East Lansing, Michigan, USA.*

**David Clarke**, *University College Cork, Ireland.*

**Daniel Doucet**, *Great Lakes Forestry Centre, Sault Ste Marie, Ontario, Canada.*

**Eric Duchaud**, *INRA, UR892, Unité Virologie et Immunologie Moléculaires, F-78350 Jouy-en-Josas, France.*

**Martin A. Erlandson**, *Agriculture and Agri-Food Canada, Saskatoon Research Centre, 107 Science Place, Saskatoon, SK S7N 0X2, Canada.*

**Shannon Escasa**, *Great Lakes Forestry Centre, Sault Ste Marie, Ontario, Canada.*

**Brian A. Federici**, *Department of Entomology and Interdepartmental Graduate Programs in Genetics, Genomics & Bioinformatics and Cell, Molecular & Developmental Biology. University of California, Riverside, CA 92521, USA.*

**Sophie Gaudriault**, *INRA and Université Montpellier II, UMR1133 Laboratoire EMIP, F-34000 Montpellier, France.*

**Heidi Goodrich-Blair**, *University of Wisconsin, Madison, Wisconsin, USA.*

**Parwinder S. Grewal**, *Ohio State University, Wooster, Ohio, USA.*

**Quan Guoxing**, *Great Lakes Forestry Centre, Sault Ste Marie, Ontario, Canada.*

**Richard A. Humber**, *USDA-ARS, Robert W. Holley Center for Agriculture and Health, Ithaca, NY 14853, USA.*

**Bradley C. Hyman**, *Department of Biology, University of California, Riverside, CA 92521, USA.*

**Hinanit Koltai**, *Department of Ornamental Horticulture, ARO, Volcani Center, Bet Dagan, 50250, Israel.*

**Tim Ladd**, *Great Lakes Forestry Centre, Sault Ste Marie, Ontario, Canada.*

**Madoka Nakai**, *Graduate School of Agriculture, Tokyo University of Agriculture and Technology, Fuchu, Tokyo, Japan.*

**Miroslav Oborník**, *University of South Bohemia, Faculty of Sciences, Department of Molecular Biology, and Biology Centre of the Academy of Sciences of the Czech Republic, Institute of Parasitology, Branišovská 31, 370 05 České Budějovice, Czech Republic.*

**Hyun-Woo Park**, *John A. Mulrennan Sr Public Health Entomology Research and Education Center, College of Engineering Sciences, Technology and Agriculture, Florida A&M University, Panama City, FL 32405, USA.*

**Scott M. Peat**, *Department of Microbiology and Molecular Biology, Brigham Young University, Provo, UT 84602-5253, USA.*

**Stephen A. Rehner**, *USDA, ARS Systematic Mycology and Microbiology Laboratory, Beltsville, MD 20705-2350, USA.*

**Jeffrey Slack**, *Great Lakes Forestry Centre, Sault Ste Marie, Ontario, Canada.*

**Ian Smith**, *Nara Institute of Science and Technology, Ikoma, Nara, Japan.*

**Raymond J. St Leger**, *Department of Entomology, University of Maryland, College Park, MD 20742, USA.*

**S. Patricia Stock**, *Department of Entomology, University of Arizona, Tucson, AZ 85721-0036, USA.*

**Patrick Tailliez**, *Institut National de la Recherche Agronomique (INRA), Unité d'Ecologie Microbienne des Insectes et Interactions Hôte-Pathogène (EMIP/UMR1133), Université Montpellier II, Place Eugène Bataillon, Case courrier 54, Bâtiment 24, 3ème étage, 34095 Montpellier CEDEX 5, France.*

**David A. Theilmann**, *Agriculture and Agri-Food Canada, Pacific Agri-Food Research Centre, 4200 Highway 97 – Summerland, British Columbia, V0H 1Z0, Canada.*

**Chengshu Wang**, *Institute of Plant Physiology and Ecology, Shanghai Institutes for Biological Sciences, Chinese Academy of Sciences, Shanghai 200032, China.*

**Brian L. Weiss**, *Department of Epidemiology and Public Health, Section of Vector Biology, Yale University School of Medicine, LEPH 606, 60 College Street, New Haven, CT 06510, USA.*

**Zhen Li**, *Great Lakes Forestry Centre, Sault Ste Marie, Ontario, Canada.*

# Preface

Like other natural enemies, insect pathogens, including bacteria, viruses, fungi, protozoa and nematodes, have demonstrated to be environmentally safe and economical alternatives for control of a wide range of arthropod pests. Because of their documented efficacy and the advantages they offer in comparison with chemical pesticides, interest in these organisms has increased exponentially over the past decades. Furthermore, their ability to reduce insect populations has also been the compelling force to further investigate their biology, physiology, ecology and molecular biology. In this respect, molecular technology has played a crucial role in many aspects related with the field of insect pathology, leading to the development of methods and techniques that have had an impact on both basic and applied research beyond the scope of insect pathology.

In this book, we have attempted to bring together a broad array of molecular techniques and approaches currently used in insect pathology. We hope to facilitate their consideration, development and success by a wide audience including students, educators and scientists in academic, private and governmental sectors.

This book is divided into four parts: (i) Identification and Diagnostics; (ii) Evolutionary Relationships and Population Genetics; (iii) Host–Pathogen Interactions; and (iv) Genomics and Genetic Engineering. Sixteen chapters have been written by leading researchers in the field which provide comprehensive and up-to-date information on each part.

Part I: *Identification and Diagnostics*

The first five chapters of this book present an overview of the current state of taxonomy and systematics of different entomopathogens. Methods and techniques considered for collection, preservation of organisms for subsequent nucleic acid extraction and markers used for the identification and/or diagnosis of these organisms are described.

Part II: *Evolutionary Relationships and Population Genetics*

This part includes three chapters discussing different approaches for assessing evolutionary relationships among three major groups of entomopathogens: fungi, bacteria and nematodes. Methods and markers currently considered or with potential application in phylogenetics, population genetics and DNA bar coding are presented. Analytical tools (software) relevant to their application in evolutionary and population genetics studies with entomopathogens are also summarized.

Part III: *Host–Pathogen Interactions*

The three chapters of this part focus on how molecular approaches have contributed to the evolving understanding of insect–pathogen interactions. They summarize and provide an overview of the central role played by these interactions in the overall progression of the disease including insect immune response to infection, pathogen development in the host and interactions between insect–host and microbial flora–entomopathogens.

Part IV: *Genomics and Genetic Engineering*

The five chapters included in this part review methods for engineering and genetic transformation of entomopathogens. Approaches and methods considered in this part also include those for the study of genomes including utilization of the genomic sequence information in functional genomics and comparative genomics.

This book is especially timely as it provides current methods and approaches for diagnosis and identification of entomopathogens. It also summarizes molecular techniques applied for the study of population genetics and evolutionary relationships of entomopathogens themselves and in relation with their arthropod hosts. Moreover, it includes the latest information on methods used to unravel pathogen–host interactions and for the study of their genomes and strains improvement (GMOs).

Some techniques and/or methods might be replaced in the future, but most of the methods included in this book should remain as classic molecular approaches and should provide a foundation for studies in the future. The challenge is great and there is still a need for expanded research and collaboration between scientists among different disciplines. Our hope is that this book will be viewed as a primer of molecular techniques in insect pathology and that it will stimulate further research for the improvement of entomopathogens in pest management and also for their consideration as biological model systems.

We wish to take the opportunity to express our gratitude to all contributing authors and the numerous reviewers who participated and/or contributed with valuable comments in the creation of this book. We are also grateful to CABI for its patience and to our families for their endless support and understanding.

**S. Patricia Stock**
**John Vandenberg**
**Noël Boemare**
**Itamar Glazer**

# Glossary of Terms

**Amplification:** Selective replication of a gene to produce more than the normal single copy in a haploid genome.

**Auxotroph:** An organism that requires one or more substances in addition to minimal medium.

**cDNA library:** A collection of cDNA molecules that were generated *in vitro* from mRNA of a single type of cell population.

**Codon:** A sequence of three adjacent nucleotides encoding an amino acid or termination of translation.

**Colony hybridization:** A procedure for selecting a bacterial clone containing a gene of interest. DNAs from a large number of clones are simultaneously tested with a labelled probe that hybridizes to the gene of interest.

**Constitutive:** A constitutively expressed gene is always turned on.

**Electrophoresis:** Separation of charged molecules, such as DNA, RNA and proteins, in an electric field.

**Exon:** A region of a gene that is ultimately represented in that gene's mature transcript.

**Gene:** The fundamental unit of heredity; it contains the information for making one RNA.

**Gene expression:** The process by which gene products are made.

**Genome:** A haploid set of chromosomes of an organism.

**Heterokaryon:** A cell containing two or more nuclei of different origin.

**Homologous recombination:** Recombination that requires extensive sequence similarity between the recombining DNAs.

**Hybridization:** A process in which double-stranded structures are formed from two nucleotides (either DNA or RNA) from different sources.

**Hybridization stringency:** The combination of factors (temperature, salt, organic solvent, detergent, etc.) that influences the ability of two polynucleotide strands to hybridize.

**Intron:** A region that interrupts the transcribed part of a gene; an intron is transcribed, but is removed by splicing during maturation of the transcript.

**Karyotype:** A pictorial representation of all the chromosomes in a given organism.

**Kilobase pair:** One thousand base pairs.

**Klenow fragment:** A fragment of DNA polymerase I, created by cleaving with protease, that lacks the 5'–3' exonuclease activity.

**Knockout:** Inactivation of specific genes.

**Lambda phage:** A phage of *Escherichia coli*; it can replicate lytically or lysogenically.

**Marker:** A gene or mutation that serves as a signpost at a known location in the genome.

**Microarray:** Also called a gene chip or a DNA chip; microarrays consist of large numbers of molecules (often, but not always, DNA) distributed in rows in a very small space. Microarrays permit scientists to study gene expression by providing a snapshot of all the genes that are active in a cell at a particular time.

**Open reading frame (ORF):** A reading frame that is uninterrupted by translation stop codons.

**Phage:** A virus capable of infecting and multiplying in bacteria.

**Plaque:** A clearing zone that a phage makes on a layer of growing bacterial cells by infecting and either killing or slowing their growth.

**Plasmid:** A circular DNA that replicates independently of the cell's DNA.

**Primer:** A polynucleotide sequence that serves as a growing point for polymerization.

**Promoter:** A region of DNA to which RNA polymerase binds prior to initiation of translation.

**Prophage:** Inactive state of a bacteriophage maintained as a part of chromosomal DNA in a host cell.

**Recombinant DNA:** The product of recombination between two or more fragments of DNA; it can either occur naturally or can be performed *in vitro*.

**Restriction endonucleases:** A class of enzymes that recognizes specific base sequences in DNA and cuts at those specific sites.

**Restriction fragment:** A piece of DNA cut from a larger DNA by the action of restriction endonucleases.

**Restriction map:** A map that shows the locations of restriction sites in a region of DNA.

**RNA interference (RNAi):** Post-transcriptional gene silencing in which double-stranded RNA mediates the destruction of messenger RNAs in a sequence-specific fashion.

**Stop codon:** One of the three codons (UAG, UAA and UGA) that code for the termination of translation.

**Transformation:** The genetic modification induced by the incorporation of a foreign DNA into a cell.

# I Identification and Diagnostics

# 1 Molecular Approaches to Virus Characterization and Detection

M.A. Erlandson[1] and D.A. Theilmann[2]

[1]*Agriculture and Agri-Food Canada, Saskatoon Research Centre, Saskatoon, Canada;* [2]*Agriculture and Agri-Food Canada, Pacific Agri-Food Research Centre, Summerland, Canada*

## 1.1. Introduction

Insects are associated with viruses in a variety of pathogenic relationships, including vectors of plant and animal viruses, and also in very specialized symbiotic relationships with insect parasitoids in which virus gene products are responsible for suppression of host-insect immune systems allowing insect parasitoids to fully develop (e.g. polydnaviruses). Historically, many insect viruses were initially described and characterized by entomologists working on specific insect groups or pest species. Initial characterizations were based on symptomology and observations of pathology at the level of resolution of light microscopy. Evans and Shapiro (1997) provide an excellent overview of symptomology and microscopy techniques used to identify insect viruses. Not surprisingly, most early work focused on those virus groups associated with impressive epizootics in insect populations (baculoviruses), those having unique and readily observable symptomology (baculoviruses and iridescent viruses) and those able to be detected by light microscopy as a consequence of being occluded in relatively large protein crystals (baculoviruses, cypoviruses and entompoxviruses) within infected host cells.

## 1.2. Current Virus Taxonomy and Classification

The current taxonomic classification of viruses is outlined in the Eighth Report of the International Committee on Virus Taxonomy (ICTV) (Fauquet *et al.*, 2005), as well as the ICTV Taxonomy and Index to Virus Classification, and the Nomenclature Taxonomic Lists and Catalogue of Viruses that can be found on the National Centre for Biotechnology Information (NCBI) web site (http://www.ncbi.nlm.nih.gov/ICTVdb/Ictv/index.htm). The ICTV catalogue can be searched in various ways and lists 22 virus families whose hosts include invertebrates for at least some members of the group. A tremendous diversity of viruses representing at least 14 virus families has been reported to be associated with invertebrates to at least some degree as pathogens (Fauquet *et al.*, 2005) (Table 1.1).

The working definition of virus species established by Van Regenmortel (1990) is 'a polythetic class of viruses that constitutes a replicating lineage and occupies a particular ecological niche'. The polythetic nature of the virus species requires that viruses within a species have several characteristics in common but they may not share a single defining characteristic (Ball, 2005). In contrast, higher viral taxa such as genera and family are 'universal' groupings in which specific characteristics are common to all members. Some of the characteristics used to classify viruses in taxonomic groups include: morphology (virion size and shape, presence or absence envelope); biochemical properties such as type and form of nucleic acid constituting the virus genome, protein complement, and lipid and carbohydrate content; genome organization and replication strategy; and finally biological properties such as host range, mode of transmission and tissue tropisms.

Details of key characteristics which define the various virus families are clearly laid out in the Eighth Report of the ICTV (Fauquet *et al.*, 2005). In the following section, we will simply highlight some of the key features of the major

**Table 1.1.** Virus families associated with insects.

| Family/genus | Nucleic acid | Virion shape | Occlusion body | Related viruses in Vertebrates | Related viruses in Plants |
|---|---|---|---|---|---|
| ***Baculoviridae*** | | | | | |
| *Alphaviridae* | dsDNA, circular | Rod | + | None | None |
| *Betabaculovirus* | dsDNA, circular | Rod | + | None | None |
| *Gammabaculovirus* | dsDNA, circular | Rod | + | None | None |
| *Deltabaculovirus* | dsDNA, circular | Rod | + | None | None |
| ***Poxviridae*** | | | | | |
| *Entomopoxvirus* | dsDNA, linear | Ovoid | + | Vaccine virus | None |
| ***Iridoviridae*** | | | | | |
| *Iridovirus* and *Chloridovirus* | dsDNA, linear | Isometric | – | Frog virus 3 | None |
| ***Ascoviridae*** | | | | | |
| *Ascovirus* | dsDNA, circular | Rod–ovoid | – | None | None |
| ***Polydnaviridae*** | | | | | |
| *Bracovirus* | dsDNA, m-circular | Rod | – | None | None |
| *Ichnovirus* | dsDNA, m-circular | Fusiform | – | None | None |
| ***Parvoviridae*** | | | | | |
| *Densovirus* | ssDNA | Isometric | – | Canine parvo virus | None |
| ***Reoviridae*** | | | | | |
| *Cypovirus* (cytoplasmic polyhedrosis) | dsRNA, linear, ten segments | Isometric | + | Reovirus (e.g. blue tongue) | Rice dwarf Wound tumour virus |
| ***Tetraviridae*** | | | | | |
| *Betatetravirus* | ssRNA+, 1-linear | Isometric | – | None | None |
| *Omegatetravirus* | ssRNA+, 2-linear | Isometric | – | None | None |
| ***Dicistroviridae*** | | | | | |
| *Cripavirus* | ssRNA+, linear | Isometric | – | None | Many |
| ***Nodaviridae*** | | | | | |
| *Alphanodavirus* | ssRNA+, 2-linear | Isometric | – | None | None |
| ***Picornaviridae*** | ssRNA+ | Isometric | – | Polio | Many |
| ***Rhabdoviridae*** | ssRNA– | Bullet-shaped | – | Rabies | Many |

virus families infecting insects, and the virus families will be grouped for convenience based on nucleic acid type without implying any phylogenetic relationship among the virus families in each section. Later in the chapter, we will review some of the molecular techniques currently used for virus identification at the genus and species level and by extension in diagnostic techniques for tracking virus infection and persistence in the environment.

There is a relatively large group of unassigned viruses that infect invertebrates (Fauquet *et al.*, 2005). These are viruses whose key characteristics do not readily fit those of existing virus genera placed within families. Many of the unassigned viruses are small ribonucleic acid (RNA) viruses (SRV) from *Drosophila* or honeybees while others are well-characterized viruses such as Oryctes rhinoceros virus and Heliothis virus 1, which were once thought to be associated with the baculoviruses (Fauquet *et al.*, 2005). The number of unassigned viruses speaks for the dynamic and complex nature of virus taxonomy.

### 1.2.1. Double-stranded DNA viruses

*Baculoviridae*: viruses in the family *Baculoviridae* have been well characterized due to their potential as biological control agents and their recent use as eukaryotic expression vector systems. The family is characterized by rod-shaped enveloped virions (hence 'baculo') whose genome consists of a single covalently closed, circular, double-stranded deoxyribonucleic acid (dsDNA) molecule of 80–180 kb (Theilmann *et al.*, 2005; Jehle *et al.*, 2006a). Typically two virion phenotypes occur in baculovirus infections, the occlusion-derived virion (ODV) that is occluded in a crystalline protein matrix during the final stages of virus replication in the nucleus of infected cells, and the budded virion (BV) that is produced as nucleocapsids bud through the plasma membrane of infected cells. BVs are responsible for virus spread to tissues throughout the host producing a systemic infection, and ODVs are responsible for horizontal transmission of baculovirus subsequently initiating virus infection in the insect midgut upon ingestion of the virus occlusion body (OB). The family *Baculoviridae* contains four genera, *Alpha-*, *Beta-*, *Gamma-* and *Deltabaculovirus*. The *Alpha-*, *Gamma-* and *Deltabaculovirus* (formally known collectively as the genus *Nucleopolyhedrovirus* [NPV]) OB are polyhedral-shaped, 0.15–15 µm in diameter and contains multiple virions. In contrast, the *Betabaculovirus* formerly known as the genus *Granulovirus* [GV] OB is capsule-shaped, 0.3 × 0.5 µm in diameter and contains a single virion. To date, *Betabaculovirus* has been isolated only from Lepidoptera, whereas *Alpha-*, *Gamma-* and *Deltabaculovirus* has been described from Lepidoptera, Hymenoptera and Diptera. Baculovirus or baculovirus-like particles have also been reported from crustaceans such as shrimp but the data are limited and further studies need to be done.

Species designations within *Baculoviridae* are based on host range and restriction endonuclease (REN) profiles of genomic DNA, and increasingly DNA sequence analysis is used to differentiate species (Theilmann *et al.*, 2005). Historically, baculovirus species have been named on the basis of the host from which they were isolated. Thus, the type species for NPV, *Autographa californica MNPV* (AcMNPV), was originally isolated from the alfalfa looper, *A. californica*, and there are currently at least seven strains of this virus including *Anagrapha falcifera MNPV* (AnfaNPV) and *Trichoplusia ni MNPV* (TnMNPV) isolated from celery and cabbage looper hosts, respectively. Many of the earliest characterized NPV isolates were designated as 'multiple nucleopolyhedrovirus' (MNPV) or 'single nucleoployhedrovirus' (SNPV) based on whether multiple or single nucleocapsids occur in each enveloped ODV; *Trichoplusia ni SNPV* (TnSNPV) is an example of the latter. However, the M and S designations have no phylogenetic significance. There are currently over 28 NPVs characterized to the extent to be recognized species. The type species for *Betabaculovirus*, *Cydia pomonella GV* (CpGV), was originally isolated from codling moth, *C. pomonella*, larvae. At present, there are over 14 recognized species of *Betabaculovirus*. However, virus taxonomy is a dynamic system and as more baculovirus genome sequences are completed and phylogenetic studies done on key genes the genus descriptions within *Baculoviridae* may need adjustment (Jehle *et al.*, 2006b). Phylogenetic studies of key baculovirus genes such as *lef-8*, *lef-9* and *polyhedrin*/*granulin* (major OB protein) indicate that *alphabaculovirus* from lepidopteran hosts fall into two subgroups (Group 1 and Group 2) (Jehle *et al.*, 2006b).

A number of recent reference texts and reviews describe the biology, molecular biology and genomics of baculoviruses (Miller, 1997; Bonning, 2005) as well as their potential use as biological control agents for insect pest control (Moscardi, 1999; Vail *et al.*, 1999). Baculoviruses are considered safe biologically based insect control agents based on an extensive database on host range (arthropod-specific) and history of use in insect control with no issues of target specificity, negative environmental or human health issues being identified. These observations are enhanced by the fact that no morphological or genetic similarity to any other virus taxa has been noted for baculoviruses.

*Poxviridae* – subfamily *Entomopoxvirinae*: entomopoxviruses, in the subfamily *Entomopoxvirinae*, are typical poxviruses having large brick-shaped enveloped virions with a single, covalently closed linear dsDNA genome of 130–375 kb (Buller *et al.*, 2005). These large complex viruses replicate in the cytoplasm of infected cells, and viral genes involved in virus DNA replication and transcriptional regulation are expressed even before the virion is completely uncoated. Although these viruses replicate in the cytoplasm, there is evidence that host nuclear factors are required for virus gene expression following viral DNA replication. Mature virions are typically occluded within an OB, referred to as a spheroid. There are three recognized genera in the *Entomopoxvirinae* distinguished by virion morphology, host range and genome size (Buller *et al.*, 2005). *Alphaentomopoxvirus* has been isolated exclusively from Coleoptera, the virions are ovoid (450 × 250 nm) and the genome size ranges from 260 to 370 kbp. There are only six species other than the type species, *Melolontha melolontha entomopoxvirus* (MMEV). *Betaentomopoxvirus* has been isolated from Lepidoptera and Orthoptera, virions are ovoid (350 × 250 nm) and the genome size is approximately 250 kbp. *Amsacta moorei entomopoxvirus 'L'* (AMEV) is the type species of this genus. Finally, *Gammaentomopoxvirus* includes entomopoxviruses isolated from Diptera, the virions are brick-shaped (320 × 230 nm) and the genome size ranges from 250 to 380 kbp. Species within the genera are currently described based on host range and virion morphology. However, gene content and gene order as well as REN profiles and serological criteria will become increasingly important for species characterization within this group.

Entomopoxviruses have been investigated for their potential use as biological control agents for Orthoptera including grasshoppers and locusts. However, morphological and biochemical similarities of entomopoxviruses to those poxviruses infecting mammals have raised potential safety concerns. As more entomopoxvirus genomic DNA sequences become available, a clearer picture of a set of common co-linear core genes distinguishing these insect viruses from the poxviruses of vertebrates is emerging (Buller *et al.*, 2005). In addition, no serological relationship exists between the insect and vertebrate poxviruses, and the evidence for host restriction of *Entomopoxvirinae* to insects is solid. A recent review (King *et al.*, 1998) provides a detailed description of entomopoxviruses.

*Iridoviridae*: iridescent viruses have been isolated only from poikilotherms such as amphibians, fish and invertebrates (Williams, 1998; Chinchar *et al.*, 2005). The virions are icosahedral, 120–200 nm in diameter, with a complex capsid structure consisting of capsid proteins incorporated in a lipid membrane. Genomes consist of a single linear dsDNA molecule, 140–300 kbp in diameter, with redundant sequences at each end. The replication strategy of iridescent viruses is complex, with initial genomic DNA synthesis and gene transcription occurring in the host-cell nucleus and a second stage of viral DNA synthesis in the cytoplasm. Mature virions are assembled in the cytoplasm of infected cells and often accumulate in paracrystalline arrays and remain cell-associated. The regular packing of the virus particles in these arrays produces the characteristic 'iridescent' blue- to green-coloured sheen of iridescent virus-infected individuals. This virus group infects dipteran insects that are vectors of animal diseases and thus have received some attention as possible biological control agents. However, transmission cycles are poorly understood and the incidence of infection induced by oral inoculation is very low. Also, the incidence of virus infection in natural populations is typically quite low and displays limited virulence.

The iridescent viruses infecting insects fall into two genera: *Iridovirus* and *Chloriridovirus*. *Iridovirus* virions are typically 120–130 nm in diameter, with genome size ranging from 140 to 210 kbp. Iridoviruses infect a wide range of arthropods but most have been isolated from insects in the orders Diptera, Coleoptera and Lepidoptera found in aquatic or soil habitats. As stated above, the transmission cycle of these viruses are poorly understood but may involve cannibalism (Williams, 1998). The natural host range of these viruses may vary but injection experiments show replication in a broad array of insect hosts. The type species for *Iridovirus*, *Invertebrate iridescent virus 6* (IIV-6), demonstrates the currently accepted nomenclature, based on numerical designations. There are only two recognized species in this genus, IIV-6 and *Invertebrate iridescent virus 1* (IIV-1), in addition to a large number of tentative species for which limited characterization data are available. The ICTV outlines detailed biochemical information to establish species within *Iridovirus* (Chinchar *et al.*, 2005) including amino acid (aa) sequence analysis of the major capsid protein (distinct species exhibit no more than 90% aa sequence identity); DNA–DNA dot-blot hybridization (less than 50% hybridization values for distinct species); restriction enzyme length polymorphism (RFLP) using a panel of not fewer than four RENs; and serological cross-reactivity among strains within the same species (Western blot analysis using antibodies raised against disrupted virions). *Chloriridovirus* contains a single species, *Invertebrate iridescent virus 3* (IIV-3) infecting aquatic Diptera, mainly mosquitoes. The virions are considerably larger (180 nm in diameter) than those of *Iridovirus* and the genome is approximately 135 kbp.

*Ascoviridae*: ascoviruses infect larvae of Lepidoptera but their transmission appears to be dependent on parasitic wasps (Hymenoptrea: Braconidae and Ichneumonidae). For most ascoviruses, wasp transmission is mechanical, but for one species there is evidence of vertical transmission in the wasp vector (Miller, 1998). Virions of *Ascoviridae* are bacilliform to ovoid in shape, 130 nm in diameter and 200–400 nm in length, with a complex outer envelope and a circular dsDNA genome of 120–180 kbp (Federici *et al.*, 2005). Ascoviruses replicate in the host-cell nucleus, which subsequently undergoes hypertrophy, ruptures and the cell becomes packed with virion-containing vesicles in a process that resembles apoptosis. Ascoviruses typically produce a chronic infection in the lepidopteran host with delayed development, reduced feeding and disruption of the moulting process. Infected larvae have milky white haemolymph due to the presence of massive numbers of virion-filled vesicles released from lysed cells.

The *Ascoviridae* contains a single genus, *Ascovirus*, and four assigned species including the type species *Spodoptera frugiperda ascovirus 1a* (SfAV-1a). The species characterization is based on virion morphology, host range, tissue tropism, specific hymenopteran parasitoid association, RFLP profile and DNA hybridization data. Recent phylogenetic analysis suggests a relationship between *Ascoviridiae* and *Iridoviridae* (Federici *et al.*, 2005).

*Polydnaviridae*: polydnaviruses are enveloped DNA viruses with a genome consisting of multiple dsDNA circular molecules ranging from 2 to 31 kbp, with total genome size ranging from 150 to 250 kbp. However, total genome size is difficult to estimate because of the redundancy of DNA sequences among different-sized genome segments. Polydnaviruses are associated with hymenopteran endoparasitoids as a symbiont that suppresses the normal 'immune' response of the lepidopteran larval host. The polydnavirus genome exists as a proviral form incorporated at multiple sites within the wasp genome. Virus replication occurs only in the nucleus of calyx cells associated with reproductive tissues of female wasp during pupal development and virus particles are assembled in these cells and bud or are released into the calyx fluid. Virus particles are injected into the host lepidopteran larva along with wasp eggs and calyx fluid. Although polydnaviruses do not replicate in the lepidopteran host, select virus genes are expressed, and these play a role in altering the host's physiology to the advantage of the developing parasitoid.

There are two genera in *Polydnaviridae* differentiated by virion morphology and parasitoid family association: *Bracovirus* is associated with braconid wasps and has cylindrical virions with a single unit membrane envelope, while *Ichnovirus* is associated with ichneumonid wasps and characterized by fusiform virions enveloped by two unit membranes (Webb *et al.*, 2005). *Cotesia melanoscela bracovirus* (CmeBV) is the type species for *Bracovirus* and is one of 32 described species in the genus. *Campletis sonorensis ichnovirus* (CsIV) is the type species for *Ichnovirus* and is one of 21 described species in the genus along with a number of tentative species. The species criteria are somewhat similar for each genus in that virions need to be isolated from the oviduct of the associated female wasp, reference specimens of the wasp host need to be identified by a specialist and deposited in an insect collection, virion morphology must fit the criteria for the respective genera, and the virus genome isolated from virions must consist of multiple dsDNA circular molecules. Other features used to identify species are REN profiles of genomic DNA and the lepidopteran host range of the virus (non-replicative host) (Webb *et al.*, 2005). Webb (1998) and Kroemer and Webb (2004) provide reviews of polydnavirus biology, molecular biology and taxonomy.

Unassigned viruses: there are a number of unassigned dsDNA viruses that have some similarity to baculoviruses and were in the past classified as non-occluded members of *Baculoviridae*. These include the Oryctes rhinoceros virus (OrV), Heliothis zea virus 1 (HzV-1), Helicoverpa zea virus-2 (HzV-2) or gonad-specific virus, as well as Gryllus rubens virus (GrV). As more DNA sequence information becomes available, it is increasingly clear that these viruses have limited distant relationships to baculoviruses based on a limited number of genes; however, HzV-1 and HzV-2 appear to be closely related to each other.

### 1.2.2. Single-stranded DNA viruses

*Parvoviridae* – subfamily *Densovirinae*: parvoviruses are characterized by small non-enveloped icosahedral virions (18–26 nm in diameter) and a single linear single-stranded deoxyribonucleic acid (ssDNA) genome of 4–6 kb (Tattersall *et al.*, 2005). All parvoviruses that infect arthropods are classified in the subfamily *Densovirinae* with four genera (*Densovirus*, *Iteravirus*, *Brevidensovirus* and *Pefudensovirus*) distinguished on the basis of genome structure (monosense or ambisense) and transcription strategy. Species designation within each genus is based on serological distinctiveness and DNA sequence (less than 95% identity between species for the non-structural genes). Some densoviruses appear to infect a single host in nature but others have broad experimental host ranges.

*Junonia coenia densovirus* (JcDNV), the type species for *Densovirus*, has a 6 kb ssDNA genome with ambisense organization and long terminal repeats. The nomenclature of these viruses follows that of most other insect virus groups, viruses being designated based on the host-insect genus and species name from which they were initially identified. The two-letter abbreviated form in some cases has been extended to four letters to avoid confusion due to similar host names. *Bombyx mori densovirus* (BmDNV), the type species for *Iteravirus*, has a 5 kb ssDNA genome with monosense organization. *Aedes aegypti densovirus* (AaeDNV), the type species for *Brevidensovirus*, has a 4 kb ssDNA genome with monosense organization. Finally, *Periplaneta fuliginosa densovirus* (PfDNV), the type species for *Pefudensovirus*, has a 5.5 kb ssDNA genome with ambisense organization.

Densoviruses were originally noted in insects reared commercially such as wax moth and silk worm where host density is high. Densoviruses typically infect larval stages and produce symptoms that include alterations in cuticular pigmentation and progressive paralysis. Some of the densoviruses infecting economically important pests are quite virulent and host-specific and have been investigated as biocontrol agents. For example, AaeDNV was developed for mosquito control (Bergoin and Tijssen, 1998). In addition, densoviruses have been investigated as expression vector systems, as plasmids containing almost the entire virus genome are 'infectious' upon transfection (Bergoin and Tijssen, 1998).

### 1.2.3. Double-stranded RNA viruses

*Reoviridae*: reoviruses are characterized by icosahedral virions (60–80 nm in diameter) and genomes consisting of 10–12 linear dsRNA molecules. The family *Reoviridae* contains a number of genera, which include *Orbivirus*, *Coltivirus* and *Seadornavirus* that replicate in both vertebrate hosts and arthropod vectors such as ticks and mosquitoes. Similarly, reoviruses include plant viruses such as *Phytoreovirus*, *Fijivirus* and *Aryzavirus* that replicate both in plant hosts and arthropod vectors such as leafhoppers. A single reovirus genus, *Cypovirus* (CPV), infects insects exclusively (Bellonick and Mori, 1998; Mertens *et al.*, 2005).

Cypoviruses like baculoviruses and entomopoxviruses are typically occluded within a crystalline proteinaceous OB formed in the cytoplasm of infected cells. Cypoviruses are typically transmitted orally with the OB dissolving in the alkaline pH of the host-insect gut. Infection is typically restricted to midgut epithelial cells and usually produces chronic disease symptoms similar to starvation. These symptoms include reduced feeding and larval growth, increased development times, and if diseased individuals successfully pupate and eclose, the adults are often malformed and have reduced longevity and fecundity. Cypoviruses have been isolated most frequently from Lepidoptera but a few isolates from Diptera and Hymenoptera have also been described (Mertens *et al.*, 2005). Although cypoviruses are highly infectious and persist in insect populations, they have received limited attention for development as viral insecticides due to the general chronic rather than acute symptomology. Readers are referred to a recent review by Belloncik and Mori (1998) for a complete description of cypovirus biology.

Cypoviruses have single-layered iscosahedral capsids with 12 surface spikes and genomes consisting of ten dsRNA linear molecules varying in size from 0.6 to 5.6 kbp and each coding for a single virus gene product (Mertens *et al.*, 2005). The virions are occluded in OB or polyhedra with cubic to polyhedral shape and range in size from 0.2 to 10 μm in diameter. Species with the genus *Cypovirus* have historically been defined by electrophoretic profiles of dsRNA segments in agarose or polyacrylamide gels. More recent but limited RNA sequence analysis comparisons, antigenic relatedness of capsid proteins and RNA–RNA hybridization studies have confirmed the validity of electrophoretic profiles for classification of species. *Cypovirus 1* (CPV-1) is the type species for the genus *Cypovirus* and the various isolates of CPV-1 are named with respect to the host species from which it was isolated, for example Bombyx mori cypovirus 1 (BmCPV-1). There are currently 16 CPV species described on the basis of dsRNA genomic electrophoretic profiles in agarose or polyacrylamide gel electrophoresis (PAGE) gels, with at least seven dsRNA segments showing similar mobility within species and at least three segments with significantly different mobility to distinguish between species (Mertens *et al.*, 2005).

### 1.2.4. Single-stranded RNA viruses

*Tetraviridae*: the host range of *Tetraviridae* is restricted exclusively to insects that have only been isolated from Lepidoptera (Gordon and Hanzlik, 1998). Tetraviruses are small (40 nm in diameter) icosahedral, non-enveloped viruses with a genome consisting of either one or two linear, positive-sense, ssRNA segments. There are two genera in the family: *Betatetravirus* has a single-segment ssRNA genome of approximately 6.5 kb, whereas *Omegatetravirus* has a two-segment ssRNA genome (5.3 and 2.45 kb) (Hanzlik *et al.*, 2005). Infection by viruses in both genera appears to be restricted to the midgut; thus, an oral route of infection is suspected and horizontal transmission is more than likely the major route of infection. However, indirect evidence for vertical transmission exists for both genera. There is great variation in pathogenicity among virus isolates, with symptoms ranging from inapparent to acutely lethal infections. The definitive criterion for species identity in both *Betatetravirus* and *Omegatetravirus* is the nucleotide sequence of the capsid protein gene, but in practice several other methods including host range, cross-reactivity of antisera to capsid proteins and size of genomic RNA upon electrophoretic analysis have been used for species characterization (Hanzlik *et al.*, 2005).

*Nudaurelia capensis β virus* (NβV), the type species for *Betatetravirus*, is one of ten species in the genus. *Nudaurelia capensis ω virus* (NωV) is the type species for *Omegatetravirus* and the only other species in the genus is the extensively studied *Helicoverpa armigera stunt virus* (HaSV). HaSV has been investigated as a potential biocontrol agent for pest species in the Helicoverpa group because it produces rapid feeding cessation, significant delays in larval growth and very characteristic stunting or shrinkage of the larval body.

*Dicistroviridae*: *Cripavirus*, the lone genus in the family *Dicistroviridae*, contains a single, positive-sense, ssRNA linear genome of 9–10 kb within small (30 nm diameter) icosahedral, non-enveloped virions (Christian *et al.*, 2005a). *Cricket paralysis virus* (CrPV), the type species for *Cripavirus*, has a very broad host range having been isolated from Orthoptera, Hymenoptera, Lepidoptera, Hemiptera and Diptera. Species in this genera are distinguished largely on the basis of serology and sequence of capsid proteins (identity >90% at species level) and to some degree natural host range and cell culture replication. Cripaviruses include Black queen cell virus and Acute bee paralysis virus associated with honeybees either as acute or inapparent infections and the degree of symptomology is often related to the presence of other pathogens or parasites (Christian and Scotti, 1998).

The *Dicistroviridae* are similar to other positive-sense ssRNA picornavirus-like viruses which include *Iflavirus* (see below). There are also a large number of small (30 nm diameter) RNA-containing viruses (SRV) that have similarities to *Dicistroviridae* but are currently unclassified. Some of these may eventually be reclassified within *Dicistroviridae*. Readers are referred to Christian and Scotti (1998) for a more complete description of picorna-like viruses associated with arthropods.

*Nodaviridae* – genus *Alphanodavirus*: alphanodaviruses are small (32–33 nm diameter) spherical, non-enveloped viruses with a bipartite genome of two positive-sense, linear ssRNA molecules (3.1 and 1.4 kb), both of which are required for infectivity and are encapsulated in the same virion (Scheemann *et al.*, 2005). Alphanodaviruses have been isolated from insects and the host range appears to be restricted to insects with the exception of the type species *Nodamura virus* (NoV). There is some evidence that NoV infects pigs and is transmissible to suckling mice by a mosquito vector (Ball and Johnson, 1998). Most of the alphanodaviruses are infectious for wax moth larvae upon injection and typically cause paralysis and death; however, little is known about their natural transmission cycle or ecology. Many of the other alphanodaviruses have been isolated from soil-dwelling beetle species in the south Pacific, for example *Flock house virus* (FHV) and *Black beetle virus* (BBV), but they have also been isolated from Diptera and Lepidoptera (Scheemann *et al.*, 2005). The definitive means of distinguishing species in this genus is by nucleotide sequence analysis of the capsid protein gene.

*Iflavirus*: this is another SRV with small (30 nm diameter), non-enveloped, spherical virions containing a single, positive-sense ssRNA molecule of 8.5–9.5 kb (Christian *et al.*, 2005b). *Iflavirus* is not currently assigned to a virus family but, as stated above, it has many features in common with picorna-like viruses. The type species is *Infectious flacherie virus*, which infects *B. mori*. Another notable member of this group is Sacbrood virus, which infects honeybees and is of economic significance (Christian and Scotti, 1998). As stated previously, there are a large number of small RNA viruses that are vectored by arthropods to plant and animal hosts, but these are beyond the scope of this chapter.

## 1.3. Preliminary Approaches to Virus Identification

Historically, those virus groups that exclusively infect arthropods and particularly insects have been of interest due to their economic impact on beneficial insects or their potential as biological control agents of pest insect. These viruses were largely diagnosed and identified based on symptomology in the host and morphological characters determined at the level of light microscopy. Thus, much of the early work focused on those viruses which are occluded in large (0.5–15 μm diameter) crystalline protein OBs detectable by light microscopy. Readers are referred to Evans and Shapiro (1997) for a review of techniques used to identify and diagnose virus infection based on host symptomology and microscopy including light (various optical systems) and electron microscopy. Their review also provides useful information on virus isolation, quantification and bioassay assessment techniques.

## 1.4. Methodology for Virus Isolation and Fractionation (Nucleic Acid and Protein Purification Techniques)

A good overview of techniques for isolating various viruses from infected hosts is provided by Tompkins (1991) and summarized in tabular form by Evans and Shapiro (1997). The ultracentrifuge is a key piece of laboratory equipment for virus purification, particularly when highly purified samples are required for biochemical characterizations. For most virus groups, virion purification requires some form of isopycnic centrifugation on CsCl or sucrose gradients to separate virions from cellular material based on particle density, as well as size and shape. For those viruses that are occluded in proteinaceous crystals or OBs, the virions need to be released from the OB typically by incubation in an alkaline buffer that disrupts the OB crystalline structure yet leaves the virions intact. Protocols for purifying different virus groups can vary substantially; thus, it is important to consult published protocols for specific virus groups where possible. For the purposes of diagnostics, some short cuts can be taken as crude preparations work for many molecular protocols, and to be of practical value high-throughput protocols may be required for diagnostic tests. We provide some examples of short-cut techniques for virus particle and nucleic acid isolation in the section on polymerase chain reaction (PCR) techniques.

Once virions have been purified, it is often of interest to extract the nucleic acid component for genome analysis. There are a number of good molecular biology manuals available that are excellent resources for nucleic acid and protein purification protocols (see Davis *et al.*, 1986; Sambrook and Russell, 2001).

Typically the first step is an incubation of virions in Tris-HCl buffer along with a protease and a detergent such as sodium dodecyl sulfate (SDS) to denature and digest the protein capsid of the virion and release the nucleic acid genome from binding proteins. This is followed by phenol/chloroform extractions to separate protein from nucleic acids which partition to the aqueous phase. For RNA viruses, a ribonuclease (RNase) inhibitor such as quanidine isothiocyanate is required in combination with phenol/chloroform extraction to avoid RNA degradation, or ready-to-use commercial products such as Trizol are available for RNA extractions. Depending on the ultimate use of the purified nucleic acid, RNA or DNA can be concentrated by precipitation in ethanol at −20°C or dialysed against Tris-HCl, ethylenediaminetetraacetic acid (EDTA) buffer at 4°C. We have found that for large DNA viruses, such as baculoviruses, dialysed DNA is preferable when high-quality preparations are required as for instance in transfection experiments in which intact genomic DNA is required.

A variety of commercial RNA and DNA extraction kits are also available and have been used to purify virus genomes ranging from small RNA viruses (Yue and Genersch, 2005) to large DNA viruses such as baculoviruses (England *et al.*, 2005; Jehle *et al.*, 2006b).

## 1.5. Biochemical/Molecular Approaches to Virus Identification and Diagnosis

In the following sections, we will briefly describe some of the biochemical and molecular biological approaches for virus identification and detection. Wiedbrauk and Farkas (1995) provide a complete manual of molecular techniques for virus detection based on nucleic acid components and it is a good resource for those readers wanting more in-depth information on some of the techniques listed below.

As PCR technology has become such a standard and important approach, we discuss its application to insect virology in a separate section. The techniques discussed below and examples cited are by no means an exhaustive review of insect virus characterization and detection using molecular techniques but are meant to serve as an introduction to the various molecular approaches that are currently available and/or are more widely used.

### 1.5.1. Virus structural proteins

The separation of denatured proteins from virus particles by sodium dodecyl sulfate-polyacrylamide gel electrophoresis (SDS-PAGE) has been used to characterize and distinguish virus isolates from a variety of insect virus families. Virus samples

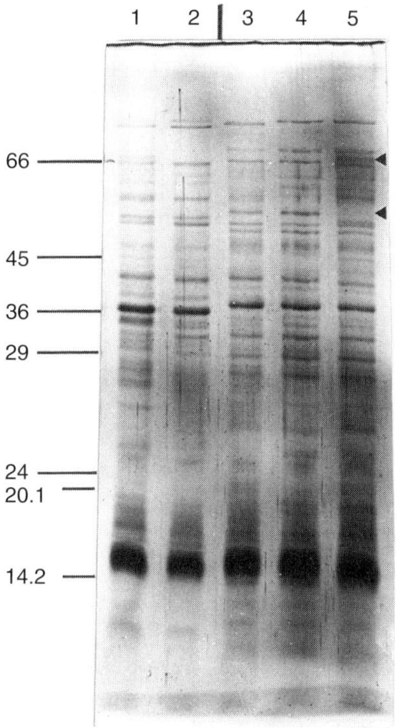


**Fig. 1.1.** Analysis of released occluded virions of: (1) MacoNPV-A1; (2) MacoNPV-A2; (3) MbNPV-NL82/1; (3) MbNPV-NL; (4) MbNPV-D. Two major differences between MbNPV-NLs and MbNPV-D are indicated on the right. Also note the size difference of the major capsid protein between MacoNPV (36.7 kDa) and MbNPV (37.2 kDa). (From Erlandson, 1990.)

are typically resuspended in SDS-PAGE sample buffer containing 1% SDS and 1% 2-mercaptoethanol and heated to 100°C in order to denature proteins. Following electrophoresis, gels are stained with Commassie blue or silver- or nickel-based stains which provide increased sensitivity. Figure 1.1 shows the comparative profiles of polypeptide components of purified ODV of several NPVs that infect *Mamestra* species electrophoresed on 10% SDS-PAGE slab gels. In this case, different species of viruses can be distinguished by size differences of the major capsid proteins (~36–39 kDa). This technique does not have the sensitivity or specificity of more recently developed techniques discussed below but still has a role in a complete characterization of new virus isolates. Recent descriptions of new virus isolates from among *Cypovirus* (Green *et al.*, 2006) and *Dicistroviridae* (Nakashima *et al.*, 2006) are examples of polypeptide analysis as part of virus characterization.

## 1.5.2. Serological detection systems

Serological analysis has played, and continues to play, an important role in insect virus characterization and diagnostic identification. A variety of techniques including complement fixation, immunodiffusion, radioimmune assays and enzyme-linked immunosorbent assay (ELISA) have been used to detect and quantify insect viruses. Of these, ELISA has been the predominant method used and has the benefit of being relatively rapid, specific for the antigen targeted and sensitive. McCarthy and Gettig (1986) provide an overview of serological techniques used to study baculoviruses and Harlow and Lane's (1988) work is a good resource for basic theory and protocols for serological methods.

Recently, a dipstick immunoassay was developed for the detection of *Orgyia pseudotsugata MNPV* (OpMNPV) and Orgyia pseudotsugata SNPV (OpSNPV) strains in infected Douglas-fir tussock moth larvae (Thorne *et al.*, 2007), and this approach appears to be applicable to determining virus prevalence in field-collected larvae. The assay is based on application of two monoclonal mouse anti-OpMNPV and OpSNPV antibodies to nitrocellulose membrane dipsticks along with an internal positive control antibody. The assay is relatively quick (several hours), uses crude larval homogenates and involves an easily scored colour change based on a horseradish peroxidase reaction (Thorne *et al.*, 2007).

### 1.5.3. Nucleic acid hybridizations: Southern blots – dot-blot

The use of DNA renaturation kinetics, in solution, to determine the relationship between insect virus isolates goes back to the work of David Kelly on iridescent viruses (Kelly and Avery, 1974) and baculoviruses (Kelly, 1977). More recently, DNA immobilization or transfer from gels on to various solid substrates such as nitrocellulose or nylon membranes, denaturation to ssDNA targets and hybridization with radio-labelled ssDNA probes following the method of Southern (1975) has been used for the detection and characterization of insect virus, as well as assisting the development of physical maps of viral genomes.

Sambrook and Russell (2001) provide an excellent resource on the theory and practice of analysis of DNA homology using the Southern technique and provide a good discussion of factors influencing the sensitivity of the technique including labelling efficiency and DNA probe length. Southern blot analysis of REN digests of virus DNA and use of $^{32}$P-labelled probes, either whole virus genome or cloned fragment DNA probes, under different hybridization conditions has been used extensively for investigating homology between different virus species and strains. The studies of DNA homology among baculoviruses (Smith and Summers, 1982) and iridoviruses (Williams, 1994) are examples of the application of Southern blot analysis to specific groups of insect viruses.

A more commonly used technique for virus detection is the simpler DNA–DNA dot-blot protocol because neither virus particles nor virus DNA need necessarily be purified for the technique to be applicable. One of the earliest studies using the DNA–DNA dot-blot technique for detection of an insect virus in infected larvae is that of Ward *et al.* (1987). These authors used $^{32}$P-labelled *Wiseana signata* NPV (WisiNPV) *Bam*HI fragment clones to detect WisiNPV DNA in infected insect homogenates that were treated by various methods prior to blotting on to nylon membranes. They found that crude insect homogenates applied directly to nylon membranes gave almost as good detection sensitivity as alkali-treated homogenates (alkali treatment to release virions from OB, disrupt virions and denature the DNA) or phenol/chloroform extracted homogenates. The crude homogenate method simply involved direct spotting of the insect homogenate on the membrane, air drying and layering on to 3 mm filter paper soaked in alkali buffer (0.5 M NaOH, 1.5 M NaCl) for 30 min and then air drying. The crude homogenate dot-blot was found to be superior to phase-contrast microscopy, differential staining or radioimmunoassays for virus detection and as little as 150 pg of DNA or 20 OBs could be detected. A similar comparative study testing DNA dot-blot hybridization for detection of *Lymantria dispar MNPV* (LdMNPV) infection in gypsy moth larvae was conducted by Keating *et al.* (1989). These authors tested an even more direct approach of larval squashes in which intact larvae were squashed between two nitrocellulose membranes prewetted in 10× saline sodium citrate (SSC) buffer (1.5 M NaCl, 0.15 M Na-citrate pH 7.0) plus 50 mM EDTA for 1–2 h resulting in the release of larval tissues on to the membranes. In this study, the estimate of infection rates by LdMNPV $^{32}P$-labelled DNA hybridization on both dot-blot and larval squash membranes was similar to mortality rates determined from a subsample of infected larvae reared until pupation or mortality. One of the disadvantages of the DNA–DNA dot-blot hybridizations described above is the handling problem associated with radioactively labelled DNA probes. Kaupp and Ebling (1993) tested DNA–DNA dot-blot systems for detection of LdMNPV and *Choristoneura fumiferana MNPV* (CfMNPV) infection in gypsy moth and spruce budworm larvae, respectively. They used horseradish peroxidase-labelled DNA probes and enhanced chemiluminescence detection techniques to overcome the problems of handling radioactive probes. They demonstrated a detection sensitivity of 0.1–0.5 ng using purified DNA or an equivalent of 5000 LdMNPV OB in infected insect tissue applied to nylon membranes. The sensitivity of this DNA probe labelling method is somewhat lower than for $^{32}P$-labelled probes but still useful for detection of virus infection. Despite the specificity and sensitivity of the dot-blot techniques, to a large degree this technology has been superseded by the availability of PCR for virus identification and detection.

### 1.5.4. Restriction endonuclease analysis

The use of Type II RENs that bind to specific nucleotide recognition sites and cleave dsDNA followed by the separation of cleaved fragments, based on size, using agarose gel electrophoresis has been extensively exploited to characterize dsDNA viruses such as baculoviruses (Possee and Rohrmann, 1997), entomopoxviruses (Erlandson and Street, 1997), iridoviruses (Williams, 1998) and Oryctes virus (Crawford *et al.*, 1986). This is a very simple and routine technique, and a wide variety of Type II REN enzymes are available commercially, typically supplied with ready-to-use buffers for digest incubation. Following incubation, the digestion mixture is typically applied to 0.5–1.4% agarose gels made up in appropriate running buffer and electrophoresed for several hours to overnight (see Sambrook and Russell, 2001, Chapter 5 for a complete explanation of gel electrophoresis). The agarose gels are then incubated with appropriate stains that intercalate into dsDNA fragments and fluoresce upon exposure to ultraviolet (UV) light so that the DNA fragments can be visualized. Ethidium bromide (0.5 μg/ml) is the most commonly used dye and 250 ng to 1.0 μg of genomic DNA is sufficient to visualize REN profiles. Recently more-sensitive and less-hazardous dyes such as SYBR Gold (Molecular Probes) that increase the detection sensitivity up to 10× over ethidium bromide have been used (e.g. Graham *et al.*, 2004).

REN fragment profile analysis can be used to estimate genome sizes when used in conjunction with appropriate standard DNA size markers, different concentration agarose gels and sufficient replication of gel runs. In conjunction with Southern blot, end sequencing of cloned REN fragments and/or double REN digestion, REN analysis can be used to produce physical maps of virus genomes that are useful in characterizing viruses and mapping potential genetic differences between virus strains. This approach has been used extensively in characterizing baculoviruses (Possee and Rohrmann, 1997). Thus, REN analysis has been helpful in defining virus species but it is particularly useful in distinguishing strains within species. Figure 1.2 shows the *Hin*dIII REN fragment profiles of different

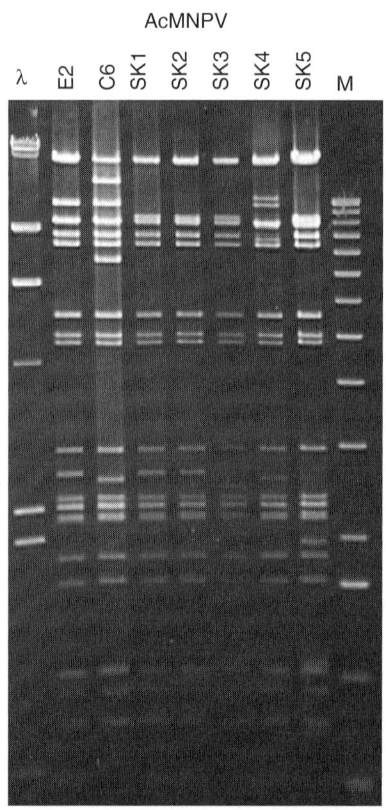


**Fig. 1.2.** Restriction endonuclease profiles of *Autographa californica* nucleopolyhedrovirus strains. One microgram of purified virus DNA was digested with *Hin*dIII, electrophoresed on a 0.7% agarose gel (1× TBE) and REN fragments visualized after staining with ethidium bromide. The DNA size markers, *Hin*dIII-digested lambda phage DNA (λ) and 1 kb marker (M), are indicated on the left and right, respectively. The AcMNPV strains include standard strains E2 and C6 as well as field isolates SK1-SK5.

strains of AcMNPV isolated from individual *T. ni* larvae from different field populations in comparison to standard strains E2 and C6. Note that although almost all of the DNA fragments have identical mobility for most of the isolates, for each isolate there are unique-sized fragments. Differences in REN profiles among virus strains may be due to single nucleotide changes at REN recognition site or result from significant insertion, deletion or genome rearrangements. The availability of a physical map makes the interpretation of alterations in REN fragment profiles easier and critical information on genetic differences among strains can be derived.

The DNA samples in Fig. 1.2 were purified by a short-cut technique based on a modified protocol of Smith and Crook (1988) in which infected larvae were homogenized in 0.1% SDS and the NPV OB concentrated by centrifugation, resuspended in 200 µl of sterile double-distilled $H_2O$ ($ddH_2O$) and 5 µl of 1 M $Na_2CO_3$ to disrupt the OB. The released ODVs were then digested with 1% SDS and proteinase K (0.2 mg/ml) for 2 h, and the DNA extracted by phenol/chloroform/isoamyl-alcohol (50:48:2) and precipitated in ice-cold 70% ethanol. This approach has been used extensively to characterize natural isolates of baculoviruses and study genotypic variation of virus populations in forest (Graham *et al.*, 2004; Cory *et al.*, 2005) and agricultural (McIntosh *et al.*, 1987; Munoz *et al.*, 1998) pest species.

### 1.5.5. Electrophoretic profiles of whole genomes

The REN analysis described above is only applicable to those insect viruses that contain dsDNA genomes. For some groups of RNA viruses a somewhat more straightforward approach has been used in that intact purified genomic RNA is electrophoresed on 1% agarose or 3–5% polyacrylamide gels to determine size and number of genomic fragments. This system has worked well for cypoviruses since it was first proposed by Payne and Mertens (1983) as a method of distinguishing CPV isolates. A recent example demonstrates the technique's utility. Green *et al.* (2006) purified a previously uncharacterized mosquito *Culex restuans* CPV (CrCPV) on Ludox gradients, isolated genomic RNA using a commercial kit (QIAampViral Mini Kit from Qiagen) and 100 ng of purified RNA was electrophoresed on a 1% agarose gel (Fig. 1.3). The RNA profile of CrCPV was compared with those of well-characterized CPVs, BmCPV-1 and TnCPV-15, for descriptive purposes. This approach has been used with a number of RNA virus families including tetraviruses and nodaviruses.

### 1.5.6. Genome nucleic acid sequence

The increasing ease and cost-effectiveness of nucleic acid sequencing has led to the elucidation of complete genome sequences for a variety of insect virus groups. Indeed, new complete genome sequences are routinely added to the GenBank database. For example, the GenBank genome database lists 28 completely sequenced baculovirus genomes, including 25 NPV and GV from lepidopteran hosts, two NPV from hymenopteran hosts and one from a dipteran host. To date there are

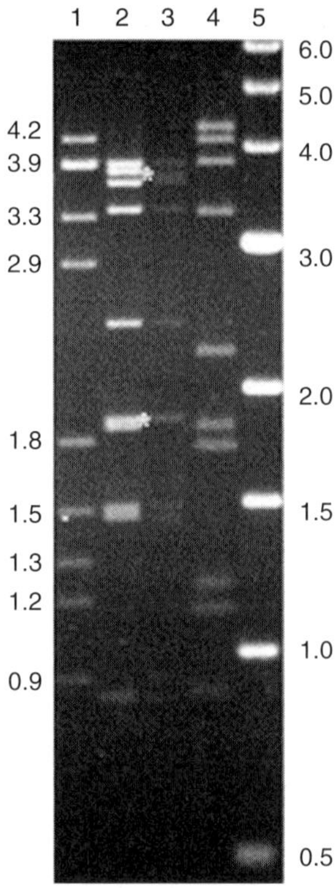


**Fig. 1.3.** Electrophoretic separation of various CPVs on a 1% agarose gel: BmCPV-1 (lane 1), CrCPV-17 (lane 2), UsCPV-17 (lane 3), TnCPV-15 (lane 4) and 1 kb marker (lane 5). The asterisks between lanes 2 and 3 indicate the visual differences between CrCPV-17 and UsCPV-17. The published segment sizes for BmCPV-1 (lane 1) are labelled to the left of the figure. The published sizes for TnCPV-15 (lane 4) are: Seg-1 = 4361 bp and Seg-10 = 897 bp. (From Green *et al.*, 2006.)

complete genome sequences available for two entomopoxviruses, one iridovirus, 14 densoviruses, 12 viruses from *Dicistroviridae*, four tetraviruses, five nodaviruses and at least four cypoviruses in which all 11 genome segments have been completely sequenced. The gene sequence and gene order information have led to more precise definitions of species within these virus groups and potentially improved methods for virus identification. The availability of complete genome sequences has also led to more thorough understanding of the phylogenetic relationships among viruses, a topic that will be covered elsewhere in this book.

Recently the use of nucleic acid sequencing technology has moved into the realm of pathogen discovery and detection in insect populations. Cox-Foster *et al.* (2007) used a metagenomics survey approach, based on high-throughput pyrosequencing technology, to survey for microbes associated with honeybee colony collapse disorder (CCD). They used total RNA extracted from honeybee samples in order to detect RNA viruses as well as other pathogens. This study detected a wide variety of insect pathogens associated with honeybee colonies but showed a high association of Israeli acute paralysis virus with bees from CCD colonies (Cox-Foster *et al.*, 2007).

## 1.6. PCR and Virus-specific Primer Development

### 1.6.1. PCR basics

The development of molecular marker systems based on DNA PCR technology has become commonplace in the identification and detection of insect viruses. One of the major advantages of this technique is its sensitivity to detect specific DNA templates at nanogram and picogram concentrations. There are several good manuals that describe standard PCR protocols (Innis *et al.*, 1990; Dieffenbach and Dveksler, 2003) and these are useful resources both for getting started and for troubleshooting PCR. Typical PCR reactions with genomic DNA from organisms involve three basic steps: (i) denaturation of double-stranded DNA to produce single-stranded target DNA; (ii) annealing or hybridization of oligonucleotide primers (15–25 nucleotides) to complimentary single-stranded target DNA sequences; and (iii) an elongation step for 5′ to 3′ extension of the primers by a thermostable DNA polymerases, such as *Taq* or *Pfu*, reading off the DNA template.

A typical PCR thermocycle condition consists of an initial denaturation step at 94°C for 2–5 min followed by 20–35 cycles of the following conditions: 94°C for 1 min (denaturation step), selected annealing temperature (50–65°C) for 0.5–1 min, 72°C for 0.5–1.5 min (elongation step duration dependent on the length of anticipated PCR product, approximately 1 min per 1000 nucleotides) and final elongation step at 72°C.

A standard PCR reaction mixture consists of a heat-stable DNA polymerase, typically *Taq*, $MgCl_2$ (a cofactor for the polymerase enzyme), the four deoxyribonucleotide triphosphates (dNTPs) required for DNA synthesis, an appropriate reaction buffer and finally the DNA template and oligonucleotide primers designed to hybridize to opposite strands of the target DNA template in a specific manner. The concentration of each component of the PCR reaction often requires optimization; for example, $MgCl_2$ (1–2 mM) and primer concentration can significantly impact the yield of PCR products as well as the fidelity of the reaction.

In some systems DNA template concentration can be important and too much template DNA can lead either to no authentic PCR product or to significant mispriming and therefore non-specific PCR products. Some trial and error may be required to determine the optimum annealing temperature for the PCR reaction, although theoretical equations can be used to estimate suitable annealing temperatures based on the melting temperatures of the PCR primers selected. As well, PCR reaction conditions cannot always be adopted directly from published protocols as the cycling conditions of thermocycler models may vary. This relates largely to the rapidity with which the temperature control block transits from one temperature to the next. The volume chosen for PCR reactions (typically from 20 to 100 μl), the type of PCR tubes, the type of thermocycler available (heated lid versus thermocyclers, which require oil overlays on the PCR reaction mixture to minimize evaporation) will all effect the consistency and ease of the PCR assay.

The quality of DNA template can impact the efficiency and success of PCR reactions. For infected tissue culture cell samples and most insect tissue, simple PCR detergent isolation procedure (see, e.g. Malitschek and Schart, 1991) or commercial DNA extraction kits (see, e.g. Lupiani *et al.*, 1999) can be used with good success, particularly for non-occluded virions. For occluded viruses such as baculoviruses and entomopoxviruses, it may be necessary to release the virions from the OB by dissolution in a high-pH carbonate buffer or proteinase K and potentially purify the virus DNA by phenol/chloroform extraction. Also, in many cases, insects contain high levels of polyphenols and other compounds that decrease the efficiency and fidelity of DNA polymerases used in PCR reactions and a single phenol/chloroform extraction followed by ethanol precipitation of DNA may be required to generate consistently good-quality DNA preparations.

The methodology outlined above is directly applicable to DNA viruses; however, for insect viruses with RNA genomes an additional step is required. Typically, to detect viral RNA a complementary DNA (cDNA) copy is first synthesized from the RNA template using reverse transcriptase and the cDNA is then amplified by *Taq* polymerase in a protocol referred to a reverse transcriptase PCR (RT-PCR). Genersch (2005) provides an example of the application of this technique using a commercial one-step RT-PCR kit to detect Deformed wing virus (DWV), a positive-sense ssRNA virus of honeybees. Recently, a more sensitive one-step real-time RT-PCR method based on CYBER Green chemistry was developed to detect DWV and Black queen cell virus in honeybees (Kukielka *et al.*, 2008). This method was several orders of magnitude more sensitive than RT-PCR and has the advantage of potentially giving quantitative estimates of these viruses in bee samples.

### 1.6.2. Primer development and virus identification strategies

The specificity and efficiency of PCR amplification are significantly impacted by the design of PCR primer pairs. There are numerous software packages to assist in primer design, but a few key factors are important in designing primers and selecting parameters to guide primer design software. Specificity is influenced by the length of primers and typically primers between 18–24 nucleotides are suitable for PCR. Specificity is also influenced by annealing (melting) temperatures (*Tm*), and selecting primers with *Tm* in the range of 54–62°C seems to give the most consistent results. Sequence at the 3′ end of a potential primer can be critical for specificity and differences of only two nucleotides between DNA templates can be enough to separate viral species based on PCR. The 3′ region of the primer is also most important in avoiding complimentary homology between primer pairs which can lead to primer dimmer formation (when the primers have a higher affinity for each other than for the target DNA) and very much reduced efficiency of PCR amplification. The length of the predicted PCR product also has an impact on the efficiency of PCR amplification. Generally for the purposes of detecting a specific DNA sequence, PCR products of 150–1000 nucleotides are ideal. Amplification of longer PCR products can be less efficient particularly if the quality of the DNA template is not good.

As more complete genome sequences become available for an increasing variety of insect viruses, the job of PCR primer design is becoming less problematic. A few examples will be discussed here to outline primer design strategies. Baculoviruses are the most thoroughly studied insect virus group and there is complete genome sequence for over 40 baculovirus species. To date there appear to be at least 30 core genes common to all baculoviruses and 62 genes common to lepidopteran baculoviruses (McCarthy and Theilmann, 2008; Herniou *et al.*, 2003). Recently, a PCR method was developed for identification of baculoviruses based on highly conserved regions of three genes, *late expression factor 8* (*lef-8*), *lef-9* and *polyhedrin*, and for which a series of degenerative primers were designed (Lange *et al.*, 2004). The degenerative primers also included 5′ extension homologous to universal sequencing primers such that the PCR products could be directly sequenced. The use of degenerative primers allows for some degree of variation in the target DNA sequence and thus Jehle *et al.* (2006) were able to generate PCR product and partial gene sequence for an additional 48 NPV and 23 GV isolates. This type of data allowed for more detailed phylogenetic analysis and produced further information on species definition among lepidopteran baculovirus groupings. For a new or previously uncharacterized baculovirus isolate, the exploitation of these degenerative PCR primers and subsequent DNA sequence analysis would provide a good first step in establishing its relationship to already characterized viruses. The *polyhedrin/granulin* gene is among the most highly conserved genes in lepidopteran baculoviruses and has been exploited extensively for development of PCR primers for virus detection (Burand *et al.*, 1992; de Moraes and Maruniak, 1997; Wang *et al.*, 2000; Woo, 2001). In some of these cases, PCR primers have been designed to detect a single virus species; however, in other studies, degenerative primers have been designed against conserved regions of the *polyhedrin* gene. For example, de Mores and Maruniak (1997) designed degenerative PCR primers that recognized the following NPVs: AcMNPV (and strain AnfaNPV, AgMNPV, SfMNPV, SeMNPV, OpMNPV and HzSNPV, and a 575 bp PCR fragment is produced with each of these virus DNA templates. Subsequently viruses could be distinguished from each other by RFLP in which the PCR product is digested with three different RENs that gave unique fragment profiles for each virus upon electrophoresis in agarose gels. This approach is generally referred to as PCR-RFLP. These authors also describe the importance of PCR optimization in terms of dNTP and $MgCl_2$ concentration for both PCR product yield and reaction specificity. Although using degenerative primers, they were able to obtain consistent PCR products with 0.5–10 ng of virus DNA template and as little as 1 pg of AcMNPV DNA template gave positive PCR amplification.

The PCR-RFLP technique targeted against *polyhedrin* sequences has been used by many researchers; for example, Wu and Wang (2005) were able to distinguish *Lymantria xylina NPV* from LdMNPV based on different sized *Bam*HI digest products of a 677 bp *polyhedrin*-based PCR product from the two species. Figure 1.4 shows an example of PCR-RFLP using PCR primers specific to a portion of the AcMNPV *chitinase* gene followed by *Hin*dIII digestion to distinguish several field strains of AcMNPV from the standard strain C6. Other baculovirus genes and PCR strategies have also been used to identify and detect specific viruses. For example, Lo and Chao (2004) have used primers specific to AcMNPV *ie-1* in a real-time quantitative PCR as a means of rapidly estimating AcMNPV titres in cell culture infection experiments. Khurad *et al.* (2004) also used PCR primers designed to *Bombyx mori MNPV* (BmMNPV) *ie-1* gene for detection of BmMNPV in a study examining vertical transmission of the virus in silkworm colonies. The use of PCR primer sets against two different genes, particularly those with less well-conserved

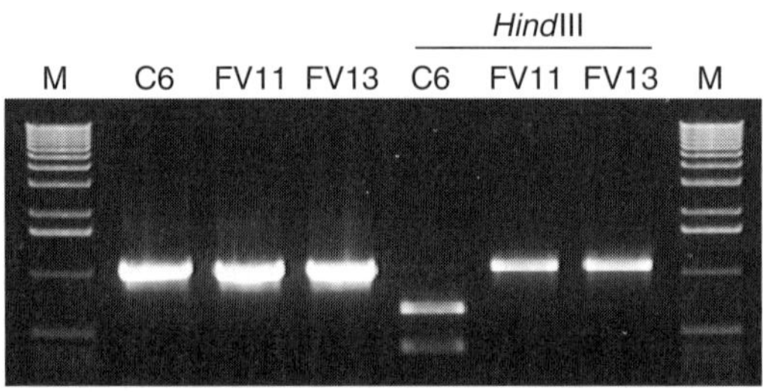


**Fig. 1.4.** Restriction endonuclease fragment polymorphisms of PCR amplicons from *Autographa californica* nucleopolyhedrovirus strains. Purified genomic DNA from AcMNPV-C6 or field strains FV11 or FV13 were subjected to PCR using primers specific to a portion of the AcMNVP-*chitinase* (10 pmol), 0.5 units of *Taq* DNA polymerase (Invitrogen, Burlington, Ontario), 0.4 mM dNPTs and 4 mM $MgCl_2$. Intact and *Hin*dIII-digested PCR products were compared on a 1.0% agarose gel (1×TAE). Note that *Hin*dIII digestion of the PCR products distinguishes the field strains from AcMNPV-C6.

sequence, can be used to differentially detect two viruses in a multiplex PCR system in which all four primers are included in a single PCR reaction. This allows the detection of either or both of the virus species in a single reaction depending on the DNA template present. We have used an AcMNPV-specific primer set developed to *pe38* gene that produces a 619 bp AcMNPV-specific product and a TnSNPV-specific primer set developed to *alkaline exonuclease* that produces a 405 bp TnSNPV-specific product in a multiplex PCR to detect the prevalence of each virus in cabbage looper populations (Erlandson *et al.*, 2007) (Fig. 1.5). In addition, primer pairs are often designed to detect recombinant baculoviruses as distinguished from the wild-type (wt) parental virus. Typically one PCR primer is designed against a portion of the foreign gene and the second primer to the baculovirus gene promoter used in recombinant construction.

A more limited variety of gene targets have been used in detection and identification of RNA viruses. A few PCR detection systems for viruses associated with honeybees will be cited as examples of genes exploited. Tentcheva *et al.* (2004) describe an RT-PCR detection system for DWV based on PCR primers corresponding to an approximately 400 nucleotide region of the *RNA-dependent RNA polymerase* gene. As described previously, a first-step reverse transcriptase reaction is necessary and in this study the authors used a Thermoscript® RT-PCR kit (Invitrogen) with random hexamer primers to generate a cDNA from the 3′ end of the DWV genome. Bakonyi *et al.* (2002) used a one-step RT-PCR system to detect acute bee paralysis virus and primers targeted to the capsid protein gene to amplify a 355 bp region from a cDNA generated by reverse transcriptase.

### 1.6.3. Confirmation of infection and persistent infection status

Many honeybee viruses produce non-apparent infections and are difficult to detect; thus, the development of RT-PCR systems to detect a variety of small RNA

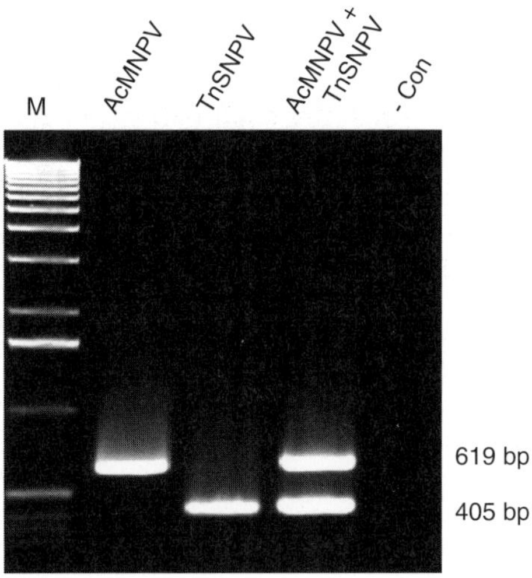


**Fig. 1.5.** Species-specific multiplex PCR assay for AcMNPV and TnSNPV. Purified genomic DNA from AcMNPV-HR3, TnSNPV-RJ or mixed DNA samples were subjected to multiplex PCR with both AcMNVP-specific *p38* and TnSNPV-specific *alkaline endonuclease* primers (10pmol), 0.5 units of *Taq* DNA polymerase (Invitrogen, Burlington, Ontario), 0.4mM dNPTs and 4mM $MgCl_2$ and electrophoresed on a 1.0% agarose gel (1×TAE). (From Erlandson *et al.*, 2007.)

viruses has been instrumental in virus screening. Tentcheva *et al.* (2004) used RT-PCR to detect DWV in adult bee, pupae and varroa mite samples from bee hives in France. Similarly Chen *et al.* (2005) used a series of six individual RT-PCR assays to detect six different viruses in queen bee samples. They demonstrated that queens could harbour multiple viruses suggesting that vertical transmission of viruses occurs and indeed DWV was detected in eggs and larval stages as well. Yue *et al.* (2006) were able to detect DWV and acute bee paralysis virus in the semen of honeybee drones suggesting an additional pathway for vertical transmission of these viruses.

PCR assays have also been used to detect insect DNA viruses in a variety of contexts and just a few examples are cited here. Burand *et al.* (1992) used primers targeted to the LdMNPV *polyhedrin* gene in PCR assays to determine the level of LdMNPV OB on gypsy moth eggs. They showed that the PCR assays could detect as little as five virus genome copies or one OB equivalent and indicated that PCR could be very useful in studies aimed at a better understanding of virus epizootiology as well as investigations of transovarial transmission of LdMNPV. Khurad *et al.* (2004) also used PCR detection of *Bombyx mori NPV* (BmNPV) in a study of vertical transmission of this virus in *B. mori*. More complex variations of PCR including real-time quantitative PCR have been used as a means to study baculovirus replication kinetics (Rosinski *et al.*, 2002) and for rapid titre determinations (Lo and Chao, 2004).

Other molecular approaches have been used to examine baculovirus infection and transmission cycles, among them RT-PCR to detect the level of expression of specific gene transcripts in various insect tissues. For example, Simón *et al.* (2004) used RT-PCR to examine immediate early (*ie-0*), early (*egt*, *DNA polymerase*), late (*chitinase*) and very late (*polyhedrin*) gene expression during SeMNPV and Spodoptera littoralis NPV (SpliNPV) infection of homologous and heterologous *Spodoptera* hosts. Hughes *et al.* (1997) used RT-PCR to examine the level of *Mamestra brassicae MNPV* (MbMNPV) *polyhedrin* expression in the messenger RNA (mRNA) pool extracted from larvae from a laboratory colony of *M. brassicae* in which there was evidence of vertical transmission of the MbMNPV virus. Their results indicated a low-level persistent MbMNPV infection in these insects by virtue of detection of MbMNPV *polyhedrin* transcripts, suggesting that the mechanism of maintaining a 'latent' infection may be similar to the measles model (Hughes *et al.*, 1997).

### 1.6.4. Detection of virus in environmental samples

For ecological studies examining virus persistence and cycling in insect populations and post-application tracking of the environmental fate of virus-based bioinsecticides, virus detection in environmental samples can be an important issue. Recovery of amplifiable AgMNPV DNA from soil samples spiked with known concentrations of AgMNPV OBs was examined by de Moraes *et al.* (1999). One key problem in purifying amplifiable DNA from soil samples is the presence of phenolic compounds and organic acids that can interfere with DNA polymerases. These authors examined two methods – phenol-ether extraction and magnetic capture-hybridization (MCH) – for extraction of AgMNPV DNA from soil. The MCH method proved superior and AgMNPV *polyhedrin*-specific PCR products could routinely be amplified from soil extracts. Similarly, England *et al.* (2005) have examined PCR-based methods for detecting recombinant baculovirus (CfMNPV $egt^-/lacZ^+$) DNA in aquatic microcosms designed to mimic forest ponds. This study utilized 0.5% Na pyrophosphate and isopropanol precipitation to concentrate virus DNA from spiked pond water samples, and DNA samples were extracted from pond sediments by incubation in 0.5% Na pyrophosphate, centrifugation and application of the supernatant to Sephadex g-75 spin columns. The extracted DNA was then detected by PCR using primers specific to the $egt^-/lacZ^+$ component of the CfMNPV recombinant. The detection limit for CfMNPV DNA in spiked water samples was 13.5 pg $ml^{-1}$ of pond water. The use of such microcosms and PCR techniques should be useful in determining the persistence of both intact virus particles as well as free DNA in the environment following application of viral-based bioinsecticides.

## 1.7. Conclusions

The increasing availability of molecular techniques has allowed more rapid and precise identification and detection of insect viruses. Continuing technical advances will allow for new approaches to ecological studies and post-application tracking of insect viruses in terms of persistence and cycling both in biological samples and the physical environment.

## References


Bakonyi, T., Farkas, R., Szendori, A., Dobos-Kovacs, M. and Rusvai, M. (2002) Detection of acute bee paralysis virus by RT-PCR in honeybee and *Varroa destructor* field samples: rapid screening of Hungarian apiries. *Apidologie* 33, 63–74.

Ball, L.A. (2005) The universal taxonomy of viruses in theory and practice. In: Fauquet, C.M., Mayo, M.A., Maniloff, M., Desselberger, U. and Ball, L.A. (eds) *Virus Taxonomy, Eighth Report of the International Committee on Virus Taxonomy*. Elsevier, San Diego, California, pp. 3–7.

Ball, L.A. and Johnson, L.K. (1998) Nodaviruses of insects. In: Miller, L.K. and Ball, L.A. (eds) *The Insect Virus.* Plenum Press. New York, pp. 225–267.

Belloncik, S. and Mori, H. (1998) Cypoviruses. In: Miller, L.K. and Ball, L.A. (eds) *The Insect Viruses.* Plenum Press, New York, pp. 337–369.

Bergoin, M. and Tijssen, P. (1998) Biological and molecular properties of densoviruses and their use in protein expression and biological control. In: Miller, L.K. and Ball, L.A. (eds) *The Insect Viruses.* Plenum Press, New York, pp. 141–169.

Bonning, B.C. (2005) Baculoviruses: biology, biochemistry, and molecular biology. In: Gilbert, L., Iatrou, K. and Gill, S. (eds) *Comprehensive Molecular Insect Science*, Vol. 6. Elsevier BV, Oxford, pp. 233–270.

Buller, R.M., Arif, B.M., Black, D.N., Dumbell, K.R., Esposito, J.J., Lefkowitz, E.J., McFadden, G., Moss, B., Mercer, A.A., Moyer, R.W., Skinner, M.A. and Tripathy, D.N. (2005) Family *Poxviridae*. In: Fauquet, C.M., Mayo, M.A., Maniloff, M., Desselberger, U. and Ball, L.A. (eds) *Virus Taxonomy, Eighth Report of the International Committee on Virus Taxonomy*. Elsevier, San Diego, California, pp. 117–132.

Burand, J.P., Horton, H.M., Retnasami, S. and Elkinton, J.S. (1992) The use of polymerase chain reaction and shortwave UV irradiation to detect baculovirus DNA on the surface of gypsy moth eggs. *Journal of Virological Methods* 36, 141–150.

Chen, Y., Pettis, J.S. and Feldlaufer, M.F. (2005) Detection of multiple viruses in queens of the honeybee *Apis mellifera* L. *Journal of Invertebrate Pathology* 90, 118–121.

Chinchar, V.G., Essbauer, S., He, J.G., Hyatt, A., Miyazaki, T., Seligy, V. and Williams, T. (2005) Family *Iridoviridae*. In: Fauquet, C.M., Mayo, M.A., Maniloff, M., Desselberger, U. and Ball, L.A. (eds) *Virus Taxonomy, Eighth Report of the International Committee on Virus Taxonomy*. Elsevier, San Diego, California, pp. 145–162.

Christian, P., Carstens, E., Domier, L., Johnson, K., Nakashima, N., Scotti, P. and van der Wilk, F. (2005a) *Dicistroviridae*. In: Fauquet, C.M., Mayo, M.A., Maniloff, M., Desselberger, U. and Ball, L.A. (eds) *Virus Taxonomy, Eighth Report of the International Committee on Virus Taxonomy*. Elsevier, San Diego, California, pp. 783–788.

Christian, P., Carstens, E., Domier, L., Johnson, K., Nakashima, N., Scotti, P. and van der Wilk, F. (2005b) *Iflavirus*. In: Fauquet, C.M., Mayo, M.A., Maniloff, M., Desselberger, U. and Ball, L.A. (eds) *Virus Taxonomy, Eighth Report of the International Committee on Virus Taxonomy*. Elsevier, San Diego, California, pp. 779–782.

Christian, P.D. and Scotti, P.D. (1998) Picornalike viruses of insects. In: Miller, L.K. and Ball, L.A. (eds) *The Insect Viruses*. Plenum Press, New York, pp. 301–336.

Crawford, A.M., Zelazny, B. and Alfiler, A.R. (1986) Genotypic variation in geographic isolates of *Oryctes* baculovirus. *Journal of General Virology* 67, 949–952.

Cory, J.S., Green, B.M., Paul, R.K. and Hunter-Fujita, F. (2005) Genotypic and phenotypic diversity of a baculovirus population within an individual insect host. *Journal of Invertebrate Pathology* 89, 101–111.

Cox-Foster, D.L., Conlan, S., Holmes, E.C., Palacios, G., Evans, J.D., Moran, N.A., Quan, P.-L., Briese, T., Horning, M., Geiser, D.M., Martinson, V., vanEngelsdorp, D., Kakstein, A.L., Drysdale, A., Hui, J., Zhai, J., Fui, L., Hutchison, S.K., Simons, J.F., Egholm, M., Pettis, J.S. and Lipkin, W.I.

(2007) A metagenomic survey of microbes in honeybee colony collapse disoreder. *Science* 318, 283–290.

Davis, L.G., Dibner, M.D. and Battey, J.F. (1986) *Basic Methods in Molecular Biology*. Elsevier, New York, 383 pp.

Dieffenbach, C.W. and Dveksler, G.S. (eds) (2003) *PCR Primer, A Laboratory Manual*, 2nd edn. Cold Spring Harbor Laboratory Press, Plainview, New York, 520 pp.

England, L.S., Pollock, J., Vincent, M., Kreutzweiser, D., Fick, W., Trevors, J.T. and Holmes, S.B. (2005) Persistence of extracellular baculoviral DNA in aquatic microcosms: extraction, purification, and amplification by polymerase chain reaction (PCR). *Molecular and Cellular Probes* 19, 75–80.

Erlandson, M.A. (1990) Biological and biochemical comparison of *Mamestra configurata* and *Mamestra brassicae* nuclear polyhedrosis virus isolates pathogenic for bertha armyworm: *Mamestra configurata* (Lepidoptera: Noctuidae). *Journal of Invertebrate Pathology* 56, 47–56.

Erlandson, M.A. and Street, D. (1997) Entomopoxviruses associated with grasshoppers and locusts: biochemical characterization. *Memoirs of the Entomological Society of Canada* 171, 131–146.

Erlandson, M.A., Newhouse, S., Moore, K., Janmatt, A., Myers, J. and Theilmann, D.A. (2007) Characterization of naturally occurring baculovirus isolates from *Trichoplusia ni* populations from vegetable greenhouses. *Biological Control* 41, 256–263.

Evans, H. and Shapiro, M. (1997) Viruses. In: Lacey, L.A. (ed.) *Manual of Techniques in Insect Pathology*. Academic Press, San Diego, California, pp. 17–53.

Fauquet, C.M., Mayo, M.A., Maniloff, J., Desselberger, U. and Ball, L.A. (2005) *Virus Taxonomy, Eighth Report of the International Committee on Virus Taxonomy*. Elsevier, San Diego, California, pp. 691–700.

Federici, B.A., Bigot, Y., Granados, R.R., Hamm, J.J., Miller, L.K., Newton, I., Stasiak, K. and Vlak, J.M. (2005) *Ascoviridae*. In: Fauquet, C.M., Mayo, M.A., Maniloff, M., Desselberger, U. and Ball, L.A. (eds) *Virus Taxonomy, Eighth Report of the International Committee on Virus Taxonomy*. Elsevier, San Diego, California, pp. 269–274.

Genersch, E. (2005) Development of a rapid and sensitive RT-PCR method for detection of deformed wing virus, a pathogen of the honeybee (*Apis mellifera*). *The Veterinary Journal* 169, 121–123.

Gordon, K.H.J. and Hanzlik, T.N. (1998) Tetraviruses. In: Miller, L.K. and Ball, L.A. (eds) *The Insect Viruses*. Plenum Press, New York, pp. 269–299.

Graham, R.I., Tyne, W.I., Possee, R.D., Sait, S.M. and Hails, R.S. (2004) Genetically variable nucleopolyhedroviruses isolated from spatially separate populations of the winter moth *Operophtera brumata* (Lepidoptera: Geometridae) in Orkney. *Journal of Invertebrate Pathology* 87, 29–38.

Green, T.B., Shapiro, A., White, S., Rao, S., Mertens, P.P.C., Carner, G. and Becnel, J.J. (2006) Molecular and biological characterization of a Cypovirus from the mosquito *Culex restuans*. *Journal of Invertebrate Pathology* 91, 27–34.

Hanzlik, T.N., Gordon, K.H.J., Gorbalenya, A.E., Hendry, D.A., Pringle, F.M., Ward, V.K. and Zeddam, J-L. (2005) *Tetraviridae*. In: Fauquet, C.M., Mayo, M.A., Maniloff, M., Desselberger, U. and Ball, L.A. (eds) *Virus Taxonomy, Eighth Report of the International Committee on Virus Taxonomy*. Elsevier, San Diego, California, pp. 873–883.

Harlow, E. and Lane, D. (1988) *Antibodies, A Laboratory Manual*. Cold Spring Harbor Laboratory Press, Plainview, New York, 726 pp.

Herniou, E.A., Olszewski, J.A., Cory, J.S. and O'Reilly, D.R. (2003) The genome sequence and evolution of baculoviruses. *Annual Review of Entomology* 48, 211–234.

Hughes, D.S., Possee, R.D. and King, L.A. (1997) Evidence for the presence of a low-level, persitent baculovirus infection of *Mamestra bassicae insects*. *Journal of General Virology* 78, 1801–1805.

Innis, M.A., Gelfand, D.H., Sninsky, J.J. and White, T.J. (eds) (1990) *PCR Protocols, A Guide to Methods and Applications*. Academic Press, San Diego, California, pp. 482.

Jehle, J.A., Blissard, G.W., Bonning, B.C., Cory, J.S., Herniou, E.A., Rohrmann, G.F., Theilmann, Thiem, S.M. and Vlak, J.M. (2006a) On the classification and nomenclature of baculoviruses: a proposal for revision. *Arctives of Virology* 151, 1257–1266.

Jehle, J.A., Lange, M., Wang, H., Hu, Z., Wang, Y. and Hauschild, R. (2006b) Molecular identification and phylogenetic analysis of baculoviruses from Lepidoptera. *Virology* 346, 180–193.

Kaupp, W.J. and Ebling, P.M. (1993) Horseradish peroxidase-labelled probes and enhanced chemiluminescence to detect baculoviruses in gypsy moth and eastern spruce budworm larvae. *Journal of Virological Methods* 44, 89–98.

Keating, S.T., Burand, J.P. and Elkinton, J.S. (1989) DNA hybridization assay for detection of gypsy moth nuclear polyhedrosis virus in infected gypsy moth (*Lymantria dispar* L.) larvae. *Applied and Environmental Microbiology* 55, 2749–2754.

Kelly, D.C. (1977) The DNA contained by nuclear polyhedrosis viruses isolated from four *Spodoptera* sp. (Lepidoptera, Noctuidae): genome size and homology assessed by DNA reassociation kinetics. *Virology* 76, 468–471.

Kelly, D.C. and Avery, R.J. (1974) The DNA content of four small iridescent viruses: genome size, redundancy, and homology determined by renaturation kinetics. *Virology* 57, 425–435.

King, L.A., Wilkinson, N., Miller, D.P. and Marlow, S.A. (1998) Entompoxviruses. In: Miller, L.K. and Ball, L.A. (eds) *The Insect Viruses*. Plenum Press, New York, pp. 1–29.

Khurad, A.M., Mahulikar, A., Rathod, M.K., Rai, M.M., Kanginakurdu, S. and Nagaraju, J. (2004) Vertical transmission of nucleopolyhedrovirus in silkworm, *Bombyx mori* L. *Journal of Invertebrate Pathology* 87, 8–15.

Kroemer, J.A. and Webb, B.A. (2004) Polydnavirus genes and genomes: emerging gene families and new insights into polydnavirus replication. *Annual Review of Entomology* 49, 431–456.

Kukielka, D., Esperón, F., Higes, M. and Sánchez-Vizcaíno, J. (2008) A sensitive one-step real-time RT-PCR method of detection of deformed wing virus and black queen cell virus in honeybee *Apis mellifera*. *Journal of Virological Methods* 147, 275–281.

Lange, M., Wang, H., Zhihong, H. and Jehle, J.A. (2004) Towards a molecular identification of and classification system of lepidopteran-specific baculoviruses. *Virology* 325, 36–47.

Lo, H-R. and Chao, Y-C. (2004) Rapid titre determination of baculovirus by quantitative real-time polymerase chain reaction. *Biotechnology Progress* 20, 354–360.

Lupiani, B., Raina, A.K. and Huber, C. (1999) Development and use of a PCR assay for detection of reproductive virus in wild populations of *Helicoverpa zea* (Lepidoptera: Noctuidae). *Journal of Invertebrate Pathology* 73, 107–112.

Malitschek, B. and Schart, M. (1991) Rapid identification of recombinant baculoviruses using PCR. *BioTechniques* 11, 177–178.

McCarthy, W.J. and Gettig, R.R. (1986) Current developments in baculovirus serology. In: Granados, R.R. and Federici, B.A. (eds) *The Biology of Baculoviruses*, Vol I. CRC Press, Boca Raton, Florida, pp. 147–148.

McCarthy, C.B. and Theilmann, D.A. (2008) AcMNPV ac143(odve 18) is essential for mediating budded virus production and is the 30th baculovirus core gene. Virology 375(1), 277–291.

McIntosh, A.H., Rice, W.C. and Ignoffo, C.M. (1987) Genotypic variants in wild-type populations of baculovirus. In: Maramorosch, K. (ed.) *Biotechnology in Invertebrate Pathology and Cell Culture*. Academic Press, New York, pp. 305–325.

Mertens, P.P.C., Rao, S. and Zhou, Z.H. (2005) *Cypovirus*. In: Fauquet, C.M., Mayo, M.A., Maniloff, M., Desselberger, U. and Ball, L.A. (eds) *Virus Taxonomy, Eighth Report of the International Committee on Virus Taxonomy*. Elsevier, San Diego, California, pp. 522–533.

Miller, L.K. (1997) *The Baculoviruses*. Plenum Press, New York, pp. 447.

Miller, L.K. (1998) Ascoviruses. In: Miller, L.K. and Ball, L.A. (eds) *The Insect Viruses*. Plenum Press, New York, pp. 91–103.

de Moraes, R.R. and Maruniak, J.E. (1997) Detection and identification of multiple baculoviruses using polymerase chain reaction (PCR) and restriction endonuclease analysis. *Journal of Virological Methods* 63, 209–217.

de Moraes, R.R., Maruniak, J.E. and Funderburk, J.E. (1999) Methods for detection of *Anticarsia gemmatalis* nucleopolyhedrovirus DNA in soil. *Applied and Environmental Mircobiology* 65, 2307–2311.

Moscardi, F. (1999) Assessment of the application of baculoviruses for control of Lepidoptera. *Annual Review of Entomology* 44, 257–289.

Muñoz, D., Castillejo, J.I. and Caballero, P. (1998) Naturally occurring deletion mutants are parasitic genotypes in a wild-type nucleopolyhedrovirus population of *Spodoptera exigua*. *Applied and Environmental Mircobiology* 64, 4372–4377.

Nakashima, N., Kawahara, N., Omura, T. and Noda, H. (2006) Characterization of a novel satellite virus and a strain of Himetobi P virus (Dicistroviridae) from the brown planthopper, *Nilaparvata lugens*. *Journal of Invertebrate Pathology* 91, 53–56.

Possee, R.D. and Rohrmann, G.F. (1997) Baculovirus genome organization and evolution. In: Miller, L.K. (ed.) *The Baculoviruses*. Plenum Press, New York, pp. 109–140.

Payne, C.C. and Mertens, P.P.C. (1983) Cytoplasmic polyherdrosis viruses. In: Joklik, W.K. (ed.) *The Reoviridae*. Plenum Press, New York, pp. 425–504.

Rosinski, M., Reid, S. and Nielsen, L.K. (2002) Kinetics of baculovirus replication and release using real-time quantitative polymerase chain reaction. *Biotechnology and Bioengineering* 77, 476–480.

Sambrook, J. and Russell, D.W. (2001) *Molecular Cloning: A Laboratory Manual*, 3rd edn, 3 vols. Cold Spring Harbor Laboratory Press, Plainview, New York, 726 pp.

Scheemann, A., Ball, L.A., Delsert, C., Johnson, J.E. and Nishizawa, T. (2005) *Nodaviridae*. In: Fauquet, C.M., Mayo, M.A., Maniloff, M., Desselberger, U. and Ball, L.A. (eds) *Virus Taxonomy, Eighth Report of the International Committee on Virus Taxonomy*. Elsevier, San Diego, California, pp. 365–369.

Simón, O., Williams, T., López-Ferber, M. and Caballero, P. (2004) Virus entry or primary infection cycles are not the principal determinants of host specificity of *Spodoptera* spp. nucleopolyhedroviruses. *Journal of General Virology* 85, 2845–2855.

Smith, G.E. and Summers, M.D. (1982) DNA homology among subgroup A, B, and C baculoviruses. *Virology* 123, 393–406.

Smith, I.R.L. and Crook, N.E. (1988) *In vivo* isolation of baculovirus genotypes. *Virology* 166, 240–244.

Southern, E.M. (1975) Detection of specific sequences among DNA fragments separated by gel electrophoresis. *Journal of Molecular Biology* 98, 503–517.

Tattersall, P., Bergoin, M., Bloom, M.E., Brown, K.E., Linden, R.M., Muzyczka, N., Parrish, C.R. and Tijssen, P. (2005) *Densovirinae*. In: Fauquet, C.M., Mayo, M.A., Maniloff, M., Desselberger, U. and Ball, L.A. (eds) *Virus Taxonomy, Eighth Report of the International Committee on Virus Taxonomy*. Elsevier, San Diego, California, pp. 365–369.

Tentcheva, D., Gauthier, L.G., Sandrine, J., Canabady-Rochelle, L., Dainat, B., Cousserans, F., Colin, M.E., Ball, B.V. and Bergoin, M. (2004) Polymerase chain reaction detection of deformed wing virus (DWV) in *Apis mellifera* and *Varroa destructor. Apidologie* 35, 431–439.

Theilmann, D.A., Blissard, G.W., Bonning, B., Jehle, J., O'Reilly, D.R., Rohrmann, G.F., Theim, S. and Vlak, J. (2005) Family *Baculoviridae*. In: Fauquet, C.M., Mayo, M.A., Maniloff, M., Desselberger, U. and Ball, L.A. (eds) *Virus Taxonomy, Eighth Report of the International Committee on Virus Taxonomy*. Elsevier, San Diego, California, pp. 177–185.

Thorne, C.M., Otvos, I.S., Conder, N. and Levin, D.B. (2007) Development of a dipstick immunoassay to detect nucleopolyhedroviruses in Douglas-fir tussock moth larvae. *Journal of Virological Methods* 146, 188–195.

Tompkins, G.J. (1991) Purification of invertebrate viruses. In: Adams, J.R. and Bonami, J.R. (eds) *Atlas of Invertebrate Viruses*. CRC Press, Boca Raton, Florida, pp. 31–40.

Vail, P.V., Hostetter, D.L. and Hoffmann, D.F. (1999) Development of multi-nucleocapsid nucleopolyhedrovirses (MNPVs) infectious to loopers (Lepidoptera: Noctuidae: Plusiinae) as microbial control agents. *Integrated Pest Management Reviews* 4, 231–257.

Van Regenmortel, M.H.V. (1990) Virus species, a much overlooked but essential concept in virus classification. *Intervirology* 31, 241–254.
Wang, C-H., Yang, H-N., Liu, H-C., Kou, G-H. and Lo, C-F. (2000) Nested polymerase chain reaction and *in situ* hybridization for detection of nucleopolyhdrosis. *Journal of Virological Methods* 84, 65–75.
Ward, V.K., Fleming, S.B. and Kalmakoff, J. (1987) Comparison of a DNA-DNA dot-blot hybridization assay with light microscopy and radioimmunoassay for detection of a nuclear polyhedrosis virus. *Journal of Virological Methods* 15, 65–73.
Webb, B.A. (1998) Polydnavirus biology, genome structure, and evolution. In: Miller, L.K. and Ball, L.A. (eds) *The Insect Viruses*. Plenum Press, New York, pp. 105–139.
Webb, B.A., Beckage, N.E., Hayakwa, Y., Lanzrein, B., Stoltz, D.B., Strand, M.R. and Summers, M.D. (2005) Family *Polydnaviridae*. In: Fauquet, C.M., Mayo, M.A., Maniloff, M., Desselberger, U. and Ball, L.A. (eds) *Virus Taxonomy, Eighth Report of the International Committee on Virus Taxonomy*. Elsevier, San Diego, California, pp. 255–267.
Wiedbrauk, D.L. and Farkas, D.H. (eds) (1995) *Molecular Methods for Virus Detection*. Academic Press, San Diego, California, p. 386.
Williams, T. (1994) Comparative studies of iridoviruses: further support for a new classification. *Virus Research* 33, 99–121.
Williams, T. (1998) Invertebrate iridescent viruses. In: Miller, L.K. and Ball, L.A. (eds) *The Insect Viruses*. Plenum Press, New York, pp. 105–139.
Woo, S.D. (2001) Rapid detection of multiple nucleopolyhedroviruses using polymerase chain reaction. *Molecules Cells* 3, 334–340.
Wu, C-Y. and Wang, C-H. (2005) Characterization and *polyhedrin* gene cloning of *Lymantria xylina* multiple nucleopolyhedrovirus. *Journal of Invertebrate Pathology* 88, 238–246.
Yue, C. and Genersch, E. (2005) RT-PCR analysis of *Deformed wing virus* (DWV) in honeybees (*Apis mellifera*) and mites (*Varroa destructor*). *Journal of General Virology* 86, 3419–3424.
Yue, C., Schröder, M., Bienefeld, K. and Genersch, K. (2006) Detection of viral sequences in semen of honeybees (*Apis mellifera*): evidence for vertical transmission of viruses through drones. *Journal of Invertebrate Pathology* 92, 105–108.

# 2 Molecular Approaches and Techniques for the Study of Entomopathogenic Bacteria

N. Boemare and P. Tailliez

*INRA, UMR1133 Laboratoire EMIP, Montpellier, France; Université Montpellier II, UMR1133 Laboratoire EMIP, Montpellier, France*



## 2.1. Introduction

Entomopathogenic bacteria have colonized diverse habitats and can be isolated from water, soils, plants, animals and humans. They are able to multiply in the insect haemocoel from small inocula composed of less than 10,000 viable cells and produce fatal septicaemia (Bucher, 1960). Furthermore, they have developed different strategies to interact with, and kill, insects, and are characterized by having a wide range of putative virulence factors, including insecticidal toxins, exoenzymes and haemolysins. Entomopathogenic bacteria are either facultative or obligate pathogens (Bucher, 1960; Krieg, 1981). After ingestion, only obligate pathogenic bacteria such as *Bacillus thuringiensis* are able to damage the healthy gut wall and act by toxaemia rather than septicaemia (Schnepf *et al.*, 1998). In the case of *Paenibacillus larvae*, the spores germinate in the midgut lumen of the insects. The vegetative bacteria massively proliferate within the midgut before starting to locally breach the epithelium using the paracellular route and then invade the haemocoel (Yue *et al.*, 2008). As a result of bacterial growth in the haemocoel, bacteraemia occurs, often followed by fatal septicaemia. In contrast, facultative pathogenic bacteria such as *Serratia marcescens* cannot overcome the gut barrier of insects except if these latter are under severe stress (Matsumoto *et al.*, 2003). Septicaemia can develop only if such bacteria are injected directly into the haemocoel. In the case of *Photorhabdus* and *Xenorhabdus*, bacterial cells penetrate the haemocoel of insects using their nematode vectors and then multiply and kill the host within 24–48 h. However, some *Photorhabdus* strains are also able to kill insects after ingestion (Bowen *et al.*, 1998).

Entomopathogenic bacteria are distributed in phylogenetically diverse groups of prokaryotes. Gram-negative entomopathogenic bacteria include: (i) members of the Enterorbacteriaceae such as *Serratia entomophila*, *S. marcescens*, *Serratia proteamaculans* and the nematophilic genera *Photorhabdus* and *Xenorhabdus*; and (ii) members of the Pseudomonadaceae such as *Pseudomonas entomophila*. Gram-positive entomopathogenic bacteria comprise the most studied group of bacteria in insect pathology and include members of the genus *Bacillus* (Bacillaceae). This genus has recently been submitted to several taxonomic revisions, and a number of new genera such as *Brevibacillus*, *Lysinibacillus* and *Paenibacillus* have been recognized from species previously described in the genus *Bacillus* (Ash *et al.*, 1993; Shida *et al.*, 1996; Ahmed *et al.*, 2007).

In this chapter, we summarize the most common molecular methods and techniques considered for identification and diagnosis of entomopathogenic bacteria. Further protocols and techniques can also be found in Priest (1993), Stackebrandt (2006) and Chapters 6 (Tailliez and Boemare, this volume), 11 (Goodrich-Blair *et al.*, this volume) and 13 (Gaudriault and Duchaud, this volume).

## 2.2. Classification of Entomopathogenic Bacteria

### 2.2.1. Phenotypic screening and identification of entomopathogenic bacteria

Isolation and identification of entomopathogenic bacteria require expertise in various fields including biochemistry, serology, physiology and molecular biology. Phenotypic screening is usually considered the first step in the identification of bacterial strains as it allows a rough comparison and placement into similar or different groups. Most commonly considered phenotypic traits include: (i) colony morphology (size and shape); (ii) colony pigmentation; (iii) cell shape and size; (iv) motility and flagellar arrangement; (v) Gram stain reaction; and (vi) aerobic or/and anaerobic metabolisms. For example, entomopathogenic bacilli (i.e. members of the family Bacillaceae) are Gram-positive, endosporeforming, aerobic or facultative rod-shaped bacteria. Members of the genus *Clostridia*, including entomopathogenic isolates of *Clostridium bifermentans*, can be distinguished from bacilli by their anaerobic metabolism. Members of the family Enterobacteriaceae including entomopathogenic nematodes symbionts are Gram-negative, oxidase-negative, asporogenous, rod-shaped bacteria generally motile by peritrichous flagellae and able to grow both aerobically and anaerobically. Diagnostic morphological characters common to Pseudomonads are: rod-shaped cells, presence of polar flagella. Pseudomonadaceae are also Gram-negative and have an aerobic metabolism.

Thiéry and Frachon (1997) provided a complete summary of the most commonly considered methods for isolation, culture and preservation of entomopathogenic bacteria, and readers should refer to this publication for further information. *Bergey's Manual of Systematic Bacteriology* (Garrity *et al.*, 2005) should

**Table 2.1.** Entomopathogenic bacteria.

| | Insect target | Interaction with insect – mode of action |
|---|---|---|
| **Gram-positive endospore-forming bacteria** | | |
| *Bacillus thuringiensis* (Berliner) | Wide spectrum | – Release of active toxin by processing of protoxin by insect midgut proteases |
| *B. thuringiensis* var. *israelensis* (Goldberg and Margalit, 1977; de Barjac, 1978) | Mosquito *Aedes*, *Culex* | – Insertion of toxin into apical membrane, creation of ion channels<br>– Leakage of intracellular contents, death of insects |
| *Brevibacillus laterosporus* (Claus and Berkeley, 1986; Shida *et al.*, 1996) | Wide spectrum but limited potential for mosquito control | – Toxin produced during the vegetative phase of the bacterial growth cycle<br>– Toxin maintained during sporulation |
| *Lysinibacillus sphaericus* (Ahmed *et al.*, 2007) | Mosquito *Anopheles*, *Culex* | – Mode of action likely similar to that of *B. thuringiensis* var. *israelensis* |
| *Paenibacillus popilliae* (Dutky, 1940; Pettersson *et al.*, 1999) | Scarab larvae | – Toxicity by injection of intact or solubilized parasporal bodies into haemocoel<br>– Toxicity following multiplication and sporulation of vegetative rods in the haemolymph |
| *Paenibacillus larvae* (White, 1906; Heyndrickx *et al.*, 1996; Genersch *et al.*, 2006) | Honeybee larvae (American foul brood) | – Rapid germination of spores in the midgut of honeybee larvae<br>– Massive proliferation in the midgut lumen of vegetative *P. larvae*<br>– Penetration of the midgut epithelium by vegetative cells mainly via the paracellular route and invasion of the haemocoel |
| *Clostridium bifermentans* serovar *malaysia* (de Barjac *et al.*, 1990) | Mosquito Blackfly larvae | – Mode of action likely similar to that of *B. thuringiensis* var. *israelensis* |
| **Enterobacteria** | | |
| *Serratia entomophila* (Grimont *et al.*, 1988) | New Zealand Grass grub *Costelytra zealandica* (Amber disease) | – Cessation of feeding and gut clearance<br>– Amber coloration of the grub<br>– Invasion of the haemocoel |
| *Serratia marcescens* (Grimont and Grimont, 2005) | Opportunistic Wide spectrum | – Invasion of insect haemocoel following injury or stress<br>– Increase apoptosis of insect brain cells by an influx of dopamine from the haemolymph |
| *Photorhabdus* sp. (Boemare *et al.*, 1993) | Wide spectrum | – Released of symbiotic bacteria in the haemocoel of infected insects by nematodes of the family Heterorhabditidae<br>– Invasion of the haemocoel by the bacteria provoking toxaemia and septicaemia |

*Continued*

**Table 2.1.** Continued

| | Insect target | Interaction with insect – mode of action |
|---|---|---|
| *Xenorhabdus* sp. (Thomas and Poinar, 1979) | Wide spectrum | – Release of symbiotic bacteria in the haemocoel of infected insects by nematodes of the family Steinernematidae<br>– Invasion of the haemocoel by the bacteria provoking toxaemia and septicaemia |
| **Pseudomonads** | | |
| *Pseudomonas entomophila* (Vodovar *et al.*, 2005) | Wide spectrum | – Strong perturbation of the midgut epithelium<br>– Food uptake blockage |

also be consulted for complete descriptions of phenotypic traits and techniques. A summary table of most studied entomopathogenic bacteria and their taxonomic position are listed in Table 2.1. The following web site also provides additional information on this subject: http://www.ebi.ac.uk/2can/genomes/bacteria.html.

### 2.2.2. Molecular approaches used for identification and diagnosis

Molecular biology revolutionized systematics and classification of bacteria. Major contributions in this field were made during the 1970s and 1980s, specifically the study of G + C (guanine plus cytosine) contents in bacterial chromosomal DNA became a valuable method for differentiating *Bacillus* spp. and strains. DNA reassociation methods were also widely applied during the 1980s, particularly for the study of species complexes. With this method, strains that have DNA relatedness greater than 70% are considered conspecific (Stackebrandt and Goebel, 1994).

Consideration of molecular diagnosis for entomopathogenic bacteria isolates is now viewed as a 'must-do-step' for accurate identification and diagnostic purposes. In particular, molecular approaches have been considered critical for studying species complexes and also to help define species. The study of the genetic diversity among bacterial species and strains has also had an impact in assessing relatedness among species and understanding their diversity.

Among the most commonly used methods are: DNA–DNA relatedness, DNA typing and nucleic acid sequencing.

#### *2.2.2.1. The bacteria genome*

Bacteria are characterized by having one single chromosome within the nucleoid. The size of this chromosome varies among species. For example, the bacterial genomes of *B. thuringiensis* (the widely studied insect pathogen), *Bacillus cereus* and *Bacillus anthracis* are around 5.4 Mb (Carlson and Kolstø, 1993; Ivanova *et al.*, 2003).

Other *Bacillus* spp. such as *Bacillus subtilis* and *Bacillus licheniformis* have a smaller chromosome (4.2 Mb) (Kunst *et al.*, 1997; Rey *et al.*, 2004). Entomopathogenic nematode symbiotic bacteria *Photorhabdus luminescens* has a genome size of 5.7 Mb (Duchaud *et al.*, 2003), whereas *Xenorhabdus bovienii* and *Xenorhabdus nematophila*, other nematode symbiotic bacteria, are estimated to have a genome size of approximately 4.3 Mb (http://www.xenorhabdus.org/).

Many bacteria also contain extrachromosomal elements (plasmids), which are smaller double-stranded DNA (dsDNA) molecules that replicate independent of the chromosomal DNA. Genes located in bacterial plasmids usually code for proteins that determine specific phenotypes, but do not code for products needed for bacterial survival and growth. Bacterial genes are organized in operons or cassettes that consist of a promoter, a series of genes and a transcription terminator.

The most important genes studied for entomopathogenic bacteria are the *Cry* genes that code for insecticidal crystal proteins (*B. thuringiensis*). These genes are usually found in large transmissible plasmids or more rarely in the chromosome. These proteins show entomopathogenic properties to insects from orders Lepidoptera, Diptera and Coleoptera (WHO, 1999). Different combinations of *Cry* genes are found in various *B. thuringiensis* strains including those with one, two or even four different genes (Lereclus *et al.*, 1993). More than 200 toxin genes with pesticidal activity have been cloned from a wide range of *B. thuringiensis* strains. The genes can be grouped into 80 classes (Crickmore *et al.*, 1998). Sequence analysis of these genes is currently considered not only relevant to the classification of bacteria species but also for interpreting evolutionary relationships among taxa (Crickmore, 2000).

Bacterial genome studies have provided extremely valuable information regarding the genetic diversity of species. Not only have they assisted in resolving taxonomies at various levels, but they have also contributed to assessing evolutionary relationships. Readers should refer to Chapters 13 (Gaudriault and Duchaud, this volume) and 14 (Slack *et al.*, this volume) for further information on bacterial genomes.

#### *2.2.2.2. Methods*

DNA–DNA HYBRIDIZATION. Bacterial species are currently defined as a collection of strains characterized by at least one distinctive diagnostic phenotypic trait and by having at least 70% DNA–DNA relatedness. Strains with a lesser DNA–DNA relatedness value are considered different species (Wayne *et al.*, 1987). Because of the comparative nature of this molecular technique, it is necessary to initiate identification of new bacterial isolates considering phenotypic traits followed by 16S ribosomal RNA (rRNA) gene sequence comparisons (see sections below).

Two standard procedures for quantitative DNA–DNA hybridization are mainly used: (i) DNA labelling with 32P by nick-translation; and (ii) DNA hybridization by the hydroxyapatite (HA) method (Brenner *et al.*, 1969).

DNA LABELLING WITH 32P BY NICK-TRANSLATION. After equilibration with 0.14 M sodium phosphate pH 6.8 (PB) and 0.1% sodium dodecyl sulfate, columns of HA are plunged into a thermoregulated ethylene glycol bath. The DNA solutions are pipetted into the columns. When the liquid level is close to the bed surface, portions of 0.14 M phosphate buffer are added in order to elute the single-stranded DNA and duplex DNA is progressively denatured and eluted with 0.14 M PB at increasing temperatures. Residual adsorbed materials (dsDNA) are then eluted with 0.40 M phosphate buffer.

DNA HYBRIDIZATION BY THE HYDROXYAPATITE (HA). In this procedure, DNA is labelled with tritium-labelled nucleotides and DNA is hybridized by the Sl nuclease-trichloroacetic acid method (Crosa *et al.*, 1973). Labelled single-stranded DNA is eliminated by action of the S1 nuclease. The labelled double-stranded which is S1-resistant is the homogeneous part of the heteroduplexes.

Both methods provide a dsDNA (homologous) and single-stranded DNA. The relative binding ratio (RBR) is the percentage of value of homologous DNA in the heteroduplex. It gives an estimate of the similarity of both hybridized DNA. It has been shown that RBR determined by the HA method and by S1 nuclease methods is similar for a given pair of DNAs (Grimont *et al.*, 1980).

Temperature at which 50% of the DNA is eluted by 0.14 M sodium phosphate buffer ($Tm(e)$) is the DNA denaturation point. $Tm(e)$ is determined graphically as the temperature at which half of the renatured DNA is eluted. The value called $\Delta Tm(e)$ is the difference between the $Tm(e)$ of the homologous reaction and the $Tm(e)$ of the heterologous reaction. So $\Delta Tm(e)$ is an estimate of the divergence between two DNAs. $\Delta Tm$ values are determined by the method of Crosa *et al.* (1973).

*Photorhabdus* and *Xenorhabdus* (Table 2.2), bacterial symbionts of entomopathogenic nematodes, represent a good example of the value of this method for delimitation of genera and species. Briefly, these two genera were originally considered in a single genus *Xenorhabdus* on the basis of basic phenotypic traits and of their intimate association with entomopathogenic nematodes (Thomas and Poinar, 1979). Subsequent studies based on DNA–DNA relatedness and phenotypic characters showed that there were important differences between *Xenorhabdus luminescens* (a symbiont of Heterorhabditid nematodes) and the other *Xenorhabdus* spp. (symbionts of Steinernematid nematodes) (Grimont *et al.*, 1984; Akhurst and Boemare, 1988; Boemare *et al.*, 1993). The formerly described *X. luminescens* was then transferred into a new genus, *Photorhabdus*, as *P. luminescens* (Boemare *et al.*, 1993).

DNA relatedness studies also assisted in the identification of species within *Photorhabdus* (Fischer-Le Saux *et al.*, 1999). *P. luminescens* sp. includes three subgroups based on DNA–DNA relatedness (more than 80% of DNA binding with $\Delta Tm$ < 1.5°C) and phenotypic characters: *P. luminescens* ssp. *akhurstii*, *P. luminescens* ssp. *laumondii* and *P. luminescens* ssp. *luminescens* (Fischer-Le Saux *et al.*, 1999). A subgroup corresponding to *Photorhabdus temperata* ssp. *temperata* was defined within the species *P. temperata* (Fischer-Le Saux *et al.*, 1999). At least, the species *Photorhabdus asymbiotica* currently comprises two subspecies *P. asymbiotica* ssp. *asymbiotica* for the American strains and *P. asymbiotica* ssp. *australis* for the Australian strains (Akhurst *et al.*, 2004).

In a similar manner, DNA–DNA relatedness methods also contributed to the identification of *Xenorhabdus* sp. initially characterized by phenotypic approaches

**Table 2.2.** Species and subspecies of the genus *Photorhabdus* and *Xenorhabdus* associated with the entomopathogenic nematodes.

| Species/subspecies | References | Type strain | Nematode host | 16S rDNA accession number |
|---|---|---|---|---|
| ***Photorhabdus* Boemare *et al.*, 1993** | | | | |
| *P. asymbiotica* ssp. *asymbiotica* | Fischer-Le Saux *et al.*, 1999 | ATCC43950$^T$ (3265–86$^T$) | *No host* | Z76755 |
| *P. asymbiotica* ssp. *australis* | Akhurst *et al.*, 2004 | CIP108025$^T$ (9802892$^T$) | *No host* | AY280572 |
| *P. luminescens* ssp. *akhurstii* | Fischer-Le Saux *et al.*, 1999 | CIP105564$^T$ (FRG04$^T$) | *Heterorhabditis indica* | AJ007359 |
| *P. luminescens* ssp. *kayaii* | Hazir *et al.*, 2004 | DSM15194$^T$ (1121$^T$) | *H. bacteriophora* | AJ560630 |
| *P. luminescens* ssp. *laumondii* | Fischer-Le Saux *et al.*, 1999 | CIP105565$^T$ (TT01$^T$) | *H. bacteriophora* | AJ007404 |
| *P. luminescens* ssp. *luminescens* | Fischer-Le Saux *et al.*, 1999 | DSM3368$^T$ (Hb$^T$) | *H. bacteriophora* | AY278640 |
| *P. luminescens* ssp. *thracensis* | Hazir *et al.*, 2004 | DSM15199$^T$ (39–8$^T$) | *H. bacteriophora* | AJ560634 |
| *P. temperata* | Fischer-Le Saux *et al.*, 1999 | CIP105563$^T$ (XINach$^T$) | *H. megidis* | AJ007405 |
| *P. temperata* ssp. *temperata* | Fischer-Le Saux *et al.*, 1999 | CIP105563$^T$ (XINach$^T$) | *H. megidis* | AJ007405 |
| ***Xenorhabdus* Thomas and Poinar, 1979** | | | | |
| *X. beddingii* | Akhurst, 1986a | DSM4764$^T$ (Q58$^T$) | Unknown | AY278675 |
| *X. bovienii* | Akhurst, 1983 | DSM4766$^T$ (T228$^T$) | *Steinernema feltiae* | X82252 |
| *X. budapestensis* | Lengyel *et al.*, 2005 | DSM16342$^T$ | *S. bicornutum* | AJ810293 |
| *X. cabanillasii* | Tailliez *et al.*, 2006 | DSM17905$^T$ (USTX62$^T$) | *S. riobrave* | AY521244 |
| *X. doucetiae* | Tailliez *et al.*, 2006 | DSM17909$^T$ (FRM16$^T$) | *S. diaprepesi* | DQ211709 |
| *X. elhersii* | Lengyel *et al.*, 2005 | DSM16337$^T$ | *S. longicaudum* | AJ810294 |
| *X. griffiniae* | Tailliez *et al.*, 2006 | DSM17911$^T$ (ID10$^T$) | *S. hermaphroditum* | DQ211710 |
| *X. hominickii* | Tailliez *et al.*, 2006 | DSM17903$^T$ (KE01$^T$) | *S. karii* | DQ211719 |
| *X. indica* | Somvanshi *et al.*, 2006 | DSM17382$^T$ | *S. thermophilum* | AM040494 |
| *X. innexi* | Lengyel *et al.*, 2005 | DSM16336$^T$ | *S. scapterisci* | AJ810292 |
| *X. japonica* | Nishimura *et al.*, 1994 | DSM16522$^T$ | *S. kushidai* | DQ202310 |
| *X. koppenhoeferi* | Tailliez *et al.*, 2006 | DSM18168$^T$ (USNJ01$^T$) | *S. scarabaei* | DQ205450 |
| *X. kozodoii* | Tailliez *et al.*, 2006 | DSM 17907$^T$ (SaV$^T$) | *S. arenarium* | DQ211716 |
| *X. mauleonii* | Tailliez *et al.*, 2006 | DSM17908$^T$ (VC01$^T$) | *Unknown* | DQ211715 |
| *X. miraniensis* | Tailliez *et al.*, 2006 | DSM17902$^T$ (Q1$^T$) | Steinernematidae | DQ211713 |
| *X. nematophila* | Poinar and Thomas, 1965 | DSM3370$^T$ (AN6$^T$) | *S. carpocapsae* | AY278674 |
| *X. poinarii* | Akhurst, 1986b | CIP109143$^T$ (G6$^T$) | *S. glaseri* | D78010 |
| *X. romanii* | Tailliez *et al.*, 2006 | DSM17910$^T$ (PR06-A$^T$) | *S. puertoricense* | DQ211717 |
| *X. stockiae* | Tailliez *et al.*, 2006 | DSM17904$^T$ (TH01$^T$) | *S. siamkayai* | DQ202309 |
| *X. szentirmaii* | Lengyel *et al.*, 2005 | DSM16338$^T$ | *S. rarum* | AJ810295 |

(*Xenorhabdus beddingii, X. bovienii, Xenorhabdus japonica, X. nematophila* and *Xenorhabdus poinarii*) (Boemare *et al.*, 1993; Nishimura *et al.*, 1994) and also assisted in the identification of novel species such as *Xenorhabdus miraniensis (isolate Q1), Xenorhabdus kozodoii* (isolate SaV) and *Xenorhabdus szentirmaii* (isolate K77) (Tailliez *et al.*, 2006).

In contrast, for the genus *Bacillus*, several studies have shown that strains of the species *B. anthracis, B. cereus* and *B. thuringiensis* have high DNA homology (greater than 70%) suggesting that these three species should be considered as a single one (Somerville and Jones, 1972; Kaneko *et al.*, 1978). In this example, the DNA–DNA hybridization technique is ineffective for accurate identification of new isolates in the species *B. thuringiensis*.

Techniques required to obtain DNA–DNA relatedness values, however, tend to be expensive, time-consuming and often require radioactive labels and type strains used as reference. An alternative approach to quantification of genome relatedness is to compare selected DNA sequences for a group of bacterial strains as proposed below.

MOLECULAR TYPING. A combination of molecular typing techniques, ERIC and RAPD fingerprinting, has been developed to group strains that belong to the same species based on the amplification of conserved genomic regions (migration of bands of the same size in different profiles). Molecular typing methods are useful for screening large collections of bacterial strains. We briefly refer to these methods and provide protocols used for entomopathogenic-nematophilic bacteria.

RANDOM AMPLIFIED POLYMORPHIC DNA (RAPD). The simplicity of the RAPD technique makes it ideal for genetic mapping and DNA fingerprinting with particular utility in the field of population genetics. In many cases, only a small number of primers are necessary to identify polymorphism within species (Williams *et al.*, 1990). Polymerase chain reaction (PCR) primers usually consist of 8–10 nucleotides. They are used individually during PCR reactions. Primers usually adhere to a specific nucleotide segment of the genomic DNA. If primers hybridize in the proper orientation and at a suitable distance from each other, the fragment between these primers is amplified and measured using gel electrophoresis. Recently, this method was used notably for the molecular characterization of *B. thuringiensis* and *Xenorhabdus* strains and isolates (Tailliez *et al.*, 2006; Konecka *et al.*, 2007; Chaves *et al.*, 2008).

ENTEROBACTERIAL REPETITIVE INTERGENIC CONSENSUS (ERIC) (HULTON ET AL., 1991). ERIC sequences, also described as intergenic repetitive units, differ from most other bacterial repeats in being distributed across a wider range of species. ERIC sequences were first described in *Escherichia coli, Vibrio cholerae, Salmonella enterica* serovar Typhimurium (=*Salmonella typhimurium*) and other members of the Enterobacteriaceae (Sharples and Lloyd, 1990; Hulton *et al.*, 1991). ERIC sequences have been found only in intergenic regions and within transcribed regions (Hulton *et al.*, 1991). The number of copies of the ERIC sequence varies among species. Initially, it was estimated that there may be about 30 copies in *E. coli* K-12 and perhaps 150 in *S. enterica* serovar Typhimurium LT2 (Hulton *et al.*,

1991), while the genome sequence of *P. luminescens* has been reported to contain over 700 copies (Duchaud *et al.*, 2003).

RAPD_ERIC PROFILES AND THEIR APPLICATION IN ENTOMOPATHOGENIC-NEMATOPHILIC BACTERIA. RAPD primers:

P1 (5′-TGCTCTGCCC-3′), P2 (5′-GGTGACGCAG-3′);
P3 (5′-TCGCTGGGAC-3′).

Primers are used in three separate PCR reactions.
ERIC primers:

ERIC1R (5′-GCTATGCTCCYGGGGRTT-3′);
ERIC2 (5′-ACTATGTGAYTGGGGTGA-3′).

We also recommend considering a modified version of the sequences of the ERIC primers proposed by Versalovic *et al.* (1991). These modified sequences are in agreement with the ERIC sequences present in the *Xenorhabdus* genome (http://maizeapache.ddpsc.org/xeno_blast/index.html).

PCR amplifications can be performed in a final volume of 50 µl containing 1 × PCR buffer (Qbiogene, http://www.qbiogene.com/), 20–100 ng of bacterial genomic DNA, 0.2 mM $MgCl_2$, 0.5 µM primer, 200 µM of each dNTP and 3.75 U *Taq* DNA polymerase (concentration of enzyme stock: 15 U/µl). PCR was carried out in a GeneAmp 2400 thermal cycler PCR system (Perkin Elmer, http://las.perkinelmer.com/).
Amplification conditions that we recommend are the following.

| | |
|---|---|
| Step 1 | 94°C for 5 min |

Then 30 cycles composed of the following.

| | |
|---|---|
| Step 2 | 94°C for 1 min |
| Step 3 | 48°C (ERIC) or 42°C (RAPD) for 1 min |
| Step 4 | 3 (ERIC) to 6 min (RAPD) of temperature ramping to 72°C |
| Step 5 | 72°C for 1 min |

Briefly, 20 µl of the PCR reactions are electrophoresed in 1% Seakem GTG agarose (Tebu, http://www.tebu-bio.com/) gels in TBE (for 1 l: Tris base, 10.8 g; boric acid, 5.5 g; EDTA, 0.5 M pH 8.0, 4 ml), alongside with the 123-bp ladder (Invitrogen). Each gel contains ten lanes of PCR products and three lanes of ladder located at both sides and in the centre. After electrophoresis, DNA gels are stained using an ethidium bromide solution (1 mg/l).

For each strain, the ERIC and RAPD profiles are combined and then compared using the Pearson similarity coefficient taking into account the number, the position and the intensity of the DNA bands. The resulting dendrogram presented in Fig. 2.1 is calculated with the Unweighted Pair Group Method Using Arithmetic Averages (UPGMA) (Sokal and Michener, 1958) module of GelCompar software (Applied-Maths, http://www.applied-maths.com/).

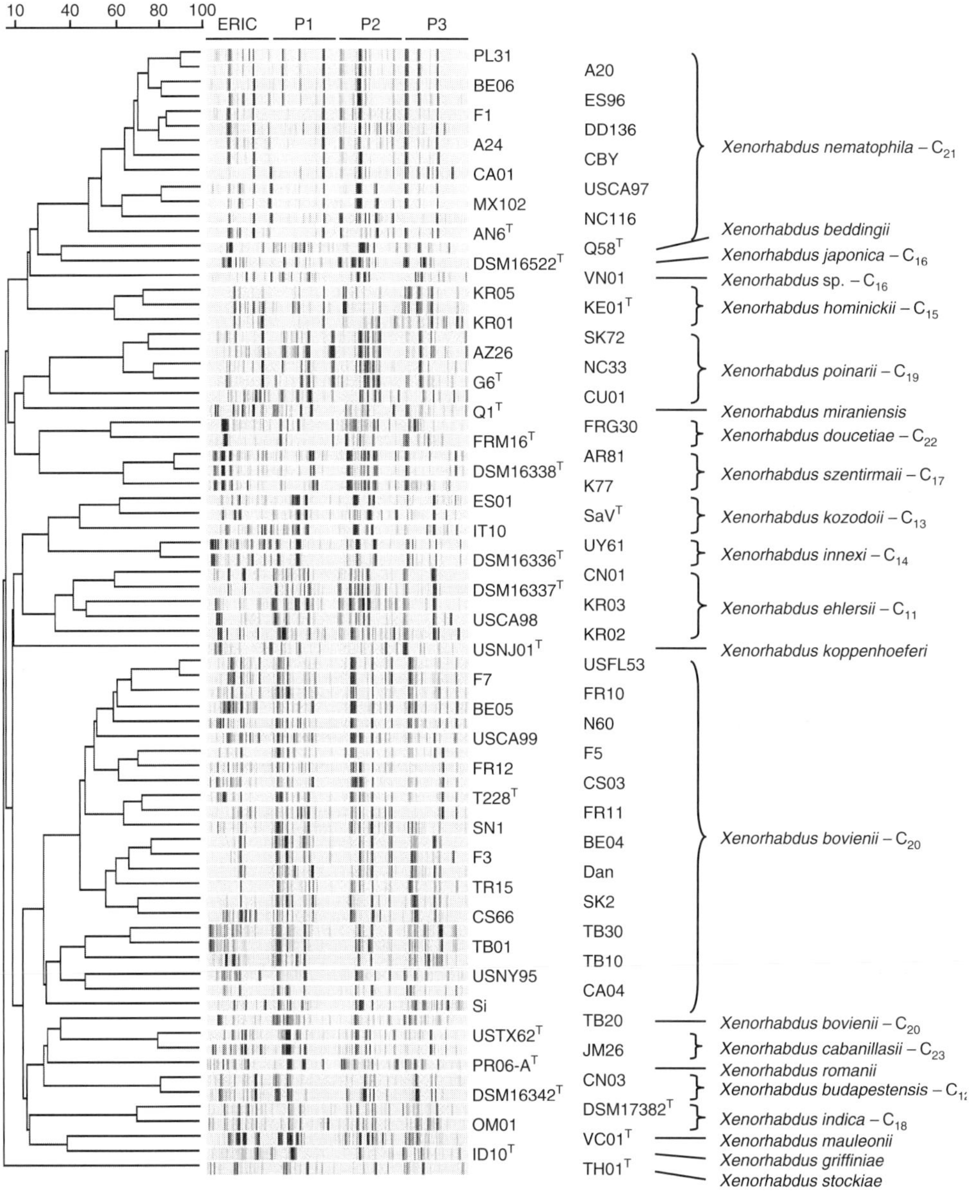


**Fig. 2.1.** Comparison of the molecular typing profiles of 76 *Xenorhabdus* strains. One ERIC profile and three RADP profiles obtained with the primers P1, P2 and P3 used in independent reactions are combined for each strain. The combined molecular typing profiles are compared using the Pearson similarity coefficient. The corresponding similarity matrix is used to generate a dendrogram using UPGMA (Sokal and Michener, 1958) module of GelCompar software (Applied-Maths). Type strains are indicated by the name of the bacterial species being in bold typeface. Clusters $C_{11}$ to $C_{23}$ correspond to 16S clusters defined in Fig. 2.2.

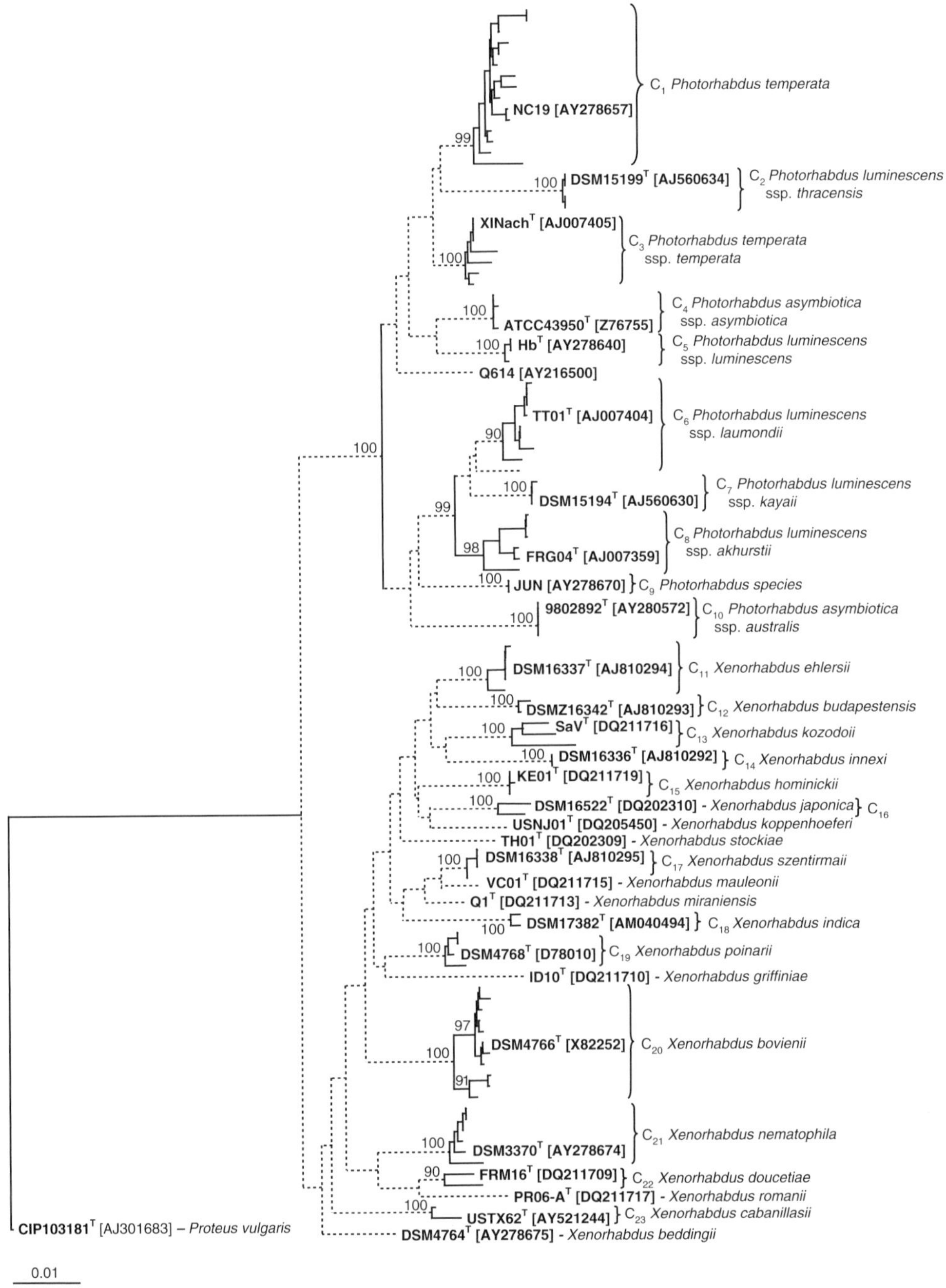


**Fig. 2.2.** Distance tree comparing the 16S rRNA gene sequences of 54 *Xenorhabdus* and 58 *Photorhabdus* strains. The tree was constructed using the 16S rRNA gene sequences (1426 nucleotides), the model of Jukes and Cantor (1969) and the neighbour-joining (NJ) (Saitou and Nei, 1987) module of PAUP software (Swofford, 2003). The 16S rRNA gene sequence of *Proteus vulgaris* type-strain CIP103181$^T$ was used as outgroup.

*Continued*

SEQUENCE ANALYSIS. One critical step for generation of good-quality sequences relies on good and reliable DNA extraction procedures and PCR amplification. Readers should be aware that a number of DNA extraction kits are available from different manufactures. In our laboratory, we currently use the QIAamp DNA mini kit (Qiagen, http://www1.qiagen.com/) for extracting the total bacterial genomic DNA of *Photorhabdus* and *Xenorhabdus*.

SMALL SUBUNIT (16S) rRNA GENE. Comparative sequence analysis of the 16S rRNA gene has been extensively used as a diagnostic method and also to determine phylogenetic relationships of novel bacterial isolates/species (see Tailliez and Boemare, Chapter 6, this volume). Moreover, this gene is characterized by having areas of secondary structure which have also proven useful for diagnostic purposes. The size of this gene is approximately 1.6 kb, and is composed of regions with different levels of variability thus providing valuable information for differentiation of taxa. Strains that show less than 97% 16S rRNA sequence are considered to be new and/or different species, as there are no examples in which strains with this extent of divergence in 16S rRNA sequence meet the criteria of more than 70% DNA–DNA relatedness (Stackebrandt and Goebel, 1994).

### Case study I: analysis of 16S rRNA sequences for identification of novel bacterial species in Xenorhabdus and Photorhabdus

16S rRNA gene sequences have been widely considered to depict new *Photorhabdus* and *Xenorhabdus* taxa. For example, Hazir *et al.* (2004), using a combination of 16S rRNA gene sequences and phenotypic traits, identified two novel *Photorhabdus* ssp.: *P. luminescens* ssp. *kayaii* and *P. luminescens* ssp. *thracensis*. Lengyel *et al.* (2005) and Somvanshi *et al.* (2006), considering similar approaches, identified new *Xenorhabdus*: *Xenorhabdus budapestensis*, *Xenorhabdus ehlersii*, *Xenorhabdus indica*, *Xenorhabdus innexi* and *X. szentirmaii*.

rRNA gene sequence comparisons often lack resolution when compared to quantitative DNA–DNA hybridization. Isolates that have more than 97% identity may or may not meet the 70% relatedness criterion for inclusion in the same species. In our example presented in Fig. 2.2, 16S rRNA gene sequences between two *Xenorhabdus* spp. or two *Photorhabdus* spp. are often less than 3% (and always

**Fig. 2.2.** (Continued) Bootstrap values (percentages of 1000 replicates) of more than 90% are shown at the nodes. Dashed lines indicate unreliable links between groups and unique sequences. Accession numbers in brackets correspond to 16S rRNA gene sequences retrieved from GenBank (http://www.ncbi.nlm.nih.gov/). Twenty-three clusters, $C_1$ to $C_{23}$, supported by high bootstrap values (>90%), were identified and assigned to described *Photorhabdus* and *Xenorhabdus* species and subspecies. Only sequences corresponding to type strains or representative of a group are indicated by the number of the strain being in bold typeface. The bar indicates 1% sequence divergence.

less than 5%), so this frequently used bacterial taxonomy threshold (Stackebrandt and Goebel, 1994) cannot be used here without precaution to support proposals for new species of *Photorhabdus* and *Xenorhabdus*. Consequently, high similarity of rRNA gene sequences does not eliminate the need to apply other methods such as DNA–DNA hybridization (or molecular typing methods) to further assess identity of isolates.

### Case study II: identification of *Bacillus* sp./strains

Ash *et al.* (1991b) considered 16S rRNA gene sequences of approximately 60 strains of *Bacillus* and related species to assess the taxonomic position and phylogenetic relationships of these taxa. Their study depicted five major clades. Members of clade I, '*Bacillus* sensu stricto group' comprised species such as *B. anthracis*, *B. cereus*, *B. thuringiensis* and *B. subtilis*. Sequence analysis of 16S rRNA genes of these first three *Bacillus* 'species' showed that they are highly similar (greater than 99%) (Ash *et al.*, 1991a) but significantly different from the sequences of the phylogenetically closely related *B. subtilis* subgroup (less than 94%); (Ash *et al.*, 1991b). Clade II included a number of round-spore forming bacilli such as *Bacillus sphaericus*. Clade III grouped *Paenibacillus* spp. such as *P. larvae* and *Paenibacillus popilliae*. *Brevibacillus* spp. were considered members of clade IV. Thermophilic bacilli such as *Bacillus stearothermophilus* and *Bacillus thermocloacae* were included in clade V. At least, a new genus, *Alicyclobacillus*, was proposed by Wisotzkey *et al.* (1992) to include acidophilic species such as *Alicyclobacillus cycloheptanicus* (formerly *Bacillus cycloheptanicus*).

## 2.3. Conclusions

Identification of new bacterial isolates using basic phenotypic tests continues to be the first identification step of bacteria. However, sequence analysis of 16S rRNA genes has shown to be useful for rapidly identifying novel isolates. Although the advantages of this approach are clear, classification by rRNA gene sequences alone is sometimes unsatisfying for several reasons (lack of resolution comparing with quantitative hybridization; horizontal gene transfer, see Tailliez and Boemare, Chapter 6, this volume). We recommend to take into consideration both molecular and phenotypic studies in a polyphasic approach to obtain a more accurate classification at the generic level. For example, for the genus *Xenorhabdus*, DNA–DNA hybridization, 16S rRNA gene sequence analysis, molecular and phenotypic typing methods allowed the precise classification of isolates into different species (Tailliez *et al.*, 2006). Sequences-based approaches such as multilocus sequence analysis (MLSA) (Maiden *et al.*, 1998) are also rapid and robust, and can be used efficiently to analyse bacterial infraspecies diversity or population structure of closely related species such as *B. anthracis*, *B. cereus* and *B. thuringiensis* (Sorokin *et al.*, 2006).

## References


Ahmed, I., Yokota, A., Yamazoe, A. and Fujiwara, T. (2007) Proposal of *Lysinibacillus boronitolerans* gen. nov. sp. nov., and transfer of *Bacillus fusiformis* to *Lysinibacillus fusiformis* comb. nov. and *Bacillus sphaericus* to *Lysinibacillus sphaericus* comb. nov. *International Journal of Systematic and Evolutionary Microbiology* 57, 1117–1125.

Akhurst, R.J. (1983) Taxonomic study of *Xenorhabdus*, a genus of bacteria symbiotically associated with insect pathogenic nematodes. *International Journal of Systematic Bacteriology* 33, 38–45.

Akhurst, R.J. (1986a) *Xenorhabdus nematophilus* subsp. *beddingii* (Enterobacteriaceae): a new subspecies of bacteria mutualistically associated with entomopathogenic nematodes. *International Journal of Systematic Bacteriology* 36, 454–457.

Akhurst, R.J. (1986b) *Xenorhabdus nematophilus* subsp. *poinarii*: its interaction with insect pathogenic nematodes. *Systematic and Applied Microbiology* 8, 142–147.

Akhurst, R.J. and Boemare, N.E. (1988) A numerical taxonomic study of the genus *Xenorhabdus* (Enterobacteriaceae) and proposed elevation of the subspecies of *X. nematophilus* to species. *Journal of General Microbiology* 134, 1835–1845.

Akhurst, R.J., Boemare, N.E., Janssen, P.H., Peel, M.M., Alfredson, D.A. and Beard, C.E. (2004) Taxonomy of Australian clinical isolates of the genus *Photorhabdus* and proposal of *Photorhabdus asymbiotica* subsp *asymbiotica* subsp nov and *P. asymbiotica* subsp *australis* subsp nov. *International Journal of Systematic and Evolutionary Microbiology* 54, 1301–1310.

Ash, C., Farrow, J.A., Dorsch, M., Stackebrandt, E. and Collins, M.D. (1991a) Comparative analysis of *Bacillus anthracis*, *Bacillus cereus*, and related species on the basis of reverse transcriptase sequencing of 16S rRNA. *International Journal of Systematic Bacteriology* 41, 343–346.

Ash, C., Farrow, J.A.E., Wallbanks, S. and Collins, M.D. (1991b) Phylogenetic heterogeneity of the genus *Bacillus* revealed by comparative analysis of small subunit – ribosomal RNA sequences. *Letters in Applied Microbiology* 13, 202–206.

Ash, C., Priest, F.G. and Collins, M.D. (1993) Molecular identification of rRNA group 3 bacilli (Ash, Farrow, Wallbanks and Collins) using a PCR probe test. *Antonie van Leeuwenhoek* 64, 253–260.

Berliner, E. (1915) Uber die Schlaffsuchtder Mehlmottenraupe (*Ephestia kühniella* Zell) und ihren Erreger *Bacillus thuringien* sis n.sp. *Zeitschrift für Angewandte Entomologie* 2, 29–56.

Boemare, N.E., Akhurst, R.J. and Mourant, R.G. (1993) DNA relatedness between *Xenorhabdus* spp. (Enterobacteriaceae), symbiotic bacteria of entomopathogenic nematodes, and a proposal to transfer *Xenorhabdus luminescens* to a new genus, *Photorhabdus* gen. nov. *International Journal of Systematic Bacteriology* 43, 249–255.

Bowen, D., Rocheleau, T.A., Blackburn, M., Andreev, O., Golubeva, E., Bhartia, R., ffrench-Constant, R.H. (1998) Insecticidal toxins from the bacterium *Photorhabdus luminescens*. *Science* 280, 2129–2132.

Brenner, D.J., Fanning, G.R., Rake, A.V. and Johnson, K.E. (1969) Batch procedure for thermal elution of DNA from hydroxyapatite. *Analytical Biochemistry* 28, 447–459.

Bucher, G.E. (1960) Potential bacterial pathogens of insects and their characteristics. *Journal of Insect Pathology* 2, 172–195.

Carlson, C.R. and Kolstø, A.B. (1993) A complete physical map of a *Bacillus thuringiensis* chromosome. *Journal of Bacteriology* 175, 1053–1060.

Chaves, J.Q., Cavados, C.F.G. and Rabinovitch, L. (2008) Phenotypic and genotypic features of new autoagglutinating *Bacillus thuringiensis* strains. *Journal of Invertebrate Pathology* 98, 85–92.

Claus, D. and Berkeley, R.C.W. (1986) *Bacillus* Cohn 1872. In: Sneath, P.H.A. (ed.) *Bergey's Manual of Systematic Bacteriology*, Vol. 2. The Williams and Wilkins, Baltimore, Maryland, pp. 1105–1140.

Crosa, J.H., Brenner, D.J. and Falkow, S. (1973) Use of single-strand-specific nuclease for analysis of bacterial and plasmid deoxyribonucleic acid homo- and heteroduplexes. *Journal of Bacteriology* 115, 904–911.

Crickmore, N. (2000) The diversity of *Bacillus thuringiensis* delta-endotoxins. In: Charles, J.-F., Delecluse, A. and Nelson-LeRoux, C. (eds) *Entomopathogenic Bacteria: From Laboratory to Field Application*. Kluwer Academic Publishers, London, pp. 65–79.

Crickmore, N., Zeigler, D.R. and Feitelson, J. (1998) Revision of the nomenclature for the *Bacillus thuringiensis* pesticidal crystal proteins. *Microbiological and Molecular Biological Reviews* 62, 807–813.

de Barjac, H. (1978) Une nouvelle variété de *Bacillus thuringiensis* très toxique pour les moustiques: *Bacillus thuringiensis* var *israelensis*, sérotype 14. *Compte-Rendu de l'Académie des Sciences Série D* 286, 797–800.

de Barjac, H., Sebald, M., Charles, J.F., Cheong, W.H. and Lee, H.H. (1990) *Clostridium bifermentans* serovar *malaysia*, une nouvelle bactérie anaérobie pathogène des larves de moustiques et simulies. *Compte-Rendu de l'Académie des Sciences Série III Sciences de la Vie* 310, 383–387.

Duchaud, E., Rusniok, C., Frangeul, L., Buchrieser, C., Givaudan, A., Taourit, S., Bocs, S., Boursaux-Eude, C., Chandler, M., Charles, J.F., Dassa, E., Derose, R., Derzelle, S., Freyssinet, G., Gaudriault, S., Medigue, C., Lanois, A., Powell, K., Siguier, P., Vincent, R., Wingate, V., Zouine, M., Glaser, P., Boemare, N., Danchin, A. and Kunst, F. (2003) The genome sequence of the entomopathogenic bacterium *Photorhabdus luminescens*. *Nature Biotechnology* 21, 1307–1313.

Dutky, S.R. (1940) Two new spore-forming bacteria causing the milky disease in Japanese beetle larvae. *Journal of Agricultural Research* 61, 57–68.

Fischer-Le Saux, M., Viallard, V., Brunel, B., Normand, P. and Boemare, N.E. (1999) Polyphasic classification of the genus *Photorhabdus* and proposal of new taxa: *P. luminescens* subsp. *luminescens* subsp. nov., *P. luminescens* subsp *akhurstii* subsp. nov., *P. luminescens* subsp *laumondii* subsp. nov., *P. temperata* sp. nov., *P. temperata* subsp. *temperata* subsp. nov. and *P. asymbiotica* sp. nov. *International Journal of Systematic Bacteriology* 49, 1645–1656.

Garrity, G.M., Brenner, D.J., Krieg, N.R. and Staley, J.T. (2005) *Bergey's Manual of Systematic Bacteriology, The Proteobacteria*, Vol. 2, 2nd edn. Springer Science, New York, 1106.

Genersch, E., Forsgren, E., Pentikäinen, J., Ashiralieva, A., Rauch, S., Kilwinski, J. and Fries, I. (2006) Reclassification of *Paenibacillus larvae* subsp. *larvae* and *Paenibacillus larvae* subsp. *pulvifaciens* as *Paenibacillus larvae* without subspecies differentiation. *International Journal of Systematic and Evolutionary Microbiology* 56, 501–511.

Goldberg, L.J. and Margalit, J. (1977) A bacterial spore demonstrating rapid larvicidal activity against *Anopheles sergentii*, *Uranotaenia unguiculata*, *Culex univittatus*, *Aedes egypti* and *Culex pipiens*. *Mosquito News* 37, 355–358.

Grimont, F. and Grimont, P.A.D. (2005) Genus XXXIV *Serratia* Bizio. In: Garrity, G.M., Brenner, D.J., Krieg, N.R. and Staley, J.T. (eds) *Bergey's Manual of Systematic Bacteriology, The Gammaproteobacteria*. Springer, New York, pp. 799–811.

Grimont, P.A.D., Popoff, M.Y., Grimont, F., Coynault, C. and Lemelin, M. (1980) Reproducibility and correlation study of three deoxyribonucleic acid hydridization procedures. *Current Microbiology* 4, 325–330.

Grimont, P.A.D., Steigerwalt, A.G., Boemare, N., Hickman-Brenner, F.W., Deval, C., Grimont, F. and Brenner, D.J. (1984) Deoxyribonucleic acid relatedness and phenotypic study of the genus *Xenorhabdus*. *International Journal of Systematic Bacteriology* 34, 378–388.

Grimont, P.A.D., Jackson, T.A., Ageron, E. and Noonan, M.J. (1988) *Serratia entomophila* sp. nov. associated with amber disease in the New Zealand grass grub, *Costelytra zealandica*. *International Journal of Systematic Bacteriology* 38, 1–6.

Hazir, S., Stackebrandt, E., Lang, E., Schumann, P., Ehlers, R.-U. and Keskin, N. (2004) Two new subspecies of *Photorhabdus luminescens*, isolated from *Heterorhabditis bacteriophora* (Nematoda:

Heterorhabditidae): *Photorhabdus luminescens* subsp *kayaii* subsp nov and *Photorhabdus luminescens* subsp *thracensis* subsp nov. *Systematic and Applied Microbiology* 27, 36–42.

Heyndrickx, M., Vandemeulebroecke, K., Hoste, B., Janssen, P., Kersters, K., de Vos, P., Logan, N.A., Ali, N. and Berkeley, R.C.W. (1996) Reclassification of *Paenibacillus* (formely *Bacillus*) *pulvifaciens* (Nakamura 1984) Ash *et al.*, 1994, a later subjective synonym of *Paenibacillus* (formely *Bacillus*) *larvae* (White, 1906) Ash *et al.*, 1994, as a subspecies of *P. larvae* subsp. *larvae* and *P. larvae* subsp. *pulvifaciens. International Journal of Systematic Bacteriology* 46, 270–279.

Hulton, C.S.J., Higgins, C.F. and Sharp, P.M. (1991) ERIC sequences: a novel family of repetitive elements in the genomes of *Escherichia coli*, *Salmonella typhimurium* and other enterobacteria. *Molecular Microbiology* 5, 825–834.

Ivanova, N., Sorokin, A., Anderson, I., Galleron, N., Candelon, B., Kapatral, V., Bhattacharyya, A., Reznik, G., Mikhailova, N., Lapidus, A., Chu, L., Mazur, M., Goltsman, E., Larsen, N., D'Souza, M., Walunas, T., Grechkin, Y., Pusch, G., Haselkorn, R., Fonstein, M., Ehrlich, S.D., Overbeek, R. and Kyrpides, N. (2003) Genome sequence of *Bacillus cereus* and comparative analysis with *Bacillus anthracis. Nature* 423, 87–91.

Jukes, T.H. and Cantor, C.R. (1969) Evolution of protein molecules. In: Munro, H.N. (ed.) *Mammalian Protein Metabolism*, Vol. 3. Academic Press, NewYork, pp. 21–132.

Kaneko, T., Nozari, R. and Aizawa, K. (1978) Deoxyribonucleic acid relatedness between *Bacillus anthracis*, *Bacillus cereus* and *Bacillus thuringiensis. Microbiology and Immunology* 22, 639–641.

Konecka, E., Kaznowski, A., Ziemnicka, J. and Ziemnicki, K. (2007) Molecular and phenotypic characterisation of *Bacillus thuringiensis* isolated during epizootics in *Cydia pomonella* L. *Journal of Invertebrate Pathology* 94, 56–63.

Krieg, A. (1981) The genus *Bacillus*: insect pathogens. In: Starr, M.P., Stolp, H., Trüper, H.G., Balows, A. and Schlegel, H.G. (eds) *The Prokaryotes*. Springer, Berlin, pp. 1743–1755.

Kunst, F., Ogasawara, N., Moszer, I., Albertini, A.M., Alloni, G., Azevedo, V., Bertero, M.G., Bessières, P., Bolotin, A., Borchert, S., Borriss, R., Boursier, L., Brans, A., Braun, M., Brignell, S.C., Bron, S., Brouillet, S., Bruschi, C.V., Caldwell, B., Capuano, V., Carter, N.M., Choi, S.-K., Codani, J.-J., Connerton, I.F., Cummings, N.J., Daniel, R.A., Denizot, F., Devine, K.M., Düsterhöft, A., Ehrlich, S.D., Emmerson, P.T., Entian, K.D., Errington, J., Fabret, C., Ferrari, E., Foulger, D., Fritz, C., Fujita, M., Fujita, Y., Fuma, S., Galizzi, A., Galleron, N., Ghim, S.-Y., Glaser, P., Goffeau, A., Golightly, E.J., Grandi, G., Guiseppi, G., Guy, B.J., Haga, K., Haiech, J., Harwood, C.R., Hénaut, A., Hilbert, H., Holsappel, S., Hosono, S., Hullo, M.-F., Itaya, M., Jones, L., Joris, B., Karamata, D., Kasahara, Y., Klaerr-Blanchard, M., Klein, C., Kobayashi, Y., Koetter, P., Koningstein, G., Krogh, S., Kumano, M., Kurita, K., Lapidus, A., Lardinois, S., Lauber, J., Lazarevic, V., Lee, S.-M., Levine, A., Liu, H., Masuda, S., Mauël, C., Médigue, C., Medina, N., Mellado, R.P., Mizuno, M., Moestl, D., Nakai, S., Noback, M., Noone, D., O'Reilly, M., Ogawa, K., Ogiwara, A., Oudega, B., Park, S.-H., Parro, V., Pohl, T.M., Portetelle, D., Porwollik, S., Prescott, A.M., Presecan, E., Pujic, P., Purnelle, B., Rapoport, G., Rey, M., Reynolds, S., Rieger, M., Rivolta, C., Rocha, E., Roche, B., Rose, M., Sadaie, Y., Sato, T., Scanlan, E., Schleich, S., Schroeter, R., Scoffone, F., Sekiguchi, J., Sekowska, A., Seror, S.J., Serror, P., Shin, B.-S., Soldo, B., Sorokin, A., Tacconi, E., Takagi, T., Takahashi, H., Takemaru, K., Takeuchi, M., Tamakoshi, A., Tanaka, T., Terpstra, P., Tognoni, A., Tosato, V., Uchiyama, S., Vandenbol, M., Vannier, F., Vassarotti, A., Viari, A., Wambutt, R., Wedler, E., Wedler, H., Weitzenegger, T., Winters, P., Wipat, A., Yamamoto, H., Yamane, K., Yasumoto, K., Yata, K., Yoshida, K., Yoshikawa, H.-F., Zumstein, E., Yoshikawa, H. and Danchin, A. (1997) The complete genome sequence of the gram-positive bacterium *Bacillus subtilis. Nature* 390, 249–256.

Lengyel, K., Lang, E., Fodor, A., Szállas, E., Schumann, P. and Stackebrandt, E. (2005) Description of four novel species of *Xenorhabdus*, family Enterobacteriaceae: *Xenorhabdus budapestensis* sp. nov., *Xenorhabdus ehlersii* sp. nov., *Xenorhabdus innexi* sp. nov., and *Xenorhabdus szentirmaii* sp. nov. *Systematic and Applied Microbiology* 28, 115–122.

Lereclus, D., Delécluse, A. and Lecadet, M.-M. (1993) Diversity of *Bacillus thuringiensis* toxins and genes. In: Entwistle, P., Cory, J.S., Bailey, M.J., Higgs, S. (eds) *Bacillus thuringiensis, An Environmental Biopesticide: Theory and Practice*. Wiley, Chichester, UK, pp. 37–69.

Maiden, M.C.J., Bygraves, J.A., Feil, E., Morelti, G., Russell, J.E., Urwin, R., Zhang, Q., Zhou, J., Zurth, K., Caugant, D.A., Feavers, I.M., Achtman, M. and Spratt, B.G. (1998) Multilocus sequence typing: a portable approach to the identification of clones within populations of pathogenic microorganisms. *Proceedings of the National Academy of Sciences of the USA* 95, 3140–3145.

Matsumoto, H., Tanaka, K., Noguchi, H. and Hayakawa, Y. (2003) Cause of mortality in insects under severe stress. *European Journal of Biochemistry* 270, 3469–3476.

Nishimura, Y., Hagiwara, A., Suzuki, T. and Yamanaka, S. (1994) *Xenorhabdus japonicus* sp. nov. associated with the nematode *Steinernema kushidai*. *World Journal of Microbiology and Biotechnology* 10, 207–210.

Pettersson, B., Rippere, K.E., Yousten, A.A. and Priest, F.G. (1999) Transfer of *Bacillus lentimorbus* and *Bacillus popilliae* to the genus *Paenibacillus* with emended descriptions of *Paenibacillus lentimorbus* comb. nov. and *Paenibacillus popillae* comb. nov. *International Journal of Systematic Bacteriology* 49, 531–540.

Poinar, G.O. Jr and Thomas, G.M. (1965) A new bacterium, *Achromobacter nematophilus* sp. nov. (Achromobacteriaceae: Eubacteriales) associated with a nematode. *International Bulletin of Bacteriological Nomenclature and Taxonomy* 15, 249–252.

Priest, F.G. (1993) Biodiversity of entomopathogenic, endospore forming bacteria. In: Charles, J.F. and Nielsen-Le Roux, C. (eds) *Entomopathogenic Bacteria: From Laboratory to Field Application*. Kluwer Academic Publishers, Dordrecht, The Netherlands, pp. 1–22.

Rey, M.W., Ramaiya, P., Nelson, B.A., Brody-Karpin, S.D., Zaretsky, E.J., Tang, M., Lopez de Leon, A., Xiang, H., Gusti, V., Groth Clausen, I., Olsen, P.B., Rasmussen, M.D., Andersen, J.T., Jørgensen, P.L., Larsen, T.S., Sorokin, A., Bolotin, A., Lapidus, A., Galleron, N., Ehrlich, S.D. and Berka, R.M. (2004) Complete genome sequence of the industrial bacterium *Bacillus licheniformis* and comparisons with closely related *Bacillus* species. *Genome Biology* 5, R77.

Saitou, N. and Nei, M. (1987) The neighbour-joining method: a new method for reconstructing phylogenetic trees. *Molecular and Biological Evolution* 4, 406–425.

Schnepf, E., Crickmore, N., Van Rie, J., Lereclus, D., Baum, J., Feitelson, J., Zeigler, D.R. and Dean, D.H. (1998) *Bacillus thuringiensis* and its pesticidal crystal proteins. *Microbiology and Molecular Biology Reviews* 62, 775–806.

Sharples, G.J. and Lloyd, R.G. (1990) A novel repeated sequence located in the intergenic regions of bacterial chromosomes. *Nucleic Acids Research* 18, 6503–6508.

Shida, O., Takagi, H., Kadowaki, K., Komagata, K. (1996) Proposal for two new genera, *Brevibacillus* gen. nov. and *Aneurinibacillus* gen. nov. *International Journal of Systematic Bacteriology* 46, 939–946.

Sokal, R.R. and Michener, C.D. (1958) A statistical method for evaluating systematic relationships. *University Kansas Science Bulletin* 22, 1409–1438.

Somerville, H.J. and Jones, M.L. (1972) DNA competition studies within the *Bacillus cereus* group of bacilli. *Journal of General Microbiology* 73, 257–265.

Somvanshi, V.S., Lang, E., Ganguly, S., Swiderski, J., Saxena, A.K. and Stackebrandt, E. (2006) A novel species of *Xenorhabdus*, family Enterobacteriaceae: *Xenorhabdus indica* sp. nov., symbiotically associated with entomopathogenic nematode *Steinernema thermophilum* Ganguly and Singh, 2000. *Systematic and Applied Microbiology* 29, 519–525.

Sorokin, A., Candelon, B., Guilloux, K., Galleron, N., Wackerow-Kouzova, N., Ehrlich, S.D., Bourguet, D. and Sanchis, V. (2006) Multiple-locus sequence typing analysis of *Bacillus cereus* and *Bacillus thuringiensis* reveals separate clustering and a distinct population structure of psychrotrophic strains. *Applied and Environmental Microbiology* 72, 1569–1578.

Stackebrandt, E. (2006) *Molecular Identification, Systematic, and Population Structure of Prokaryotes*. Springer, Berlin, 320.

Stackebrandt, E. and Goebel, B.M. (1994) Taxonomic note: a place for DNA-DNA reassociation and 16S rRNA sequence analysis in the present species definition in bacteriology. *International Journal of Systematic Bacteriology* 44, 846–849.

Swofford, D.L. (2003) *PAUP*: Phylogenetic Analysis Using Parsimony (*and Other Methods)*, Version 4.0b10. Sinauer Associates, Sunderland, Massachusetts.

Tailliez, P., Pagès, S., Ginibre, N. and Boemare, N. (2006) New insight into diversity in the genus *Xenorhabdus*, including the description of ten new species. *International Journal of Systematic and Evolutionary Microbiology* 56, 2805–2818.

Thiéry, I. and Frachon, E. (1997) Identification, isolation, culture and preservation of entomopathogenic bacteria. In: Lacey, L.A. (ed.) *Manual of Techniques in Insect Pathology*. Academic Press, London, 409.

Thomas, G.M. and Poinar, G.O. Jr (1979) *Xenorhabdus* gen. nov., a genus of entomopathogenic nematophilic bacteria of the family Enterobacteriaceae. *International Journal of Systematic Bacteriology* 29, 352–360.

Versalovic, J., Koeuth, T. and Lupski, J.R. (1991) Distribution of repetitive DNA sequences in eubacteria and application to fingerprinting of bacterial genomes. *Nucleic Acids Research* 19, 6823–6831.

Vodovar, N., Vinals, M., Liehl, P., Basset, A., Degrouard, J., Spellman, P., Boccard, F. and Lemaitre, B. (2005) *Drosophila* host defense after oral infection by an entomopathogenic *Pseudomonas* species. *Proceedings of the National Academy of Sciences of the USA* 102, 11414–11419.

Wayne, L.G., Brenner, D.J., Colwell, R.R., Grimont, P.A.D., Kandler, O., Krichevsky, M.I., Moore, L.H., Moore, W.E.C., Murray, R.G.E., Stackebrandt, E., Starr, M.P., Trüper, H.G. (1987) Report of the ad hoc committee on reconciliation of approaches to bacterial systematics. *International Journal of Systematic Bacteriology* 37, 463–464.

White, G.F. (1906) The bacteria of the apiary with special reference to bee disease. In: US Department of Agriculture (ed.) *Bureau of Entomology, Technical Series n°14*. Washington, DC, pp. 1–50.

WHO (World Health Organization) (1999) International programme on chemical safety (IPCS): microbial pest control agent *Bacillus thuringiensis*. *Environmental Health Criteria* 217, 1–105.

Williams, J.G.K., Kubelik, A.R., Livak, K.J., Rafalski, J.A. and Tingey, S.V. (1990) DNA polymorphisms amplified by arbitrary primers are useful as genetic markers. *Nucleic Acids Research* 18, 6531–6535.

Wisotzkey, J.D., Jurtshuk, P., Fox, G.E., Deinhard, G. and Poralla, K. (1992) Comparative sequence analyses on the 16S rRNA (rDNA) of *Bacillus acidocaldarius*, *Bacillus acidoterrestris* and *Bacillus cycloheptanicus* and proposal for the creation of a new genus, *Alicyclobacillus* gen. nov. *International Journal of Systematic Bacteriology* 42, 263–269.

Yue, D., Nordhoff, M., Wieler, L.H. and Genersch, E. (2008) Fluorescent *in situ* hybridization (FISH) analysis of the interactions between honeybee larvae and *Paenibacillus larvae*, the causative agent of American foulbrood of honeybees (*Apis mellifera*). *Environmental Microbiology* doi:10.111/j.1462–2920.

# 3 Molecular Methods for Identification and Diagnosis of Fungi

L.A. Castrillo[1] and R.A. Humber[2]

[1]Department of Entomology, Cornell University, Ithaca, USA; [2]USDA-ARS, Robert W. Holley Center for Agriculture and Health, Ithaca, USA



## 3.1. Introduction

The use of PCR-based molecular techniques to determine relationships among organisms enables new phylogenetically based revisions of taxonomies and classifications (see Chapter 7, this volume). The emphasis of this chapter, however, is on more applied uses of molecular techniques on the fungi affecting insects and other invertebrates. Despite the indispensable uses of molecular data for the taxonomy and systematics of fungal entomopathogens, the identifications and diagnoses of these fungi offer separate problems and are increasingly dependent on molecular – and particularly PCR-based – techniques. 'Identification' here refers to assigning appropriate and correct scientific names to organisms; fungal identifications are usually needed at the rank of species or below (subspecies, variety, etc.). Identifications only to genus are usually used for ecologically based studies. Accurate generic identifications are often easily done by morphology without depending on complex, expensive and time-consuming molecular approaches. However, molecular methods for even generic identifications are becoming vital for fungi whose latest taxonomies are based on genomic criteria: Phylogenetic revisions for insect fungi include those segregating the family Clavicipitaceae (Hypocreales) into three families (Sung *et al.*, 2007) with its attendant revision of *Cordyceps*, and the reclassifications of *Verticillium* section *Prostrata* into *Lecanicillium* and other related genera (Gams and Zare, 2001) and of *Paecilomyces* section *Isarioidea* (Luangsa-ard *et al.*, 2005; all species are in Hypocreales but while most common entomopathogens are now placed in *Isaria* in the Cordycipitaceae, other invertebrate pathogens were referred to Ophiocordycipitaceae or Clavicipitaceae but not transferred into other anamorphic genera).

The techniques treated here that are not PCR-based are the oldest and are now rarely used. The strong current preference for PCR-based methods does not, however, mean that these older methods cannot still be useful if carefully applied.

In this chapter 'diagnosis' involves detecting whether a fungus is present in a population of possible hosts or in environmental substrates (e.g. soil or leaf litter). Diagnoses are mainly useful for broadly ecological levels of study completely apart from any need to find infected individuals or to do a detailed identification. Diagnostic tests can monitor fungal introductions to, or dispersals into, or losses from, natural habitats. The use of a given molecular identification or diagnostic method depends on the rank in the taxonomic hierarchy (Fig. 3.1) being considered. The total amount of (molecular) data already generated for a given fungus and its relatives also affects the probable success in using a given molecular technique for identifications or diagnoses. However, devising phylogenetic taxonomies is much less dependent on data from extensively and repetitively sampled populations of a given taxon than are accurate identifications and diagnoses.

This chapter does not provide fully detailed protocols for the included methods but seeks to familiarize readers with the nature and applicabilities of these methods. Discussions of techniques cite references to studies using them primarily with entomopathogenic fungi or that which provide further details on materials, methodologies or appropriate means to analyse or to interpret their results.

## 3.2. General Considerations

### 3.2.1. The fungal genome

Identification and diagnosis of fungi using molecular methods rely on DNA polymorphisms that occur at high frequency. Most of these polymorphisms are in the non-coding regions of the genome, made up of introns, spacers and repeated sequences. In contrast, the functional coding regions are more conserved because of selection pressure. Non-coding regions of the fungal nuclear DNA may be up to 30%, a relatively small percentage compared to the genome of most other eukaryotes, and accounts for fungi's compact genome (Moore and Frazer, 2002).

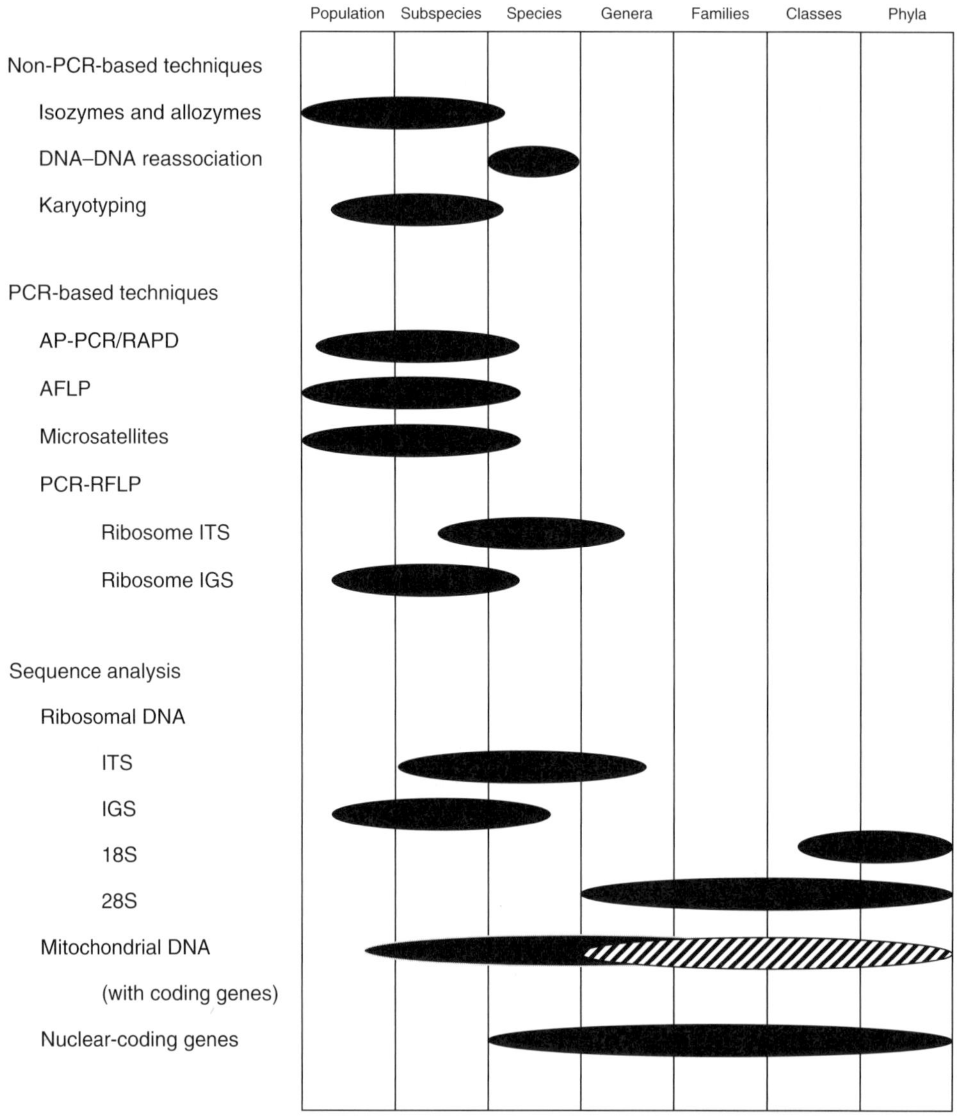


**Fig. 3.1.** Taxonomic resolution of some of the available molecular techniques for fungi. (Modified from Bruns *et al.*, 1991.)

DNA polymorphisms may arise from deletions, insertions, transitions, transversions and errors in replication. More than 95% of DNA polymorphisms are based on a single base change (Moore and Frazer, 2002), a feature utilized in strain identification by most of the techniques to be discussed in this chapter.

The fungal genome is comprised primarily of nuclear and mitochondrial DNA (mtDNA). Although fungi may harbour a number of extra chromosomal elements like plasmids and double-stranded RNA (dsRNA) that may contribute to intraspecies polymorphisms, they will not be covered in this chapter.

### 3.2.2. Selecting a correct technique

For any study on identification and diagnosis, defining the objectives and its specifics precedes and determines which molecular techniques are appropriate to resolve questions on identity (see Rehner, Chapter 7, this volume, for techniques considered for higher-taxa relationships). A few of the key questions one should consider before deciding on which molecular techniques or markers should be employed include: (i) the taxonomic level of identification (i.e. strain versus species identification); (ii) purity of samples (i.e. pure or monosporic cultures versus mixed samples from the field); and (iii) availability of samples (i.e. obligate pathogens of limited quantity versus facultative pathogens).

The taxonomic level (species, genus, family, etc.) at which an identification is desired is of primary importance because the utility of a given molecular technique is limited to only a few levels of the taxonomic hierarchy (see Fig. 3.1). Depending on the location and genetic stability of their target regions, some molecular markers are ideal for strain identification and population studies but are too variable for higher taxa relationship studies. Conversely, molecular markers for highly conserved regions may not provide information that is significant at, or below, the species level but are ideal for phylogenetic studies.

Another important consideration in the use of molecular techniques to identify strains and/or species for studies on epizootics, tracking introductions or identifying origins is the fungus life cycle (Taylor *et al.*, 1999). The mode of reproduction (asexual, sexual or mixed reproduction strategy) of a fungus needs to be considered because some molecular markers are dominant. They identify alleles of a locus only as either present or absent, and are of limited usefulness for diploid fungi.

Other criteria for technique selection, keeping in mind the number of samples you need to process, are the cost of equipment and reagents and the ease of application. Given that there may be different ways to address the same questions, one can choose an appropriate technique or combination of techniques while considering practical aspects of a given study.

### 3.2.3. Sample preparation

For studies on strain/species differentiation and identification using either arbitrary primers or primers for targeted genome regions (e.g. telomere, ribosomal spacers), DNA should be obtained from monosporic cultures, especially if further studies are planned, to obtain valid data for analysis. One should never assume that isolates, especially hypomycetous fungi, obtained from an insect cadaver or from a colony forming unit are clonal even after serial dilution. For techniques to isolate single-spore isolates see Veen (1967) and Castrillo *et al.* (2004).

For detection in environmental samples or infected hosts by use of strain- or species-specific markers, mixed DNA samples (with background DNA from soil microorganisms, plant or insect host) may be utilized after testing the specificity and sensitivity of the primers developed. Note, however, that mixed DNA from environmental samples also contain contaminants (e.g. humic acid, polysaccharides) that can inhibit assay reactions, and tests should be conducted using additives (e.g., bovine serum albumin (BSA), dimethyl sulfoxide (DMSO), and T4 gene 32 protein) to relieve inhibition without significantly affecting sensitivity and efficacy of assays.

Isolation of DNA of high molecular weight and of high purity is central to any molecular study of fungal entomopathogens. Although there are various ways of extracting genomic DNA, the principles are the same: the fungal cell wall is disrupted or fungal cell lysed to allow release of nucleic acids; the nucleic acids are solubilized and then separated from proteins and other contaminants prior to precipitation for collection and storage. Readers should consult Chapter 16 (this volume) for detailed information on nucleic acids extraction methods. Unless the RNA component needs to be examined (for dsRNA studies), samples are usually treated with RNase during the extraction process. There are various protocols published in print and online and DNA extraction kits are also available from various manufacturers.

DNA extraction kits adaptable to fungal samples or developed specifically for plants and fungi have been utilized for different entomogenous fungi (e.g. Enkerli *et al.*, 2001; Aquino de Muro *et al.*, 2005; Nielsen *et al.*, 2005). Depending on availability of fungal material, the nature of the study and case of extraction, DNA may be obtained from spores, mycelia, hyphal bodies or a mixture of these stages (see Chapter 16, this volume). Additionally, for certain PCR applications, extraction of DNA from fungal tissues may not be necessary if crushed spores or mycelium can be used directly (Aufauvre-Brown *et al.*, 1992).

## 3.3. Genetic Fingerprinting

Molecular techniques used for identification and diagnosis of fungal strains and species can be grouped into non-PCR-based and PCR-based methods. Non-PCR-based methods were used in early (pre-PCR) studies of fungal strain differentiation and species identification. However, the use of these methods is now limited or has been abandoned due to advantages offered by PCR techniques such as use of small quantity of DNA, non-radioactive probes, high sensitivity and ease of use.

### 3.3.1. Non-polymerase chain reaction (PCR)-based techniques

#### *3.3.1.1. Multilocus enzyme electrophoresis (MLEE)*

Protein electrophoresis of isozymes (multiple forms of an enzyme due to variations in primary structure) and allozymes (allelic variants of an enzyme) was the first molecular technique broadly used for fungal strain and species identification. This technique is based on the differential rate of migration of non-denatured proteins due to their net charge during electrophoresis in either starch or polyacrylamide gels. A protein's net charge (based on its amino acid composition), size and shape determine its migration properties (Avise, 2004). Soluble proteins from whole-cell extracts are separated by electrophoresis or isoelectric focusing, and the gel is incubated in a histochemical stain specific for a given enzyme to reveal the comparative locations of the various allozymes in their separate lanes. Phenotypes are assessed by the collective banding positions of all enzyme systems tested.

This relatively easy technique produces robust data. In addition, isozyme and allozyme markers are codominant. However, a major drawback in using MLEE is that inter-isolate variation at the nucleotide level cannot be detected unless they change primary amino acid composition. Generally, data from at least ten enzymes need to be combined to show variability (Soll, 2000). In a similar manner, even changes in amino acid composition are also undetectable unless electrophoretic mobility is affected, thereby causing different alleles from different samples to be considered identical. Furthermore, these markers may not be neutral as some enzymes may be under selection pressure.

Readers should consult Hames and Rickwood (1990) for more information on MLEE methods. MLEE has been used to assess genetic variation among isolates of entomopathogenic fungi like *Beauveria* spp. (e.g. St Leger *et al.*, 1992), *Hirsutella thompsonii* (Boucias *et al.*, 1982) and *Metarhizium anisopliae* (Riba *et al.*, 1986).

#### *3.3.1.2. Electrophoretic karyotyping*

Fungal karyotypes (or chromosome number and sizes) can vary among strains of a given species and have been used to assess intraspecies diversity, especially among asexual fungi (Kistler and Miao, 1992). Karyotype polymorphisms arise from genomic rearrangements such as deletions, duplications, insertions or translocations, and from the presence and accumulation of supernumerary chromosomes (Covert, 1998; Walz, 2004) and from polyploidization. Supernumerary chromosomes (also called B chromosomes) are believed to be dispensable for normal growth, containing DNA material not present in other isolates of the same species (Covert, 1998). They are generally smaller than the normal chromosomes and are meiotically unstable (Goosen and Debets, 1996). Since fungal chromosomes are too small to see, polymorphism is indirectly visualized using pulse field gel electrophoresis (PFGE), where intact chromosomes, prepared as protoplast plugs, are run in agarose gels and forced to change direction during electrophoresis allowing different size fragments (above 30–50 kb) to separate from each other. With standard electrophoresis, these large fragments migrate with the same mobility regardless of size. By using two alternating fields, smaller fragments move to new directions faster than larger DNA allowing separation in agarose gels and detection with ethidium bromide stain visualized under ultraviolet (UV) light. Chromosome size is estimated by running standards, like *Saccharomyces cerevisiae* and *Schizosaccharomyces pombe* chromosomes, alongside samples of interest. Several systems of alternating or PFGE have been developed (Walz, 2004), with the contour-clamped homogeneous electric field (CHEF) being the most commonly used.

Critical factors in karyotype analysis are the preparation of chromosome samples and selection of appropriate electrophoretic conditions (field strength, pulse time, reorientation angle, buffer, agarose and temperature). Karyotype analysis requires considerable technical expertise and is costly to conduct. For more details on karyotype electrophoresis of filamentous fungi readers should refer to Chapter 16 (this volume). Additionally, Walz (2004) may be consulted for further readings on this subject. For protocols on preparation of intact chromosomes from entomogenous fungi refer to Chapter 16 (this volume) and Shimizu *et al.* (1992), Pfeifer and Khachatourians (1993) and Viaud *et al.* (1996).

Because karyotype analysis is affected by electrophoretic conditions used to separate chromosome bands, this technique is often complemented with telomere fingerprinting, which provides an estimate of telomere number and, consequently, chromosome number. Telomere fingerprinting utilizes restriction fragment length polymorphism (RFLP; see below) and probes based on available sequences or chromosome ends (telomeres) of related fungi. For example, Viaud *et al.* (1996) used a telomeric probe containing eight telomeric repeats $(TTAGGG)_8$, and 159 base pairs (bp) of a *Botrytis cinerea* telomere-associated region, along with CHEF electrophoresis of chromosomal DNA, to determine karyotype polymorphism among nine *Beauveria bassiana* isolates. Telomere fingerprinting has also been used independently as a marker to differentiate strains of entomopathogenic fungi such as *M. anisopliae* var. *acridum* (= *Metarhizium flavoviride*) (Inglis *et al.*, 1999) and *Nomuraea rileyi* (Boucias *et al.*, 2000).

### *3.3.1.3. Restriction fragment length polymorphism (RFLP)*

This technique is based on the digestion of total genomic DNA with one or more restriction enzymes, endonucleases that recognize specific 4–6 or more base pairs of DNA sequences and cut the DNA in or near the recognition sequence, followed by hybridization with specific genomic probes. Polymorphisms are based on the presence or absence of a recognition site for a particular restriction enzyme resulting from base substitutions within the cleavage site, addition or deletion of DNA, or sequence rearrangements (Avise, 2004). The restriction phenotype produced is visualized by running DNA fragments in agarose gels or by Southern blot hybridization. In the latter method, DNA restriction fragments are separated by gel electrophoresis, denatured and transferred to nylon or nitrocellulose membrane, hybridized with probes labelled with radioactive or fluorescent markers and exposed to X-ray films to obtain an autoradiograph of the restriction pattern. For more details on Southern blotting see Sambrook *et al.* (1989). Probes are labelled to allow detection of bands on X-ray films because restriction digests of chromosomal DNA from complex organisms result in numerous small fragments appearing as a 'smear' rather than in distinct, separate bands in agarose gels stained with ethidium bromide. Probes are generated from highly purified DNA, cloned DNA fragments or from cDNA to detect specific fragments of interest. They may vary from single-gene probes that generate patterns with only one to two bands to probes based on repeat sequences (i.e. ribosomal DNA (rDNA), transposable elements and satellite DNA) that produce more informative complex patterns.

Restriction enzymes are sequence-specific and allow for reliable differentiation and identification of fungal strains and species. Furthermore, since RFLP markers can detect variations in multiple alleles, they can be used for analysis of diploids or recombinants from haploid parents. RFLP analysis, however, requires extraction of large amounts of DNA (5–10 µg) for restriction digests and utilizes strictly regulated radioactive materials, which not only require special storage, handling and disposal protocols, but also increase the cost of analysis.

RFLP markers for fungi can usefully differentiate intraspecific to interspecific diversity depending on the probe used and the organisms studied. For example, Typas *et al.* (1992) found a combination of mitochondrial and ribosomal RFLP useful for species to subspecies studies of the genus *Verticillium*, and Pipe *et al.* (1995) found rDNA useful for the genus *Metarhizium*. Walsh *et al.* (1990), however, observed that rDNA RFLP was useful within Entomophthorales only in delineation of genera and species but not at the subspecies level. Hajek *et al.* (1990) used RFLP of the rDNA to confirm that the fungal pathogen *Entomophaga maimaiga* was responsible for epizootics in gypsy moth in seven contiguous north-eastern states in the USA.

#### *3.3.1.4. DNA–DNA hybridization*

This technique involves pairwise comparison of isolates using DNA–DNA re-association kinetics for species identification. It is based on the double-stranded nature of DNA and that complimentary strands are held together by hydrogen bonds (Avise, 2004). Samples are heated to break the hydrogen bonds, generating single-stranded DNA that is labelled with radioactive iodine. This sample is then mixed with single-stranded DNA of a known species, and allowed to reanneal into double-stranded DNA; the mixture then is heated gradually and monitored for dissociation to single strands. High-sequence homology between two samples will result in a stable duplex even at a high temperature. Consequently, closely related species or samples of the same species show a high melting temperature ($T_m$) value, the temperature at which 50% of DNA strands are dissociated. Although a standard procedure for bacterial taxonomy, this technique was not widely adapted for fungi because its sensitivity declines with divergence. This technique is also labour-intensive because each assay is limited to a single pair of samples.

### 3.3.2. PCR-based techniques

The PCR technique utilizes multiple cycles of different temperature profiles programmed on a thermocycler to denature, anneal to primers and elongate template DNA (or RNA). The process is exponential, with the amplified products or amplicons of the previous cycle serving as templates for the next cycle. This feature circumvents the problem of small amounts of the starting DNA sample and allows for a highly sensitive detection system. Typical PCR assays consist of 25–45 cycles, resulting in enough amplicons for visualization of products in agarose gels stained with ethidium bromide.

The denaturation step breaks hydrogen bands between the double-stranded DNA, generating single-stranded templates to which a pair of primers (or a single primer in case of random amplified polymorphic DNA (RAPD)) anneal to on opposite strands of the template DNA, flanking the sequence to be amplified. The primer binding sites determine the size and number of amplification products. A heat stable DNA polymerase, like *Taq* DNA polymerase, recognizes the free 3′ hydroxyl end of the primer-template DNA duplex and catalyses DNA synthesis of the region between the two primers. Included in PCR mix are reaction buffer, deoxynucleotide triphosphates (dNTPs) to incorporate in the extension step and magnesium, a co-factor in the enzymatic reaction. Interaction of the different components and cycling parameters determine specificity, sensitivity and fidelity of PCR reactions. Optimal assay conditions for different studies will vary and need to be determined empirically.

Primers are short chains of synthetic oligonucleotides, usually 18–24 nucleotides, which bind to complimentary sequences on the template DNA and direct synthesis to specific segments of the genome. They may be random or specific, based on conserved regions of the genome or on known sequences of the organism of interest. The design of target-specific primers is critical in the development of any strain- or species-specific detection assays. Although optimizing various reagent concentrations and cycling parameters can further enhance specificity of PCR assays, considerable time and effort are saved by starting with a good primer design. Primer sequence and length also determine annealing temperature. Generally, longer primers generate more specific products, requiring complementation for longer sequences with the template DNA and having higher annealing temperature that further minimize mispriming (binding to unintended template). Primer design programs to optimize primer sequences for PCR reactions are available commercially. There are also online sites for checking primer sequences for possible primer–dimer formation (primers bind to each other rather than to the template DNA) and calculating annealing temperature. For more on PCR, see Palumbi (1996) and Cold Spring Harbor Protocols (www.cshprotocols.org). For PCR optimization and primer design see Grunenwald (2003) and Hyndman and Mitsuhashi (2003), respectively.

Additives to PCR mixes such as BSA, DMSO, glycerol, formamide or T4 Gene 32 protein may enhance efficiency and specificity of binding. The type and concentration of additive to use depend on the sample DNA (e.g. BSA helps relieve inhibition of PCR reaction by environmental contaminants like humic acid, while DMSO helps amplify GC-rich template). As with other reagents in the PCR mix, the optimal concentration for a given additive needs to be determined empirically.

PCR products are usually visualized by running a small aliquot (usually 5 μl) in agarose gel stained with ethidium bromide (final concentration of 0.5 μg/ml) and photographed under UV light. DNA fragment size is estimated by running samples alongside molecular standards of the appropriate size range. For efficient separation of PCR products, the percentage of agarose (w/v) and type used will vary with the expected DNA fragment size. Different types of agarose with different resolving ranges are available. For recipes and guidelines on gel electrophoresis, see Sambrook *et al.* (1989) and Cold Spring Harbor Protocols (www.cshprotocols.org).

The development of PCR techniques has resulted in a proliferation of PCR-based methods currently used for fungal identification and diagnosis. These are often the methods of choice if genome sequences and/or primers are available. Depending on the primer sequence, target binding regions may be specific to rDNA or mtDNA or may span the whole nuclear genome. For an overview of the general steps in PCR-based fingerprinting (see Fig. 3.2).

### *3.3.2.1. Nuclear genome amplification*

ARBITRARY PRIMED PCR (AP-PCR) OR RANDOM AMPLIFIED POLYMORPHIC DNA (RAPD). The RAPD technique utilizes short arbitrary primers (usually ten nucleotides) that are expected to complement to some or a few parts of the target DNA and, thus, does not require prior information on the fungal genome or its sequences (Welsh and McClelland, 1990). Polymorphisms are based on the number and location of binding sites, and are visualized on ethidium bromide-stained agarose gels as the number and size of bands generated by RAPD primers. Variation detected may be as simple as a single base substitution that alters a binding site. The ease of application and its low cost made this technique popular for initial studies on entomopathogenic fungal systematics but most especially for intraspecies and interspecies diversity.

Typically, a preliminary screening is conducted on a number of primers to select for ten or more that produce a reasonable number of distinct,

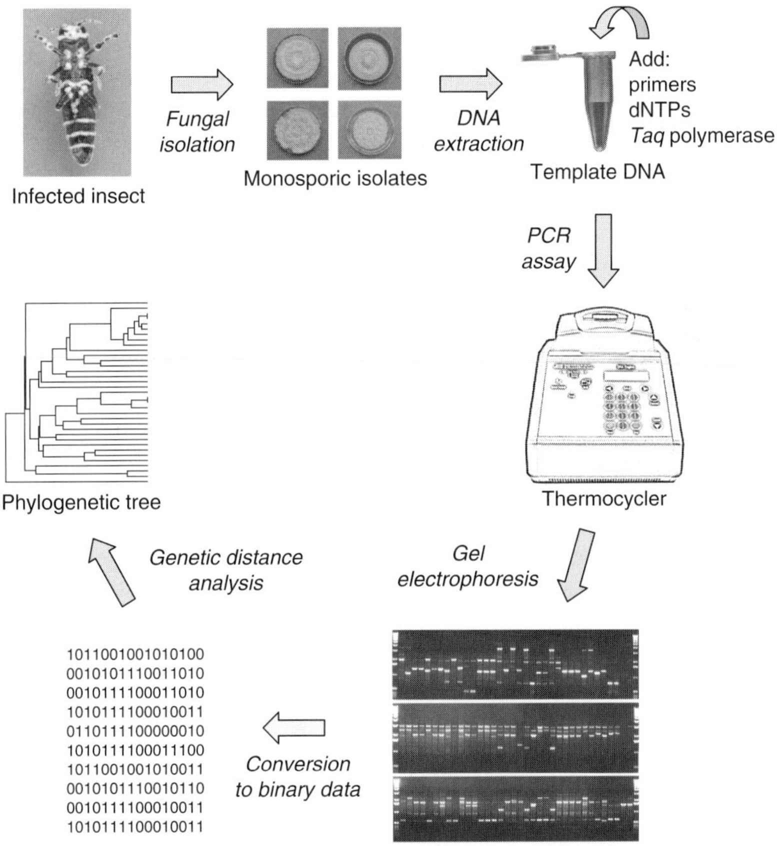


**Fig. 3.2.** Overview of the general steps in PCR-based DNA fingerprinting of fungi.

reproducible and well-separated amplified fragments easy to score. For each primer, the presence or absence of a band of a given molecular size, considered a locus, is scored for each isolate. Thus, RAPD markers are dominant (i.e. alleles are either absent or present). They cannot differentiate between dominant homozygotes and heterozygotes and are not applicable for analysis of diploid fungi (McDonald, 1997).

The requirement for small DNA samples and the generation of multiple PCR products for analysis in a short period of time allow for rapid screening of numerous isolates compared to RFLP. However, since DNA fragments of comparable size migrate at the same rate despite differences in sequence, the apparent level of polymorphism may be underestimated. This problem may be alleviated by combining RAPD with single-strand conformation polymorphism analysis, which is based on the different electrophoretic mobility of fragments of the same size with different DNA sequences (Dowling *et al.*, 1996). Another important consideration when using the RAPD technique is the reproducibility of data generated between, and even within, laboratories. Since slight variations in concentration of reagents (i.e. primer and $MgCl_2$) can alter results, reproducibility of RAPD data is a major concern; it is generally conceded, however, that well-standardized techniques can allow reliable reproducibility of results within a given laboratory.

AMPLIFIED FRAGMENT LENGTH POLYMORPHISM (AFLP). The AFLP technique effectively combines RAPD and RFLP techniques and offers advantages of both techniques. The genomic DNA is digested with one or more restriction enzymes (e.g. *Eco*RI and *Mse*I) followed by selective amplification (Vos *et al.*, 1995). The restriction fragments are ligated to oligonucleotide adapter molecules that act as primer binding sites. Primers with homology to the adapters are used to amplify the ligated fragments. The primer length and sequence vary by the addition of one to three nucleotides to the primer 3′ end. Longer primers are used for larger genomes to restrict the number of fragments for analysis. Variation in sequence of the additional bases also results in the detection of more loci. Compared to RAPD, AFLP produces more robust data because it utilizes longer primers and more stringent reaction conditions (i.e. higher annealing temperature). This technique also screens more loci and, thus, generates more complex patterns that allow for better separation of similar isolates. PCR products can be numerous and are best resolved on denaturing polyacrylamide gels for adequate separation. AFLP has been used for intraspecies study of entomopathogenic fungi like *B. bassiana* (Aquino *et al.*, 2005), *N. rileyi* (Boucias *et al.*, 2000), and *E. maimaiga* (Nielsen *et al.*, 2005). The use of AFLP does require a larger initial DNA sample than RAPD, as well as greater technical expertise, and is more costly. Furthermore, it suffers the same drawback as RAPD since AFLP markers are dominant.

MICROSATELLITES. Microsatellites (also known as simple sequence repeats (SSR)) are tandem repeats of short DNA sequences (one to six nucleotides) found in many prokaryotes and eukaryotes, including fungi. They are generally highly polymorphic based on the number of reiterations of the repeated sequence, and are widely used for population studies. Detectable polymorphism is comparable to, or higher than, other fingerprinting methods because strand slippage during DNA replication can change the number of short repeats (Taylor *et al.*, 1999). In fungi, repeats from one to six nucleotides have been recorded from nine completely sequenced haploid fungal genomes (Karaoglu *et al.*, 2005). The most abundant of such repeats are mononucleotide, dinucleotide (e.g. AT, AG or CT) and trinucleotide repeats (e.g. AAG, GAG or TTC) (Karaoglu *et al.*, 2005). Microsatellite markers are more reliable than RAPD markers and are codominant with up to 50 detectable alleles per locus (Jarne and Lagoda, 1996). A pair of primers for a specific locus can detect many different alleles from one or two up to several hundred base pairs in length. Primer development, however, requires information on the target genome. In the absence of genome sequence information, microsatellite analysis requires cloning, detection of microsatellites and sequencing to determine regions flanking microsatellite-rich areas that may be used for PCR primer design. Among entomogenous fungi, microsatellite primers have been developed for *Beauveria brongniartii* (Enkerli *et al.*, 2001) and *B. bassiana* (Rehner and Buckley, 2003). Because of the range in size and number of amplicons generated by microsatellite primers, results are best resolved in polyacrylamide or high-resolution agarose gels (e.g. MetaPhor® agarose) stained with ethidium bromide and visualized under UV light.

REPETITIVE ELEMENT-BASED PCR (REP-PCR). Transposable elements (TEs) are mobile elements that cause spontaneous genetic changes in fungi and other eukaryotes via insertion, excision and transposon-mediated genome rearrangements (Daboussi and Capy, 2003). They are characterized by their mode of transposition and by their structural organization. Fungal TEs, reported mostly from asexual filamentous fungi, are classified as either class I (retrotransposons or retroposons), which transpose via an RNA intermediate, or class II (DNA transposons), which transpose on the DNA level via a 'cut-and-paste' mechanism (Daboussi and Capy, 2003). Since TEs are often confined to specific fungal strains or populations, they may be used as diagnostic tools (Kempken, 1999). Based on this feature of TEs, rep-PCR utilizes a pair of outwardly directed primer sequences from known TEs to determine their presence and copy number and location among strains. DNA sequences lying between TEs are amplified and fragment length polymorphisms are analysed by gel electrophoresis. While PCR markers based on TEs have been useful diagnostic tools for fungal plant pathogens at intraspecific levels such as formae speciales or races, their utility for entomopathogenic fungi still has to be evaluated.

PCR-BASED RFLP. PCR-based RFLP technique permits analysis of the same target genome regions for strain/species identification as does traditional RFLP, but obviates the need for large amounts of DNA samples or the use of radioactivity. Target regions of the genome are first amplified using primers based on known or conserved sequences flanking the region of interest. Amplified fragments are then digested with one or more restriction endonucleases and separated by gel electrophoresis for analysis. Since restriction products are fewer compared to total genomic digests, restriction patterns can be visualized by gel electrophoresis. PCR-based RFLP serves as another method of analysing rDNA and mtDNA in addition to sequencing.

### 3.3.2.2. Ribosomal markers

Part of the nuclear genome in fungi and other eukaryotes is the rDNA, which codes for the RNA component of the ribosomal RNA (rRNA). The rDNA genes occur in tandem repeats of up to 100–200 copies. Each rDNA gene unit has a copy of the 28S (large subunit), 18S (small subunit) and 5.8S of the rRNA, separated by internal transcribed spacers (ITS). The region between each gene unit is the non-transcribed intergenic spacer (IGS), which often contains the separately transcribed 5S rRNA (Fig. 3.3). The 28S, 18S and 5.8S are conserved in comparison to the ITS and IGS regions and are used to study relationships at the upper levels of the taxonomic hierarchy. In contrast, variability in the ITS and IGS regions has been utilized for specific to generic and for population to subspecific recognitions, respectively (Carlile *et al.*, 2001). Since the 18S and 28S regions are strongly conserved across distantly related organisms, and their sequences have been determined from different eukaryotes, the primers complementary to sections of these regions are used to amplify and to obtain the polymorphic ITS and IGS regions for analysis (Fig. 3.3).

The amplified fragments are directly analysed on agarose gels for length polymorphism, cut with restriction endonucleases, and the resulting fragments run in agarose gels (amplified rDNA restriction analysis (ARDRA) ), or sequenced for differentiation at the base-pair level. ITS and IGS polymorphisms have been used to look at intraspecific and interspecific diversity in entomogenous fungi (e.g. Rohel *et al.*, 1997; Jensen and Eilenberg, 2001; Coates *et al.*, 2002; Pantou *et al.*, 2003) and to develop species-specific primers for diagnostic studies (e.g. Entz *et al.*, 2005; Castrillo *et al.*, 2007).

### 3.3.2.3. Mitochondrial markers

Mitochondria are small organelles that serve as sites of oxidative phosphorylation in eukaryotes, whose DNA is, because of their endosymbiotic bacterial origins, wholly distinct from the nuclear DNA of the cells in which they occur.

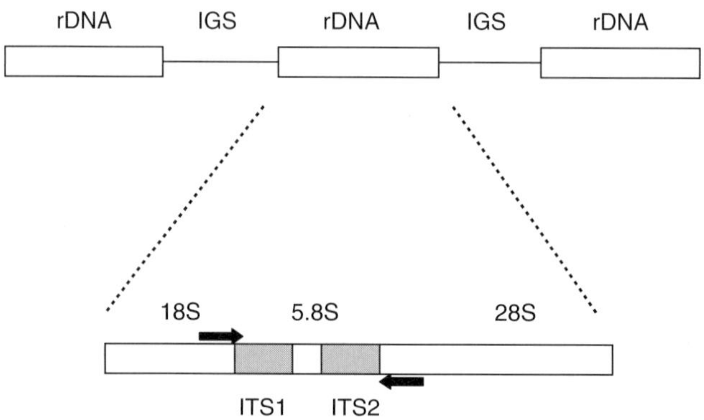


**Fig. 3.3.** Diagram of ribosomal DNA structure. Primers complimentary to sections of conserved 18S and 28S are used to amplify and obtain polymorphic ITS regions for analysis.

Although their genomes are much smaller compared to the nuclear genome, they make up a considerable part of the total DNA in a cell because of their numbers. The mtDNA is also characterized by its distinctively high (70–80%) AT content, lack of methylation and conserved gene function (Paquin *et al.*, 1997). The genome contains genes for mitochondrial ribosomal and transfer RNAs and enzymes involved in oxidative phosphorylation. The highly conserved regions are useful for phylogenetic studies, while the highly variable regions are used for intraspecific and interspecific identification (Kouvelis *et al.*, 2004; Ghikas *et al.*, 2006). Although mitochondrial genomes can evolve at their own rate relative to the nuclear genomes of the organisms in which they occur, most comparison studies of the two genomes in fungi have shown concordance (e.g. Kurdyla *et al.*, 1995; Sommerhalder *et al.*, 2007). Cases of discordant evolution (i.e. comparatively low variation in mtDNA) are likely due to selection pressure and because all mitochondrial genes are linked (Sommerhalder *et al.*, 2007). Thus, a comparative study of genetic variation of both genomes may be ideal for classification studies (e.g. Kouvelis *et al.*, 2008).

Studies on diversity of the mitochondrial genome of entomopathogenic fungi have utilized RFLP analysis (e.g. Mavridou and Typas, 1998) and DNA sequencing analyses (e.g. Typas *et al.*, 1992; Hegedus and Khachatourians, 1993). The complete sequence of mitochondrial genomes of the entomopathogenic fungi *Lecanicillium muscarium* and *M. anisopliae* var. *anisopliae* have also been published (Kouvelis *et al.*, 2004; Ghikas *et al.*, 2006). For more on mitochondrial genomes, with focus on entomopathogenic fungi see Rodriguez *et al.* (2004).

### 3.3.3. DNA fingerprint data analyses

DNA fingerprinting data are meaningful only with proper quantitative analyses and knowledgeable interpretation. For this review, we focus on the analysis of two-state data sets, where bands generated and visualized on agarose gels are converted to binary values of 1 and 0, indicating presence or absence of band, respectively, for a given product generated by a given primer. This type of analysis is applicable to most of the methods discussed, including markers that generate bands without defined loci and those that provide allelic information.

Banding patterns resolved by electrophoresis may be inspected or 'scored' either manually, by visual examination, or automatically, by use of image processing software that detects, scans and classifies bands. Manual scoring is possible only for markers that generate a few bands or relatively simple banding patterns (i.e. microsatellite markers or AP markers). For markers that generate complex banding patterns such as AFLP, computer-assisted band detection is necessary. Data generated by different primers for all isolates are combined, and a measure of similarity for every possible pair of isolates is calculated using appropriate coefficients (i.e. Jaccard or Dice coefficient for two-state data) available in software packages such as NTSYS-PC (EXETER software). The most used coefficient is Jaccard, which treats all bands equally regardless of the speed of marker evolutionary clock; this coefficient assumes that different patterns need not contain the same number of bands and that increasing pattern complexity increases resolution (Sneath and Sokal, 1973). Coefficient values range from 0 to 1, where 0 reflects no common bands and increasing values reflect increasing degrees of similarity. A value of 1 indicates all common bands or identical/clonal isolates. For more information on Jaccard, Dice and other coefficients, see Sneath and Sokal (1973).

Following calculation of similarity coefficient values for all pairs of isolates, a matrix of coefficients is computed to generate dendrograms or phylogenetic trees using unweighted pair – group arithmethic averages method (UPGMA), neighbour-joining method or other appropriate tree-building methods. UPGMA is the simplest distance matrix method, and can be used when the rate of gene substitution is more or less constant (Nei and Kumar, 2000). In cases where the rate of evolution varies from lineage to lineage, other distance methods such as least square, minimum evolution or neighbour-joining method should be used. For more on principle, algorithm and calculations for each method, see Nei and Kumar (2000). Reliability of tree topology obtained by UPGMA or other distance matrix methods may be tested by use of interior branch test (Nei *et al.*, 1985) or bootstrap test (Felsenstein, 1985). In cases where two or more trees are generated from the same matrix, a bootstrap consensus tree is generated.

Ideally, quantitative data from DNA fingerprinting with appropriate stable markers of numerous isolates of a species should be stored in a database. The fingerprint profiles could be used as reference standard among different laboratories for identification of unknowns or novel strains. Databases could also be combined with current systematic studies for species placement, especially within recognized species complexes such as *B. bassiana, M. anisopliae* and *L. lecanii* (Humber, 1997). Although DNA sequence data would provide the most stable database, sequencing of numerous isolates for strain identification may not be practical for most laboratories.

## 3.4. DNA Sequencing

A variety of coding and non-coding regions can be screened for utility, and data from multiple sequences analysed to resolve taxonomic questions. For example, Rehner *et al.* (2006) found two nuclear intergenic regions (a length of DNA containing no, or very few, genes) in *B. bassiana* species that contain sequence variation useful for studies on biogeography and epidemiology. Examples of coding genes that have been utilized for genetic diversity studies include elongation factor 1-α, α and β tubulin, and histone genes, in addition to ribosomal and mitochondrial genes. Since genes that code for proteins may have different evolutionary clocks and may be prone to bias due to selection pressure, more than one gene should be sequenced to assess genetic diversity.

It is beyond the scope of this chapter to review DNA sequencing techniques (i.e. Maxam-Gilbert versus Sanger dideoxy, the latter is the basis of cycle and automated sequencing techniques) and sequence data analysis. Readers are referred to books on these topics.

## 3.5. Diagnosis and Detection

### 3.5.1. Species- and strain-specific identification

When the primary objective of the study is the identification of a specific strain or species (e.g. to track the establishment or dispersal of a biological control agent in the field), specific primers in combination with sensitive assay protocols, need to be designed for detection and diagnosis. These primers are based on previous knowledge of target DNA sequences and may detect specific genes (e.g. rDNA or pathogenicity genes) or unique sequences may be generated from DNA fingerprints by AP-PCR or AFLP. An example of the latter is sequence-characterized amplified region (SCAR) markers based on sequences unique to a strain or species detected by use of AP-PCR (Paran and Michelmore, 1993). This strategy has been utilized to develop SCAR primers for detection of *B. bassiana* strain GHA, the active ingredient in commercially available mycoinsecticides in the USA, in environmental samples for field efficacy and persistence studies (Castrillo *et al.*, 2003). The advantage of these specific markers is that they allow detection of the target strain/species from mixed environmental samples, especially when other indigenous strains or closely related species may be present. For example, *E. maimaiga*-specific primers, based on the ITS sequence of several *E. maimaiga* strains, were developed for detection of the fungus in forest soil where other *Entomophaga* spp. occur (Castrillo *et al.*, 2007).

### 3.5.2. Real-time PCR

Strain- and species-specific primers can also be utilized in real-time PCR assays for quantitative detection of the organism of interest. Real-time PCR is an advancement of the standard PCR technique, where amplified fragments at the assay end point are visualized by gel electrophoresis. The results are evaluated as presence or absence of a given band and do not resolve variations in yield (gel electrophoresis cannot detect less than tenfold change in yield). In contrast, real-time PCR detects amplification products at the exponential phase, when reactions are specific and precise and while the PCR products are doubled at every cycle. Quantification is based on the relationship between the amount of starting target sample and the amount of amplified products at any given cycle during this phase. Detection is accomplished by the use of different chemistries, or fluorescent reporters, that are either amplicon sequence-specific (i.e. Taqman®, Molecular Beacons and Scorpion® probes) or non-specific (i.e. SYBR® Green). These are used in conjunction with proper instrumentation that monitors the accumulation of these reporter molecules as the PCR reaction proceeds. At some point during the exponential phase, the accumulated products generate a measurable fluorescence above the background. This point, called the threshold cycle, is used to calculate the starting template DNA of an unknown by running parallel reactions using a standard. Non-specific chemistry utilizes molecules such as SYBR Green that binds to the minor groove of *any* double-stranded DNA. SYBR Green is easy to use and inexpensive but can result in overestimation of the target sequence due to primers–dimers or misprimed non-specific products (Mackay *et al.*, 2002). In contrast, the TaqMan assay utilizes a labelled fluorogenic probe that anneals to a specific sequence between those targeted by the forward and reverse primers. This specificity permits the use of multiple primers-probe sets, each with a unique fluorophore, to detect and to quantify different targets in the same reaction tube. This could be useful for studies on mixed infections and possible interactions between different pathogens in a given host. Species- and strain-specific TaqMan-based real-time PCR assays have been developed for *E. maimaiga* and *B. bassiana* GHA, respectively, by Castrillo *et al.* (2007, 2008).

Although real-time PCR assays are very sensitive, and capable of detecting less than 1.0 pg of pure spore DNA, care should be taken prior to their use for mixed environmental samples. Variations in sample recovery, DNA extraction, background DNA present and PCR-inhibiting contaminants could affect assay sensitivity or limit of detection and, consequently, the derived estimate of fungal titer present. Once these factors have been taken into consideration, real-time PCR assay could be used for field samples, thereby eliminating the need to culture the fungus prior to molecular analysis. Another advantage of real-time PCR over standard PCR is that data analysis is performed automatically, saving the researcher considerable work time. Unfortunately, start-up cost for the required equipment can be prohibitively high for many laboratories. For more on real-time PCR detection chemistries, primers and probe design, and data analysis see Mackay (2007) and Dorak (2006).

## 3.6. Conclusions and Future Prospects

Many molecular techniques may help identify or diagnose fungi rank but these techniques are currently useful only with taxa for which at least an outline of a PCR-based taxonomy exists or for which several published molecular studies can be compared. In a practical sense, then, molecular identifications or diagnoses are still useful for only a limited range of taxa in the order Hypocreales even though these fungi – species of *Beauveria*, *Metarhizium*, *Lecanicillium*, *Isaria*/*Paecilomyces* and *Nomuraea* – are worldwide the most widely distributed, common, and most frequently applied insect fungi. Other important fungal entomopathogens whose phylogenetic foundations are too incomplete to be used for identifications or diagnoses include zygomycetes in *Entomophthorales*, blastocladialeans in *Coelomomyces*, and rust-like basidiomycetes in *Septobasidium*. Diagnoses tend to focus on generic or familial ranks and may need different technical approaches than for identifications that are usually desired at or below the rank of species.

Molecular (sequence) data are increasingly necessary to identify species from such genera as *Beauveria*, *Metarhizium*, and *Cordyceps* (in the broad sense). As increasingly detailed phylogenetic revisions of fungal genera appear, primary identifications will depend more on sequence matching than on traditional characters. Even while the number of taxa for which sequence-based data allow accurate identifications continually rises, the total number of entomopathogens for which sequence-based data allow accurate generic, specific or infraspecific identifications still remains low. Into the indefinite future it will be true that no matter how many sequences from rising numbers of genes and of taxonomically diverse fungi become available, the traditional, morphological characters will still remain the only possible means to identify a vast range of these fungi.

The computational analyses of molecular data powerfully reveal patterns of relationships among organisms. However, the reality of a living, reactive organism presents the ultimate integration of its whole genome and a level of integrated expression that comparative molecular analyses can only aspire to achieve. Traditional taxonomies need to recognize that their phylogenetically based equivalents often show that morphologically similar organisms have distinctly different genomes and also that traditional taxonomies may be incapable of distinguishing rich patterns of diversity and relationships detectable only by DNA-based analyses. Synergistic cooperation between traditional and molecular approaches to systematics will allow the integration of all available knowledge into a conceptually unified body of biological wisdom.

## References


Aquino de Muro, M., Elliot, S., Moore, D., Parker, B.L., Skinner, M., Reid, W. and El Bouhssini, M. (2005) Molecular characterization of *Beauveria bassiana* isolates from overwintering sites of Sunn Pests (*Eurygaster* and *Aelia* species). *Mycological Research* 109, 294–306.

Aufauvre-Brown, A., Tang, C.M. and Holden, D.W. (1992) Detection of gene disruption events in *Aspergillus* transformants by polymerase chain reaction directly from conidiospores. *Current Genetics* 24, 177–178.

Avise, J.C. (2004) *Molecular Markers, Natural History and Evolution.* Sinauer Associates, Massachusetts.

Boucias, D.G., McCoy, C.W. and Joslyn, D.J. (1982) Isozyme differentiation among 17 geographical isolates of *Hirsutella thompsonii. Journal of Invertebrate Pathology* 39, 329–337.

Boucias, D.G., Tigano, M.S., Sosa-Gomez, D.R., Glare, T.R. and Inglis, P.W. (2000) Genotypic properties of the entomopathogenic fungus *Nomuraea rileyi. Biological Control* 19, 124–138.

Bruns, T.D., White, T.J. and Taylor, J.W. (1991) Fungal molecular systematics. *Annual Reviews of Ecology and Systematics* 22, 525–564.

Carlile, M.J., Watkinson, S.C. and Gooday, G.W. (2001) *The Fungi.* Academic Press, San Diego, California.

Castrillo, L.A., Vandenberg, J.D. and Wraight, S.P. (2003) Strain-specific detection of introduced *Beauveria bassiana* in agricultural fields by use of sequence-characterized amplified region markers. *Journal of Invertebrate Pathology* 82, 75–83.

Castrillo, L.A., Griggs, M.H. and Vandenberg, J.D. (2004) Vegetative compatibility groups in indigenous and mass-released strains of the entomopathogenic fungus *Beauveria bassiana*: likelihood of recombination in the field. *Journal of Invertebrate Pathology* 86, 26–37.

Castrillo, L.A., Thomsen, L., Juneja, P. and Hajek, A.E. (2007) Detection and quantification of the gypsy moth fungal pathogen *Entomophaga maimaiga* resting spores in forest soil using real-time PCR. *Mycological Research* 111, 324–331.

Castrillo, L.A., Griggs, M.H. and Vandenberg, J.D. (2008) Quantitative detection of *Beauveria bassiana* GHA (Ascomycota: Hypocreales), a potential microbial control agent of the emerald ash borer, by use of real-time PCR. *Biological Control* 45, 163–169.

Coates, B.S., Hellmich, R.L. and Lewis, L.C. (2002) *Beauveria bassiana* haplotype determination based on nuclear rDNA internal transcribed spacer PCR-RFLP. *Mycological Research* 106, 40–50.
Covert, S.F. (1998) Supernumerary chromosomes. *Current Genetics* 33, 311–319.
Daboussi, M.J. and Capy, P. (2003) Transposable elements in filamentous fungi. *Annual Review of Microbiology* 57, 275–299.
Dorak, M.T. (2006) *Real-time PCR*. Taylor & Francis, New York.
Dowling, T.E., Moritz, C., Palmer, J.D. and Riesenberg, L.H. (1996) Nucleic acids III: analysis of fragments and restriction sites. In: Hillis, D.M., Moritz, C. and Mable, B.K. (eds) *Molecular Systematics*. Sinauer Associates, Massachusetts, pp. 249–320.
Enkerli, J., Widmer, F., Gessler, C. and Keller, S. (2001) Strain-specific microsatellite markers in the entomopathogenic fungus *Beauveria brongniartii*. *Mycological Research* 105, 1079–1087.
Entz, S.C., Johnson, D.L. and Kawchuk, L.M. (2005) Development of a PCR-based diagnostic assay for the specific detection of the entomogenous fungus *Metarhizium anisopliae* var. *acridum*. *Mycological Research* 109, 1302–1312.
Felsenstein, J. (1985) Confidence limits on phylogenies: an approach using the bootstrap. *Evolution* 39, 783–791.
Gams, W. and Zare, R. (2001) A revision of *Verticillium* sect. *Prostrata*. III. Generic classification. *Nova Hedwigia* 72, 329–337.
Ghikas, D.V., Kouvelis, V.N. and Typas, M.A. (2006) The complete mitochondrial genome of the entomopathogenic fungus *Metarhizium anisopliae* var. *anisopliae*: gene order and *trn* gene clusters reveal a common evolutionary course for all Sordariomycetes, while intergenic regions show variation. *Archives of Microbiology* 185, 393–401.
Goosen, T. and Debets, F. (1996) Molecular genetic analysis. In: Bos, C.J. (ed.) *Fungal Genetics: Principles and Practice*. Marcel Dekker, New York, pp. 97–117.
Grunenwald, H. (2003) Optimization of polymerase chain reactions. In: Bartlett, J.M.S. and Stirling, D. (eds) *PCR Protocols*. Humana Press, New Jersey, pp. 89–100.
Hajek, A.E., Humber, R.A., Elkinton, J.S., May, B., Walsh, S.R.A. and Silver, J.C. (1990) Allozyme and restriction fragment length polymorphism analyses confirm *Entomophaga maimaiga* responsible for 1989 epizootics in North American gypsy moth populations. *Proceedings of the National Academy of Sciences of the USA* 87, 6979–6982.
Hames, B.D. and Rickwood, D. (1990) *Gel Electrophoresis of Proteins: A Practical Approach*. Oxford University Press, New York.
Hegedus, D.D. and Khachatourians, G.G. (1993) Identification of molecular variants in mitochondrial DNAs of members of the genera *Beauveria, Verticillium, Paecilomyces, Tolypocladium*, and *Metarhizium*. *Applied and Environmental Microbiology* 59, 4283–4288.
Humber, R.A. (1997) Fungi: identification. In: Lacey, L. (ed.) *Manual of Techniques in Insect Pathology*. Academic Press, San Diego, California, pp. 153–186.
Hyndman, D.L. and Mitsuhashi, M. (2003) PCR primer design. In Bartlett, J.M.S. and Stirling, D. (eds) *PCR Protocols*. Humana Press, New Jersey, pp. 81–88.
Inglis, P.W., Magahaes, B.P. and Valadares-Inglis, M.C. (1999) Genetic variability in *Metarhizium flavoviride* revealed by telomeric fingerprinting. *FEMS Microbiology Letters* 179, 49–52.
Jarne, P. and Lagoda, P.J.L. (1996) Microsatellites, from molecules to populations and back. *Trends in Ecology and Evolution* 10, 424–429.
Jensen, A.B. and Eilenberg, J. (2001) Genetic variation within the insect pathogenic genus *Entomophthora*, focusing on the *E. muscae* complex, using PCR-RFLP of the ITS II and the LSU rDNA. *Mycological Research* 105, 307–312.
Karaoglu, H., Lee, C.M.Y. and Meyer, W. (2005) Survey of simple sequence repeats in completed fungal genomes. *Molecular Biology and Evolution* 22, 639–649.
Kempken, F. (1999) Fungal transposons: from mobile elements towards molecular tools. *Applied Microbiology and Biotechnology* 52, 756–760.

Kistler, H.C. and Miao, V.P.W. (1992) New modes of genetic change in filamentous fungi. *Annual Review of Phytopathology* 30, 131–152.

Kouvelis, V.N., Ghikas, D.V. and Typas, M.A. (2004) The analysis of the complete mitochondrial genome of *Lecanicillium muscarium* (synonym *Verticillium lecanii*) suggests a minimum common gene organization in mtDNAs of Sordariomycetes: phylogenetic implications. *Fungal Genetics and Biology* 41, 930–940.

Kouvelis, V.N., Sialakouma, A. and Typas, M.A. (2008) Mitochondrial gene sequences alone or combined with ITS region sequences provide firm molecular criteria for the classification of *Lecanicillium* species. *Mycological Research* 112, 829–844.

Kurdyla, T.M., Guthrie, P.A.I., McDonald, B.A. and Appel, D.N. (1995) RFLPs in mitochondrial and nuclear DNA indicate low levels of genetic diversity in oak wilt pathogen *Ceratocystis fagacearum*. *Current Genetics* 27, 373–378.

Luangsa-ard, J.J., Hywel-Jones, N.L., Manoch, L. and Samson, R.A. (2005) On the relationships of *Paecilomyces* sect. *Isarioidea* species. *Mycological Research* 109, 581–589.

Mackay, I.M. (2007) *Real-time PCR in Microbiology. From Diagnosis to Characterization.* Caister Academic Press, Norfolk, UK.

Mackay, I.M., Arden, K.E. and Nitsche, A. (2002) Real-time PCR in virology. *Nucleic Acid Research* 30, 1292–1305.

Mavridou, A. and Typas, M.A. (1998) Intraspecific polymorphism in *Metarhizium anisopliae* var. *anisopliae* revealed by analysis of rRNA gene complex and mtDNA RFLPs. *Mycological Research* 102, 1233–1241.

McDonald, B.A. (1997) The population genetics of fungi: tools and techniques. *Phytopathology* 87, 448–453.

Moore, D. and Frazer, L.N. (2002) *Essential Fungal Genetics*. Springer, New York.

Nei, M. and Kumar, S. (2000) *Molecular Evolution and Phylogenetics*. Oxford University Press, Oxford.

Nei, M., Stephens, J.C. and Saitou, N. (1985) Methods for computing the standard errors of branching points in evolutionary tree and their application to molecular data from humans and apes. *Molecular Biology and Evolution* 2, 66–85.

Nielsen, C., Milgroom, M.G. and Hajek, A.E. (2005) Genetic diversity in the gypsy moth fungal pathogen *Entomophaga maimaiga* from founder populations in North America and source populations in Asia. *Mycological Research* 109, 941–950.

Palumbi, S.R. (1996) Nucleic Acids II: the polymerase chain reaction. In: Hillis, D.M., Moritz, C. and Mable, B.K. (eds) *Molecular Systematics*. Sinauer Associates, Massachusetts, pp. 205–248.

Pantou, M.P., Mavridou, A. and Typas, M.A. (2003) IGS sequence variation, group I-introns and the complete nuclear ribosomal DNA of the entomopathogenic fungus *Metarhizium*: excellent tools for isolate detection and phylogenetic analysis. *Fungal Genetics and Biology* 38, 159–174.

Paquin, B., LaForest, M.-J., Forget, L., Roewer, I., Wang, Z., Longcore, J. and Lang, F.B. (1997) The fungal mitochondrial genome project: evolution of fungal genomes and their gene expression. *Current Genetics* 31, 380–395.

Paran, I. and Michelmore, R.W. (1993) Development of a reliable PCR-based markers linked to downy mildew resistance gene in lettuce. *Theoretical and Applied Genetics* 85, 985–993.

Pfeifer, T.A. and Khachatourians, G.G. (1993) Electrophoretic karyotype of the entomopathogenic deuteromycete *Beauveria bassiana*. *Journal of Invertebrate Pathology* 61, 231–235.

Pipe, N.D., Chandler, D., Bainbridge, B.W. and Heale, J.B. (1995) Restriction fragment length polymorphism in the ribosomal RNA gene complex of isolates of the entomopathogenic fungus *Metarhizium anisopliae*. *Mycological Research* 99, 485–491.

Rehner, S.A. and Buckley, E.P. (2003) Isolation and characterization of microsatellite loci from the entomopathogenic fungus *Beauveria bassiana* (Ascomycota: Hypocreales). *Molecular Ecology Notes* 3, 409–411.

Rehner, S.A., Posada, F., Buckley, E.P., Infante, F., Castillo, A. and Vega, F.E. (2006) Phylogenetic origins of African and neotropical *Beauveria bassiana* s. l. pathogens of the coffee berry borer, *Hypothenemus hampei*. *Journal of Invertebrate Pathology* 93, 11–21.

Riba, G., Bouvier-Fourcade, I. and Caudal, A. (1986) Isoenzymes polymorphism in *Metarhizium anisopliae* (Deuteromycotina: Hypomycetes) entomogenous fungi. *Mycopathology* 96, 161–169.

Rodriguez, R.J., Cullen, D., Kurtzman, C.P., Khachatourians, G.G. and Hegedus, D.D. (2004) Molecular methods for discriminating taxa, monitoring species, and assessing fungal diversity. In: Mueller, G.M., Bills, G.F. and Foster, M.S. (eds) *Biodiversity of Fungi. Inventory and Monitoring Methods.* Elsevier, San Diego, California, pp. 77–102.

Rohel, E., Couteadier, Y., Papierok, B., Cavelier, N. and Dedryver, C.A. (1997) Ribosomal internal transcribed spacer size variation correlated with RAPD-PCR pattern polymorphism in the entomopathogenic fungus *Erynia neoaphidis* and some closely related species. *Mycological Research* 101, 573–579.

Sambrook, J., Fritsch, E.F. and Maniatis, T. (1989) *Molecular Cloning*. Cold Spring Harbor Laboratory Press, New York.

Shimizu, S., Arai, A. and Matsumoto, T. (1992) Electrophoretic karyotype of *Metarhizium anisopliae*. *Journal of Invertebrate Pathology* 60, 165–187.

Sneath, P.H.A. and Sokal, R.R. (1973) *Numerical Taxonomy*. W. H. Freeman, California.

Soll, D.R. (2000) The ins and outs of DNA fingerprinting the infectious fungi. *Clinical Microbiology Reviews* 13, 332–370.

Sommerhalder, R.J., McDonald, B.A. and Zhan, J. (2007) Concordant evolution of mitochondrial and nuclear genomes in the wheat pathogen *Phaeosphaeria nodorum*. *Fungal Genetics and Biology* 44, 764–772.

St Leger, R.J., Allee, L.L., May, B., Staples, R.C. and Roberts, D.W. (1992) World wide distribution of genetic variation among isolates of *Beauveria* spp. *Mycological Research* 96, 1007–1015.

Sung, G.-H., Hywel-Jones, N.L., Sung, J.-M., Luangsa-ard, J.J., Shrestha, B. and Spatafora, J.W. (2007) Phylogenetic classification of *Cordyceps* and the clavicipitaceous fungi. *Studies in Mycology* 57, 5–59.

Taylor, J.W., Geiser, D.M., Burt, A. and Koufopanou, V. (1999) The evolutionary biology and population genetics underlying fungal strain typing. *Clinical Microbiology Reviews* 12, 126–146.

Typas, M.A., Griffen, A.M., Bainbridge, B.W. and Heale, J.B. (1992) Restriction fragment length polymorphism in mitochondrial DNA and ribosomal RNA gene complexes as an aid to the characterization of species and sub-species populations in the genus *Verticillium*. *FEMS Microbiology Letters* 95, 157–162.

Veen, K.H. (1967) Monospore culture and determination of nucleus numbers. *Journal of Insect Pathology* 9, 276–278.

Viaud, M., Couteaudier, Y., Levis, C. and Riba, G. (1996) Genome organization in *Beauveria bassiana*: electrophoretic karyotype, gene mapping, and telomeric fingerprint. *Fungal Genetics and Biology* 20, 175–183.

Vos, P., Hogers, R., Bleeker, M., Reijens, M., Vandelee, T., Hornes, M., Frijiters, A., Pot, J., Kuiper, M. and Zabeau, M. (1995) AFLP: a new technique for DNA fingerprinting. *Nucleic Acids Research* 23, 4407–4414.

Walsh, S.R.A., Tyrrell, D., Humber, R.A. and Silver, J.C. (1990) DNA restriction fragment length polymorphisms in the rDNA repeat unit of *Entomophaga*. *Experimental Mycology* 14, 381–392.

Walz, M. (2004) Electrophoretic karyotyping. In Kück, U. (ed.) *The Mycota II Genetics and Biotechnology*. Springer, Berlin-Heidelberg, Germany, pp. 53–70.

Welsh, J. and McClelland, M. (1990) Fingerprinting genomes using PCR with arbitrary primers. *Nucleic Acid Research* 18, 7213–7224.

# 4 Molecular Approaches and the Taxonomy of Insect-parasitic and Pathogenic Nematodes

S.P. Stock

*Department of Entomology, University of Arizona, Tucson, USA*



## 4.1. Introduction

Advances made in molecular biology and software technology have revitalized the discipline of taxonomy. Indeed, taxonomy, the science of naming and classifying organisms, has become a trendy topic for research in recent years. For example, existing diversity and inventory projects of living organisms heavily rely on the development and application of accurate diagnostic methods to help with the identification of organisms (and their by-products) with beneficial properties for medical research and pharmaceutical bioprospecting. Improved analytical molecular methods and data analyses have also changed routine diagnostics and identification procedures, making them more accurate and exciting than they had been before. Moreover, currently ongoing 'tree of life' projects depend on the continuous improvement of methods for inferring, evaluating and using phylogenies to test evolutionary hypotheses.

With respect to insect-parasitic and pathogenic nematodes, molecular techniques have been developed to distinguish species, races and biotypes, as well as to study genetic variability and phylogenetic relationships of these organisms (Adams *et al.*, 1998; Nguyen *et al.*, 2001; Stock *et al.*, 2001; Perlman *et al.*, 2003; Spiridonov *et al.*, 2004). More than a decade ago, a workshop on taxonomy and systematics of entomopathogenic nematodes was organized at the CAB International Institute, St Albans, UK, to establish a solid foundation for future work. A benchmark document was created to address topics such as: (i) revised lists of described species (including synonyms, species *inquirenda* and *nomina dubia*); (ii) protocols for preservation and molecular characterization of species; (iii) development of taxonomic keys; and (iv) guidelines for species descriptions (Hominick *et al.*, 1997). Indeed, outcomes from this workshop set a yardstick that helped taxonomists and non-taxonomists around the world describe various novel entomopathogenic nematode species. However, advances made in molecular biology over the past decade have made these guidelines out of date and current markers, genes and methods previously considered must be reassessed and new technologies incorporated to keep up with the discovery of new nematode species (Stock and Hunt, 2005).

In this chapter, I summarize past and current molecular methods considered for diagnosis of insect-parasitic and pathogenic nematodes. Readers should be aware that the practice of developing molecules as taxonomic tools is not a trivial and quick task. On the contrary, it requires the same rigour and attention that has been applied for the past centuries to morphologically based taxonomy. Molecular taxonomy is herein considered as an additional step for species descriptions. It is only because of the focus of this book that emphasis is put on molecular methods. Comprehensive taxonomic studies aided with biological and behavioural observations should be undertaken to assess the genetic diversity and delimit species boundaries of novel nematode species.

## 4.2. Nematode Diagnosis and the Barcode System

The DNA bar-code or a 'bar-code of life' is a concept that was developed in 2003 at Cold Spring Harbor Laboratory (Stoeckle *et al.*, 2003). The DNA bar-code is considered analogous to the universal product code (UPC) commonly used on retail products and also known as the 'zebra code'. Instead of this being a numeric code, the molecular bar-code has nucleotide sequence information from a common gene and serves as a unique identifier for every species in the planet. The bar-code system is not a useful tool to infer phylogenetic relationships, but information gained from bar-coding, can be used with pre-existing phylogenies. There is overlap between DNA bar-coding, taxonomy and systematics, and together they can work to build a better system for species identification, measure their biological diversity and assess evolutionary relationships between and among various taxonomic ranks.

In spite of the existing arguments on the utility and universality of the DNA bar-code system, numerous research projects are under way to test the concept. Nematodes are among the first organisms used to test the bar-code concept (Floyd *et al.*, 2002a,b; Blaxter, 2003; Hebert *et al.*, 2003). For example, Floyd *et al.* (2002b) placed unknown nematodes sampled from Scottish upland grasslands by interpreting the signal from 18S ribosomal DNA (rDNA) bar-code. More recently, 18S sequences were also considered as a 'coarse diagnostic tool' to identify 360 nematode species from a tropical rainforest in Costa Rica (Mullin *et al.*, 2006; Powers *et al.*, 2009). In this study, 18S rDNA sequences were generated via direct sequencing of the polymerase chain reaction (PCR) products, and sorted into *m*olecular *o*perational *t*axonomic *u*nits (MOTUs) on the basis of primary sequence. A total of 167 unique nematode MOTUs were identified and compared with small subunit (SSU) sequences archived in GenBank to assess putative identifications and likely relationships.

Currently, there is insufficient information in the nematode databases for extensive species identification based on the 18S bar-code, but one glance at the tree permits an educated guess regarding the taxonomic affinities of their unknown nematode samples (Powers *et al.*, 2004). Expansion of the 18S nematode tree of life through collaboration of projects such as NemATOL (National Science Foundation (NSF)-funded nematode branch of the Tree of Life Project, http://nematol.unh.edu/) will undoubtedly become a valuable resource to the DNA bar-code system of this Nematoda.

## 4.3. A Review of Molecular Approaches Considered for Insect-parasitic and Pathogenic Nematode Taxonomy

Insect-parasitic and pathogenic nematodes show significant variation in behaviour, host range, infectivity, reproduction and environmental tolerances. This biological variation has stimulated interest to more fully characterizing their genetic diversity, specifically because new species and/or isolates may prove more useful than those currently used as biological control agents against agriculturally important pests (Stock and Reid, 2003; Stock, 2005).

Accurate identification of these insect parasites and pathogens is of critical importance. For example, matching the right nematode species with the appropriate target insect pest is relevant for the success of biological control and integrated pest management (IPM) programmes. Proper nematode diagnosis is particularly related to the importation of exotic species and/or strains which could eventually outcompete native species/strain therefore compromising local biodiversity. Furthermore, correct diagnosis of nematodes species and/or strains has also implications in the commercialization of this organisms and proprietary rights for use of patents and other legal issues regarding the use and application of insect-parasitic and pathogenic nematodes (Stock, 2005, 2007).

Molecular characters were quickly adopted and became useful tools for species identification and systematics of nematodes. Specifically, molecular approaches became essential tools when dealing with taxonomic ambiguities and helped resolving problems such as identification of members of a species complex and in the differentiation of species that are morphologically similar (Cutler and Stock, 2003; Stock *et al.*, 2004).

This section reviews the most common molecular techniques and markers that have been considered for taxonomic studies of insect-parasitic and pathogenic nematodes. Techniques are briefly outlined and discussed. Detailed methods can be found elsewhere (Hussey, 1979; Curran, 1991; Curran and Robinson, 1993; Avise, 1994; Hillis *et al.*, 1996; Powers and Fleming, 1998; Stock and Hunt, 2005; Stock, 2006). Chapter 8 (Peat *et al.*, this volume) should also be consulted as a reference for approaches and methods considered in nematode phylogenetics and population genetic studies.

### 4.3.1. Randomly amplified polymorphic DNA (RAPD)

The RAPD technique uses 10 base pair (bp) random primers to detect random segments of genomic DNA to depict polymorphisms (Williams *et al.*, 1990). These primers adhere to a specific nucleotide segment of the genomic DNA. The DNA is cut into several segments of a specific length which can be measured using gel electrophoresis. For a mutation to change the RAPD pattern, it must occur in the priming region or must change the length of the DNA between priming regions. In this way the RAPD analysis can provide a simple method for measuring genomic variation (Lynch and Milligan, 1994).

The RAPD technique has some advantages over other systems of genetic documentation because it has a universal set of primers, no preliminary work such as probe isolation, filter preparation or nucleotide sequencing is necessary (Williams *et al.*, 1990). The ease and simplicity of the RAPD technique made it ideal for genetic mapping, plant and animal breeding programmes and DNA fingerprinting, with particular utility in the field of population genetics. Nevertheless, theoretical and technical concerns arose regarding the use of RAPD methods: for example, reproducibility of results could be a problem, especially due to weakly amplified bands or poor quality and concentration of primer and/or template and PCR cycling conditions (including type of PCR machine used) (Muralidharan and Wakeland, 1993; Schierwater and Ender, 1993). With this method it is also difficult to accurately measure genetic variability at interspecific and intraspecific levels, and it may lead to possible misdiagnosis.

RAPDs have typically been used for fingerprinting and population genetic structure studies of various nematode groups (Liu and Berry, 1995; Schwenk *et al.*, 1996; Blok *et al.*, 1997; Randig *et al.*, 2002). This method has also been considered to aid species identifications and/or to infer phylogenetic affinities of members of Heterorhabditidae and Steinernematidae (Gardner *et al.*, 1994; Liu and Berry, 1995; Hashmi *et al.*, 1996; Liu and Berry, 1996; see Peat *et al.*, Chapter 8, this volume). In spite of these efforts, RAPDs have fallen into disfavour within systematics, mostly because of all the issues outlined above and principally because they are not of direct utility for phylogenetic analysis.

### 4.3.2. Restriction fragment length polymorphism (RFLP)

A decade ago, restriction enzymes and PCR-RFLP methods were considered a useful diagnostic tool for identification of species in the Steinernematidae and Heterorhabditidae (Reid and Hominick, 1992; Reid *et al.*, 1997; Hussaini *et al.*, 2001; Phan *et al.*, 2001a,b; Anis *et al.*, 2002). Moreover, it was thought they were a good complement to morphological traits, specially when dealing with descriptions of new undescribed species in *Steinernema* (Joyce *et al.*, 1994; Nasmith *et al.*, 1996; Stock *et al.*, 1998; Luc *et al.*, 2000; Stack *et al.*, 2000; Phan *et al.*, 2001a,b).

PCR-RFLP approach was also applied to interpret evolutionary relationships among entomopathogenic nematodes (Reid, 1994; Reid *et al.*, 1997; Stock and Hunt, 2005). However, it is now recognized that even for large sequences or entire genomes, restriction enzymes are variable in their efficiency for generating RFLPs (Whitkus *et al.*, 1994). Moreover, without restriction site maps, fragment patterns cannot reliably produce homologous characters required to infer phylogenetic relationships or delimit species. Without a priori knowledge of cleavage site homology, interpretation of fragment patterns can be complicated or misleading (Hillis *et al.*, 1996).

Current technological advances, such as DNA sequencing, have made RFLP-PCR methods only of limited value. They can be applied for the characterization of large number of individuals in a sample, but of course having a pre-existing knowledge of their sequence variation.

### 4.3.3. DNA sequence analysis

DNA sequence analysis has been incorporated and is now widely accepted in nematode systematics. Currently, it is considered the most suitable approach not only for assessing phylogenetic relationships at different taxonomic levels but also for species delimitation (Powers *et al.*, 1994; Meldal *et al.*, 1997; Blaxter *et al.*, 1998; Iwahori *et al.*, 1998; Szalanski *et al.*, 2000; Nguyen *et al.*, 2001; Stock *et al.*, 2001; Perlman *et al.*, 2003; Stock and Koppenhöfer, 2003; Nadler *et al.*, 2006a,b; Wang *et al.*, 2007; Ye *et al.*, 2007).

## 4.4. Techniques Considered for Obtaining DNA Sequences

### 4.4.1. Specimen collection and preservation

One of the most critical aspects to consider when dealing with molecular systematics is that samples must be kept in a structurally intact physiologically active state (Dessauer *et al.*, 1996). A number of techniques and strategies have been considered for preservation of nematodes to both maintain their morphology and allow extraction of nucleic acids for molecular diagnostics.

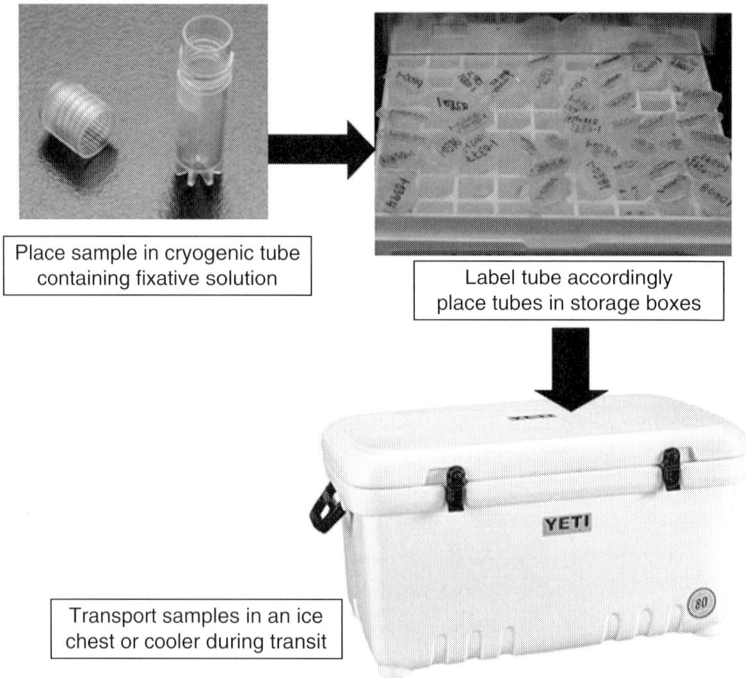


**Fig. 4.1.** Handling of samples in the field for future nucleic acid extraction.

No special method is required for collection of insect-parasitic and pathogenic nematodes for future molecular analysis that is currently known. Readers should refer to Kaya and Stock (1997) for further details on sampling methods. However, one aspect to keep in mind is that samples should be packed and transported as fast and clean as possible to avoid tissue degradation and/or contamination. If nematodes are collected from dissected insects in the field, tissues should be kept in plastic cryogenic tubes or bags (Fig. 4.1). Cryogenic storage vials should be properly labelled indicating samples names or numbers and any other useful information. Samples placed in cryogenic tubes should have space to allow tissue expansion during freezing. Coolers with dry ice or blue-ice packs should be used to maintain samples at appropriate temperature (4°C to −20°C) (Fig. 4.1).

If nematodes are abundant and are amenable for rearing or maintenance in the laboratory with *in vivo* or *in vitro* culturing methods, efforts should be placed in maintaining live cultures and also in preserving specimens for future molecular biology research and/or for exchange of samples with other laboratories.

#### *4.4.1.1. Fresh samples*

DNA extractions methods are more amenable when dealing with live samples. Fresh unfrozen tissues give the highest yield of mitochondrial DNA (mtDNA) (Dowling *et al.*, 1996). Particularly, if dealing with adult stages which usually have their mouth opened and are feeding, exposure of specimens to lysis buffer solution (i.e. containing proteinase K or other lysis enzymes) facilitates digestion of tissues. When dealing with immature stages, particularly with third-stage infective juveniles of entomopathogenic nematodes, digestion process will normally require additional time, to allow lysis solution digest the nematode's double cuticle. This process can be sped-up by grinding samples with a pestle and also by exposing nematodes to alternating 10 min freezing and thawing periods.

#### *4.4.1.2. Formalin-fixed specimens*

Nematodes are traditionally preserved in formalin or formalin-based fixatives because they are effective in maintaining morphological integrity of specimens. However, specimens fixed in formalin for long periods often cannot be used for molecular studies. This is mostly because formalin preservation makes DNA poorly available for amplification and subject to amplification errors, besides other direct and indirect effects such as irreversible denaturation, modification and formations of covalent bonds between nucleotides (Chang and Loew, 1995; Schander and Halanych, 2003).

Thomas *et al.* (1997) demonstrated that successful DNA extraction and subsequent molecular analyses can be carried out with formalin-fixed specimens for 2 days. In another study, Bhadury *et al.* (2005) tested un-buffered formalin as a preservative for short period of time. In their study, these authors concluded that specimens fixed for a period of 7 to 11 days showed no ambiguity in DNA extraction and subsequent amplification of samples. This study showed that formalin can be used to fix nematodes for a short period of time allowing also preservation of their morphological traits.

#### *4.4.1.3. Ethanol-preserved samples*

Organic solvents such as absolute alcohol for molecular studies have also been used to preserve nematode tissues for subsequent nucleic acid extraction, despite the problems these methods may have in maintaining a specimen's morphological integrity. Ethanol preserves specimens by the inhibition of cellular enzymes. This process is assumed to be complete and irreversible and thus samples must be rehydrated in increasing concentrations of $H_2O$ before protease digestion and DNA extraction. According to Dorris *et al.* (2002), DNA has been notoriously difficult to extract from ethanol-stored nematodes. Tissues are difficult to digest with protease K in comparison with either fresh or formalin-fixed specimens. Alcohol preservation of nematodes results in shrinkage of specimens, often making accurate morphological identification impossible. A minimum of 95% ethanol has been recommended for long-term storage. For nematode material fixed in 70% ethanol, no PCR amplification of any sized fragment has been possible from nematode material stored at room temperature in 70% ethanol. Dorris *et al.* (2002) have suggested that this problem is due perhaps to DNases present in nematodes which are active in 70% ethanol (See section 4.4.3.6).

#### *4.4.1.4. Frozen samples*

Samples that need long-term preservation are usually stored in liquid nitrogen storage tanks or in ultracold (−70°C) freezers. Generally, a variety of cardboard or plastic cryogenic storage containers are available for this purpose. Cryogenic tubes can be stored and organized in these boxes with corresponding identifica-

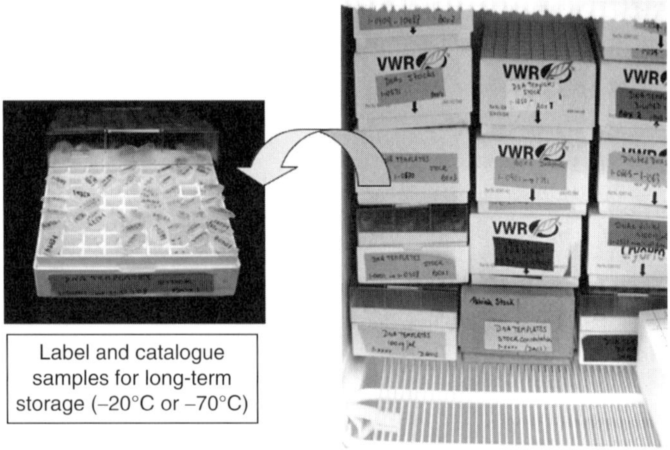


**Fig. 4.2.** Storage of frozen tissues and nucleic acids.

tion of contents (Fig. 4.2). It is advisable to keep an inventory of frozen tissues indicating whenever possible sample number, storage date, taxonomic status (e.g. family, genus, species, strain name, etc.), removal date (from the storage box) and purpose for which the sample was used/removed (i.e. DNA extraction).

For a better access to samples and to avoid damage of samples due to freezer malfunctioning due to recurrent opening of freezer doors for extraction of samples, it is recommended to have a 'working freezer', this is a freezer where samples that are being processed are placed (a −20°C freezer can be used for this purpose), and an ultracold freezer for long-term storage of samples. Readers should consult Dessauer *et al.* (1996) for further details on sample storage and preservation.

## 4.4.2. Selection of appropriate genes

### *4.4.2.1. Nuclear genes*

Ribosomal genes have been used extensively to study nematodes at the different molecular level. These genes are found in high copy numbers, and each gene evolves at different rates. Ribosomal genes include the external non-transcribed spacer (NTS), SSU or 18S, the internal transcribed spacer 1 (ITS1), the 5.8S, the internal transcribed spacer 2 (ITS2) and the large subunit (LSU) or 28S. Each of these genes are also characterized by having variable and conserved regions. This variability from gene to gene across rDNA has been considered useful for taxonomic studies at different taxonomic levels, including delimitation of nematode taxa (e.g. Stock, 2005; Curran and Driver, 1994; Blaxter *et al.*, 1998; Nadler and Hudspeth, 1998, 2000; Power *et al.*, 1997) (Fig. 4.3). All rDNA sequences are compromised to a greater or lesser degree due to ambiguity of inferring positional homology of characters. Accurate positional homology determinations are critical for construction of characters used to estimate nematode molecular phylogenies and therefore delimiting species based on the phylogenetic species concept.

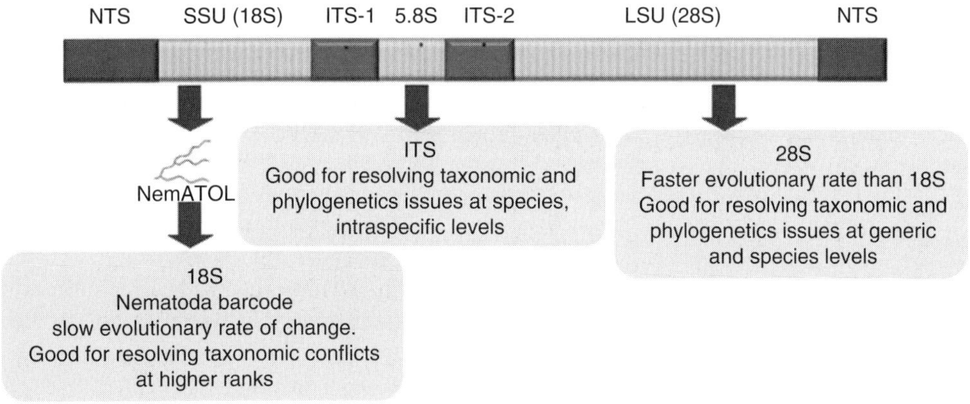


**Fig. 4.3.** Schematic representation of nematode ribosomal genes.

### *4.4.2.2. Small subunit (SSU) gene of rDNA*

This gene is characterized by having a slow evolutionary rate of change. And because of its conservative nature, it has been used to resolve taxonomic conflicts at higher ranks (Fig. 4.3). For example, Blaxter *et al.* (1998) utilized SSU sequences from 53 nematode taxa to construct the first molecularly based phylogenetic framework of Nematoda. In this study, major nematode groups were differentiated, and the newly constructed evolutionary framework was considered to assess monophyletic origins of nematodes.

The SSU of RNA is considered suboptimal for solving taxonomic conflicts at the species level, at least for entomopathogenic nematodes (Liu *et al.*, 1997; Stock *et al.*, 2001; Nadler *et al.*, 2006a). The conserved nature of this gene hinders resolution of nematode taxa at the specific or infraspecific levels.

### *4.4.2.3. Internal transcribed spacer (ITS) region and 5.8S gene of rDNA*

The internal transcribed spacer (ITS) region of rDNA is composed of three gene regions, ITS-1, 5.8S and ITS-2. The 5.8S ribosomal RNA (rRNA) gene is a short and highly conserved region. Contrarily, ITS-1 and ITS-2 rRNA are regions that evolved at a much higher rate than the 18S and 28S genes, making these regions ideal for phylogenetic studies at the species and population levels, population genetic studies and also for taxonomic purposes (Ferris *et al.*, 1993; Chilton *et al.*, 1995; Cherry *et al.*, 1997); (Fig. 4.3).

With particular reference to insect-pathogenic and parasitic nematodes, this variable gene has revealed numerous diagnostic utilities. It has been used to identify species of entomopathogenic nematodes and also to asses their evolutionary history (Nguyen *et al.*, 2001; Perlman *et al.*, 2003; Spiridonov *et al.*, 2004). For example, ITS-1 region has revealed sufficient genetic variation for differentiating *Heterorhabditis* spp. and has also been considered valuable for assessment of evolutionary relationships between species of this genus (Adams *et al.*, 1998). The entire ITS region has also been used to assess phylogenetic relationships and to delimit species of *Steinernema* spp. (Nguyen *et al.*, 2001). However, the length of sequences is highly variable (>100 bp difference between some species), and nucleotide composition is also very variable. Inference of positional homology across extensive regions of sequence of ITS is also quite dubious.

### *4.4.2.4. Large subunit (LSU) of rDNA*

The LSU (28S) of rRNA is more variable than SSU rRNA (Fig. 4.3) and has fewer ambiguously aligned positions than ITS. LSU sequence data have been used to assess phylogenetic relationships among *Steinernema* spp. (Stock *et al.*, 2001). In the study by Stock *et al.* (2001), 28S rDNA proved to be a suitable and informative region for interpreting evolutionary relationships among *Steinernema* spp. (see Peat *et al.*, Chapter 8, this volume, for further information). This region is also considered an effective and reliable approach for delimitation of terminal taxa in *Steinernema* as well as for diagnostic purposes (Stock *et al.*, 2001; Stock and Hunt, 2005).

### *4.4.2.5. Mitochondrial genes*

Mitochondrial genes have different evolution rates; therefore, knowledge on the evolutionary speed of a given locus is critical for the type of work to be done. Mitochondrial DNA loci such as the cytochrome c oxidase subunit 1 (*cox*1) evolve slowly; therefore, they are best suited to deeper lineage phylogeny (i.e. genera).

At present, a few mitochondrial genes have been considered in studies of genetic variation within and among nematodes with potential as biological control agents (Fig. 4.4). Powers *et al.* (1986) studied the molecular structure of nematode mtDNA using the mermithid *Romanomermis culicivorax*. In a later study, Powers *et al.* (1994) compared several mtDNA genes (e.g. NADH dehydrogenase subunit 3 (ND3), large rRNA and cytochrome b genes) to measure the genetic divergence from several nematode species, including *R. culicivorax*. Blouin *et al.* (1999) and Liu *et al.* (1999) studied the genetic variation among several *Heterorhabditis marelatus* populations using the ND4 gene of mtDNA, and found limited intraspecific variation. Other mtDNA genes studied include COX II and 16S rDNA (Szalanski *et al.*, 2000). These loci showed variation at the species level and proved useful for discrimination between selected *Steinernema* spp. However, they failed to show variation at the intraspecific level when tested with several *Steinernema feltiae* populations.

More recently mtDNA genes have been considered to assess evolutionary relationship of entomopathogenic nematodes in *Steinernema* (Nadler *et al.*, 2006a) and also for mermithid nematodes phylogenetic studies (Tang and Hyman, 2007). Readers should also refer to Chapter 8 (Peat *et al.*, this volume) for further details on this subject.

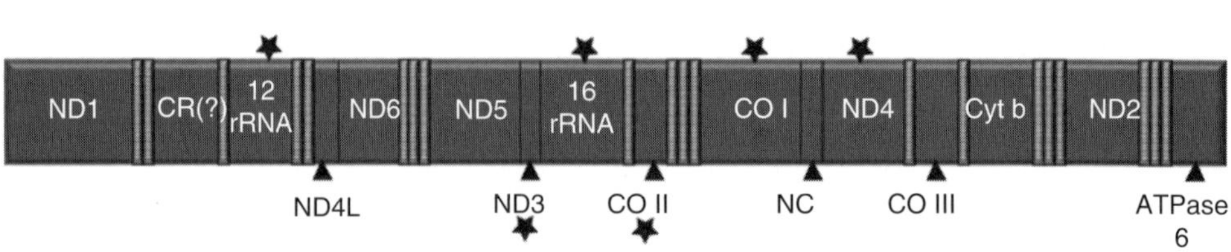


**Fig. 4.4.** Schematic representation of nematode mitochondrial genes. "Stars" indicate mitochondrial genes studied until now.

## 4.4.3. DNA extraction methods

### *4.4.3.1. DNA extraction from bulk specimens*

PHENOL-CHLOROFORM EXTRACTION (MODIFIED FROM AUSBEL, 1989). Phenol extraction is a common technique used to purify DNA samples. Usually, an equal volume of transposable elements (TE)-saturated phenol is added to an aqueous sample of DNA in a microcentrifuge tube. The mixture is vigorously vortexed, and then centrifuged to enact phase separation. The upper, aqueous layer is then carefully removed to a new tube, avoiding the phenol interface. This supernatant can then be subjected to two additional extractions to remove residual phenol. An equal volume of water-saturated ether is added to the tube, the mixture is vortexed and the tube is centrifuged to allow phase separation. The upper, ether layer is removed and discarded, including phenol droplets at the interface. After this extraction is repeated, the DNA is concentrated by ethanol precipitation.

For extraction of nucleic acids from nematodes, either a bulk of infective juveniles or adult stages can be considered for this purpose. Nematodes can be individually collected with the help of an 'L'-shaped needle (see Kaya and Stock, 1997) and placed in a small watch glass in TE buffer (pH 8.0). Sample should then be carefully rinsed (at least three times) with TE buffer (pH 8.0). After the last rinse, nematodes in buffer can be transferred to a cryotube for storage (long- or short-term), or placed in a microcentrifuge tube for nucleic acid extraction.

A basic phenol-chloroform extraction procedure is described below:

- Place nematodes in a 1.5 microcentrifuge tube and add approximately 500 µl of TE pH 8.0 and add 15 µl of 30% sodium dodecyl sulfate (SDS) and 20 µl of proteinase K (1 ng/µl). Note: if sample contains, for example, 100 µl (nematode suspension), add only 400 µl of TE. This is to adjust amount of TE added so final volume is 500 µl.
- Incubate sample at 50°C in a water bath (incubation time may vary according to the sample). Note: check periodically for digestion. Digestion time will vary according to the amount of nematodes in the sample and the nature of the nematodes' cuticle. For example, because of the double cuticle present in third-stage infective juveniles of Steinernematidae or Heterorhabditidae, additional time is needed for digestion. If sample is not fully digested you may after approximately a 7–10 h period add 15 µl more of proteinase K and leave overnight.
- If digestion is completed sample should be treated with 10 µl of RNase (vortexing well after addition of RNase).
- Incubate sample at 37°C for 1 h, then centrifuge it at lowest speed (if possible at 4°C) for approximately 2 min.
- Transfer supernatant to a new cryogenic tube avoiding collection debris from the bottom of the microcentrifuge tube.
- Add equal volume of phenol to the sample, vortex briefly and spin for 5 min (13,000 rpm).
- Remove supernatant carefully (make sure you do not touch the phenol interface) and place it in a new microcentrifuge tube.

- Add equal volume of 24:1 chloroform/isoamyl alcohol, vortex well and spin sample for 5 min (13,000 rpm).
- Remove upper interface carefully and transfer to a new microcentrifuge tube.
- Discard the isoamyl alcohol, centrifuge again the remaining chloroform to recover more of the upper interface (5 min at 13,000 rpm).
- Add sodium acetate (pH 5.2); (to help precipitation) 10 µl per 100 µl. For example if you have 400 µl in the tube add 40 µl of Na acetate (1 M).
- Add 100% ethanol to cover full volume of the microcentrifuge tube and mix gently by hand.
- Place tubes in freezer and leave them overnight, spin at 4°C for 10 min at full speed.
- Discard liquid and allow samples to dry in desiccators or dry pellet in a speedy vac.
- When samples are fully dried resuspend the pellet in 25 µl of TE.
- Samples are now ready for spectrophotometry reading.

#### *4.4.3.2. DNA extraction from single specimen*

Obtaining DNA template from single specimens is sometimes needed to avoid erroneous interpretations of variation within and between individual organisms. Various methods and extraction kits are currently available for single nematode or small samples of DNA extraction procedures. For nematodes, either immature (mermithids juvenile stages) or adult stages (hermaphrodites, males or females) can be used for single individual DNA extraction. When considering females or hermaphrodites, non-fertilized individuals or individuals without eggs should be selected. I briefly described the most commonly used methods.

#### *4.4.3.3. Chelex DNA extraction*

Make 5% chelex (Chelex$^{100}$, BioRad Laboratories) solution (e.g. for a 10 µl solution) as follows:

- Place stir bar in 50 µl microcentrifuge tube held upright in a beaker.
- Add 0.5 g chelex resin into conical and complete volume to 10 µl with sterile water. Note: This solution can be stored in the refrigerator for up to 1 month.
- With stir bar going, mix up chelex solution (make sure chelex beads are spinning in the water) and take 20 µl and add to tubes.
- Add 1 µl of proteinase K solution to each tube (20 mg/µl).
- Add nematode to microcentrifuge tube. You may cut each in half with a clean razor blade to facilitate digestion.
- Incubate at 56°C for 1 h.
- Boil at 100°C for 8 min.
- Cool down for a few seconds (if using thermocycler, bring down to 40°C for about 30 s), and vortex for additional 30 s.
- Use 4 µl of this mix in your PCR reactions, making sure no beads get in your PCR mix.

### *4.4.3.4. DNAzol® DNA isolation (Molecular Research Center)*

- Place individual nematode in 100 µl digestion solution in a 0.5 ml microcentrifuge tube.
- Add 100 µl of digestion solution. See recipe below.

| | |
|---|---|
| 100 mM Tris HCL pH 7.6 | 200 µl |
| 200 mM NaCl | 200 µl |
| 0.5 M EDTA pH 8.0 | 400 µl |
| 10% Sarkosyl | 200 µl |
| Proteinase K (10 mg/ml) | 20 µl (proteinase K should be made fresh – usually 0.005 g/0.5 ml) |
| Sterile distilled water | 980 µl |

- Sample should be digested overnight in 56°C water bath.
- Heat-kill proteinase by placing tube with sample in the PCR machine for 15 min at 95°C.
- Freeze and then thaw tubes with samples four times.
- Centrifuge sample for 5 min at 10,000 rpm.
- Remove 95 µl of solution, leaving bottom 5 µl, and add it to 1 ml of DNAzol isolation reagent in a 1.7 ml microcentrifuge tube.
- Add 4 µl of polyacryl carrier solution, vortex well (invert tube up and down five times).
- Add 0.5 ml 100% ethanol and mix by inverting ten times.
- Let sample sit at room temperature for 5 min and then pellet DNA by centrifuging sample at 7000 rpm for 5 min.
- Pour off ethanol and then wash DNA twice with 800 µl of 75% ethanol, spinning again if pellet breaks loose. Ethanol should be poured off, carefully removing the last of it with a pipetter.
- Allow visible ethanol to evaporate but make sure pellet does not dry.
- Resuspend pellet in 6 µl of TE buffer.

### *4.4.3.5. 'Worm lysis buffer' extraction (after Williams et al., 1990)*

This method can also be applied to bulk of specimens. Volume of lysis buffer will therefore have to be adjusted accordingly. A volume of 50 µl is recommended.

Worm lysis solution (for 1 ml of solution).

| | |
|---|---|
| KCl (stock: 1 M) | 1 µl |
| Gelatin (Dicto Bacto, stock: 1%) | 50 µl (Note: Prepare 100 mg gelatin in 10 ml water and heat in microwave. Make fresh every time is needed.) |
| Tris pH 8.2 (stock concentration: 1 M) | 10 µl |
| Tween 20 (stock concentration: 100%) | 4.5 µl |
| Proteinase K (stock concentration: 20 µg/ml) | 3.3 µl |
| $MgCl_2$ (stock concentration:1 M) | 2.5 µl |
| Double-distilled $H_2O$ | 880 µl |

- Add 15 µl lysis buffer to worm in a 0.5 ml PCR tube.
- Place at −70°C > 15 min. Sample can store for several days.
- Warm sample to room temperature and add mineral oil.
- Incubate at 60°C > 1 h.
- Heat to 95°C for 15 min.
- Cool to 4°C.
- Pipette sample up and down to mix well.

Use 2.5 µl of supernatant as template for PCR amplification for 25 µl reaction.

#### *4.4.3.6. DNeasy® DNA extraction for ethanol-fixed nematodes*

DNA has been notoriously difficult to extract from both ethanol and formalin-fixed specimens (Dorris *et al.*, 2002). A minimum of 70% ethanol has been recommended for long-term storage of nematodes to allow successful nucleic acid extraction. DNeasy® DNA tissue extraction kit (Qiagen) has shown positive results for this purpose (Graustein *et al.*, 2002; Perlman *et al.*, 2003). Manufacturer's protocol can be followed including the following modifications:

- Place single nematode specimen in a 1.5 ml microcentrifuge tube and macerate it with the help of a sterile needle or pipette tip.
- Add 20 µl proteinase K, mix by vortexing, and incubate between 57°C and 60°C until the tissue is completely lysed. Vortex every 20 min during incubation period.
- Add 4 µl of RNase A (100 mg/ml) to the sample, mix by vortexing and incubate for 2 min at room temperature.
- Vortex for 15s, add 400 µl Buffer AL–ethanol mixture to the sample and mix vigorously by vortexing.
- Pipette the mixture from step 3 into the DNeasy mini column sitting in a new 2 ml collection tube (provided with the kit). Centrifuge at 8000 rpm for 1 min.
- Discard flow-through and collection tube.
- Place the DNeasy mini column in a new 2 ml collection tube (provided with the kit), add 500 µl Buffer AW1, and centrifuge for 1 min at 8000 rpm.
- Discard flow-through and collection tube.
- Place the DNeasy mini column in a clean 1.5 ml microcentrifuge tube and pipette 200 µl. Buffer AE directly on to the DNeasy membrane.
- Incubate at room temperature for 1 min, and then centrifuge for 1 min 8000 rpm to elute.
- Repeat elution once as described in above step. A new microcentrifuge tube can be used for the second elution step to prevent dilution of the first eluate.
- Dry sample in a vacuum speed dryer for 1.5–2 h or until dry.
- Resuspend DNA in 20 µl of TE buffer.

### 4.4.4. PCR methods

For complete PCR techniques, stock solutions and procedures, readers should refer to Palumbi (1996). PCR reactions need to be carefully optimized and adjusted for different templates and primer sets. Therefore, testing of suitable conditions is a key step for successful amplification results. One of the most important aspects to consider is the temperature at which primers anneal to a given template. The number of cycles in a PCR reaction is also a critical factor that will also require some preliminary testing and should be adapted according to the type of primers considered (e.g. universal versus custom-designed primers).

In addition to these factors, the chemical components (deoxyribonucleotide triphosphates (dNTPs), $MgCl_2$, buffer, enzymes) of a PCR reaction also need to be carefully optimized and tested to enhance amplification results. A number of PCR reaction kits are available, and readers should opt for those ones that fit best their needs in terms of costs and quality of PCR products. Below is an example of PCR mix for a total volume of 25 µl.

| | |
|---|---|
| 10X buffer | 2.50 µl |
| dNTPs | 1.00 µl |
| $MgCl_2$ | 2.00 µl |
| Primer A | 1.25 µl |
| Primer B | 1.25 µl |
| Polymerase | 0.50 µl |
| MQ $H_2O$ | 14.50 µl |
| PCR mix total volume | 23.00 µl |
| DNA template | 2.00 µl |

Some researchers recommend the use of additives to enhance PCR reactions such as bovine serum albumin (BSA), dimethyl sulfoxide (DMSO), Triton X-100, Tween-20, etc. They are thought to enhance annealing of primers, stabilize enzymes and reduce secondary structure problems. However, too much additive can also make a reaction fail (Palumbi, 1996).

#### *4.4.4.1. Primers*

A number of primers sets have been considered for amplification of nuclear and mitochondrial genes considered for diagnosis/identification of insect-parasitic and entomopathogenic nematodes. However, some of these primers have also been successfully applied to other nematode taxa. See Table 4.1.

Primers considered for PCR reactions should be diluted to a concentration of 10 mM. I recommend resuspended dry primers first to a stock concentration of 100 mM. The stock concentration can then be used for future dilutions. TE buffer (pH 7.0) should be used for these dilutions. Diluted primers can be stored at either −20°C or −70°C.

PCR primers may be diluted to make sequencing primers (i.e. the primers that will be sent together with your PCR template for sequencing). For this step, double-distilled water should be considered, as TE will inhibit cycle sequencing.

**Table 4.1.** A selection of primers considered for sequencing of insect-parasitic and pathogenic nematodes.

| Primer | Orientation R = reverse F = forward | Amplified gene | Sequence | Comments | References |
|---|---|---|---|---|---|
| D2aF | F | 28S rDNA | 5′-ACAAGTACCGTGAGGGAAAsGT | Nematoda | Nunn, 1992 |
| D3bR | R | 28S rDNA | 5′- TGCGAAGGAACCAGCTACTA | Nematoda | Nunn, 1992 |
| D2bF | F | 28S rDNA | 5′-GACCCGTCTTGAAACACGGA | Nematoda | Nunn, 1992 |
| D3aR | R | 28S rDNA | 5′-TCCGTGTTTCAAGACGGGTC | Nematoda | Nunn, 1992 |
| 391 | F | 28S rDNA | 5′-AGCGGAGGAAAAGAAACTAA | *Steinernema* spp. | Stock *et al.*, 2001 |
| 501 | R | 28S rDNA | 5′-TCGGAAGGAACCGCTACTA | *Steinernema* spp. | Stock *et al.*, 2001 |
| Ferg-ID2B | F | 28S rDNA | 5′- AGTAACCTCTTGCACCAAAC | *Fergusobia* spp. | Ye *et al.*, 2007 |
| D2F1-Ferg | R | 28S rDNA | 5′-AGTACCGTGAGGGAAAGTTGAA | *Fergusobia* spp. | Ye *et al.*, 2007 |
| D2F2-Ferg | F | 28S rDNA | 5′-GGAAAGTTGAAAAGCACTTTG | *Fergusobia* spp. | Ye *et al.*, 2007 |
| D2R-Ferg | R | 28S rDNA | 5′- GATAGTTCGATTAGTCTTTCGCCC | *Fergusobia* spp. | Ye *et al.*, 2007 |
| Ferg5 | F | 28S rDNA | 5′-GAAGAGAGAGTTAAAGAGCACG | *Fergusobia* spp. | Ye *et al.*, 2007 |
| Ferg3 | R | 28S rDNA | 5′-GATAGTTCGATTAGTCTTTC | *Fergusobia* spp. | Ye *et al.*, 2007 |
| 93 | F | ITS rDNA | 5′-TTGAACCGGGTAAAAGTCG | *Steinernema*, *Heterorhabditis* spp. | Stock *et al.*, 2001 |
| 94 | R | ITS rDNA | 5′- TTAGTTTCTTTTCCTCCGCT | *Steinernema*, *Heterorhabditis* spp. | Stock *et al.*, 2001 |
| AB28 | F | ITS rDNA | 5′-ATATGCTTAAGTTCAGCGGGT | *Steinernema*, *Heterorhabditis* spp. | Curran *et al.*, 1994 |
| TW81 | R | ITS rDNA | 5′-GTTTCCGTAGGTGAACCTGC | *Steinernema*, *Heterorhabditis* spp. | Curran *et al.*, 1994 |
| 18S | F | ITS rDNA | 5′-TTGATTACGTCCCTGCCCTTT | *Steinernema* spp. | Vrain *et al.*, 1992 |
| 28S | R | ITS rDNA | 5′-TTTCACTCGCCGTTACTAAGG | *Steinernema* spp. | Vrain *et al.*, 1992 |
| 505 | F | Mitochondrial 12S rDNA | 5′-GTTCCAGAATAATCGGCTAGAC | *Steinernema* spp. | Nadler *et al.*, 2006a |
| 506 | R | Mitochondrial 12S rDNA | 5′-TCTACTTTACTACAACTTACTCCCC | *Steinernema* spp. | Nadler *et al.*, 2006a |
| 507 | F | Mitochondrial *cox*1 | 5′-AGTTCTAATCATAA(A/G)GATAT(C/T)GG | *Steinernema* spp. | Nadler *et al.*, 2006a |

| | | | | | |
|---|---|---|---|---|---|
| 588 | R | Mitochondrial *cox* 1 | 5′-TAAACTTCAGGGTGACCAAAAAATCA | *Steinernema* spp. | Nadler *et al.*, 2006a |
| 527 | F | 18S rDNA | 5′-CTAAGGAGTGTGTAACAACTCACC | Cephalobina including *Steinernema* sp. | Nadler *et al.*, 2006b |
| 532 | R | 18S rDNA | 5′-AATGACGAGGCATTTGGCTACCTT | Cephalobina including *Steinernema* sp. | Nadler *et al.*, 2006b |
| 18S-G18S4 | F | 18S rDNA | 5′- GCTTGTCTCAAAGATTAAGCC | Nematoda | De Ley and Blaxter, 2002 |
| 18S-18P | F | 18S rDNA | 5′-TGATCCWKCYGCAGGTTCAC | Nematoda | De Ley and Blaxter, 2002 |
| COI-F1 | F | Mitochondrial *cox*1 | 5′-CCTACTATGATTGGTGGTTTTGGTAATTG | Tylenchina | Kanzaki and Futai, 2002 |
| COI-R2 | R | Mitochondrial *cox*1 | 5′-GTAGCAGCAGTAAAATAAGCACG | Tylenchina | Kanzaki and Futai, 2002 |
| 537 | F | 18S rDNA | 5′-GATCCGTAACTTCGGGAAAAGGAT | Cephalobina including *Steinernema* sp. | Nadler *et al.*, 2006b |
| 531 | R | 18S rDNA | 5′-CTTCGCAATGATAGGAAGAGCC | Cephalobina including *Steinernema* sp. | Nadler *et al.*, 2006b |
| 5F | F | 18S rDNA | 5′-GCGAAAGCATTTGCCAAGAA | Mermithidae | Vandergast *et al.*, 2003 |
| 18S-9R | R | 18S rDNA | 5′- GATCCTTCCGCAGGTTCACCT | Mermithidae | Vandergast *et al.*, 2003 |

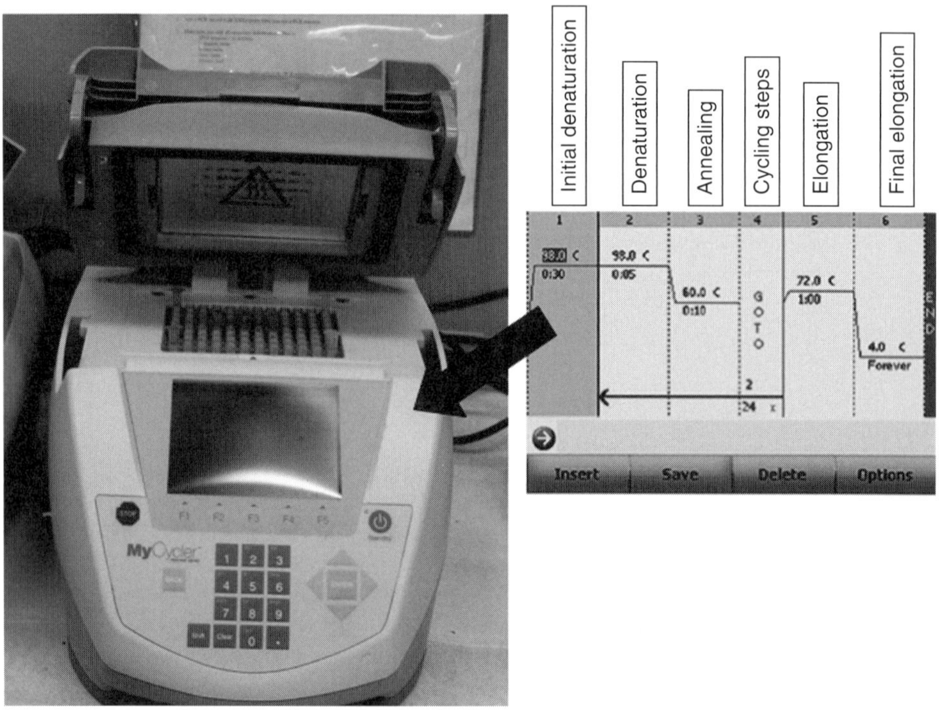


**Fig. 4.5.** PCR set-up and cycling parameters.

### *4.4.4.2. PCR cycling parameters*

Cycling parameters used will depend on the targeted gene and primers considered. A number of generic parameters have been published (Palumbi, 1996) (Fig. 4.5). Below I provide examples of PCR cycling parameters considered in my laboratory for amplifying *Heterorhabditis* and *Steinernema* ribosomal and mithocondrial genes that are considered for research in my laboratory.

Gene amplified: 28S – *Steinernema* Primer set: 391/501

| | Temperature | Time |
|---|---|---|
| Step 1 – initial denaturation | 94°C | 3 min |
| Step 2 – denaturation | 94°C | 30 s |
| Step 3 – annealing | 52°C | 30 s |
| Step 4 – elongation | 72°C | 1 min |
| Step 5 – cycling step: repeat steps 2–4 for 34 cycles | | |
| Step 6 – final elongation | 72°C | 7 min |
| Step 5 – cycle termination | 4°C | Hold |

Gene amplified: ITS – *Heterorhabditis* Primer set: 93/94

| | Temperature | Time |
|---|---|---|
| Step 1 – initial denaturation | 95°C | 3 min |
| Step 2 – denaturation | 95°C | 30 s |
| Step 3 – annealing | 60°C | 30 s |
| Step 4 – elongation | 72°C | 1 h 15 min |
| Step 5 – cycling step: repeat steps 2–4 for 32 cycles | | |
| Step 6 – final elongation | 72°C | 5 min |
| Step 5 – cycle termination | 4°C | Hold |

Gene amplified: ITS – *Steinernema* and *Heterorhabditis*
Primer set: AB28/TW81

| | Temperature | Time |
|---|---|---|
| Step 1 – initial denaturation | 94°C | 2 min |
| Step 2 – denaturation | 94°C | 30 s |
| Step 3 – annealing | 45°C | 60 s |
| Step 4 – elongation | 72°C | 90 s |
| Step 5 – cycling step: repeat steps 2–4 for 40 cycles | | |
| Step 6 – final elongation | 72°C | 15 min |
| Step 5 – cycle termination | 4°C | Hold |

Gene amplified: mt 12S – *Steinernema* Primer set: 505/506

| | Temperature | Time |
|---|---|---|
| Step 1 – initial denaturation | 94°C | 30 s |
| Step 2 – denaturation | 94°C | 3 min |
| Step 3 – annealing | 50°C | 30 s |
| Step 4 – elongation | 72°C | 45 s |
| Step 5 – cycling step: repeat steps 2–4 for 37 cycles | | |
| Step 6 – final elongation | 72°C | 7 min |
| Step 5 – cycle termination | 4°C | Hold |

Gene amplified: mt-cox 1 – *Steinernema* Primer set: 507–588

| | Temperature | Time |
|---|---|---|
| Step 1 – initial denaturation | 94°C | 1 min |
| Step 2 – denaturation | 94°C | 3 min |
| Step 3 – annealing | 40°C | 1 min |
| Step 4 – elongation | 72°C | 1 min |
| Step 5 – cycling step: repeat steps 2–4 for 36 cycles | | |
| Step 6 – final elongation | 72°C | 7 min |
| Step 5 – cycle termination | 4°C | Hold |

### 4.4.5. Agarose gel electrophoresis (AGE)

Agarose gel electrophoresis (AGE) is one of the most common ways of separating and analysing DNA. The purpose of the gel is to look at the DNA, quantify it and/or isolate a particular band. The DNA is visualized in the gel by addition of ethidium bromide. This binds strongly to DNA by intercalating between the bases, and its fluorescence means that it absorbs invisible ultraviolet (UV) light and transmits the energy as visible orange light. Most agarose gels are made between 0.7% and 2%. In my laboratory, I use 1.0% or 1.3% gels. The size of the gel and combs considered will depend on the purpose of the study. Also another critical question is the amount of DNA (or PCR product containing DNA from your sample) needed to load in a gel. This will also depend on the purpose of your research: analytical (i.e. a gel to just look for DNA of your sample) or a quantitative/preparative gel (i.e. a preparative gel to separate a DNA fragment before cutting it out of the gel for further treatment). For either option, a band should be easily visible if it contains about 20 ng of DNA = 5 μl of PCR product.

A 1% agarose gel should contain the following.

| | |
|---|---|
| Agarose | 0.5 g |
| 1X tris-borate (TBE) buffer | 50 ml |

Mix well and microwave until agarose is fully dissolved. Allow agarose mix to cool down (bowling agarose will create bubbles in your gel) and then pour into gel mould Some users add ethidium bromide (4 μl) right after the agarose has cooled down. We do not add ethidium bromide at this step but only after running the gel (see below).

Gel is ready to be loaded when it has completely solidified. The comb is removed and the gel is placed in gel box containing 1X TBE buffer. Gels should be completely covered by the buffer (Fig. 4.6). Usually, a 0.5–1.0 cm layer of buffer should be above the gel.

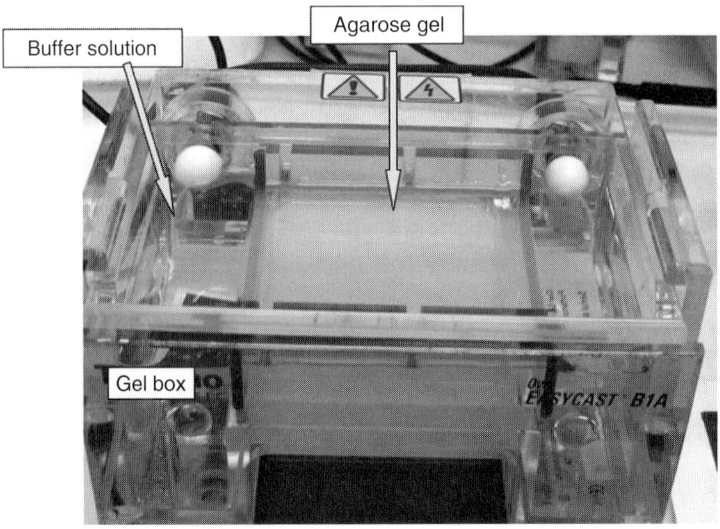


**Fig. 4.6.** Loading of agarose gel.

Volume of PCR product and ladder to be loaded in a gel will vary depending on the size and depth of the teeth of the chosen comb. We usually consider a 7 µl volume: 5 µl of PCR product plus 2 µl of tracking dye. A DNA size marker (ladder), 1 µl of ladder + 1 µl of tracking dye + 5 µl of $H_2O$ = 7 µl should be added to the gel. There are many different kinds of DNA size markers and users should make their selection based on the size of DNA fragment they wish to amplify.

Monitor the progress of the gel by reference to the marker dye. Stop the gel when the bromophenol blue has run three-fourths the length of the gel. We usually run gels at 90 V for approximately 30–45 min. Power of gel box should be switched off or unplugged. If ethidium was added to the gel, the gel is ready for visualization on a UV light box.

If ethidium bromide was not added to the gel before, it can be added at this step. For this, remove the gel from the gel box and placed it in a flat container with distilled water to which 4 µl of ethidium bromide is added. The container is covered with a lid and placed in a horizontal rotation shaker for 10–15 min. This step is followed by a distaining phase where water containing ethidium bromide is carefully poured off and replaced with clean sterile distilled water. The gel is placed again in the shaker for 10 additional min. Bands should be visualized using UV light box.

An alternative to ethidium bromide is the consideration of SYBR® Green I nucleic acid gel stain (Invitrogen). This dye has several advantages over ethidium bromide: (i) it is less mutagenic than ethidium bromide (it degrades faster in UV light); (ii) it is more sensitive (up to 50X more sensitive), and less PCR sample can be used; and (iii) gels can be reused (SYBR green stains the sample not the gel) for up to 20 times. One aspect to consider is that fidelity of SYBR green may not be as high as that of ethidium bromide.

Protocol:

- Make a thick gel (~2%): 1.0 g agarose: 50 ml tris-acetate-EDTA (TAE).
- Mix 3 µl of your sample with 1 µl of SYBR green (1 µl dye optional).
- Mix sample thoroughly and pipette directly into the wells.
- Run gel and visualize it on a gel box using SYBR green filter.

## 4.4.6. Cloning of PCR products

In rare instances, when PCR amplicons cannot be successfully sequenced directly (typically due to repeated sequence motifs), PCR products can be cloned, and sequences are then obtained from multiple clones. Several cloning methods have been developed for PCR products (see Hilllis *et al.*, 1996). Preparation of templates for sequencing can be performed with minipreps, of which there are several methods available. We herein describe a method used for cloning *Steinernema* PCR product described by Nadler *et al.* (2006a).

### *4.4.6.1. Spin filtration*

- Use ultrafree-MC spin filtration columns (Millipore 30,000 mwt filter) to clean up PCR product. Follow manufacturer's protocol for details.

- Wash product with 300 μl of diethyl pyrocarbonate (DEPC)-treated water twice. (Note: Filtered product should not be spun in excess of 5000 × g – do 5000 rpms for 5 min on a microcentrifuge.
- Remove product from filter by rinsing surface of membrane with DEPC-treated water using pipettor (but do not touch membrane surface with pipette tip).
- Adjust volume to ~20 μl (or concentrate on a speedvac centrifuge if original gel band was light).

#### 4.4.6.2. *GEMT vector ligation*

| | |
|---|---|
| 2× ligation buffer (pre-aliquotted) | 5 μl |
| Filtered PCR product (see step 1 above) | 3 μl |
| pGEMT vector – in blue freezer box | 1 μl |
| T4 ligase | 1 μl |

Incubate ligation overnight at 4°C in refrigerator.

#### 4.4.6.3. *Transformations*

LB plates should be dry and at room temperature before use.

- Thaw competent cells (JM109 *Escherichia coli*) on ice, mixing very gently.
- Add 4 μl of ligation to a 2 ml eppendorf tube – freeze remaining 6 μl.
- Add 50 μl of thawed cells to eppendorf tube.
- Gently mix.
- Place tube on ice for 90 min.
- Heat-shock cells for 50 s at 42°C in water bath.
- Place transformation on ice for 2 min.

### 4.4.7. Preparation of PCR products for sequencing

A number of protocols are currently available for purification of PCR products including silica gel membranes (Qiaquick, or other bind/elute membrane methods), size exclusion membranes (Millipore Multiscreen Eppendorf Perfect Prep, etc.) and enzymatic treatment (Exosap-IT®, USB laboratories). I summarize below two of the most commonly used methods in my laboratory.

#### 4.4.7.1. *Exosap-IT® PCR purification method*

This method is used for sequencing and further analysis of PCR products. It is also inexpensive and easy to use.
Protocol:

- Transfer 5 μl of PCR product into a 0.2 ml microcentrifuge tube.
- Add 2 μl of Exosap-IT® for a total volume of 7 μl.
- Vortex carefully and place in a thermal cycler programmed with the following steps.

| Step | Temperature/time |
|---|---|
| 1 – enzyme activation | 37°C/15 min |
| 2 – enzyme denaturation | 80°C/15 min |
| 3 – cycle termination | 4°C/indefinite |

#### *4.4.7.2. QIAquick PCR purification (Qiagen)*

This protocol is designed to purify single- or double-stranded DNA fragments from PCR and other enzymatic reactions. Fragments ranging from 100 bp to 10 kb are purified from primers, nucleotides, polymerases and salts using QIAquick spin columns in a microcentrifuge.
Protocol:

- Add 5 volumes of buffer PB to 1 volume of the PCR sample and mix.
- Place a QIAquick spin column in a provided 2 ml collection tube.
- Apply the sample to the column (i.e. to bind DNA) and centrifuge for 30–60 s.
- Discard flow-through and place the column back into the same tube.
- Wash by adding 0.75 ml buffer PE to the column and centrifuge for 30–60 s.
- Discard flow-through and place the column back in the same tube.
- Centrifuge the column for an additional 1 min at maximum speed. (Note: residual ethanol from buffer PE will not be completely removed unless the flow-through is discarded before this additional centrifugation.)
- Place the column in a clean 1.5 ml microcentrifuge tube.
- Add 50 µl of $H_2O$ to the centre of the QIAquick membrane to elute DNA and centrifuge the column for 1 min.

## 4.5. Sequencing

At present, most sequencing reactions are performed using dye-terminator sequencing chemistry, and reaction products are usually separated and detected using automated sequencers such as the ABI 3730 capillary DNA sequencer (PE Applied Biosystems) or other similar types. Sequences should be completely double-stranded for verification using reactions primed from the PCR or vector primers. Two or more additional internal sequencing primers should be considered as needed.

Site polymorphisms in directly sequenced PCR products are recorded only when both alternative nucleotide peaks are present in all sequence reactions representing both DNA strands. If the heights of the alternative nucleotide peaks at polymorphic sites are not equal, the height of the minor peak is required to significantly exceed background terminations and comprise at least 25% of the major peak to be scored as a polymorphism.

## 4.6. Sequence Manipulation and Analysis

Manipulation (editing) of sequences is a critical step for interpretation of these data and subsequent taxonomic actions that will be taken based on this information. Visualization and manual editing of electropherographs is the first and most critical step for obtaining accurate sequences. Researchers should always carefully edit sequences prior and/or after their assembly.

As mentioned above, most nucleic acid sequences are currently obtained from automated DNA sequencers. These sequencers occasionally produce poor-quality reads, particularly near the sequencing primer site, and towards the end of longer sequence runs. Also, sequences of clones from DNA libraries often contain vector sequence, polyA tails, or other unrelated sequence. Introns and primer sequence frequently flank the sequence of amplified exons and are usually included in the generated sequences. Unless removed by trimming, any of these artefacts will distort sequence assembly and downstream sequence analysis. A number of software options (e.g. EDIT VIEW (Ibis, Biosciences), SEQUENCHER (Gen Codes Corporation), LASERGEN EDITSEQ and SEQBUILDER (DNAStar)) are currently available and provide simple-to-use tools that help users visualize electropherographs and trim poor-quality or ambiguous data.

Sequence assembly refers to the alignment and merging of several fragments of a much longer nucleic acid sequence in order to reconstruct the original sequence. Usually, sequence assembly is performed with the aid of computer-based programs (see above). A wide range of software is currently available for fast and accurate sequence assembly with preset (default) parameters that allow adjustment of sequences within seconds. Many programs automatically compare the forward and the reverse-complement orientation of the primers to assemble the best possible contigs, so users can assemble DNA sequences regardless of orientation. However, other programs do not have this automated option and users need to manually orient primers (or change software default) for proper assembly of sequences.

Another crucial aspect in sequence manipulation and analysis is the alignment of sequences. Multiple alignment of sequences is often viewed as a 'central problem' in molecular systematics because both taxonomic and phylogenetic inferences are dependent on this first and challenging step. Any critical analysis of a phylogenetic hypothesis will include examination of this multiple alignment. For sequences such as rDNA, alignment ambiguity can have profound effects on phylogenetic inference (Morrison and Ellis, 1997; Chenna *et al.*, 2003; Nadler *et al.*, 2006a). Readers should refer to Chapter 8 (Peat *et al.*, this volume) for details on sequence alignment parameters and available software.

## 4.7. Conclusions

The long-term goal of molecular diagnostics is to develop protocols for the accurate and rapid identification of all nematode species. As we advance our knowledge on insect-parasitic and pathogenic nematodes and gain new insights on their diversity and evolutionary relationships, it is important that we expand and update protocols and methods. A decade ago, the concept of 'phylogentic species concept' (Nelson and Plamick, 1981; Cracraft, 1983, 1989; Nixon and Wheeler, 1990), was introduced in nematology (Adams, 1998). Since then, the concept has gained almost universal acceptance requiring taxonomists to incorporate molecularly based methods to complement morphological traits for species diagnosis and accurate identifications. Indeed, molecular techniques have provided a tremendous amount of objective data towards nematode systematics. However, readers should also be aware that they too can produce bogus results, even when care is taken to use and analyse data appropriately.

It would be a mistake to replace classical morphological approaches with molecular methods. Both morphological and molecular approaches have provided greatest explanatory power to nematode taxonomy and systematics, and together, they will continue to provide a more comprehensive view of nematode evolution, and will continue to generate robust taxonomic classifications.

## Acknowledgements

I am thankful to student and colleague members of my laboratory for their contributions to my research programme. Research in the S.P. Stock laboratory is supported by awards from National Science Foundation (awards DEB-0822631, DEB-0640899, DEB-0640807, DEB-0733729, IOS-0744336, IOS-0817592) and the University of Arizona Advance and Seed grant programmes.

## References

Adams, B.J. (1998) Species concepts and the evolutionary paradigm in modern nematology. *Journal of Nematology* 30, 1–21.

Adams, B.J., Burnell, A.M. and Powers, T.O. (1998) A phylogenetic analysis of *Heterorhabditis* (Nemata: Rhabditidae) based on internal transcribed spacer 1 DNA sequence data. *Journal of Nematology*, 30, 22–39.

Anis, M., Shahina, F., Reid, A.P. and Rowe, J. (2002) *Steinernema asiaticum* sp. n. (Rhabditida: Steinernematidae) from Pakistan. *International Journal of Nematology* 12, 220–231.

Avise, J.C. (1994) *Molecular Markers, Natural History and Evolution*. Chapman and Hall, London, 511.

Ausbel, F.M. (1989) *Current Protocols in Molecular Biology*. Wiley, New York.

Bhadury, P., Austen, M.C., Bilton, D.T., *et al.* (2006) Development and evaluation of a DNA-barcoding approach for the rapid identification of nematodes. *Marine Ecology Progress Series* 320, 1–9.

Blaxter, M.L. (2003) Counting angels with DNA. *Nature* 421, 122–124.

Blaxter, M.L., De Ley, P., Garey, J.R., Liu, L.X., Scheldeman, P., Vierstraete, A., Vanfleteren, J.R., Mackey, L.Y., Dorris, M., Frisse, L.M., Vida, J.T. and Thomas, W.K. (1998) A molecular evolutionary framework for the phylum Nematoda. *Nature* 392, 71–75.

Blok, V.C., Phillips, M.S., Mc Nicol, J.W. and Fargette, M. (1997) Genetic variation in tropical *Meloidogyne* spp. as shown by RAPDs. *Fundamental and Applied Nematology* 20, 127–133.

Blouin, M.S., Liu, J. and Berry, R.E. (1999) Life cycle variation and the genetic structure of nematode populations. *Heredity* 83, 253–259.

Chang, Y.T. and Loew, G.H. (1995) Reaction mechanisms of formaldehyde with endocyclic imino groups of nucleic acid bases. *Journal of the American Chemistry Society* 116, 3548–3555.

Chenna, R., Sugawara, H., Koike, T., Lopez, R., Gibson, T.J., Higgins, D.G. and Thompson, J.D. (2003) Multiple sequence alignment with the Clustal series of programs. *Nucleic Acids Research*, 31, 3497–3500.

Cherry, T., Szalanski, A.L., Todd, T.C. and Powers, T.O. (1997) The internal transcribed spacer region of *Belonolaimus* (Nemata: Belonolaimidae). *Journal of Nematology* 29, 23–29.

Chilton, N.B., Gasser, R.B. and Beveridge, I. (1995) Differences in a ribosomal DNA-sequence of morphologically indistinguishable species within the *Hypodontus-macropi* complex (Nematoda, Strongyloidea). *International Journal for Parasitology* 25, 647–651.

Cracraft, J. (1983) Species concepts and speciation analysis. In: Johnston, R.F. (ed.) *Current Ornithology*. Plenum Press, New York, pp. 159–187.

Cracraft, J. (1989) Speciation and its ontology: the empirical consequences of alternative species concepts for understanding patterns and process of differentiation. In: Otte, D. and Endler, J.A. (eds) *Speciation and Its Consequences*. Sinauer Associates, Sunderland, Massachusetts, pp. 28–59.

Curran, J. (1991) Application of DNA analysis to nematode taxonomy. In: Nickle, W.R. (ed.) *Manual of Agricultural Nematology*. Marcel Dekker, New York, pp. 125–143.

Curran, J. and Driver, F. (1994) Molecular taxonomy of *Heterorhabditis*. In: Burnell A.M., Ehlers, R.-U. and Masson J.-P. (eds) European Commission Publication, Luxembourg, pp. 41–48.

Curran, J. and Robinson, M.P. (1993) Molecular aids to nematode diagnosis. In: Evans, K., Trudgill, D.L. and Webster, J.M. (eds) *Plant Parasitic Nematodes in Temperate Agriculture*. CAB International, Wallingford, UK, pp. 545–564.

Cutler, C.G. and Stock, S.P. (2003) *Steinernema websteri* n. sp. (Rhabditida: Steinermeatidae), a new entomopathogenic nematode from China. *Nematologia Mediterranea* 31, 215–224.

De Ley, P. and Blaxter, M. (2002) Systematic position and phylogeny. In Lee, D.L. (ed). *The Biology of Nematodes*. Taylor & Francis, London, pp. 1–30.

Dessauer, H.C., Cole, C.J. and Hafner, M.S. (1996) Collection and storage of tissues. In: Hillis, D.M., Moritz, C. and Mable, B.K. (eds) *Molecular Systematics*. Sinauer Associates, Sunderland, Massachusetts, pp. 29–47.

Dorris, M., Viney, M.E. and Blaxter, M.L. (2002) Molecular phylogenetic analysis of the genus *Strongyloides* and related nematodes. *International Journal for Parasitology* 32, 1507–1517.

Dowling, T.E., Moritz, C., Palmer, J.D. and Rieseberg, L.H. (1996) Nucleic acids III: analysis of fragments and restriction sites. In: Hillis, D.M., Moritz, C. and Mable, B.K. (eds) *Molecular Systematics*. Sinauer Associates, Sunderland, Massachusetts, pp. 249–320.

Ferris, V.R., Ferris, J.M. and Faghihi, J. (1993) Variation in spacer ribosomal DNA in some cyst-forming species of plant parasitic nematodes. *Fundamental and Applied Nematology* 16, 177–184.

Floyd, R., Abebe, E., Papert, A. and Blaxter, M.L. (2002a) The NaOH single-nematode DNA extraction method. Available at: http://nema.cap. ed.ac.uk/npg/naoh.html.

Floyd, R., Abebe, E., Papert, A. and Blaxter, M.L. (2002b) Molecular barcodes for soil nematode identification. *Molecular Ecology* 11, 839–850.

Gardner, S.L., Stock, S.P. and Kaya, H.K. (1994) A new species of *Heterorhabditis* from the Hawaiian islands. *Journal of Parasitology* 80, 100–106.

Graustein, A., Gaspar, J.M., Walters, J.R. and Palopoli, M.F. (2002) Levels of DNA polymorphism vary with mating system in the nematode genus Caenorhabditis. *Genetics* 161, 99–107.

Hashmi, G., Glazer, I. and Gaugler, R. (1996) Molecular comparisons of entomopathogenic nematodes using randomly amplified polymorphic DNA (RAPD) markers. *Fundamental and Applied Nematology* 19, 399–406.

Hebert, P.D.N., Cywinska, A., Ball, S.L. and de Waard, J.R. (2003) Barcoding animal life: cytochrome c oxidase subunit 1 divergences among closely related species. *Proceedings of the Royal Society London Series B* 270, 96–99.

Hillis, D.M., Moritz, C. and Mable, B.K. (1996) *Molecular Systematics*, 2nd edn. Sinauer Associates, Sunderland, Massachusetts, 655.

Hominick, W.M., Briscoe, B.R., del Pino, F.G., Heng, J., Hunt, D.J., Kozodoi, E., Mracek, Z., Nguyen, K.B., Reid, A.P., Spiridonov, S., Stock, S.P., Sturhan, D., Waturu, C. and Yoshida, M. (1997) Biosystematics of entomopathogenic nematodes: current status, protocols and definitions. *Journal of Helminthology* 71, 271–298.

Hussaini, S.S., Ansari, M.A., Ahmad, W. and Subbotin, S.A. (2001) Identification of some Indian populations of *Steinernema* species (Nematoda) by RFLP analysis of the ITS region of rDNA. *International Journal of Nematology* 11, 73–76.

Hussey, R.S. (1979) Biochemical systematics of nematodes: a review. *Helminthological Abstracts (Series B)* 48, 141–148.

Iwahori, H., Tsuda, K., Kanzaki, N., Izui, K. and Futai, K. (1998) PCR-RFLP and sequencing analysis of ribosomal DNA of *Bursaphelenchus* nematodes related to pine wilt disease. *Fundamental and Applied Nematology* 21, 655–666.

Joyce, S.A., Burnell, A.M. and Powers, T.O. (1994) Characterization of *Heterorhabditis* isolates by PCR amplification of sequences of mtDNA and rDNA genes. *Journal of Nematology* 26, 260–270.

Kanzaki, N. and Futai, K. (2002) A PCR primer set for determination of phylogenetic relationships of *Bursaphelenchus* species within the *xylophilus* group. *Nematology* 4, 35–41.

Kaya, H.K. and Stock, S.P. (1997) Techniques in insect nematology. In: Lacey, L.A. (ed.) *Manual of Techniques in Insect Pathology*. Biological Techniques Series, Academic Press, San Diego, California, pp. 281–324.

Liu, J. and Berry, R.E. (1995) Differentiation of isolates in the genus *Steinernema* (Nematoda: Steinernematidae) by random amplified polymorphic DNA fragments and morphological characters. *Parasitology* 111, 119–125.

Liu, J. and Berry, R.E. (1996) Phylogenetic analysis of the genus *Steinernema* by morphological characters and randomly amplified polymorphic DNA fragments. *Fundamental and Applied Nematology* 19, 463–469.

Liu, J., Berry, R.E. and Moldenke, A.F. (1997) Phylogenetic relationships of entomopathogenic nematodes (Heterorhabditdae and Steinernematidae) inferred from partial 18S rRNA gene sequences. *Journal of Invertebrate Pathology* 69, 246–252.

Liu, J., Berry, R.E. and Blouin, M.S. (1999) Molecular differentiation and phylogeny of entomopathogenic nematodes (Rhabditida: Heterorhabditidae) based on ND4 gene sequences of mitochondrial DNA. *Journal of Parasitology* 85, 709–715.

Luc, V.P., Nguyen, K.B., Reid, A.P. and Spiridonov, S.E. (2000) *Steinernema tami* sp. n. (Rhabditida: Steinernematidae) from Cat Tien Forest, Vietnam. *Russian Journal of Nematology* 8, 33–43.

Lynch, M. and Milligan, B.G. (1994) Analysis of population genetic structure within RAPD markers. *Molecular Ecology* 3, 91–99.

Meldal, B.M.H., Debenhamb, N.J., De Ley, P., De Ley, I.T., VanXeteren, J.R., Vierstraete, A.R., Bert, W., Borgonie, G. and Moens, T. (1997) A improved molecular phylogeny of the Nematoda with emphasis on marine taxa. *Molecular Phylogenetics and Evolution* 42, 622–636.

Morrison, D. A. and Ellis, J. T. (1997) Effects of nucleotide sequence alignment on phylogeny estimation: a case study of 18S rDNAs of Apicomplexa. *Molecular Biology and Evolution*, 14, 428–441.

Mullin, P., Powers, T., Harris, T., Esquivel, A., Giblin-Davis, R., Neher, D. and Stock, S.P. (2006) Applying a ribosomal DNA sequence 'barcode' to assess nematode biodiversity in a Costa Rican rainforest preserve. Journal of Nematology, 38, 284.

Muralidharan, K. and Wakeland, E.K. (1993) Concentration of primer and template qualitatively affects products in random-amplified polymorphic DNA PCR. *BioTechniques* 14, 362–364.

Nadler, S.A. and Hudspeth, D.S.S. (1998) Ascaridoidea (Nemata: Secernentea): implications for morphological evolution and classification. *Molecular Phylogenetics and Evolution* 10, 221–236.

Nadler, S.A. and Hudspeth, D.S.S. (2000) Phylogeny of the Ascaridoidea (Nematoda: Ascaridida) based on three genes and morphology: hypotheses of structural and sequence evolution. *Journal of Parasitology* 86, 380–393.

Nadler, S.A., Bolotin, E. and Stock, S.P. (2006a) Phylogenetic relationships of *Steinernema* (Cephalobina, Steinernematidae) based on nuclear, mitochondrial, and morphological data. *Systematic Parasitology* 63, 159–179.

Nadler, S.A., De Ley, P., Mundo-Ocampo, M., Smythe, A.B., Stock, S.P., Bumbarger, D., Adams, B.J., De Ley, I.T., Holovachov, O. and Baldwin, J.G. (2006b) Phylogeny of Cephalobina (Nematoda): molecular evidence for recurrent evolution of probolae and incongruence with traditional classifications. *Molecular Phylogenetics and Evolution*, 40, 696–711.

Nasmith, C.G., Speranzini, D., Jeng, R. and Hubbes, M. (1996) RFLP analysis of PCR amplified ITS and 26S ribosomal RNA genes of selected entomopathogenic nematodes (Steinernematidae and Heterorhabditidae). *Journal of Nematology* 28, 15–25.

Nelson, G.J. and Plamick, N.I. (1981) *Systematics and Biogeography: Cladistics and Vicariance*. Columbia University Press, New York.

Nguyen, K.B., Maruniak, J. and Adams, J.B. (2001) Diagnostic and phylogenetic utility of the rDNA internal transcribed spacer sequences of *Steinernema*. *Journal of Nematology* 33, 73–82.

Nixon, K.C. and Wheeler, Q.D. (1990) An amplification of the phylogenetic species concept. *Cladistics* 6, 211–223.

Nunn, G. (1992) Nematode molecular evolution. An investigation of evolutionary patterns among nematodes based upon DNA sequences. PhD Dissertation, University of Nottingham, UK.

Palumbi, S.R. (1996) Nucleic cids II: the polymerase chain recation. In: Hillis, D.M., Moritz, C. and Mable, B.K. (eds) *Molecular Systematics*, 2nd edn. Sinauer Associates, Sunderland, Massachusetts, pp. 205–248.

Perlman, S.J., Spicer, G.S., Shoemaker, D. and Jaenike, J. (2003) Associations between mycophagous *Drosophila* and their *Howardula* nematode parasites: a worldwide phylogenetic shuffle. *Molecular Biology* 12, 237–249.

Phan, K.L., Nguyen, N.C. and Moens, M. (2001a) *Steinernema sangi* sp. n. (Rhabditida: Steinernematidae) from Vietnam. *Russian Journal of Nematology* 9, 1–7.

Phan, K.L., Nguyen, N.C. and Moens, M. (2001b) *Steinernema loci* sp. n. and *Steinernema thanhi* sp. n. (Rhabditida: Steinernematidae) from Vietnam. *Nematology* 3, 503–514.

Powers, T.O. (2004) Nematode molecular diagnostics: from Bands to Barcodes. *Annual Review of Phytopathology* 42, 367–383.

Powers, T.O. and Fleming, C.C. (1998) Biochemical and molecular characterization. In: Perry, R.N. and Wright, D.J. (eds) *The Physiology and Biochemistry of Free-Living and Plant-Parasitic Nematodes*. CAB International, Wallingford, UK, pp. 355–380.

Powers, T.O., Platzer, E.G. and Hyman, B.C. (1986) Large mitochondrial genome and mitochondrial DNA size polymorphism in the mosquito parasite, *Romanomermis culicivorax*. *Current Genetics* 11, 71–77.

Powers, T.O., Harris, T.S. and Hyman, B.C. (1994) Mitochondrial DNA sequence divergence among *Meloidogyne incognita, Romanomermis culicivorax, Ascaris suum* and *Caenorhabditis elegans*. *Journal of Nematology* 25, 564–572.

Powers, T.O., Todd, T.C., Burnell, A.M., Muray, P.C.B., Fleming, C.C., Szlanski, A.L., Adams, B.A. and Harris, T.S. (1997) The rDNA internal transcribed spacer region as a taxonomic marker for nematodes. *Journal of Nematology* 29, 441–450.

Powers, T.O., Neher, D.A., Mullin, P., Esquivel, A., Giblin-Davis, R.M., Kanzaki, N., Stock, S.P., Mora, M.M., and Uribe-Lorio, L. (2009) Tropical nematode diversity: vertical stratification of nematode communities in a Costa Rican humid lowland rainforest. *Molecular Ecology* (in press).

Randig, O., Leroy, R., Bongiovanni, M. and Castagnone-Sereno, P. (2002) RAPD characterization of single females of the root-knot nematodes, *Meloidogyne* spp. *European Journal of Plant Pathology* 107, 639–643.

Reid, A.P. (1994) Molecular taxonomy of *Steinernema*. In: Burnell, A.M., Ehlers, R.-U. and Masson, J.P. (eds) *Genetics of Entomopathogenic Nematode-Bacterium Complexes*. European Commission Publications, Luxembourg, pp. 49–58.

Reid, A.P. and Hominick, W.M. (1992) Restriction fragment length polymorphisms within ribosomal DNA repeat unit of British entomopathogenic nematodes (Rhabditida: Steinernematidae). *Parasitology* 105, 317–323.

Reid, A.P., Hominick, W.M. and Briscoe, B.R. (1997) Molecular taxonomy and phylogeny of entomopathogenic nematode species (Rhabditida: Steinernematidae) by RFLP analysis of the ITS region of rDNA repeat unit. *Systematic Parasitology* 37, 187–193.

Schander, C. and Halanych, M.K. (2003) DNA, PCR and formalinized animal tissue – a short review and protocols. *Organisms Diversity and Evolution* 3, 195–205.

Schwenk, A.E.K., Stadler, T., Streit, B. and Schierwater, B. (1996) RAPD identification of microsatellites in *Daphnia. Molecular Ecology* 5, 437–441.

Spiridonov, S.E., Reid, A.P., Podrucka, K., Subbotin, S.A. and Moens, M. (2004) Phylogenetic relationships within the genus *Steinernema* (Nemtada: RHbaditida) as inferred from analyses of sequences of the ITS-5.8S-ITS2 region of rDNA and morphological features. *Nematology* 6, 547–566.

Stack, C.M., Easwaramoorthy, S.G., Metha, U.K., Downes, M.J., Griffin, C.T. and Burnell, A.M. (2000) Molecular characterisation of *Heterorhabditis indica* isolates from India, Kenya, Indonesia and Cuba. *Nematology* 2, 477–487.

Stock, S.P. (2005) Insect-parasitic nematodes: more than model organisms. *Journal of Invertebrate Pathology* 89, 57–66.

Stock, S.P. (2007) Molecular approaches for diagnostics and phylogenetics of entomopathogenic nematodes, applications and implications for pest management. In: Papierok, B. (ed.) *Proceedings of the 10th European Meeting of the IOBC/WPRS Working Group 'Insect Pathogens and Insect Parasitic Nematodes'*. Locorotondo, Bari, Italy, 23–29 June 2005, pp. 1–5.

Stock, S.P. and Hunt, D.J. (2005) Nematode morphology and systematics. In: Grewal, P.S., Ehlers, R.U. and Shapiro-Ilan, D.I. (eds) *Nematodes as Biological Control Agents*. CAB International, Wallingford, UK, pp. 3–43.

Stock, S.P. and Koppenhöfer, A.M. (2003) *Steinernema scarabaei* n. sp. (Rhabditida: Steinernematidae), a natural pathogen of scarab beetle larvae (Coleoptera: Scarabaeidae) from New Jersey, USA. *Nematology* 5, 191–204.

Stock, S.P. and Reid, A.P. (2003) Biosystematics of entomopathogenic nematodes (Steinernematidae, Heterorhabditidae): current status and future directions. In: Cook, R. and Hunt, D.J. (eds) *Proceedings of the Fourth International Congress of Nematology, 8–13 June 2002, Tenerife, Spain. Nematology Monographs and Perspectives* 2, 435–446.

Stock, S.P., Somsook, V. and Reid, A.P. (1998) *Steinernema siamkayai* n. sp. (Rhabditida: Steinernematidae), an entomopathogenic nematode from Thailand. *Systematic Parasitology* 41, 105–113.

Stock, S.P., Campbell, J.F. and Nadler, S.A. (2001) Phylogeny of *Steinernema* Travassos, 1927 (Cephalobina: Steinernematidae) inferred from ribosomal DNA sequences and morphological characters. *Journal of Parasitology* 87, 877–889.

Stock, S.P., Griffin, C.T. and Chaenari, R. (2004) Morphological and molecular characterization of *Steinernema hermaphroditum* n. sp. (Nematoda, Steinernematidae), an entomopathogenic nematode from Indonesia, and its phylogenetic relationship with other closely related taxa. *Nematology* 6, 401–412.

Stoeckle, M., Janzen, D., Hallwachs, W., Hanken, J. and Baker, J. (2003) Taxonomy, DNA, and the Barcode of Life. Draft Conference Report. Available at: http://phe.rockefeller.edu/BarcodeConference/index.html

Szalanski, A.L., Taylor, D.B. and Mullin, P.G. (2000) Assessing nuclear and mitochondrial DNA sequence variation within *Steinernema* (Rhabditida: Steinernematidae). *Journal of Nematology* 32, 229–233.

Tang, S. and Hyman, B.C. (2007) Mitochondrial genome haplotype hypervariation within the isopod parasitic nematode *Thaumamermis cosgrovei. Genetics* 176, 1139–1150.

Thomas, W.K., Vida, J.T., Frisse, L.M., Mundo, M. and Baldwin, J. (1997) DNA sequences from formalin-fixed nematodes: integrating molecular and morphological approaches to taxonomy. *Journal of Nematology* 29, 255–267.

Vandegast, A.G. and Riderick, G.K. (2003) Mermithid parasitism of Hawaiian *Tetragnatha* spiders in a fragmented landscape. *Journal of Invertebrate Pathology*, 85, 128–136.
Vrain, T.C., Wakarchuk D.A., Levesque, A.C., and Hamilton, R.I. (1992) Intraspecific rDNA restriction fragments length polymorphisms in the *Xiphinema americanum* group. Fundamental and *Applied Nematology* 15, 563–573.
Wang, J.-Y., Xu, F., Liu, X.-S. and Wang, G.-X. (2007) Molecular phylogeny of entomopathogenic nematodes (Mermithidae) inferred from DNA sequences of 18S rDNA, 28S rDNA and COI genes. *Acta Zoologica Sinica* 53, 835–844.
Whitkus, R., Doebley, J. and Wendel, J.F. (1994) Nuclear DNA markers in systematics and evolution. In: Phillips, L. and Vasil, I.K. (eds) *DNA-based Markers in Plants*. Kluwer Academic Publishers, Dordrecht, The Netherlands, pp. 116–141.
Williams, J.G.K, Anne, R.K., Kenneth, J.L., Antoni, J.R. and Scott, V.T. (1990) DNA polymorphisms amplified by arbitrary primers are useful as genetic markers. *Nucleic Acids Research* 18, 6531–6535.
Ye, W., Giblin-Davis, R.M., Davies, K.A., Purcell, M.F., Scheffer, S.J., Taylor, G.S., Center, T.S., Morris, K. and Thomas, W.K. (2007) Molecular phylogenetics and the evolution of host plant associations in the nematode genus Fergusobia (Tylenchida: Fergusobiinae). *Molecular Phylogenetics and Evolution* 45, 123–141.

# 5 Identification and Diagnostics of Entomopathogenic Protozoa

M. Oborník

*University of South Bohemia, Faculty of Sciences, Department of Molecular Biology and Biology Centre of the Academy of Sciences of the Czech Republic, Institute of Parasitology, České Budějovice, Czech Republic*



## 5.1. Introduction

Molecular methods allow us to recognize and identify target organisms from a variety of environments including laboratory cultures, host tissues or environmental samples, through a detection of specific molecular markers. Ascending progress in the field of molecular evolution during last 15 years greatly influenced high-order taxonomy. It has been shown that the proposed kingdom Protozoa comprising unicellular heterotrophic eukaryotes is polyphyletic in its nature and its taxonomic relevance is no longer sustainable. Here I show the two current proposals dividing eukaryotic organisms into five and six major groups (kingdoms), respectively. These are currently the most accepted views on protist and eukaryotic high-order taxonomy referred by Simpson and Roger (2004) and Keeling *et al.* (2005) (Fig. 5.1), which fully reflect latest knowledge in molecular phylogeny and that actually differ each other only in placing Opisthokonta and Amoebozoa in single or separated kingdoms (see Fig. 5.1 for

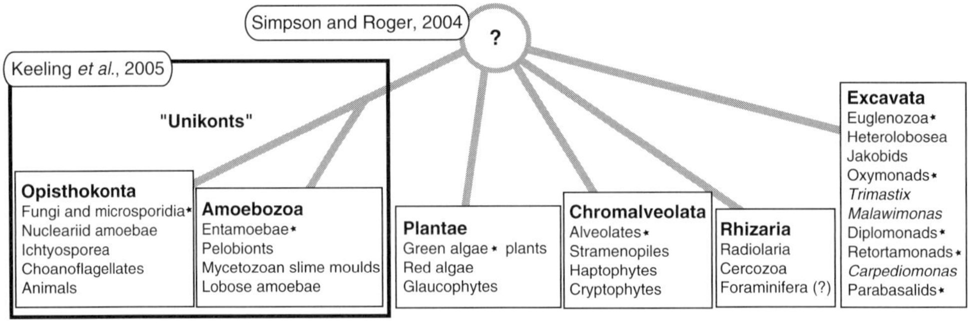


**Fig. 5.1.** Current taxonomy status of protozoa. Black stars indicate groups containing organisms traditionally classified as Protozoa associated with insects.

details). Recently, taxonomic system based on the above mentioned proposals has been published (Adl *et al.*, 2005, 2007) and is used for classification in this work.

## 5.2. Molecular Identification of Species and Strains

Entomopathogenic protozoa belonging to various eukaryotic supergroups are associated with various insect orders and thus they display extreme diversity (Fig. 5.1). Therefore, methods used to molecularly characterize particular species or strains vary and need to be adjusted and/or modified to address such diversity. Varieties of molecular markers are thus considered for each group of entomopathogenic protozoans. For example, mitochondrial minicircles can be used to identify exclusively kinetoplastid flagellates (Excavata: Euglenozoa), while plastid sequences can characterize only those protozoa that contain plastid, such as apicomplexan parasites (Chromalveolata: Alveolata) or parasitic green algae (Plantae: Chlorophyta).

### 5.2.1. Nucleic acid extraction and template preparation

Obtaining nucleic acids from a sample is a crucial step for the correct identification and diagnostics of protozoa. Proper nucleic acid extraction methods should be chosen in relation to the target organism's availability and the type of analytical method (e.g. random fragment length polymorphism (RFLP), polymerase chain reaction (PCR), sequencing) to be later considered.

#### *5.2.1.1. Live samples*

The study of insect pathogenic protozoa includes much outdoor work including collection of samples in distant places; therefore, it is very important to store samples so that collected material is preserved in a suitable manner for application of molecular diagnostic methods. For example, it is recommended to place samples into 70% ethanol or any kind of DNA-protective reagent. In most cases, it is recommended to store samples at low temperatures. For this purpose, thermoboxes with dry ice should be considered. Alternative methods include keeping samples at temperatures slightly under 0°C. For example, filling a thermoflask with a mixture of ice and $NaCl_2$ generates temperatures below −15°C that can be maintained over 10 h (J.Vávra, 2005, personal communication).

Life stages with easily breakable cell walls can be stored directly in a lysis buffer and low temperature, and should be used for nucleic acid extraction immediately after arrival of samples into the laboratory. Readers should also bear in mind that for many protozoa, samples fixed for molecular analyses cannot later be used for morphological characterization and vice versa (Undeen and Vávra, 1997).

#### *5.2.1.2. Fixed and museum specimens*

When dealing with fixed material or museum specimens, special attention needs to be placed in the quality and purity of the nucleic acids to be extracted. It has been shown that DNA from microsporidia is more successfully isolated from methanol-fixed, Giemsa-stained smears, while attempts to extract DNA from non-fixed slides have almost failed. Although the DNA from old slides (since 1956) has been highly fragmented, it has been possible to amplify fragments sufficiently to help in the diagnosis of species (Hyliš *et al.*, 2005). Two basic methods for DNA extraction routinely considered for the identification of protozoans are described below.

#### *5.2.1.3. Method 1 (Fig. 5.2.; according to Jirků* et al., *1995, modified)*

DNA extraction methods start with cell lysis in a buffer solution containing detergents (non-ionic, e.g. Tween 20 and Triton X100; or ionic, e.g. sodium dodecyl sulfate) and a protease (e.g. proteinase K or pronase E), to brake cell walls and inactivate enzymes degrading DNA. Proteins are then extracted by phenol. The water phase containing diluted DNA is separated by centrifugation and purified of the remaining phenol by chloroform–isoamyl alcohol (24:1) extraction. The water phase is then separated and the DNA precipitated, dried and diluted (Fig. 5.2).

Various modifications of this procedure can be found in the literature (e.g. Kirby, 1957; Leadon and Cerutti, 1982; Longmire *et al.*, 1987; Subrungruang *et al.*, 2004). A variety of commercial DNA extraction kits are available as well. These kits are also based on the principles described above, but have some modifications including purification of cell lysates in DNA affinity columns, wash off of proteins and cell remnants from the fixed DNA with a buffer solution, and subsequent elution from the column with an elution buffer or water (e.g. QIAGEN(USA), Invitrogen (USA), JetQuick (Genomed, Germany)).

During the process of DNA extraction of several protists, remarkable amounts of various RNAs are isolated too. Although RNA is generally very quickly degraded, it can be surprisingly tough. For example, RNA from *Gregarina garnhami*, apicomplexan parasite of locusts, usually remains very stable during extraction of the total DNA (M. Oborník and J. Lukeš, 2002, unpublished data). Since higher RNA concentration may cause problems when DNA template is used for PCR, it

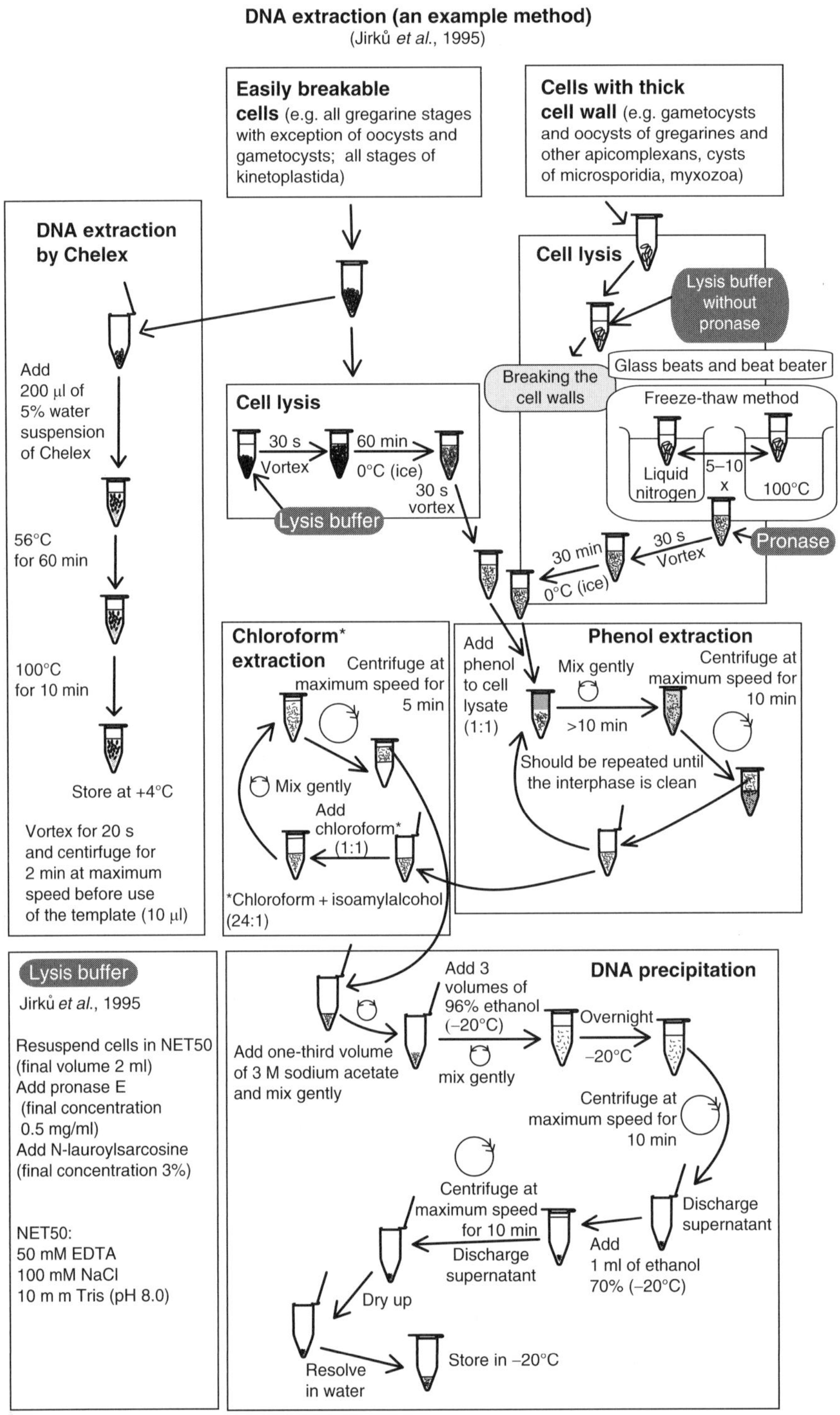


**Fig. 5.2.** See caption on the facing page.

is recommended to treat the templates with RNase (enzyme degrading RNA, the working concentration for RNase A is 1–100 μg/ml depending on the application) before use. Extracted DNA should be stored in water solution or appropriate buffer (e.g. TE buffer; 10mM Tris, 1mM EDTA, pH 7.5) in −20°C.

#### *5.2.1.4. Method 2 (according to Walsh et al., 1991, modified)*

The second DNA isolation method considered for protozoa relates to those situations when the amount of nucleic acids is low or only a very small sample (i.e. fragile life stages with easily breakable cells) is available. This is the case for some of entomogenous protists which cannot be cultivated, and only a few cells can usually be obtained from the host or environment. When dealing with this situation, cells are broken in water containing suspension of an additive such as Chelex® 100 Resin (Sigma) that is affinitive to metal ions. In this method, small gel particles capture all the metal ions from the lysate. Chelex-isolated DNA is supposed to serve only as a template for PCR amplification (Fig. 5.2). It is necessary to mention that eventual contamination of the PCR reaction mixture by Chelex particles will decrease efficiency of the enzymatic reaction. Therefore, it is necessary to homogenize the sample by vortexing and centrifuge it (2 min at 13,000 rpm) before use. Chelex particles concentrate at the bottom of the tube and the supernatant can be used as a template. Chelex-based templates can be stored in the refrigerator at +4°C for 6 months. Their freezing is not recommended.

### 5.2.2. DNA sequencing and its application in protozoa identification

DNA sequencing has been the most frequently used procedure so far for molecular diagnostics of protozoa as well as for the study of their evolutionary relationships. Such analysis provides genotypic data that can specifically identify the organism of study, if the polymorphic target region was properly chosen. PCR is the method usually used to obtain the defined DNA region for further sequencing. To get enough template for sequencing, PCR should be optimized (Fig. 5.3).

DNA sequences can be obtained either from purified PCR products (direct sequencing), or by cloning amplicons into PCR cloning vector, which is sequenced by vector-specific sequencing primers (Fig. 5.4). Generally, the cloning of a PCR product is needed when the DNA region is used as a DNA probe. Cloning the product may also help to sequence genetically distant templates. On the other side, direct sequencing is cheaper, and it overcomes possible sequencing errors caused by DNA polymerase.

---

**Fig. 5.2.** DNA extraction. This is an example method for DNA extraction referred by Jirků *et al.* (1995). It is slightly modified, and it actually has to be such for each specimen. It is recommended to break cells with thick cell walls by glass beads using Mini BeadBeater (Biospect Priducts) or Freeze-Thaw method. Cells can also be broken by deep freezing (in liquid nitrogen) and mortar and pestle. It is also advised to repeat phenol and chloroform extractions several times to get better purified templates. Each part of the procedure can be combined with commercial kits.

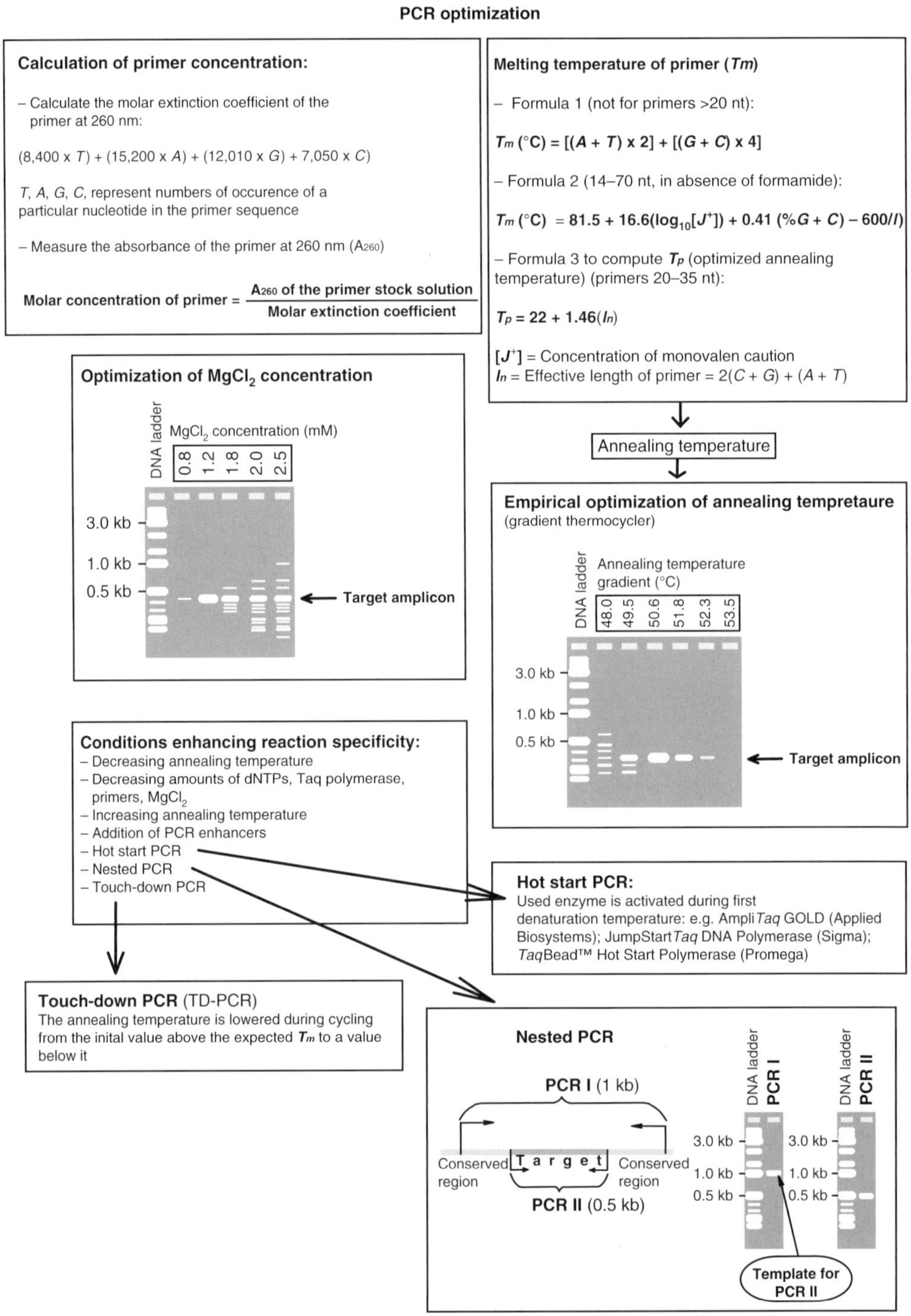


**Fig. 5.3.** PCR optimization. PCR should be optimized for each primer-template combination. Concentration of PCR components and profile of the thermal cycle can be varied to find an optimum.

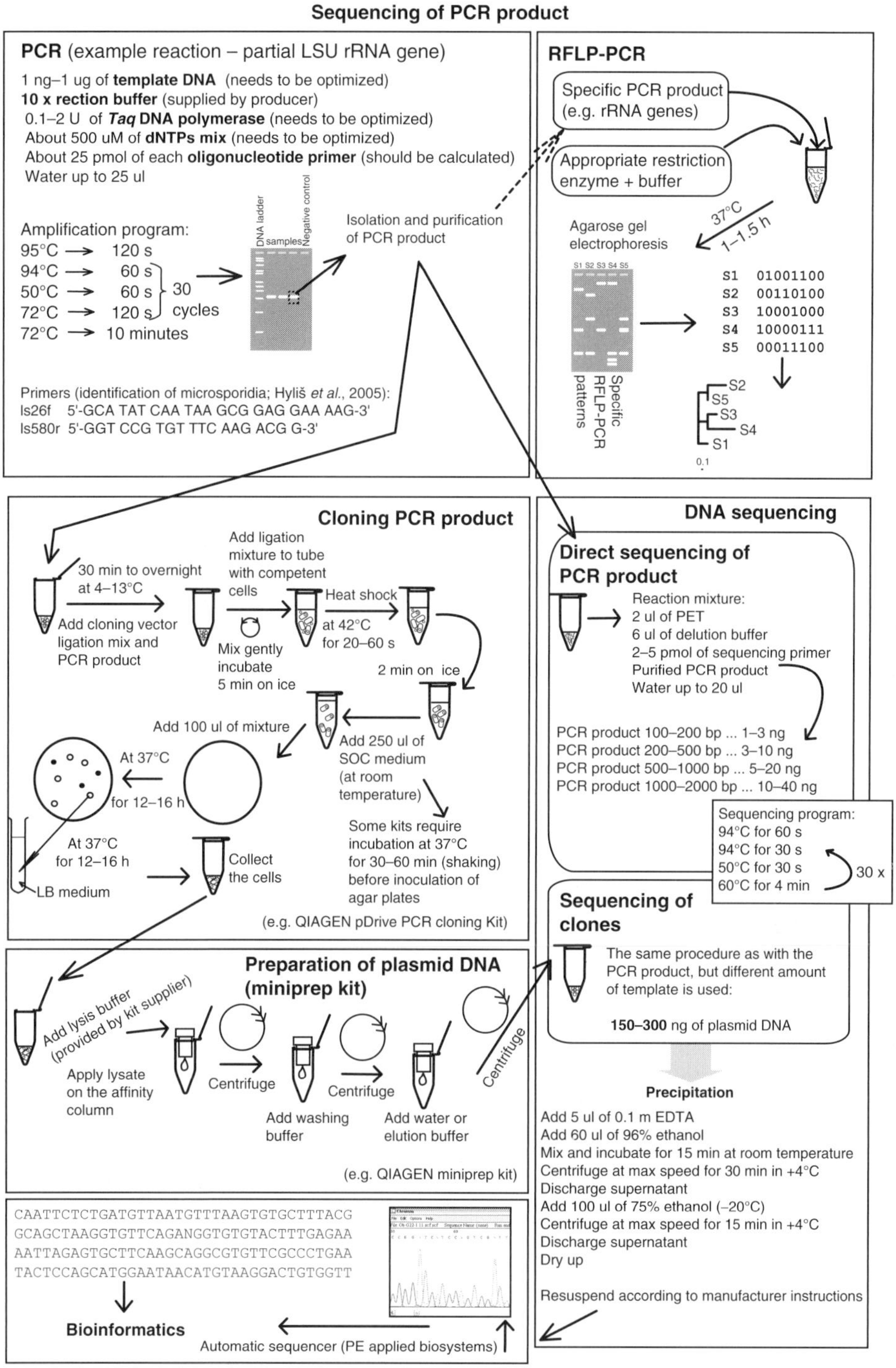


**Fig. 5.4.** Sequencing of PCR product. Genes coding for rRNA are the most frequently used markers for molecular identification. Although target genes are usually sequenced and used for phylogenetic analysis such as in microsporidia (Vossbrinck and Debrunner-Vossbrinck, 2005), RFLP-PCR represents a possible alternative to investigate polymorphism within the target. RFLP-PCR fingerprint data can give sufficient base for development of species-specific molecular marker. For all molecular methods, follow instructions of supplier of reagents.

## 5.2.3. Fingerprinting methods to identify strains

### *5.2.3.1. RFLP-based fingerprints*

RFLP is one of the oldest approaches to characterize genotypes. It is based on the ability of restriction endonucleases to specifically digest DNA within the target sequence, usually from 4 to 8 nucleotides (nt) in length. One enzyme may recognize one or more target sequences. Digested DNA is then separated by gel electrophoresis. Since restriction of total DNA often leads to unreadable or smeared DNA pattern, it is completed by Southern blot and hybridization with a specific DNA probe (Sertsrivanich and Yuthavong, 1985; McDonald and Martinez, 1990). RFLP analysis provides molecular fingerprint data that can be very well used for both phylogenetic analysis, and identification and characterization of the isolate. However, it requires much higher amount of DNA with no foreign DNA contaminants, when compared to the PCR-based methods (PCR-RFLP). Thus, RFLP based on total DNA digestion can be used only for organism available in axenic culture or those possible to purify from the host tissues, otherwise digested DNA contamination could be incorrectly scored as a marker. If performed correctly, RFLP provides high number of repeatable molecular markers that can be very well used to identify Apicomplexa (Shields and Olson, 2003; Elsheikha *et al.*, 2006) as well as Kinetoplastida (Lukeš *et al.*, 2007). However, RFLP has not been used to study protists in insects, except insect-transmitted parasites.

### *5.2.3.2. PCR-based fingerprints*

PCR-based fingerprinting utilizes single short arbitrary primers (usually 10 nt long) to produce template-specific DNA fragments. Arbitrary means that primers are not designed to anneal to the known target sequence; however, the sequence of the primer must respect general rules for primer designing to avoid artefacts, such as forming dimmers. This method (e.g. Williams *et al.*, 1990) is usually referred as arbitrary primed PCR (AP-PCR) or random amplified polymorphic DNA (RAPD). RAPD has many advantages, when compared to classical RFLP. Amplification with random primers allows us to analyse genomes with no prior knowledge of its DNA sequence. The amount of the DNA template needed for RAPD reaction is more than 1000 times lower when compared to RFLP. RAPD analysis can be very fast and usually does not require additional procedures such as Southern blot and DNA–DNA hybridizations. So far, RAPD has been used not only to characterize isolates or strains of protists, but also for gene mapping and related purposes (Reiseberg *et al.*, 1993). However, RAPD patterns are often not fully reproducible. This means that even very little variance within the reaction conditions may lead to the appearance of minor PCR products, which can be absent when the RAPD reaction is run repeatedly. Some of repeatedly obtained DNA fragments may be amplified to a level below the detection limits and are not scored. For trustable molecular identification, primers producing ambiguous amplicons should be excluded for further analysis. It is also very important to use primers that produce optimal number of RAPD amplicons leading to well-readable patterns. The readability of RAPD products depends on the particular conditions and level of DNA polymorphism. It is known that RAPD patterns also depend on the DNA polymerase (Loudon *et al.*, 1995), concentrations of reaction components and even the thermocycler used (Williams *et al.*, 1990; Carlson *et al.*, 1991; Pérez *et al.*, 1998).

Moreover, interpretation of RAPD data represents a second obstacle. RAPD products of the same size may not necessarily represent homologous sequences. In those cases where identity of products is ambiguous, it should be verified by Southern blots and DNA–DNA hybridization or the particular primer should be excluded from analysis. According to the nature of RAPD amplicons, the data cannot be analysed by unweighted maximum parsimony to construct phylogenetic trees (Backeljau *et al.*, 1995). However, various approaches based on distance methods can be applied to the RAPD-inferred 01 matrices (e.g. NeiLi and UpHolt models implemented in PAUP*; Swofford, 2000; or FREE TREE PROGRAM; Pavlíček *et al.*,1999).

RAPD markers have been used to identify various protists and to study their relationships (Lun and Desser, 1996; Vaňáčová *et al.*, 1997; Čepička *et al.*, 2005), including entomogenous microsporidia with high economical importance (Rao *et al.*, 2005, 2007).

### *5.2.3.3. Combined fingerprinting methods*

PCR-RFLP. Both the above-mentioned approaches can be combined in many ways. When a small amount of DNA is available or the DNA is contaminated, a selected target region can be amplified by group-specific primers, and the polymorphisms within the target region are then detected by enzymatic digestion of the PCR product. The resulting PCR-RFLP pattern is visualized by gel electrophoresis. Since primers designed for PCR-RFLP are specific, digestion of amplicons leads to production of highly efficient molecular markers, if appropriate target and restriction enzymes were used (e.g. Carreno *et al.*, 2001; Leng *et al.*, 1996). PCR-RFLP is frequently used to identify protists (Hopkins *et al.*, 1996; Marfurt *et al.*, 2003; Gadisa *et al.*, 2007) including entomopathogenous ones (Paxton *et al.*, 2007).

AMPLIFICATION FRAGMENT LENGTH POLYMORPHISM (AFLP). AFLP is based on the amplification of restriction fragments of enzymatically digested total DNA. After restriction of the total DNA, oligonucleotide adapters are ligated to the restriction fragments. The primers for following amplification are designed to be complementary to both the enzyme recognition sequence and the adapter sequence. The selectivity of amplification is controlled by portion of the primer sequence extending a short way into the restriction fragments. If thermostable polymerase with the absence of 3′ to 5′ exonuclease activity is used, only primers matching to the restriction fragment at their 3′end are extended and the fragments are amplified (Newton and Graham, 2000). AFLP analysis always produces very high numbers of polymorphic molecular markers that can be radioactively labelled and separated by electrophoresis on polyacrylamide gel, or labelled primers are used and AFLPs are analysed by capillary automatic sequencer (e.g. Mueller and Wolfenbarger, 1999). AFLP technique is not very much used to identify protists, mainly because of higher costs; however, if used properly, it provides high number of relevant markers (Elsheikha *et al.*, 2006).

MICROSATELLITES OR SIMPLE SEQUENCE REPEATS (SSRS). SSRs are short tandem repeats (from 10 to 50 copies) composed from mono- to tetra-nucleotide repeats (such as

$(AT)_n$ and $(CAG)_n$), which are supposed to be randomly distributed throughout the genome. Primers are designed to conserve regions flanking the SSRs. SSRs detect changes in the number of repeat units to which stepwise mutation models can be applied. These markers are codominant and it is possible to detect both nuclear and organellar sequence polymorphisms. However, the initial identification of the SSR is expensive requiring cloning and sequencing of particular SSR. In general, SSRs are frequently used to estimate population structure or gene diversity (Newton and Graham, 2000; Lowe *et al.*, 2004). SSRs are currently used to study evolution and variability of apicomplexan genus *Cryptosoridium* (Tanriverdi and Widmer, 2006) and kinetoplastid parasites *Trypanosoma rangelii* (Grisard *et al.*, 1999). Entomogenous protists have not been studied by this technique yet.

DENATURING GRADIENT GEL ELECTROPHORESIS (DGGE). This method is based on the fact that if DNA heteroduplexes differ in a single base pair, they display slightly different melting characteristics. It is necessary to bare in mind that one PCR product on a gel can represent a set of homologous molecules of the same length, which can differ in their particular sequence. Therefore, if such PCR product is separated on a gel with denaturation gradient (thermal gradient or gradient obtained by chemical denaturant such as formamide or urea), it will produce separate bands each corresponding to the particular variation within the amplified target. Such separated PCR products are then cut from the gel, cloned and sequenced, and the particular mutation which caused different melting and mobility on the gradient gel can be detected (Newton and Graham, 2000). DGGE allows detection of an occurrence of different paralogues of the target within the investigated genome, or it can be used to analyse mixed populations of closely related organisms (Nocker *et al.*, 2007). This method has been used to analyse variability of the kinetoplastid *Trypanosoma cruzi* (Stothard *et al.*, 1998).

### 5.2.4. Development of strain-specific PCR marker for molecular identification and diagnostics

To identify isolates or strains from infected host tissues or from any other complex environment, specific primers for PCR diagnostics should be constructed (Figs 5.5 and 5.6). Two ways to design such primers are addressed here.

Specific primers for PCR diagnostics can be selected based on a multiply alignment of an appropriate set of target sequences. Availability of related sequences together with the level of their polymorphism represents the main factors influencing selection of the target. Since ribosomal RNA (rRNA) genes represent the region most frequently used for phylogeny and molecular taxonomy of eukaryotes and provide various levels of DNA polymorphism, they can serve well as targets for diagnostic PCR in proposed taxonomy level (e.g. Šlapeta *et al.*, 2002; Klee *et al.*, 2006). Construction of multiply alignment of target sequences is the first step, which allows us to recognize polymorphic regions specific for the taxonomy level of interest and to design specific PCR primers. All general guidelines for primer design should be respected. It is necessary to note that having a sufficient number of sequences well representing diversity of the diagnosed group is an essential condition (Fig. 5.5).

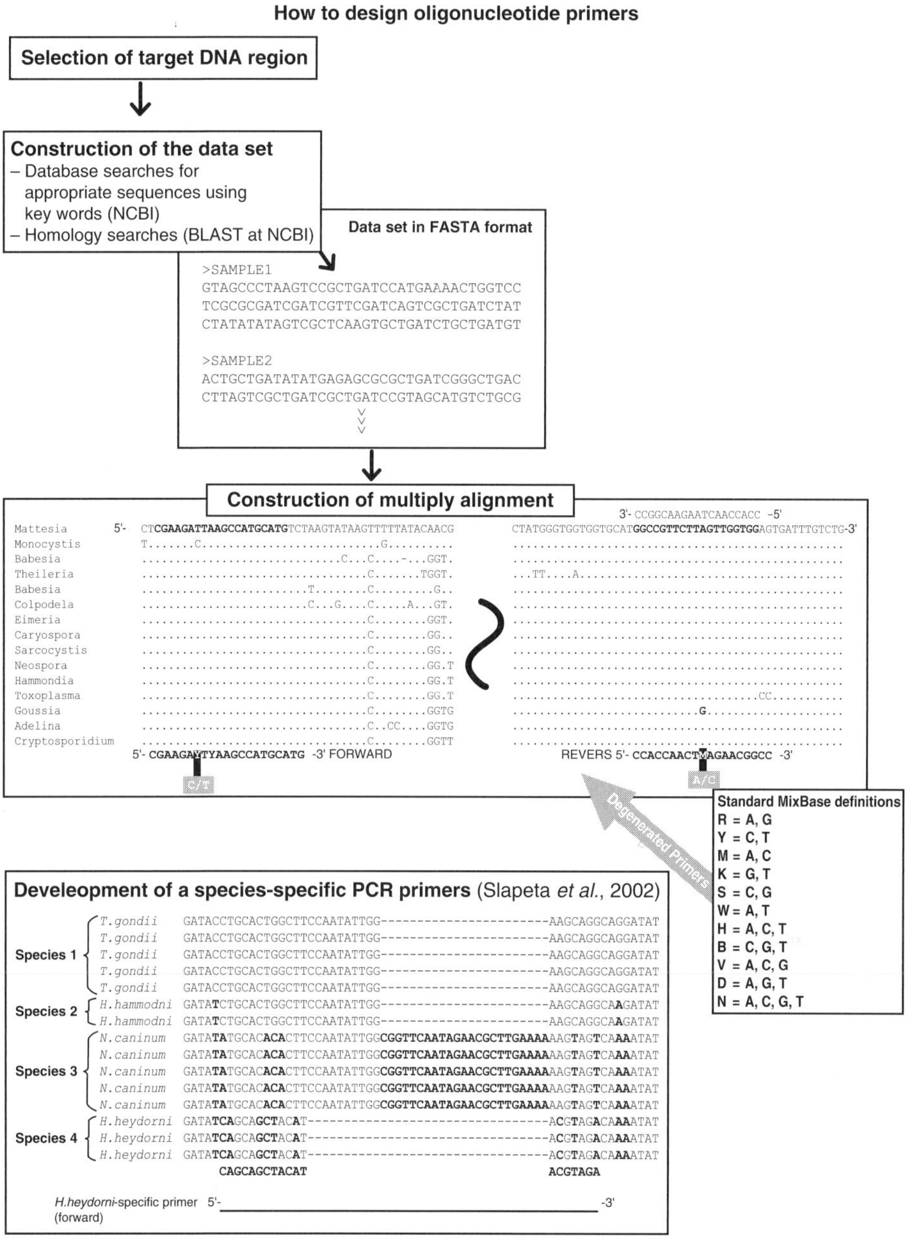


**Fig. 5.5.** How to design oligonucleotide primers. In general, two types of PCR primers are usually requested. First, primers are designed to anneal to conserved parts of the gene and to amplify the target gene from various organisms. Such primers are used to obtain data sets for phylogenetic studies. When the target is so polymorphic (i.e. no conserved region in the organism of interest can be found), degenerated primers should be designed. Species or strain-specific primers represent the second type. Based on multiple alignment of related sequences (obtained by PCR with previous type of primers), the annealing region specific for species or other taxonomic category can be found and used to design primers.

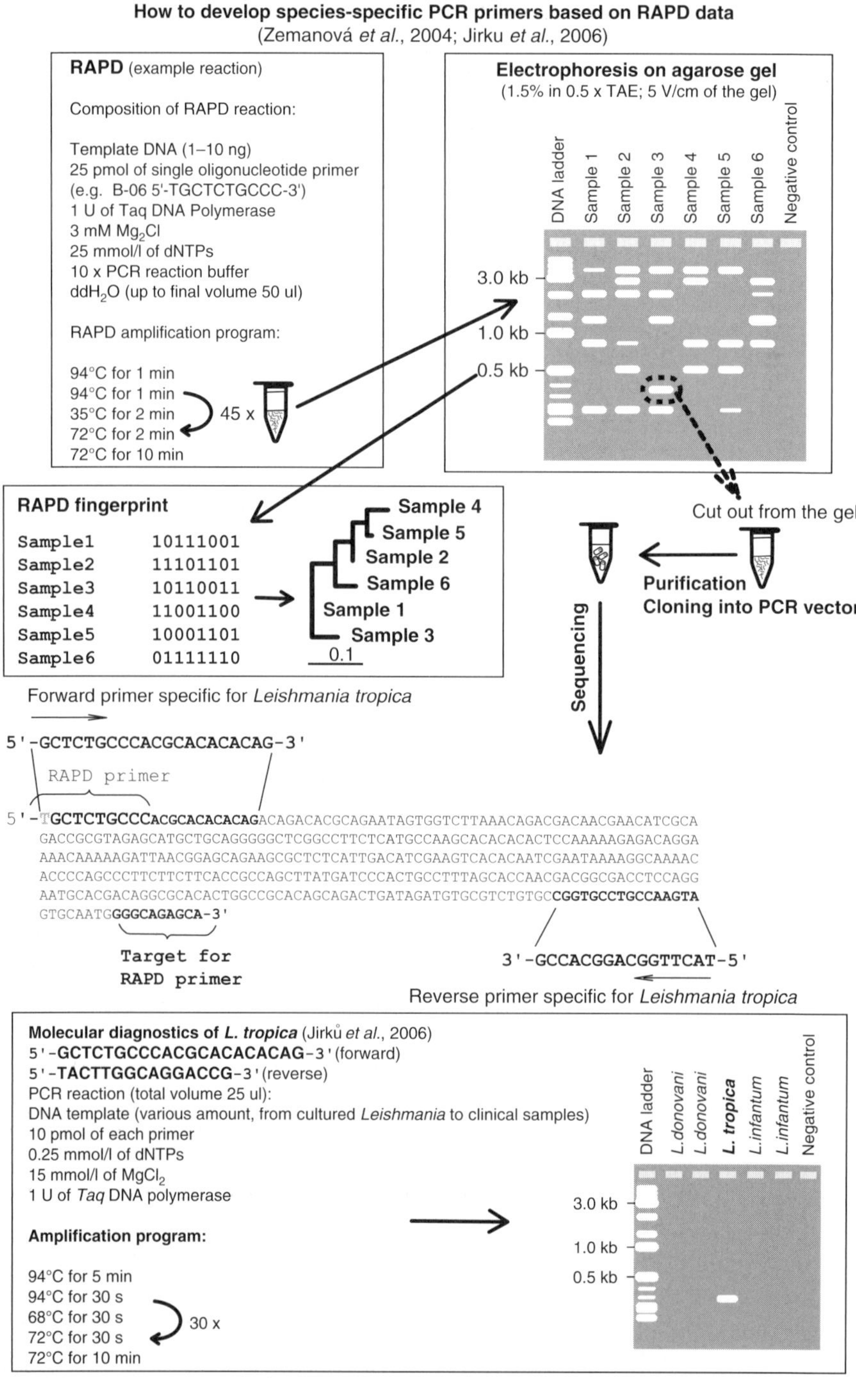


**Fig. 5.6.** How to develop species-specific PCR primers based on RAPD data. RAPD analysis can be used not only to characterize the genotype by obtaining RAPD fingerprint, but it can also be used, if the level of polymorphism detected is appropriate, to construct species-specific PCR primers. This figure shows the procedure in a easy way, but it has to be taken into account that dozens of primers have to be tested to get species-specific RAPD product. PCR primers for diagnosis of *Leishmania tropica*, which was developed from RAPD analysis has recently been published (Jirků *et al.*, 2006).

RAPD patterns can be utilized to characterize isolates of entomogenous eukaryotes, although they can be used only if axenic cultures or purified organism are available. However, RAPD technology can also be considered to construct groups or strain-specific PCR primers for molecular diagnostics. The procedure, which has been referred to identify kinetoplastid parasites of humans transmitted by insect vector (e.g. Jirků *et al.*, 2005), can be applied to other insect pathogens as well (Fig. 5.6).

### 5.2.5. Currently available molecular data for protozoa associated to insects

AMOEBAE (EUKARYOTA: AMOEBOZOA). Most insect-associated amoeba species are commensals in digestive tracts of their hosts. However, some of them, such as the families Amoebidae (*Malamoeba*, *Malpighamoeba*, *Malpigiella*; strictly entomogenous genera), and Entamoebidae, are pathogens producing amoebiasis in insects. Until now, no sequences from insect-associated amoebae have been deposited in the GeneBank database.

APICOMPLEXA (EUKARYOTA: CHROMALVEOLATA). Apicomplexans are strictly parasitic unicellular eukaryotes belonging to Alveolates (supergroup Chromalveolates). Current studies suggest that Apicomplexa is a monophyletic taxon, which forms together with ciliates (Ciliophora), dinoflagellates (Dinophyceae) (e.g. Van de Peer et al., 1996) and chromerids (Chromerida) (Moore et al., 2008) a clade called Alveolata. In addition to typical features such as apical complex and subsurface alveoli, apicomplexans also contain non-photosynthetic secondary plastid called apicoplast, where several essential metabolic pathways take place.

At present research is focused on apicomplexan parasites of humans, such as the causative agent of malaria *Plasmodium falciparum* (transmitted by *Anopheles*); however, other apicomplexans with importance in veterinary and human medicine (e.g. genera *Theileria*, *Babesia*, *Toxoplasma*, *Neospora*, *Sarcocystis*) have also been extensively studied (e.g. Šlapeta *et al.*, 2002; Kopečná *et al.*, 2006).

Apicoplast genes encoding small subunit (SSU) rRNA and ORF470 have been used to investigate apicomplexan phylogeny (Zhao and Duszynski, 2001; Oborník *et al.*, 2002). Chloroplast-specific genes in particular, such as ORF470, represent excellent target for molecular diagnostics, because they have no homologues in metazoan host. It should be noted that not all apicomplexans possess the relic plastid. It has been shown that this organelle is absent from the genus *Cryptosporidium* (Zhu *et al.*, 2000) and probably also from gregarines (Toso and Omoto, 2007).

At least there are two groups of Apicomplexa that colonize insects and may cause diseases. The first group is represented by gregarines (Apicomplexa: Gregarina), a large group of early branching unicellular parasites infecting exclusively invertebrate hosts. Gregarines have been relatively well studied by molecular techniques. This interest is motivated by the fact that they (particularly archigregarines) represent the most ancestral group within Apicomplexa (Leander and Keeling, 2004). This fact as well as the possible relationship between gregarines and the apicomplexan genus *Cryptosporidium* lead to intensive sequencing of rRNA genes and some protein coding genes from various gregarines (see Table 5.1). Thus,

**Table 5.1.** Selected entomogenous apicomplexans and sequences usable for their molecular identifications.

| Species | Gene | Accession |
|---|---|---|
| *Gregarinasina* | | |
| *Ascogregarina culicis* | 18S rRNA gene, partial sequence; internal transcribed spacer 1, 5.8S rRNA gene and ITS 2, complete sequence; and 28S rRNA gene, partial sequence | AY327258 |
| *Ascogregarina barreti* | 18S rRNA gene, partial sequence; internal transcribed spacer 1, 5.8S rRNA gene and ITS 2, complete sequence; and 28S rRNA gene, partial sequence | AY327259 |
| *Ascogregarina taiwanensis* | Actin (act1) mRNA, partial cds | AF254449 |
| *Gregarina caledia* | SSU rRNA gene, partial sequence | L31799 |
| *Gregarina chortiocetes* | SSU rRNA gene, partial sequence | L31841 |
| *Gregarina niphandrodes* | DNA-dependent RNA polymerase II largest subunit | AY168016 |
| *Gregarina polymorpha* | SSU rRNA gene, partial sequence | AF457129 |
| | Myosin A mRNA, complete cds | AY382895 |
| | myosin B mRNA, complete cds | AY382896 |
| | Actin mRNA, complete cds | AY382894 |
| *Lecudina tuzetae* | SSU rRNA gene, partial sequence | AF457128 |
| *Lecudina polymorpha* | SSU rRNA gene, partial sequence | AY196707 |
| *Leidyana migrator* | SSU rRNA gene, partial sequence | AF457130 |
| | Beta-tubulin (btub1) gene, partial sequence | AF457131 |
| *Mattesia* sp. | External transcribed spacer and 18S rRNA gene, partial sequence | AY334569 |
| *Mattesia geminata* | External transcribed spacer and 18S rRNA gene, partial sequence | AY334569 |
| *Monocystis agilis* | SSU rRNA gene, partial sequence | AF457127 |
| | Actin gene, partial cds | AY391264 |
| | Heat shock protein 90 gene, partial cds | AY391262 |
| | Actin (act2) mRNA, partial cds | AF254450 |
| | Actin (act3) mRNA, partial cds | AF254451 |
| | 18S rRNA gene, partial sequence; internal transcribed spacer 1, 5.8S rRNA gene and ITS 2, complete sequence; and 28S rRNA gene, partial sequence | AY326461 |
| *Ophriocystis elektroscirrha* (*Solenopsis invicta*) | SSU rRNA gene | AF129883 |
| *Pseudomonocystis lepidiota* | SSU rRNA gene, partial sequence | L31843 |
| *Adeleorina* | | |
| *Adelina bambarooniae* | SSU rRNA gene | |
| *Adelina dimidiata* | SSU rRNA gene | DQ096837 |
| *Adelina grylli* | SSU rRNA gene | DQ096836 |

they probably represent, together with microsporidia, the group of entomopathogenic protozoa most studied by molecular methods.

Similar to other groups, SSU rRNA gene sequences represent molecular marker most frequently used to identify gregarines (e.g. Leander, 2007; Leander *et al.*, 2003, 2006), even within environmental samples (e.g. Takishita *et al.*, 2007). Also protein coding sequences, such as β-tubulin, actin and chaperonin hsp90, have been used to infer phylogenetic relationships among early branching Apicomplexa (Leander *et al.*, 2003; Leander and Keeling, 2004). However, gregarines represent a large, diverse and unexplored group and designing group-specific primers is a very difficult task. So far, universal eukaryotic primers have not produced amplicons for all gregarine taxa that have been investigated, and for *Gregarina polymorpha*, different primers were designed to get the product (Leander *et al.*, 2003). Amplification of protein coding genes is even more complicated. Nested PCR has been considered with two pairs of highly degenerated primers to get a single β-tubulin gene coding sequence (Leander *et al.*, 2003).

The suborder Adeleorina (Apicomplexa: Eucoccidiorida) represents the second group of apicomplexans colonizing insects that has been investigated by molecular methods so far. Particularly, some entomogenous members of the genus *Adelina* (Table 5.1) have been characterized by the sequencing of nuclear SSU rRNA gene and, their phylogenetic position as a sister group to haemogregarines of the genus *Hepatozoon* within the suborder Adeleorina was suggested (Kopečná *et al.*, 2006).

CILIOPHORA (EUKARYOTA: CHROMALVEOLATA). Most of ciliates are free-living; however, many parasitic ciliates exist. Only two genera, *Lambornella* and *Tetrahymena*, have been described to parasitize on insects and cause diseases. Sequence data available for these genera are very limited. For example, only a very short fragment of gene coding for nuclear large subunit (LSU) rRNA is available (190 bp, AF010389) for *Lambornella clarkii* a pathogen of mosquitoes (Egerter *et al.*, 1986). The complete sequence of nuclear SSU rRNA gene is referred for *Lambornella* sp. (AF364043).

Contrary to *Lambornella*, *Tetrahymena pyriformis* is probably one of the most frequently sequenced ciliates found in insects (Jerome *et al.*, 1996). There are about 188 nt and 225 protein sequences available, including the complete sequence of mitochondrial genome (NC000862) (Burger *et al.*, 2000). Complete genomes from two ciliates, *Tetrahymena thermophila* (Eisen *et al.*, 2006) and *Paramecium tetraurelium* (http://paramecium.cgm.cnrs-gif.fr/db/index; Aury *et al.*, 2006), are currently available. Knowledge gained from these two genomes should serve as a base for designing specific molecular markers.

EUGLENOZOA (EUKARYOTA: EXCAVATA). Euglenozoa comprise euglenids (Euglenida), diplonemids (Diplonemea) and kinetoplastids (Kinetoplastea). Some of euglenids (e.g. *Euglena gracilis*) are photoautotrophs bearing secondary green plastids. Therefore, in the view of insect pathology, heterotrophic kinetoplastids represent organisms that may colonize insects. Although there are free-living kinetoplastids as well, most of described species are parasitic, and many of them parasitize insect hosts. Kinetoplastids got their name thanks to the uniquely structured mitochondrial (mt) DNA, which represents over 40% of the total cellular DNA.

It has been shown that kinetoplastid mitochondrial genome is composed of two different types of circular DNA molecules, minicircles (several thousands) and maxicircles (dozens). Minicircles are circular non-supercoiled molecules, which are usually about 1 kbp in size, and by their catenation constitute a network resembling chainmail armour. Maxicircles are large relaxed circular molecules that encode most of the mitochondrial genetic information and all mitochondrial genes. However, such genes are not translatable and their pre-transcripts are modified to translatable form by insertion of uridines (U) (RNA editing; see Lukeš *et al.*, 2005). These unique molecular structures are well suitable for construction of specific markers, because complex structures such as kinetoplast DNA cannot be found in any other eukaryotes. Thus, various properties of the kinetoplast (Votýpka *et al.*, 2002) as well as minicircles (e.g. Yurchenko *et al.*, 2000) can be used to characterize and identify isolates or strains.

rRNA genes (5S and 18S), small spliced leader (mini-exon) RNA gene and gene coding for glycosomal glyceraldehyde-3-phosphate dehydrogenase (gGAPDH) have been the most frequently used markers for identification of entomogenous kinetoplastids so far (Table 5.2). Since heteroxenous kinetoplastids of humans are transmitted by insects, they have been more exhaustively studied. Genomes of *Trypanosoma brucei*, *T. cruzi* and *Leishmania major* have already been sequenced (El-Sayed *et al.*, 2005), and can be searched for prospective molecular markers.

OXYMONADIDA (EUKARYOTA: EXCAVATA). Only one partial SSU rRNA gene sequence from unspecified *Oxymonas* sp. JF2002 living in *Neotermes jouteli* has been determined (Table 5.3).

TRICHOMONADIDA (EUKARYOTA: EXCAVATA: PARABASALIDEA). Trichomonads are usually referred as parasites of vertebrates, as only few have been found in invertebrates. Only SSU rRNA gene sequences from *Pseudotrypanosoma giganteum* and *Calonympha* spp. are available in the GeneBank (Table 5.3).

HYPERMASTIGIDA (EUKARYOTA: EXCAVATA). Hypermastigidia are mutualistic symbionts of insects. They live in the gut of termites, woodroaches and cockroaches (Undeen and Vávra, 1997). In addition to SSU rRNA genes from *Trichonympha agilis* and *Trichonymfa magna*, *T. agilis*, alpha-tubulin, beta-tubulin, elongation factor 1 alpha, enolase and glyceraldehyde-3-phosphate dehydrogenase (GAPDH) genes have been sequenced and are available in GeneBank (Table 5.3).

MICROSPORIDIA (EUKARYOTA: UNIKONTA: OPISTHOKONTA: FUNGI). Microsporidia are single-celled intracellular parasites that had long time been considered to be primitive amitochondrial protists. However, molecular phylogenies have suggested that microsporidia are just curious fungi and that they still possess a remnant of mitochondria (Williams *et al.*, 2002). Many microsporidia infect insects, and they represent the most extensively studied group of unicellular eukaryotic parasites of Metazoa. The complete genome of the microsporidian human pathogen *Encephalitozoon cuniculi* (2 MB; Project ID: 9545; Katinka *et al.*, 2001) has been

**Table 5.2.** Selected entomogenous kinetoplastids and sequences usable for their molecular identifications (part 1).

| Species (host) | Gene | Accession |
|---|---|---|
| *Blastocrithidia culicis* | SSU rRNA gene | L29266 |
| | 5S rRNA gene | AY547472 |
| *Blastocrithidia gerricola* | SSU rRNA gene | AF153036 |
| | Glycosomal glyceraldehyde-3-phosphate dehydrogenase (GAPDH) gene | AF322391 |
| | Spliced leader RNA gene (SL) | AY547495 |
| *Blastocrithidia triatoma* | SSU rRNA gene | AF153037 |
| | Spliced leader RNA gene | AY547467 |
| | 5S rRNA gene | AY547471 |
| *Crithidia fasciculate* | SSU rRNA gene | Y00055 |
| | 5S rRNA gene | M28975 |
| | Spliced leader RNA gene (SL) | J03470 |
| | Glycosomal glyceraldehyde-3-phosphate dehydrogenase (GAPDH) gene | AF047493 |
| *Crithidia oncopelti* | SSU rRNA gene | AF038025 |
| | 5S rRNA gene | AY547475 |
| | Spliced leader RNA gene (SL) | U96172 |
| *Endotrypanum monterogeii* | SSU rRNA gene | X5391 |
| | Spliced leader RNA gene (SL) | AB092602 |
| *Herpetomonas mariadeanei* | SSU rRNA gene | U01013 |
| | 5S rRNA gene | AY547489 |
| | Spliced leader RNA gene (SL) | AY547468 |
| *Herpetomonas megaseliae* | SSU rRNA gene | U01014 |
| | 5S rRNA gene | AY547485 |
| *Herpetomonas muscarum* | SSU rRNA gene | U01015 |
| | 5S rRNA gene | AY547484 |
| *Herpetomonas samuelpessoai* | SSU rRNA gene | L38605 |
| | 5S rRNA gene | X62331 |
| | Spliced leader RNA gene (SL) | X62331 |
| | Glycosomal glyceraldehyde-3-phosphate dehydrogenase (GAPDH) gene | AF047494 |
| *Herpetomonas roitmani* | SSU rRNA gene | AF038023 |
| *Herpetomonas ztiplika* | SSU rRNA gene | AF416560 |
| | 5S rRNA gene | AY308759 |
| *Leptomonas costaricensis* | Polymerase II LSU gene | DQ383651 |
| | SSU rRNA gene | DQ383648 |
| *Leptomonas podlipaevi* | SSU rRNA gene | DQ383649 |
| | Small spliced leader (mini-exon) RNA gene | |
| *Leptomonas* sp. | Glycosomal glyceraldehyde-3-phosphate dehydrogenase (GAPDH) gene | AF375664 |
| *Leptomonas peterhoffi* | Glycosomal glyceraldehyde-3-phosphate dehydrogenase (GAPDH) gene | AF322390 |
| *Leptomonas pyrrhocoris* | Glycosomal glyceraldehyde-3-phosphate dehydrogenase (GAPDH) gene | AY029072 |
| *Leptomonas ctenocephali* | Partial vppa gene for putative proton-translocating inorganic pyrophosphatase | AJ251218 |

*Continued*

**Table 5.2.** Continued

| Species (host) | Gene | Accession |
|---|---|---|
| *Leptomonas* sp. F6 | SSU rRNA gene | AF153042 |
| | 5S rRNA gene | AY547478 |
| | Spliced leader RNA gene (SL) | AY547461 |
| | Geranyl-geranyl diphosphate synthase-like gene, partial cds | AY029073 |
| *Leptomonas seymouri* | Maxicircle divergent region 12S rRNA-proximal repeat supercluster (kinetoplast) | DQ239761 |
| | U5 snRNA gene | AJ243569 |
| | p57 protein gene, complete cds | AY160128 |
| | p46 protein gene, complete cds | AY160127 |
| | Kinetoplast 12S rRNA gene, 9S rRNA gene, ITS 1 and ITS 2 | AJ511865 |
| | Mini-exon with insertion element LINS1 | X07488 |
| | RNA polymerase II largest subunit (RNAPII) gene, complete cds | AF338253 |
| | Kinetoplast partial 9S rRNA gene, IGS and NADH8 gene | |
| | tRNA-Gly gene, partial sequence; and tRNA-Pro gene, complete sequence; and small nuclear RNA U4 gene, complete sequence | AF204671 |
| | Glycosomal glyceraldehyde-3-phosphate dehydrogenase (GAPDH) gene, complete cds | AF047495 |
| | Alpha-tubulin mRNA 5′ end | X14005 |
| | SLA gene | X82217 |
| | Mini-exon donor RNA (medRNA) | X07487 |
| *Leptomonas collosoma* | Maxicircle divergent region containing long and short repeats; kinetoplast | AH015822 |
| | C/D box and H/ACA box small nucleolar RNA clusters, complete sequence | AY046598 |
| | C/D box small nucleolar RNA cluster, complete sequence | AF331656 |
| | Nucleolar RNA snoRNA-2, complete sequence | AF050095 |
| | Polymerase delta catalytic chain (dnap) gene, partial cds | AF351198 |
| | SmE protein gene, complete cds | AF126283 |
| | tRNA-Arg and 7SL RNA genes, complete sequences | AF006750 |
| | tRNA-Cys and U5 small nuclear RNA genes, complete sequences | AF006632 |
| | U2I RNA gene | X56453 |
| | U2III RNA gene | X56455 |
| | U2II RNA gene | X56454 |
| | Transfer RNA-Val and U3 snRNA genes | L32919 |
| | Trans-spliceosomal U2 snRNA sequence | U23406 |
| | Small spliced leader (mini-exon) RNA gene from reiteration unit | K02633 |
| *Phytomonas serpens* | SSU rRNA gene | AF016323 |
| | Spliced leader RNA gene (SL) | X87137 |
| *Trypanosomatid* G755 | SSU rRNA gene | U59491 |
| *Trypanosomatid* EVA | SSU rRNA gene | AF071866 |
| *Trypanosoma* sp. *CUL1* | SSU rRNA gene | AF416561 |

**Table 5.3.** Selected entomogenous Oxymonadida, Hypermastigida and Trichomonadida and sequences usable for their molecular identifications.

| Species | Gene | Accession |
|---|---|---|
| *Oxymonadida* | | |
| *Oxymonas* sp. JF2002 | Partial 18S rRNA gene | AJ429101 |
| *Hypermastigida* | | |
| *Trichonymfa agilis* | Beta-tubulin | AB107789 |
| | Enolase | AB107787 |
| | Glyceraldehyde-3-phosphate dehydrogenase | AB107786 |
| | Alpha tubulin | AF230348 |
| | Elongation factor 1 alpha (EF-1a) | AF230353 |
| *Trichonympha* sp. Hs10 | SSU rRNA, partial sequence | AB032230 |
| *Trichonympha* sp. Hs8 | SSU rRNA, partial sequence | AB032229 |
| *Trichonympha* sp. Hs3 | SSU rRNA, partial sequence | AB032227 |
| *Trichonymfa magna* | SSU rRNA, partial sequence | AF052714 |
| *Trichomonadida* | | |
| *Pseudotrypanosoma giganteum* | SSU rRNA gene, partial sequence | AF052707 |
| | | AF052706 |
| | | AF052705 |
| | | AF052704 |
| | | AF052703 |
| *Calonympha* sp. B14 | 16S-like SSU rRNA gene | X97976 |
| *Calonympha grassii* | SSU rRNA gene, partial sequence | AY063296 |
| | | AY063295 |
| | | AY063294 |

sequenced; sequencing of *Antonospora* (also known as *Nosema*) *locustae* genome is in progress (estimate of 2.9 MB; Project ID: 12823).

Numerous approaches have been used to diagnose microsporidia within their hosts (Klee *et al.*, 2006). Of them PCR amplification of rRNA genes (rDNA) or their fragments has been the most sensitive and specific method tested. Selected insect pathogenic microsporidian sequences available are shown in Table 5.4. Klee *et al.* (2006) recently published diagnostic PCR primers that can detect *Nosema bombi* in infected honeybees with more accuracy than light microscopy investigations. The marker is based on internal transcribed spacer (ITS) sequences. Primers specific for *N. bombi* amplify region 118–122 bp comprising ITS, 3′ end of SSU rRNA and 5′ of LSU rRNA genes (Klee *et al.*, 2006).

HAPLOSPORIDIA (EUKARYOTA: FUNGI: ZYGOMYCOTA: UNCLASSIFIED ZYGOMYCOTA). One partial (1807 bp) sequence of the SSU rRNA from *Nephridiophaga blattellae* is available in the GeneBank.

HELICOSPORIDIUM SPP. (EKARYOTA: PLANTAE: CHLOROPHYTA). This curious insect-parasitic green alga has been classified as either protozoa or fungi. It has been the

**Table 5.4.** Selected entomogenous microsporidians and sequences considered in molecular diagnostics (part1: Nosematidae).

| Species | Gene | Accession |
|---|---|---|
| *Nosematidae* | | |
| *Nosema apis* | 16S SSU rRNA gene | DQ235446 |
| | 16S rRNA gene, partial sequence; 5.8S rRNA gene, complete sequence; and 23S rRNA gene, partial sequence | U76706 |
| | SSU rRNA gene, ITS and LSU rRNA gene, complete sequence | U97150 |
| *Nosema bombii* | (62 GeneBankrecords) | |
| | 16S rRNA gene, partial sequence; ITS, complete sequence; and 23S rRNA gene, partial sequence | DQ472179 |
| | SSU rRNA gene, partial sequence; ITS, complete sequence; and LSU rRNA gene, partial sequence | AY741120 |
| | 16S rRNA gene, complete sequence | AY008373 |
| | 16S SSU rRNA gene, V4 variable region | U26158 |
| *Nosema bombycis* | (55 nt and 11 protein records) e.g. | |
| | Retrotransposon Nbr1 complete sequence | DQ444465 |
| | Alpha-tubulin gene, partial cds | DQ091252 |
| | Gene for fragmented SSU rRNA, partial sequence, 1.3 kb amplicon | AB125666 |
| | SSU rRNA, IGS, 5S rRNA | AB125664 |
| | Transposon-like element, IGS, 3′ end of LSU rRNA, ITS, fragmented SSU rRNA, inserted sequence, fragmented SSU rRNA, IGS, 5S rRNA | AB097401 |
| | Gene for putative spore surface protein, partial cds | AB107590 |
| | Elongation factor 1 alpha, partial cds | AB009600 |
| | Pseudo rRNA gene, partial sequence | D14632 |
| | DNA repair protein mRNA, complete cds | AY037305 |
| | Surface-antigen protein P30.4 (sap30.4) mRNA, | AF245278 |
| | partial cds | L28962 |
| | rRNA LSU | |
| *Nosema chrysorrhoeae* | SSU rRNA gene, partial sequence; ITS 1 and 5.8S rRNA gene, complete sequence; and LSU rRNA gene, partial sequence | AY940657 |
| | SSU rRNA gene, partial sequence | AY940656 |
| *Nosema spodopterae* | Alpha-tubulin gene, partial cds | DQ091251 |
| | SSU rRNA gene, complete sequence | AY211392 |
| | ITS, partial sequence | AY211391 |
| | LSU rRNA gene, complete sequence | AY211390 |
| | LSU rRNA gene, ITS, SSU rRNA gene, intergenic spacer, and 5S rRNA gene, complete sequence | AY747307 |
| *Gurleyidae* | | |
| *Episeptum* | SSU rRNA gene, partial sequence | AY880954 |
| *Hazardia milleri* | SSU rRNA gene, partial sequence | AY090067 |
| *Hazardia* sp. | SSU rRNA gene, partial sequence | AY090066 |
| Culicosporidae | | |
| *Pyrotheca* sp. | SSU rRNA gene, partial sequence | AY880955 |

*Continued*

**Table 5.4.** Continued

| Species | Gene | Accession |
|---|---|---|
| *Endoreticulatus* | | |
| *Endoreticulatus bombycis* | SSU rRNA gene, partial sequence | AY009115 |
| *Endoreticulatus schubergi* | SSU rRNA gene | L39109 |
| *Endoreticulatus* sp. | LSU rRNA gene, complete sequence | AY960111 |
| | Alpha-tubulin gene, partial cds | AY960112 |
| | SSU rRNA gene, complete sequence | AY502944 |
| *Burenellidae* | | |
| *Vairimorpha necatrix* | SSU rRNA gene | Y00266 |
| | rRNA LSU | L28975 |
| | Mitochondrial Hsp70 homologue mRNA, complete cds | AF008215 |
| | Largest subunit of RNA polymerase II (RPB1) gene, complete cds | AF060234 |
| | Actin gene, partial cds | AF031818 |
| *Vairimorpha lymantriae* | SSU rRNA gene sequence | AF033315 |
| | rRNA LSU | L28974 |
| *Vairimorpha imperfecta* | 16S rRNA gene | AJ131645 |
| *Vairimorpha ephistiae* | rRNA LSU | L28972 |
| *Vairimorpha heterosporum* | rRNA LSU | L28973 |
| *Caudosporidae* | | |
| *Caudospora simulii* | SSU rRNA gene, partial sequence | AY973624 |
| *Culicospora magna* | 16S rRNA gene, partial sequence | AY326269 |
| *Culicosporella lunata* | SSU rRNA gene, complete sequence | AF027683 |
| *Weiseria palustris* | 16S SSU rRNA gene, complete sequence | AF132544 |
| *Antonospora* | | |
| *Antonospora locustae* | 2562 nt and 167 protein sequences available | Project ID: 12823 |
| *Amblyospora* | | |
| *Amblyospora salinaria* | 16S rRNA gene, partial sequence | AY326270 |
| *Amblyospora bracteata* | 16S rRNA gene, partial sequence | AY090068 |
| *Amblyospora ferocious* | 16S rRNA gene, partial sequence | AY090062 |
| *Amblyospora crenifera* | 16S rRNA gene, partial sequence | AY090061 |
| *Amblyospora cinerei* | 16S rRNA gene, partial sequence | AY090060 |
| *Amblyospora canadensis* | 16S rRNA gene, partial sequence | AY090056 |
| *Amblyospora opacita* | 16S rRNA gene, partial sequence | AY090052 |
| *Amblyospora indicola* | 16S rRNA gene, partial sequence | AY090051 |
| *Amblyospora stimuli* | 16S rRNA gene, partial sequence | AY090050 |
| *Amblyospora stictici* | 16S rRNA gene, partial sequence | AY090049 |
| *Amblyospora weiseri* | 16S rRNA gene, partial sequence | AY090048 |
| *Amblyospora khaliulini* | 16S rRNA gene, partial sequence | AY090047 |
| *Amblyospora excrucii* | 16S rRNA gene, partial sequence | AY090044 |
| *Amblyospora californica* | 16S rRNA gene, partial sequence | U68473 |
| *Amblyospora connecticus* | 16S rRNA gene, partial sequence | AF025685 |

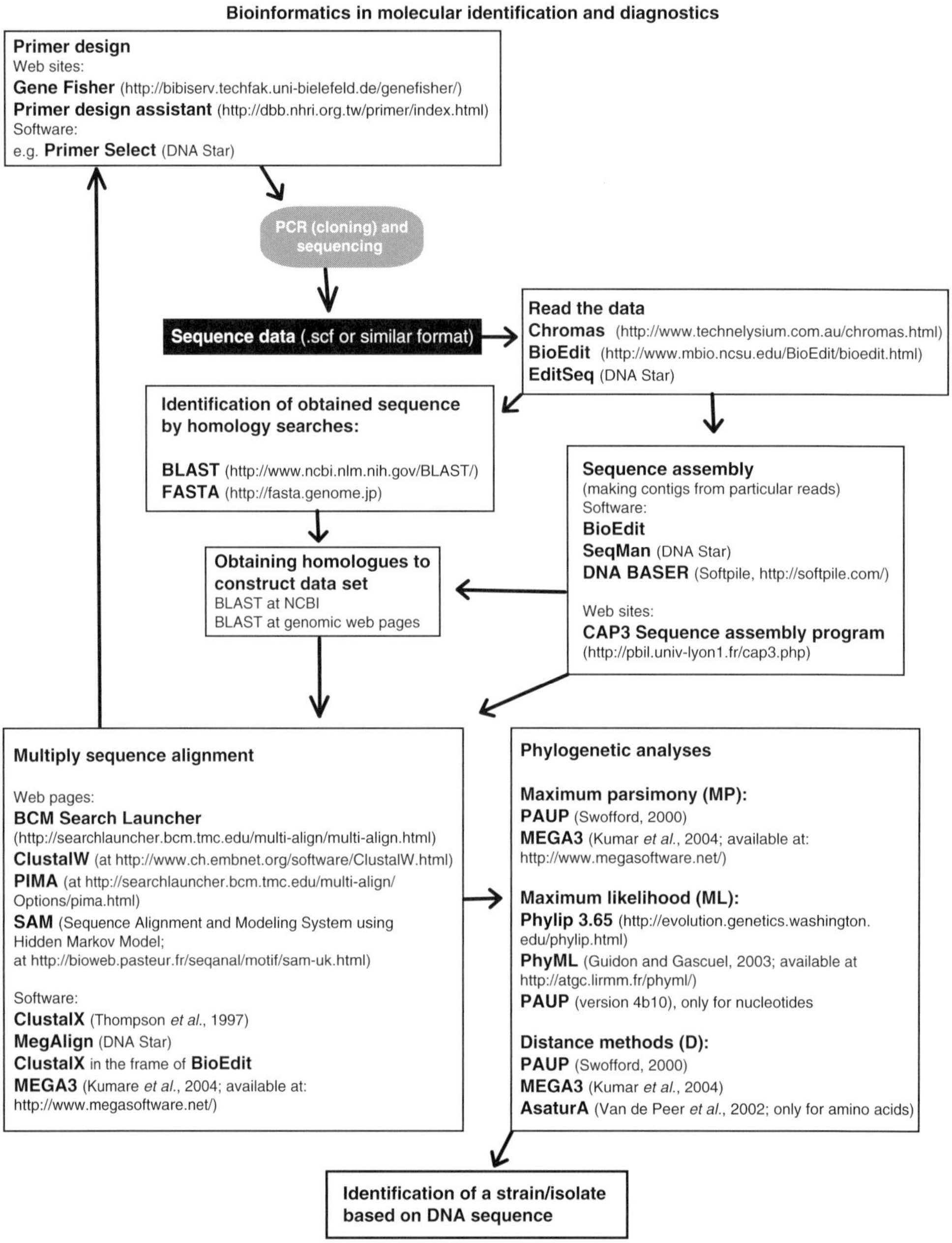


**Fig. 5.7.** Bioinformatic tools in molecular identification and diagnostics. Most bioinformatic tools are used to read and edit sequences, to search their homologues in databases, to create multiple alignments of sequences and to perform phylogenetic analysis. They can be found as portals on the internet or downloaded as freeware or shareware. There are commercial software packages that can also be used to analyse data. Although they usually work in more user-friendly environments, they rarely offer better computational methods.

subject of extensive research over the last years. Currently, numerous expressed sequence tags (EST) libraries have been sequenced and analysed. These sequence data have unequivocally confirmed the origin of this parasite within green algae (de Koning *et al.*, 2005). Additionally, complete sequences of the chloroplast genome from *Helicosporidium* sp. ex *Simulium jonesii* are available in Genbank (de Koning and Keeling, 2006).

### 5.2.6. Molecular diagnostics and bioinformatics

Bioinformatic tools are used to identify sequences, to search for their available counterparts, to construct multiply alignments, to design primers and to perform various *in silico* analyses such as search for restriction sites, phylogenetic analyses, or prediction of secondary and higher structures and targeting presequences. Figure 5.7 summarizes basic bioinformatic tools and procedures that can be used in molecular identification of entomopathogenic protozoa.

## 5.3. Conclusion and Future Perspectives

Entomogenous protist can be well identified using molecular diagnostic methods. However, the range of methods that have been already used for this purpose is limited. To apply methods of molecular diagnostics of protists to those inhabiting insects is thus highly recommended. Various methods can be applied, from the sequencing of the target DNA region, through fingerprinting methods, to the design of taxon-specific primers for PCR diagnostics. Since more and more organisms are being completely sequenced, they can serve as models to search for particular molecular markers. Thus, a wide use of molecular diagnostics of protists colonizing insects can be expected in a short time perspective.

## References

Adl, S.M., Simpson, A.G.B., Farmer, M.A., Andersen, R.A., Anderson, O.R., Barta, J.R., Bowser, S.S., Brugerolle, G., Fensome, R.A., Fredericq, S., James, T.Y., Karpov, S., Kugrens, P., Krug, J., Lane, C.E., Lewis, L.A., Lodge, J., Lynn, D.H., Mann, D.G., McCourt, R.M., Mendoza, L., Moestrup, O., Mozley-Standridge, S.E., Nerad, T.A., Shearer, C.A., Smirnov, A.V., Spiegel, F.W. and Taylor, M.F.J.R. (2005) The new higher level classification of eukaryotes with emphasis on the taxonomy of protists. *Journal of Eukaryotic Microbiology* 52(5), 399–451.

Adl, S.M., Leander, B.S., Simpson, A.G.B., Archibald, J.M., Anderson, O.R., Bass, D., Bowser, S.S., Brugerolle, G., Farmer, M.A., Karpov, S., Kolisko, M., Lane, C.E., Lodge, D.J., Mann, D.G., Meisterfeld, R., Mendoza, L., Moestrup, O., Mozley-Standridge, S.E., Smirnov, A.V. and Spiegel, F. (2007) Diversity, nomenclature, and taxonomy of protists. *Systematic Biology* 56, 684–689.

Aury, J.M., Jaillon, O., Duret, L., Noel, B., Jubin, C., Porcel, B.M., Segurens, B., Daubin, V., Anthouard, V., Aiach, N., Arnaiz, O., Billaut, A., Beisson, J., Blanc, I., Bouhouche, K., Camara, F., Duharcourt, S., Guigo, R., Gogendeau, D., Katinka, M., Keller, A.M., Kissmehl, R., Klotz, C., Koll, F., Le Mouel, A., Lepere, G., Malinsky, S., Nowacki, M., Nowak, J.K., Plattner, H., Poulain, J., Ruiz, F., Serrano, V., Zagulski, M., Dessen, P., Betermier, M., Weissenbach, J., Scarpelli, C., Schachter, V., Sperling, L.,

Meyer, E., Cohen, J. and Wincker, P. (2006) Global trends of whole-genome duplications revealed by the ciliate *Paramecium tetraurelia*. *Nature* 444(7116), 171–178.

Backeljau, T., De Bruyn, L., De Wolf, H., Jordaens, K., Van Fongen, S., Verhagen, R. and Winnepernnickx, B. (1995) Random amplified polymorphic DNA (RAPD) and parsimony methods. *Cladistics* 11, 119–130.

Burger, G., Zhu, Y., Littlejohn, T.G., Greenwood, S.J., Schnare, M.N., Lang, B.F. and Gray, M.W. (2000) Complete sequence of the mitochondrial genome of *Tetrahymena pyriformis* and comparison with *Paramecium aurelia* mitochondrial DNA. *Journal of Molecular Biology* 297, 365–380.

Carlson, J.E., Tulsieram, L.K., Glaubitz, J.C., Luk, V.W.K., Kauffeldt, C. and Rutledge, R. (1991) Segregation of random amplified polymorphic DNA markers in F1 progeny of conifers. *Theoretical and Applied Genetics* 83, 194–200.

Carreno, R.A., Pokorny, N.J., Lee, H., Trevors, J.J. and De Grandis, D.E. (2001) Phenotypic and genotypic characterization of *Cryptosporidium* species and isolates. *Journal of Industrial Microbiology and Biotechnology* 26, 95–106.

Čepička, I., Kutišová, K., Tachezy, J., Kulda, J. and Flégr, J. (2005) Cryptic species within the *Tetratrychomonas gallinatum* species complex revealed by molecular polymorphism. *Veterinary Parasitology* 128, 11–21.

de Koning, A.P. and Keeling, P.J. (2006) The complete plastid genome sequence of the parasitic green alga, *Helicosporidium* sp. is highly reduced and structured. *BMC Biology* 4, 12.

de Koning, A.P., Tartar, A., Boucias, D.G. and Keeling, P.J. (2005) Expressed sequence tag (EST) survey of the highly adapted green algal parasite, *Helicosporidium*. *Protist* 156, 181–190.

Egerter, D.E., Anderson, J.R. and Washburn, J.O. (1986) Dispersal of the parasitic ciliate *Lambornella clarki*: implications for ciliates in the biological control of mosquitoes. *Proceedings of National Academy of Sciences of the USA*, 83, 7335–7339.

Eisen, J.A., Coyne, R.S., Wu, M., Wu, D., Thiagarajan, M., Wortman, J.R., Badger, J.H., Ren, Q., Amedeo, P., Jones, K.M., Tallon, L.J., Delcher, A.L., Salzberg, S.L., Silva, J.C., Haas, B.J., Majoros, W.H., Farzad, M., Carlton, J.M., Smith, R.K., Garg, J., Pearlman, R.E., Karrer, K.M., Sun, L., Manning, G., Elde, N.C., Turkewitz, A.P., Asai, D.J., Wilkes, D.E., Wang, Y., Cai, H., Collins, K., Stewart, B.A., Lee, S.R., Wilamowska, K., Weinberg, Z., Ruzzo, W.L., Wloga, D., Gaertig, J., Frankel, J., Tsao, C.C., Gorovsky, M.A., Keeling, P.J., Waller, R.F., Patron, N.J., Cherry, J.M., Stover, N.A., Krieger, C.J., Del Toro, C., Ryder, H.F., Williamson, S.C., Barbeau, R.A., Hamilton, E.P. and Orias, E. (2006) Macronuclear genome sequence of the ciliate *Tetrahymena thermophila*, a model eukaryote. *PLoS Biology* 4, E286.

El-Sayed, N.M., Myler, P.J., Blandin, G., Berrima, M., Crabtree, J., Aggarwal, G., Caler, E., Renauld, H., Worthey, E.A., Hertz-Fowler, C., Ghedin, E., Peacock, C., Bartholomeu, D.C., Haas, B.J., Tran, A.N., Wortman, J.R., Alsmark, U.C.M., Angiuoli, S., Anupama, A., Badge,r J., Bringaud, F., Cadag, E., Carlton, J.M., Cerqueira, G.C., Creasy, T., Delcher, A.L., Djikeng, A., Embley, T.M., Hauser, C., Ivens, A.C., Kummerfeld, S.K., Pereira-Leal, J.B., Nilsson, D., Peterson, J., Salzberg, S.L., Shallom, J., Silva, J.C., Sundaram, J., Westenberger, S., White, O., Metville, S.E., Donelson, J.E., Andersson, B., Stuart, K.D. and Hall, N. (2005) Comparative genomics of trypanosomatid parasitic protozoa. *Science* 309(5733), 404–409.

Elsheikha, H.M., Schott, H.C. and Mansfield, L.S. (2006) Genetic variation among isolates of *Sarcocystis neurona*, the agent of protozoal myeloencephalitis, as revealed by amplified fragment length polymorphism markers. *Infection and Immunity* 74, 3448–3454.

Gadisa, E., Genetu, A., Kuru, T., Jirata, D., Dagne, K., Aseffa, A. and Gedamu, L. (2007) Leishmania (Kinetoplastida): species typing with isoenzyme and PCR-RFLP from cutaneous leishmaniasis patients in Ethiopia. *Experimental Parasitology* 115, 339–343.

Grisard, E.C., Campbell, D.A. and Romanha, A.J. (1999) Mini-exon gene sequence polymorphism among *Trypanosoma rangeli* strains isolated from distinct geographical regions. *Parasitology* 118, 375–382.

Guindon, S. and Gascuel, O. (2003) A simple, fast, and accurate algorithm to estimate large phylogenies by maximum likelihood *Systematic Biology* 52, 696–704.

Hopkins, R.M., Constantine, C.C., Groth, D.A., Wetherall, J.D., Reynoldson, J.A. and Thompson, R.C.A. (1996) PCR-based DNA fingerprinting of Giardia duodenalis isolates using the intergenic rDNA spacer. *Parasitology* 118, 531–539.

Hyliš, M., Weiser, J., Oborník, M. and Vávra, J. (2005) DNA isolation from museum and type collection slides of microsporidia. *Journal of Invertebrate Pathology* 88, 257–260.

Jerome, C.A., Simon, E.M. and Lynn, D.H. (1996) Description of *Tetrahymena empidokyrea* n sp, a new species in the *Tetrahymena pyriformis* sibling species complex (Ciliophora, Oligohymenophorea), and an assessment of its phylogenetic position using small-subunit rRNA sequences. *Canadian Journal of Zoology* 74, 1798–1906.

Jirků, M., Kolesnikov, A.A., Benada, O. and Lukeš, J. (1995) Marine fish and raytrypanosomes have large kinetoplast minicircle DNA. *Molecular and Biochemical Parasitology* 73, 279–283.

Jirků, M., Zemanová, E., Al-Jawabreh, A., Schönian, G. and Lukeš, J. (2006) Development of a direct species-specific PCR assay for differential diagnostics of *Leishmania tropica. Diagnostic Microbiology and Infectious Disease* 55, 75–79.

Katinka, M.D., Duprat, S., Comillot, E., Metenier, G., Thomarat, F., Prensier, G., Barbe, V., Peyretaillade, E., Brottier, P., Wincker, P., Delbac, F., El Alaoui, H., Peyret, P., Saurin, W., Gouy, M., Weissenbach, J. and Vivares, C.P. (2001) Genome sequence and gene compaction of the eukaryote parasite *Encephalitozoon cuniculi. Nature* 414(6862), 450–453.

Kazmer, D.J., Hopper, K.R., Coutinot, D.M. and Heckel, D.G. (1995) Suitability of random amplified polymorphic DNA for genetic markers in the aphid parasitoid, *Aphelinus asychis* Walker. *Biological Control* 5, 503–512.

Keeling, P.J., Burger, G., Durnford, D.G., Lang, B.F., Lee, R.W., Pearlman, R.E., Roger, A.J. and Gray, M.W. (2005) The tree of eukaryotes. *Trends in Ecology and Evolution* 20, 670–676.

Kirby, K.S. (1957) The new method for the isolation of deoxyribonuclic acids evidence on the nature of bonds between deoxyribo nucleic acid and protein. *Biochemical Journal* 66, 495–504.

Klee, J., Tay, W.T and Paxton, R.J. (2006) Specific and sensitive detection of *Nosema bombi* (Microsporidia: Nosematidae) in bumble bees (*Bombus* spp.; Hymenoptera: Apidae) by PCR of partial rRNA gene sequences. *Journal of Invertebrate Pathology* 91, 98–104.

Kopečná, J., Jirků, M., Oborník, M., Tokarev, Y.S., Lukeš, J. and Modrý, D. (2006) Phylogenetic analysis of coccidian parasites from invertebrates: search for missing links. *Protist* 157, 173–183.

Kumar, S., Tamura, K. and Nei, M. (2004) MEGA3: integrated software for molecular evolutionary genetic analysis and sequence alegment. *Briefings in Bioinformatics* 5, 150–160.

Leadon, S. and Cerutti, P. (1982) A rapid and mild procedure for isolation of DNA from mammalian cells. *Analytical Biochemistry* 120, 282–288.

Leander, B.S. (2007) Molecular phylogeny and ultrastructure of *Selenidium serpulae* (Apicomplexa, Archigregarinia) from the calcareous tubeworm *Serpula vermicularis* (Annelida, Polychaeta, Sabellida). *Zoologica Scripta* 36, 213–227.

Leander, B.S. and Keeling, P.J. (2004) Early evolutionary history of dinoflagellates and apicomplexans (Alveolata) as inferred from hsp90 and actin phylogenies. *Journal of Phycology* 40, 341–350.

Leander, B.S., Harper, J.T. and Keeling, P.J. (2003) Molecular phylogeny and surface morphology of marine aseptate gregarines (Apicomplexa): *Selenidium* spp. and *Lecudina* spp. *Journal of Parasitology* 89(6), 1191–1205.

Leander, B.S., Lloyd, S.A.J., Marshall, W. and Landers, S.C. (2006) Phylogeny of marine gregarines (Apicomplexa) – *Pterospora, Lithopystis* and *Lankesteria* – and the origin(s) of coelomic parasitism. *Protist* 157, 45–60.

Leng, X.G. and Mosier, D.A. (1996) Oberst RD differentiation of Cryptosporidium par vum, C-mirus, and C-baileyi by PCR-RFLP analysis of the 18s rRNA gene. *Veterinary Parasitology* 62, 1–7.

Longmire, J., Albright, K.L., Lewis, A.K., Meincke, L.J. and Hildebrand, C.E. (1987) A rapid and simple method for the isolation of high molecular weight cellular and chromosome-specific DNA in solution without the use of organic solvents. *Nucleic Acids Research* 15, 859.

Loudon, K.W., Coke, A.P. and Burnie, J.P. (1995) Pseudoclusters and typing by random amplified polymorphic DNA of *Aspergillus fumigatus. Journal of Clinical Pathology* 48, 183–184.

Lowe, A.L., Harris, S. and Ashton, P. (2004) *Ecological Genetics: Design, Analysis, and Application.* Blackwell Publishing, Oxford, 326.

Lukeš, J., Hashimi, H. and Zíková, A. (2005) Unexplained complexity of the mitochondrial genome and transcriptome in kinetoplastid flagellates. *Current Genetics* 48, 277–299.

Lukeš, J., Mauricio, I.L., Schonian, G., Dujardin, J.C., Soteriadou, K., Dedet, J.P., Kuhl, K., Tintaya, K.W.Q., Jirků, M., Chocholová, E., Haralambous, C., Pratlong, F., Oborník, M., Horák, A., Ayala, F.J. and Miles, M.A. (2007) Evolutionary and geographical history of the Leishmania donovani complex with a revision of current taxonomy. *Proceedings of the National Academy of Sciences of the USA* 104, 9375–9380.

Lun, Z.R. and Desser, S.S. (1996) Comparisons of molecular karyotype and RAPD patterns of anuran trypanosome isolates during long-term *in vitro* cultivation. *Folia Parasitologica* 43, 241–248.

Marfurt, J., Niederwieser, I., Makia, N.D., Beck, H.P. and Felger, I. (2003) Diagnostic genotyping of old and new world Leishmania species by PCR-RFLP. *Diagnostic Microbiology and Infectious Disease* 46, 115–124.

McDonald, B.A. and Martinez, J.P. (1990) DNA restriction-fragment-length-polymorphisms among *Mycosphaerella Graminicola* (anamorph Septoria tritici) isolates collected from a single wheat field. *Phytopathology* 80, 1368–1373.

Moore, R.B., Oborník, M., Janouškovec, J., Chrudimský, T., Vancová, M., Green, D.H., Wright, S.W., Davies, N.W., Bolch, C.J.S., Heimann, K., Šlapeta, J., Hoegh-Guldberg, O., Logsdon, J.M. and Carter, D.A.A. (2008) A photosynthetic alveolate closely related to apicomplexan parasites. *Nature* 451, 959–963.

Mueller, U.G. and Wolfenbarger, L. (1999) AFLP genotyping and fingerprinting. *Trends in Ecology and Evolution* 14, 389–394.

Newton, C. and Graham, A. (2000) *PCR*, 2nd edn. The introduction to biotechniques series, Bios Scientific Publishers, Oxford, 192.

Nocker, A., Burr, M. and Camper, A.K. (2007) Genotypic microbial community profiling: a critical technical review. *Microbial Ecology* 54, 276–289.

Oborník, M., Jirků, M., Šlapeta, J.R., Modrý, D., Koudela, B. and Lukeš, J. (2002) Notes on coccidian phylogeny as inferred from partial apicoplast SSU rRNA gene sequences. *Parasitology Research* 88, 360–363.

Pavlíček, A., Hrdá, S. and Flégr, J. (1999) FreeTree-freeware program for construction of phylogenetic trees on the basis of distance data and bootstrap jackknife analysis of the tree robustness. Application in the RAPD analysis of genus *Frenkelia. Folia Biologica* 45, 97–99.

Paxton, R.J., Klee, J., Korpela, S. and Fries, I. (2007) *Nosema carenae* has infected *Apis mellifera* in Europe since at least 1998 and may be more virulent than *Nosema apis. Apidologie* 38, 558–565.

Pérez, T., Albornoz, J. and Domínguez, A. (1998) An evaluation of RAPD fragment reproducibility and nature. *Molecular Ecology* 7, 1347–1357.

Rao, S.N., Nath, B.S. and Saratchandra, B. (2005) Characterization and phylogenetic relationships among microsporidia infecting silkworm, *Bombyx mori*, using inter simple sequence repeat (ISSR) and small subunit rRNA (SSU-rRNA) sequence analysis. *Genome* 48, 355–366.

Rao, S.N., Nath, B.S., Bhuvaneshwari, G. and Urs, S.R. (2007) Genetic diversity and phylogenetic relationships among microsporidia infecting the silkworm, *Bombyx mori*, using random amplification of polymorphic DNA: morphological and ultrastructural characterization. *Journal of Invertebrate Pathology* 96, 193–204.

Reiseberg, L.H., Choi, H.C., Chan, R. and Spore, C. (1993) Genomic map of a diploid hybrid species. *Heredity* 70, 285–293.

Sertsrivanich, R. and Yuthavong, Y. (1985) Restriction nuclease fingerprinting. In: Panyim, S., Wilairat, P. and Yuthavong, Y. (eds) *Application of Genetic Engineering to Research on Tropical Disease Pathogens with Special Reference to Plasmodia.* A laboratory manual of selected techniques. UNDP/WORLD BANK/WHO Special Programme for Research and Training in Tropical Diseases and Mahidol University, Bangkog, Thailand, pp. 451–454.

Shields, J.M. and Olson, B.H. (2003) PCR-restriction fragment length polymorphism method for detection of *Cyclospora cayetanensis* in environmental waters without microscopic confirmation. *Applied and Environmental Microbiology* 69, 4662–4669.

Simpson, A.G.B. and Roger, A.J. (2004) The real kingdoms of eukaryotes. *Current Biology* 14, R693–R696.

Šlapeta, J.R., Koudela, B., Votýpka, J., Modrý, D., Hořejš, R. and Lukeš, J. (2002) Coprodiagnostics of *Hammondia heydorni* in dogs by PCR based amplification of ITS1 rRNA: differentiation from morphologically indistinguishable oocysts of *Neospora caninum. Veterinary Journal* 163, 147–154.

Stothard, J.R., Frame, I.A., Carrasco, H.J. and Miles, M.A. (1998) Temperature gradient gel electrophoresis (TGGE) analysis of riboprints from *Trypanosoma cruzi. Parasitology* 117, 249–253.

Subrungruang, I., Mungthin, M., Chavalitshewinkoon-Petmitr, P., Rangsin, R., Naaglor, T. and Leelayoova, S. (2004) Evaluation of DNA extraction and PCR methods for detection of *Enterocytozoon bienuesi* in stool specimens. *Journal of Clinical Microbiology* 42, 3490–3494.

Swofford, D.L. (2000) *PAUP* Phylogenetic Analysis Using Parsimony (*and Related Methods).* Sinauer Associates, Sunderland, Massachusetts.

Takishita, K., Tsuchiya, M., Kawato, M., Oguri, K., Kitazato, H. and Maruyama, T. (2007) Genetic diversity of microbial eukaryotes in anoxic sediment of the saline meromictic lake Namako-ike (Japan): on the detection of anaerobic or anoxic-tolerant lineages of eukaryotes. *Protist* 158, 51–64.

Tanriverdi, S. and Widmer, G. (2006) Differential evolution of repetitive sequences in *Cryptosporidium parvum* and *Cryptosporidium hominis. Infection and Evolution* 6, 113–122.

Thompson, J.D., Gibson, T.J., Plewniak, F., Jeanmougin, F. and Higgins, D.G. (1997) The CLUSTAL_X windows interface: flexible strategies for multiple sequence alignment aided by quality analysis tools. *Nucleic Acids Research* 25, 4876–4882.

Toso, M.A. and Omoto, C.K. (2007) *Gregarina niphandrodes* may lack both a plastid genome and organelle. *Journal of Eukaryotic Microbiology.* 54, 66–72.

Undeen, A.H. and Vávra, J. (1997) Research methods for entomopathogenic protozoa. In: Lacey, L. (ed.) *Manual of Techniques in Insect Pathology.* Academic Press, San Diego, California, pp. 118–151.

Vaňáčová, Š., Tachezy, J., Kulda, J. and Flégr, J. (1997) Characterization of trichomonas species and strains by PCR fingerprinting. *Journal of Eukaryotic Microbiology* 44, 545–552.

Van de Peer, Y., Van der Auwera, G. and De Wachter, R. (1996) The evolution of stramenopiles and alveolates as derived by substitution rate calibration of small ribosomal subunit RNA. *Journal of Molecular Evolution* 42, 201–210.

Van de Peer, Y., Frickey, T., Taylor, J.S. and Meyer, A. (2002) Dealing with saturation at the amino acid level: a case study based on anciently duplicated zebrafish genes. *Gene* 295, 205–211.

Vossbrinck, C.R. and Debrunner-Vossbrinck, B.A. (2005) Molecular phylogeny of the microsporidia: ecological, ultrastructural and taxonomic considerations. *Folia Parasitologica* 52, 131–142.

Votýpka, J., Oborník, M., Volf, P., Svobodová, M. and Lukeš, J. (2002) *Trypanosoma avium* of raptors (Falconiformes): phylogeny and identification of vectors. *Parasitology* 125, 253–263.

Walsh, P.S., Metzeger, D.A. and Higuchi, R. (1991) Chelex-100 as a medium for simple extraction of DNA for PCR-based typing from forensic material. *Biotechniques* 10(4), 506–513.

Williams, B.A.P., Hirt, R.P., Lucocq, J.M. and Embley, T.M. (2002) A mitochondrial remnant in the microsporidian *Trachipleistophora hominis. Nature* 418(6900), 865–869.

Williams, J.G.K., Kubelik, A.R., Livak, K.J., Rafalski, J.A. and Tingey, S.V. (1990) DNA polymorphism amplified by arbitrary primers are useful as genetic markers. *Nucleic Acids Research* 18, 6531–6535.

Yurchenko, V., Kolesnikov, A.A. and Lukeš, J. (2000) Phylogenetic analysis of Trypanosomatina (Protozoa: Kinetoplastida) based on minicircle conserved regions. *Folia Parasitologica* 47, 1–5.

Zemanová, E., Jirků, M., Mauricio, I.L., Miles, M.A. and Lukeš, J. (2004) Genetic polymorphism within the *Leishmania donovani* complex: correlation with geographical origin. *American Journal of Tropical Medicine and Hygiene* 70, 613–617.

Zhao, X.M. and Duszynski, D.W. (2001) Phylogenetic relationships among rodent Eimeria species determined by plastid ORF470 and nuclear 18S rDNA sequences. *International Journal for Parasitology* 31, 715–719.

Zhu, G., Marchewka, M.J. and Keithly, J.S. (2000) *Cryptosporidium parvum* appears to lack a plastid genome. *Microbiology* 146, 315–321.

# II Evolutionary Relationships and Population Genetics

# 6 Phylogenetic Studies with Entomopathogenic Bacteria with Special Emphasis on Symbionts of Entomopathogenic Nematodes

P. Tailliez and N. Boemare

*INRA, UMR1133 Laboratoire EMIP, F-34095 Montpellier, France; Université Montpellier II, UMR1133 Laboratoire EMIP, F-34095 Montpellier, France*



## 6.1. Introduction

Since the 1970s, ribosomal RNA (rRNA) molecules have been considered for studying bacterial molecular genealogies. These molecules are universally distributed, easily sequenced and carry generally useful phylogenetic information. Comparison of rRNA sequences enabled notably the discovery of the domain Archaea (Balch *et al.*, 1977; Woese *et al.*, 1978), a group of prokaryotes that is more closely related to Eukarya than to the other prokaryotes (Eubacteria) (Iwabe *et al.*, 1989). However, for some bacterial groups such as *Photorhabdus* and *Xenorhabdus*, the symbiotic entomopathogenic bacteria associated with *Heterorhabditis* and *Steinernema* nematodes, respectively, 16S rRNA gene sequences have shown to be highly conserved (similarity of more than 95%), therefore giving low levels of phylogenetic information. Particularly, the deepest nodes of 16S phylogenetic trees within these two genera are poorly resolved (Rainey *et al.*, 1995; Akhurst *et al.*, 2004; Tailliez *et al.*, 2006; see also Boemare and Tailliez, Chapter 2, this volume).

Other genes have been and are currently used as alternative options to study evolutionary origins of bacteria. For example, protein-coding genes such as *recA* which encodes a bacterial DNA recombination protein and *gyrB*, encoding DNA gyrase ß-subunit, are being evaluated to assess bacterial phylogenies and to overcome problems (i.e. poor resolution) linked to 16S rRNA gene sequences (Eisen, 1995; Brendel *et al.*, 1997; Yamada *et al.*, 1999; Akhurst *et al.*, 2004; Thompson *et al.*, 2004). These genes have not been used as widely as the 16S rDNA gene and consequently databases are yet scarce.

More recently, multilocus sequence analysis (MLSA) has been proposed to assess phylogenetic relationships among bacteria. This approach is based on the analysis of a concatenation of core metabolic (housekeeping) gene sequences and has been applied to assess evolutionary relationships of *Bacillus cereus* group (Priest *et al.*, 2004; Rasko *et al.*, 2005), *Pseudomonas* spp. (Hilario *et al.*, 2004) and entomopathogenic nematode symbionts, *Photorhabdus* and *Xenorhabdus* (P. Tailliez, 2008, France, personal communication). This approach also offers a better alternative for resolving taxonomic conflicts at the genus and species levels (Gevers *et al.*, 2005).

In the future, it is expected that with the development of rapid genome sequencing techniques such as pyrosequencing (Elahi and Ronaghi, 2004; Margulies *et al.*, 2005; Shendure *et al.*, 2005), more accurate bacterial phylogenies will be developed taking into account exhaustive sets of genomic sequences (Lerat *et al.*, 2003; Rokas *et al.*, 2003).

## 6.2. Genes Considered

### 6.2.1. 16S rRNA gene sequences

#### *6.2.1.1. Serratia species*

Sequences of 16S rRNA gene have been considered to assess evolutionary relationship of *Serratia* spp. (Dauga *et al.*, 1990; Spröer *et al.*, 1999; Ashelford *et al.*, 2002). They have also helped resolve taxonomic conflicts, particularly for identification of new isolates, previously classified in this genus using basic phenotypic characters (see Boemare and Tailliez, Chapter 2, this volume). Bacteria of the genus *Serratia* are often associated with insects of many orders (Grimont and Grimont, 1978). *Serratia plymuthica* has been isolated from the intestine of healthy crickets, *Neombius fasciatus* (Steinhaus, 1941), but no infection has been attributed to this species. *Serratia marcescens* and *Serratia liquefaciens* may also be associated with insects (Grimont *et al.*, 1979a); however, these two species have been traditionally regarded as facultative pathogens, lacking the ability to invade the haemocoel of healthy insects (Steinhaus, 1959). *Serratia ficaria* is a bacterial species found in fig tree ecosystems and also isolated from figs, fig wasps (*Blastophaga psenes)* and ants visiting figs (Grimont *et al.*, 1979b). Until now, no *S. ficaria* isolate has been considered as a true insect pathogen. Contrarily, *Serratia entomophila* and *Serratia proteamaculans* are true entomopathogenic bacteria responsible for causing amber disease in grass grubs, *Costelytra zealandica*, in New Zealand (Jackson *et al.*, 2001). The disease determinants are encoded on a 115 kb plasmid designated pADAP, for amber disease-associated plasmid (Glare *et al.*, 1993). Interpretation of evolutionary origins of *Serratia* spp. considering 16S rRNA sequences suggests that entomopathogenic *S. entomophila* isolates share a common ancestor with *S. ficaria* (Fig. 6.1).

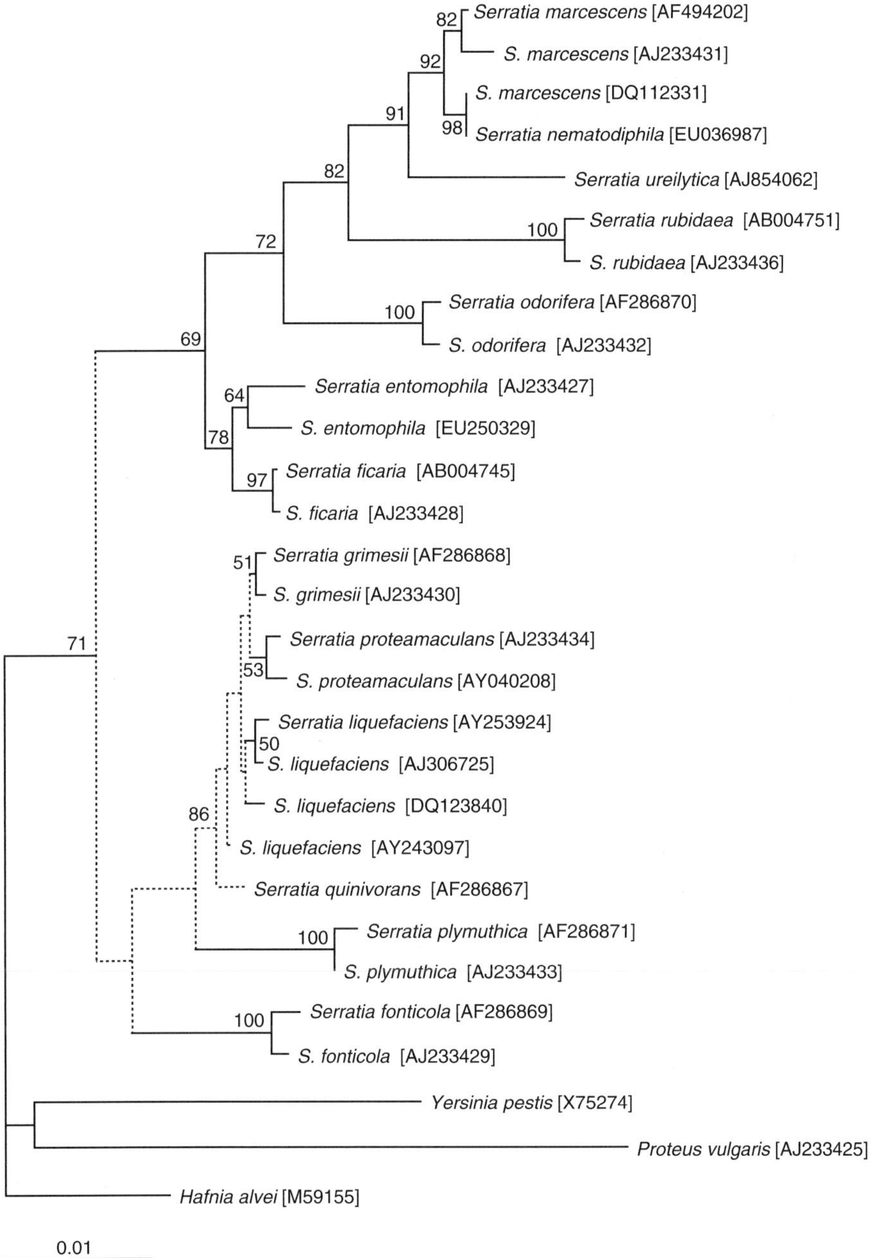


**Fig. 6.1.** Position of entomopathogenic *Serratia* in a *Serratia* phylogenetic distance tree. The distance tree is constructed using the 16S rRNA gene sequences (1318 nucleotides), the Kimura two-parameter model (1980) and the neighbour-joining (NJ) module (Saitou and Nei, 1987) of PAUP software (Swofford, 2003). The 16S rRNA gene sequences of *Hafnia alvei*, *Proteus vulgaris* and *Yersinia pestis* are used as outgroup. Bootstrap values (percentages of 1000 replicates) of more than 50% are shown at the nodes. Dashed lines indicate unreliable links between groups and unique sequences. Accession numbers in brackets correspond to 16S rRNA gene sequences retrieved from GenBank (http://www.ncbi.nlm.nih.gov/). The bar indicates 1% sequence divergence.

*S. marcescens* and *Serratia nematodiphila*, symbiotically associated with entomopathogenic nematode *Heterorhabditisoides chongmingensis* (C. Zhang and S. Yang, 2007, China, Unpublished), are grouped together (Fig. 6.1). Sequences of both taxa are similar to each other. In the same way, *Serratia grimesii*, *S. liquefaciens* and *S. proteamaculans* are three *Serratia* spp. called *S. liquefaciens*-like species (Grimont *et al.*, 1982); however, they cannot be distinguished from each other based on comparisons of their 16S rRNA gene sequences.

These results suggest that the common ancestor of all the *Serratia* spp. might have been a bacterial species which lacked virulence factors against insects and that acquisition of these genes likely happened over time and by lateral gene transfer (Hurst *et al.*, 2000).

#### *6.2.1.2. Pseudomonas species*

For over a century, the genus *Pseudomonas* Migula 1894 served as a repository for polarly flagellated strictly aerobic rods with a respiratory type of metabolism in which oxygen is used. The genus was thus very heterogeneous. Later, DNA–DNA and rRNA–DNA hybridizations split the genus *Pseudomonas* into five groups called rRNA groups I to V (Palleroni *et al.*, 1972, 1973). The genus *Pseudomonas sensu stricto* contains all species which corresponded to the rRNA group I (Palleroni, 1984).

Analysis of 16S rRNA gene sequences played an important role in resolving taxonomic conflicts in this genus. Many species originally considered in *Pseudomonas* were transferred to either existing or new genera (for a review, see Kersters *et al.*, 1996). Currently, the genus *Pseudomonas sensu stricto* includes more than 170 species (http://www.bacterio.cict.fr/p/pseudomonas.html). Phylogenetic relationships within this genus were previously studied on the basis of the 16S rRNA gene sequence comparison (Ait Tayeb *et al.*, 2005).

In this chapter, we provide analysis of 16S rRNA sequences of a selection of *Pseudomonas* spp. including *Pseudomonas entomophila* (Fig. 6.2). This true entomopathogenic species shares a common ancestor with *Pseudomonas putida* biovar A (bootstrap value of 94% indicating a high robustness of the node), which can be isolated from rhizopheric and environmental samples. The genomes of these two species are also very similar (Vodovar *et al.*, 2006). *P. entomophila* may have also acquired insect virulence factors through lateral gene transfer. Furthermore, analysis of the *P. entomophila* genome has confirmed the presence of genes that encode insecticidal toxin complexes also found in entomopathogenic enterobacteria such as *Photorhabdus luminescens*, *S. entomophila*, *Xenorhabdus nematophila* and *Yersinia* spp. (Bowen *et al.*, 1998; Waterfield *et al.*, 2001).

#### *6.2.1.3. Photorhabdus and Xenorhabdus species*

To date, only partial 16S rDNA sequences have been considered to estimate evolutionary relationships of *Xenorhabdus* and *Photorhabdus* spp. (Liu *et al.*, 1997; Tailliez *et al.*, 2006). The first study by Liu *et al.* (1997) included both genera for the analysis. In this study, isolates of *Xenorhabdus* and *Photorhabdus* claded together and formed two distinct groups. However, taxon sampling in this study was limited to only six strains representing three currently recognized *Xenorhabdus* spp. In a more comprehensive study, Tailliez *et al.* (2006) included 76 *Xenorhabdus* strains

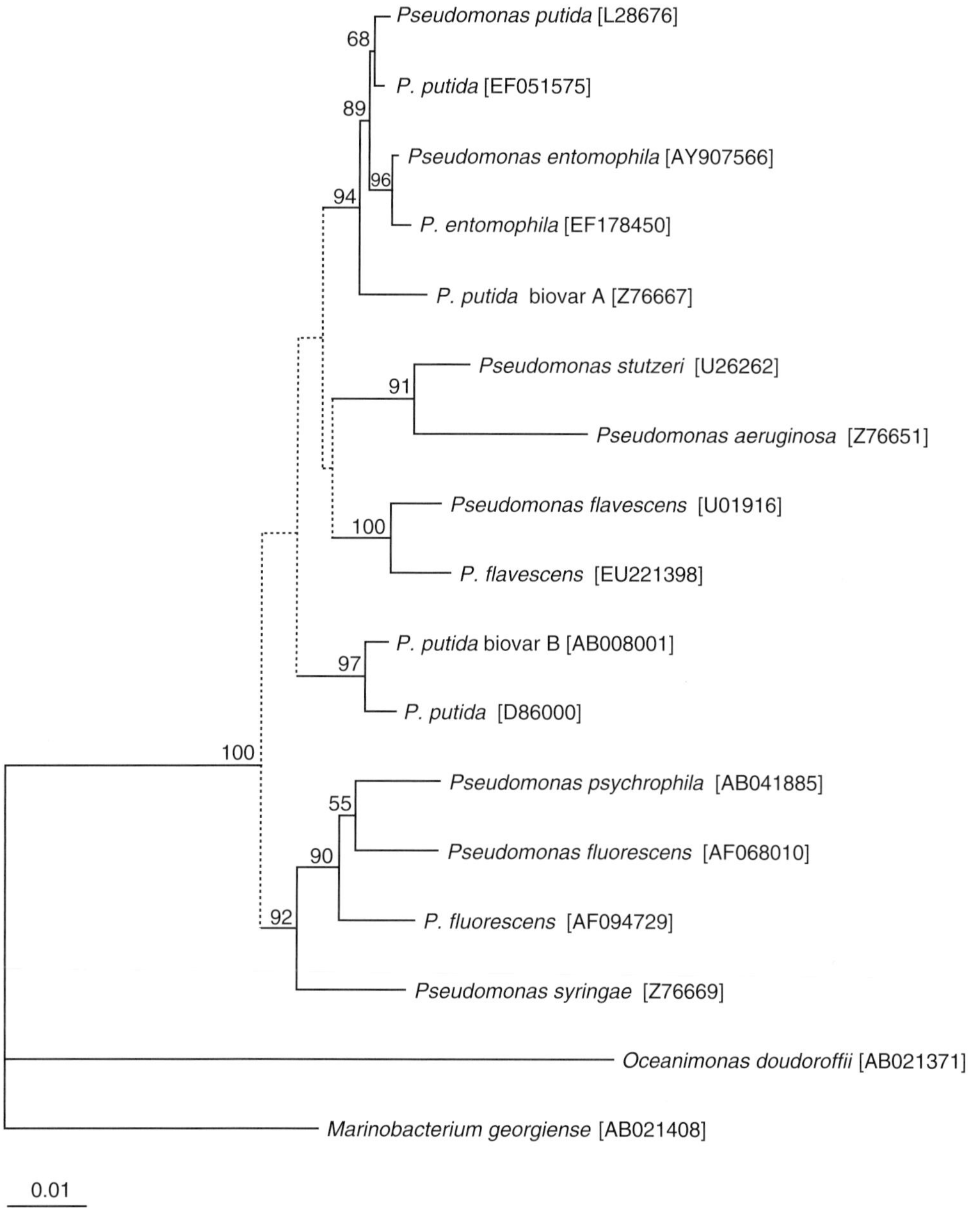


**Fig. 6.2.** Position of the entomopathogenic *Pseudomonas entomophila* in a *Pseudomonas* phylogenetic distance tree. The distance tree is constructed using the 16S rRNA gene sequences (1318 nucleotides), the Kimura two-parameter model (1980) and the neighbour-joining (NJ) module (Saitou and Nei, 1987) of PAUP software (Swofford, 2003). The 16S rRNA gene sequences of *Marinobacterium georgiense* and *Oceanimonas doudoroffii* are used as outgroup. Bootstrap values (percentages of 1000 replicates) of more than 50% are shown at the nodes. Dashed lines indicate unreliable links between groups and unique sequences. Accession numbers in brackets correspond to 16S rRNA gene sequences retrieved from GenBank (http://www.ncbi.nlm.nih.gov/). The bar indicates 1% sequence divergence.

representing known and novel species. Results from this study indicated that 16S rDNA gene is highly conserved (similarity coefficient greater than 95%), and does not provide sufficient resolution for deeper nodes of the tree. Moreover, this study also suggests the simultaneous emergence of various lineages of *Xenorhabdus* from a unique common ancestor (see Boemare and Tailliez, Chapter 2, this volume).

#### *6.2.1.4. Bacillus species*

Analysis of 16S rRNA gene sequences allows classification of novel isolates within the *Bacillus anthracis*, *B. cereus* and *Bacillus thuringiensis* groups. In a second step, it is necessary to test the virulence of the isolates on insects to classify entomopathogenic isolates in the species *B. thuringiensis*.

### 6.2.2. Protein-coding genes

Protein-coding genes have been suggested as an alternative option to avoid some of the problems encountered in phylogenetic studies with prokaryotes (Hedegaard *et al.*, 1999; Daubin *et al.*, 2002; Lerat *et al.*, 2003). Recently, two protein-coding genes *gyrB* and *recA* gene sequences have been considered for assessing evolutionary relationships in various groups of bacteria, including the entomopathogenic types (Eisen, 1995; Dauga, 2002).

#### *6.2.2.1. gyrB gene sequences and the phylogeny of Photorhabdus*

Akhurst *et al.* (2004) considered *gyrB* gene sequences to assess evolutionary relationships among *Photorhabdus* spp. In this chapter, we have compiled sequence information available in GenBank (http://www.ncbi.nlm.nih.gov/) and have also included new taxa for a more complete analysis of *Photorhabdus* phylogenies based on this gene. Below, we briefly describe some of the procedures involved in generating *gyrB* sequences, including primer information and PCR cycling parameters.

PRIMERS CONSIDERED.

| | |
|---|---|
| Forward primer | gyrBP1, 5′-TACACGAAGAAGAAGGTGTTTCAG-3′, position 245 (*P. luminescens* strain TT01 numbering, GenBank accession number BX571859); |
| Reverse primer | gyrBP2, 5′-TACTCATCCATTGCTTCATCATCT-3′, position 1641. |

PCR MASTER MIX. We suggest making a PCR master mix containing a final volume of 100 µl. Mix should contain $MgCl_2$.

| |
|---|
| 20–100 ng of DNA |
| 3 mM $MgCl_2$ (Invitrogen, http://www.invitrogen.com/ |
| 0.2 µM of each primer |
| 200 µM of each deoxynucleoside triphosphate (Invitrogen) |
| 2.5 U of *Taq* DNA polymerase (Invitrogen) in the buffer supplied with the enzyme |

PCR CYCLING PARAMETERS.

| | |
|---|---|
| Step 1 | initial denaturation: 94°C for 5 min |
| Step 2 | denaturation: 94°C for 1 min |
| Step 3 | annealing: 60°C for 1 min |
| Step 4 | elongation: 72°C for 2 min |
| Step 5 | cycling step: repeat steps 2–4 for 30 cycles |
| Step 6 | final elongation: 72°C for 7 min |
| Step 7 | cycle termination: hold at 4°C |

PURIFICATION OF PCR PRODUCTS. We usually purify PCR products using a Montage™ PCR device (Millipore, http://www.millipore.com/). However, readers should be aware that many other PCR purification kits are available and have been also cited and/or described in other chapters of Part I of this book.

SEQUENCING PRIMERS. Sequences overlapping the *gyrB* gene are then obtained using four sequencing primers.

| | |
|---|---|
| Forward primer | gyrBSP1, 5′-GATAACTCTTATAAAGTTTCCG-3′, position 316 |
| Reverse primer | gyrBSP2, 5′-CGGGTTGTATTCGTCACGGCC-3′, position 1435 |
| Forward primer | gyrBSP3, 5′-CTCTACTTAGTGGAAGGGGA-3′, position 1258 |
| Reverse primer | gyrBSP4, 5′-GCAGTAAATATTTTCCTGGA-3′, position 785 |

A number of sequence assemblage and alignment programs are available and have been discussed in Chapters 4 (Stock, this volume), 5 (Obornik, this volume) and 15 (Koltai, this volume). These programs can be applied for the assembly and alignment of *Photorhabdus* and other entomopathogenic bacteria.

#### 6.2.2.2. Interpretation of Photorhabdus phylogeny based on gyrB sequences

Phylogenetic trees based on the *gyrB* gene sequences are presented in Figs 6.3 (distance), 6.4 (likelihood) and 6.5 (parsimony). Three *Photorhabdus* groups can be recognized based on this analysis. The *P. luminescens* group is represented by four subspecies: *P. luminescens* ssp. *laumondii*, *P. luminescens* ssp. *kayaii*, *P. luminescens* ssp. *luminescens* and *P. luminescens* ssp. *akhurstii*, and an unidentified strain C8404. The *Photorhabdus asymbiotica* group comprises two subspecies: *P. asymbiotica* ssp. *asymbiotica*, *P. asymbiotica* ssp. *australis*, *Photorhabdus* strain Q614 isolated from an uncharacterized *Heterorhabditis* spp. from Queensland, Australia (Akhurst and Boemare, 1986) and strains Cbkj163 and Onlr40 isolated from *Heterorhabditis indica* isolates from Japan. The *Photorhabdus temperata* group is composed of two subspecies: *P. temperata* ssp. *temperata* and *P. luminescens* ssp. *thracensis* (strain DSM15199$^{T}$). The phylogenetic position of this latter strain, whatever the method of reconstruction used, is not congruent with its taxonomic classification as *P. luminescens* based on 16S rRNA gene sequences comparison (Hazir *et al.*, 2004). The distance and maximum likelihood analyses suggest a common ancestor for *P. asymbiotica* and the *P. luminescens* groups; whereas maximum parsimony analysis suggests a common ancestor for *P. asymbiotica* and the

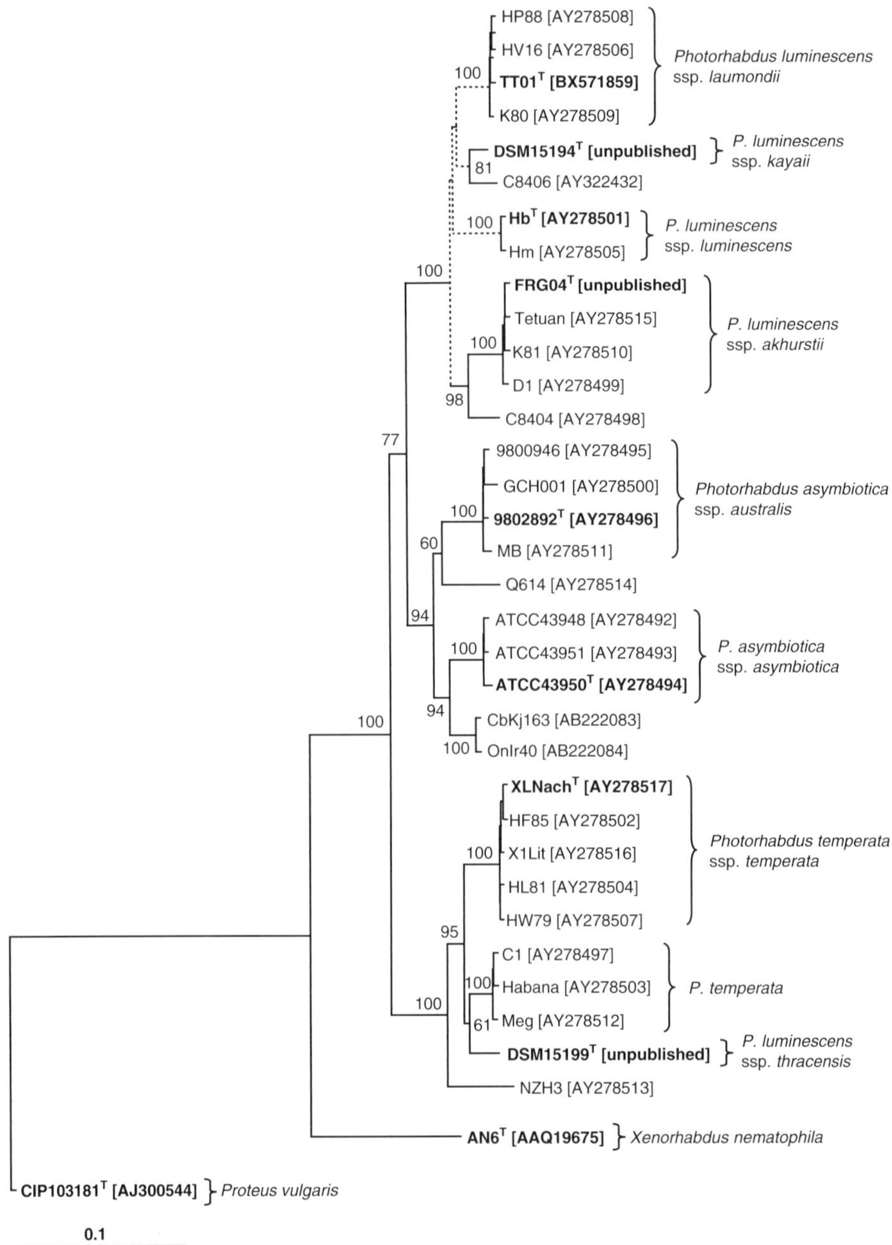


**Fig. 6.3.** Distance tree of the genus *Photorhabdus* based on the alignment of *gyrB* gene sequences. The distance tree is constructed using the *gyrB* gene sequences (889 nucleotides), the Kimura two-parameter model (1980) and the neighbour-joining (NJ) module (Saitou and Nei, 1987) of PAUP software (Swofford, 2003). The *gyrB* gene sequences of *Xenorhabdus nematophila* and *Proteus vulgaris* are used as outgroup. Bootstrap values (percentages of 1000 replicates) of more than 50% are shown at the nodes. Dashed lines indicate unreliable links between groups and unique sequences. Sequences corresponding to type strains are indicated by the number of the strain being in bold typeface. Accession numbers in brackets correspond to *gyrB* gene sequences retrieved from GenBank (http://www.ncbi.nlm.nih.gov/). The bar indicates 10% sequence divergence.

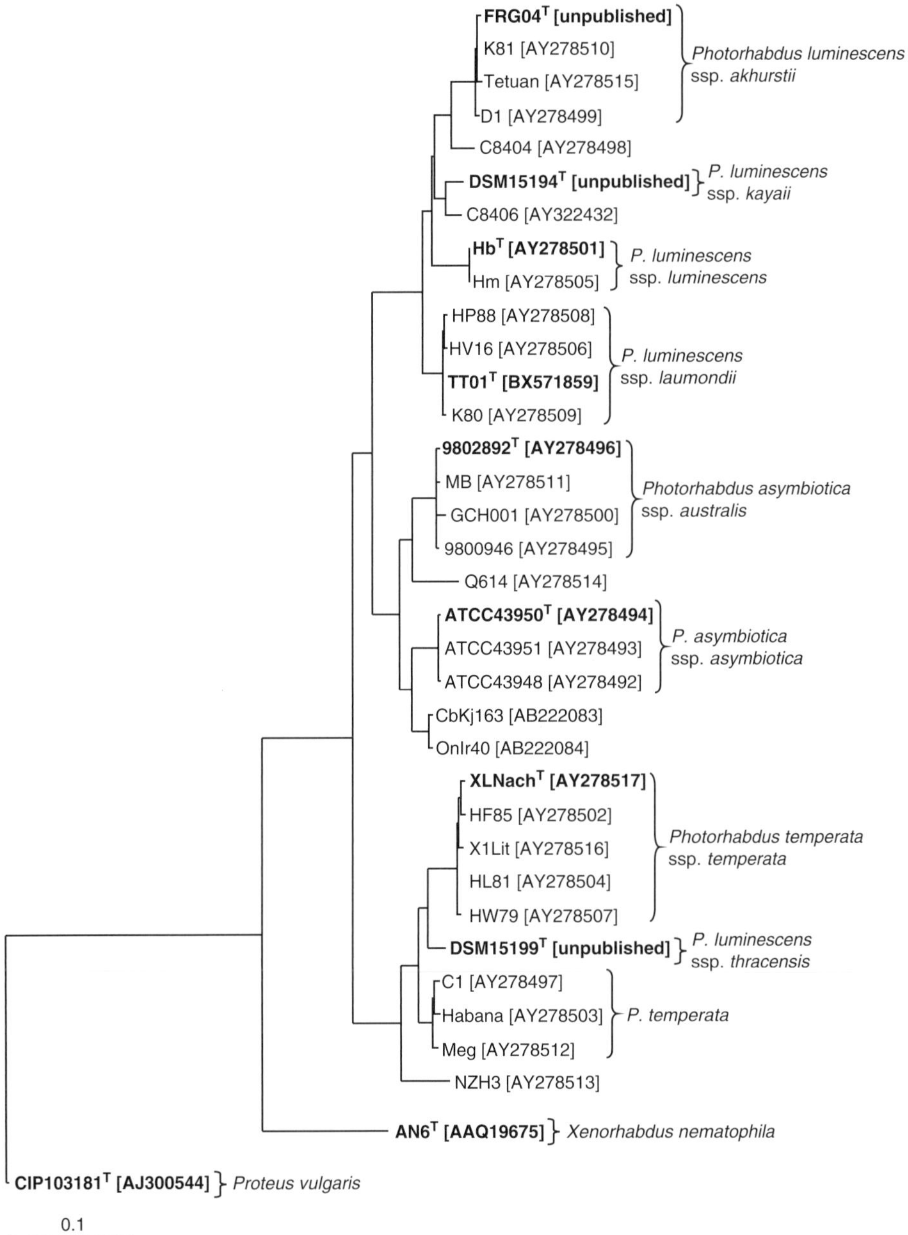


**Fig. 6.4.** Maximum likelihood tree of the genus *Photorhabdus* based on the alignment of *gyrB* gene sequences. The tree is constructed using the *gyrB* gene sequences (889 nucleotides) and the Kimura two-parameter distances (1980) included in the PAUP software (Swofford, 2003). Tree searches are performed using heuristic methods with tree-bisection reconnection (TBR) branch swapping and a minimum of 100 replicates of random stepwise addition. The *gyrB* gene sequences of *Xenorhabdus nematophila* and *Proteus vulgaris* are used as outgroup. Accession numbers in brackets correspond to *gyrB* gene sequences retrieved from GenBank (http://www.ncbi.nlm.nih.gov/). Sequences corresponding to type strains are indicated by the number of the strain being in bold typeface.

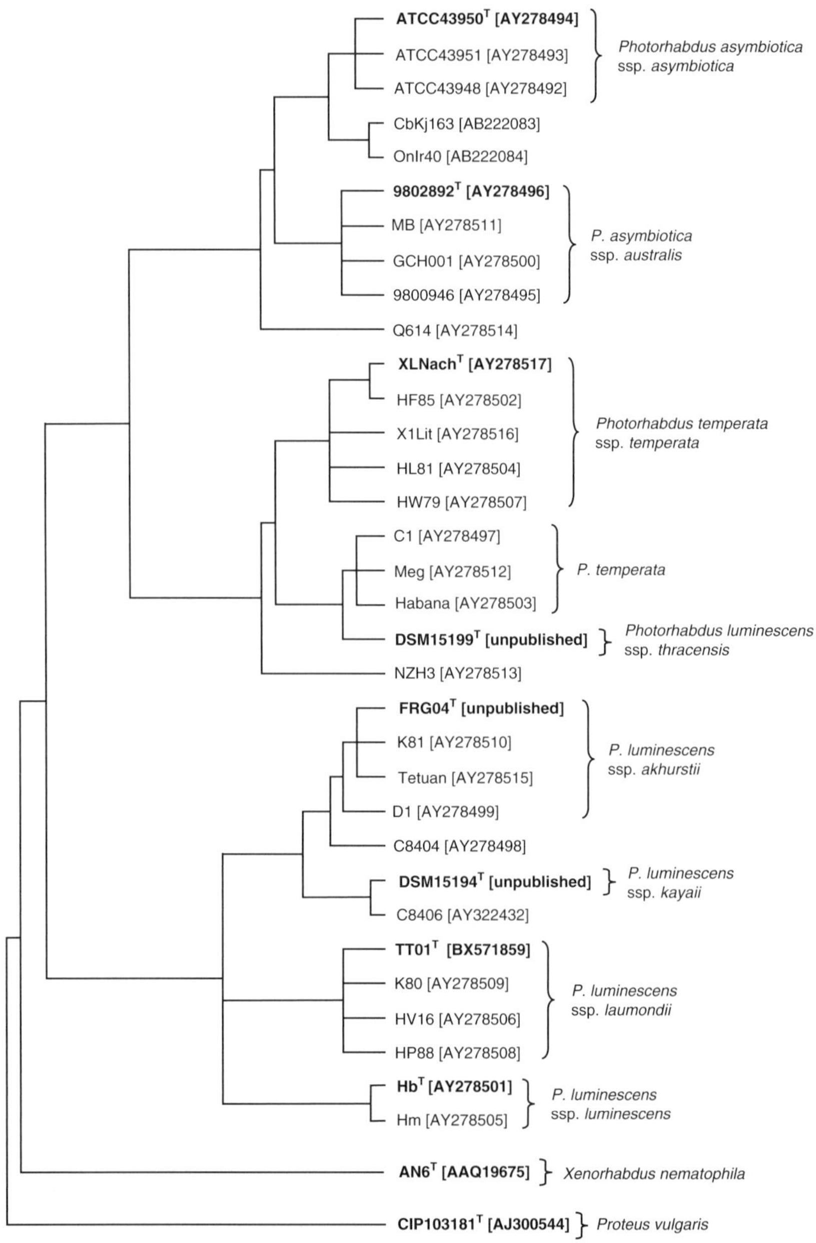


**Fig. 6.5.** Majority rule consensus of 64 equally parsimonious trees inferred by maximum parsimony analysis of an alignment of *gyrB* gene sequences of *Photorhabdus* strains. The tree is constructed using the *gyrB* gene sequences (889 nucleotides) and the maximum parsimony module included in the PAUP software (Swofford, 2003). Tree searches are performed using heuristic methods with tree-bisection reconnection (TBR) branch swapping and a minimum of 100,000 replicates of random stepwise addition. The *gyrB* gene sequences of *Xenorhabdus nematophila* and *Proteus vulgaris* are used as outgroup. Accession numbers in brackets correspond to *gyrB* gene sequences retrieved from GenBank (http://www.ncbi.nlm.nih.gov/). Sequences corresponding to type strains are indicated by the number of the strain being in bold typeface.

*P. temperata* groups. Thus, at the deepest nodes of the trees, the analysis does not allow to define precisely a hierarchical relationship between the three groups. This polytomy may be related to multiple speciation events which may have happened at the same time.

## 6.3. Conclusions and Perspectives

Sequences of 16S rRNA gene have been extensively considered to assess phylogenetic relationships of known and novel bacterial isolates, including entomopathogenic types. However, the lack of sufficient resolution of this gene has prompted investigation of other genes and methods to further explore evolutionary relationships within Eubacteria.

In the case of entomopathogenic *Photorhabdus* and *Xenorhabdus*, the key question in their evolutionary origins is the consideration of the simultaneous emergence of different lineages within each genus. 16S rRNA gene sequences have not provided strong evidence for interpretation of evolutionary origins within these two bacterial genera. Recently, consideration of other molecular markers and genes has expanded the repertoire for studying evolutionary relationships of bacteria. For example, analysis of *gyrB* gene sequences has shown more precision and has also aided in the taxonomy of *Photorhabdus* as proposed by Boemare *et al.* (1993) and Fischer-Le Saux *et al.* (1999). However, a multigene approach should be the path to follow to gather more phylogenetic information and increase robustness of evolutionary hypothesis. It would also be valuable to study the evolution and origins of pathogenic genes and investigate the hypothesis of lateral gene transfer during their evolution (Wertz *et al.*, 2003).

## References


Ait Tayeb, L., Ageron, E., Grimont, F. and Grimont, P.A.D. (2005) Molecular phylogeny of the genus *Pseudomonas* based on *rpoB* sequences and application for the identification of isolates. *Research in Microbiology* 156, 763–773.

Akhurst, R.J. and Boemare, N.E. (1986) A non-luminescent strain of *Xenorhabdus luminescens* (Enterobacteriaceae). *Journal of General Microbiology* 132, 1917–1922.

Akhurst, R.J., Boemare, N.E., Janssen, P.H., Peel, M.M., Alfredson, D.A. and Beard, C.E. (2004) Taxonomy of Australian clinical isolates of the genus *Photorhabdus* and proposal of subspecies *Photorhabdus asymbiotica* subsp. *asymbiotica* subsp. nov. and *Photorhabdus asymbiotica* subsp. *australis* subsp. nov. *International Journal of Systematic and Evolutionary Microbiology* 54, 1301–1310.

Ashelford, K.E., Fry, J.C., Bailey, M.J. and Day, M.J. (2002) Characterization of *Serratia* isolates from soil, ecological implications and transfer of *Serratia proteamaculans* subsp. *quinovora* Grimont *et al.* 1983 to *Serratia quinivorans* corrig., sp. nov. *International Journal of Systematic and Evolutionary Microbiology* 52, 2281–2289.

Balch, W.E., Magrum, L.J., Fox, G.E., Wolfe, R.S. and Woese, C.R. (1977) An ancient divergence among bacteria. *Journal of Molecular Evolution* 9, 305–311.

Boemare, N.E., Akhurst, R.J. and Mourant, R.G. (1993) DNA relatedness between *Xenorhabdus* spp. (Enterobacteriaceae), symbiotic bacteria of entomopathogenic nematodes and a proposal to transfer *Xenorhabdus luminescens* to a new genus, *Photorhabdus* gen. nov. *International Journal of Systematic Bacteriology* 43, 249–255.

Bowen, D., Rocheleau, T.A., Blackburn, M., Andreev, O., Golubeva, E., Bhartia, R. and ffrench-Constant, R.H. (1998) Insecticidal toxins from the bacterium *Photorhabdus luminescens*. *Science* 280, 2129–2132.

Brendel, V., Brocchieri, L., Sandler, S.J., Clark, A.J. and Karlin, S. (1997) Evolutionary comparisons of RecA-like proteins across all major kingdoms of living organisms. *Journal of Molecular Evolution* 44, 528–541.

Daubin, V., Gouy, M. and Perrière, G. (2002) A phylogenomic approach to bacterial phylogeny: evidence of a core of genes sharing a common history. *Genome Research* 12, 1080–1090.

Dauga, C. (2002) Evolution of the *gyrB* gene and the molecular phylogeny of Enterobacteriaceae: a model molecule for molecular systematic studies. *International Journal of Systematic and Evolutionary Microbiology* 52, 531–547.

Dauga, G., Grimont, F. and Grimont, P.A.D. (1990) Nucleotide sequences of 16S rRNA from ten *Serratia* species. *Research in Microbiology* 141, 1139–1149.

Eisen, J.A. (1995) The RecA protein as a model molecule for molecular systematic studies of bacteria: comparison of trees of RecAs and 16S rRNAs from the same species. *Journal of Molecular Evolution* 41, 1105–1123.

Elahi, E. and Ronaghi, M. (2004) Pyrosequencing, a tool for DNA sequencing analysis. In: Zhao, S. and Stodolsky, M. (eds) *Methods in Molecular Biology, Volume 255, Bacterial Artificial Chromosomes, Volume 1: Library Construction, Physical Mapping, and Sequencing*. Humana Press, Totowa, New Jersey, pp. 211–219.

Fischer-Le Saux, M., Viallard, V., Brunel, B., Normand, P. and Boemare, N.E. (1999) Polyphasic classification of the genus *Photorhabdus* and proposal of new taxa: *P. luminescens* subsp. *luminescens* subsp. nov., *P. luminescens* subsp *akhurstii* subsp. nov., *P. luminescens* subsp *laumondii* subsp. nov., *P. temperata* sp. nov., *P. temperata* subsp. *temperata* subsp. nov. and *P. asymbiotica* sp. nov. *International Journal of Systematic Bacteriology* 49, 1645–1656.

Gevers, D., Cohan, F.M., Lawrence, J.G., Spratt, B.G., Coenye, T., Feil, E.J., Stackebrandt, E., Van de Peer, Y., Vandamme, P., Thompson, F.L. and Swings, J. (2005) Re-evaluating prokaryotic species. *Nature Reviews Microbiology* 3, 733–739.

Glare, T.R., Corbett, G.E. and Sadler, A.J. (1993) Association of a large plasmid with amber disease of the New Zealand grass grub, *Costelytra zealandica*, caused by *Serratia entomophila* and *Serratia proteamaculans*. *Journal of Invertebrate Pathology* 62, 165–170.

Grimont, P.A.D. and Grimont, F. (1978) The genus *Serratia*. *Annual Review of Microbiology* 32, 221–248.

Grimont, P.A.D., Grimont, F. and Lysenko, O. (1979a) Species and biotype identification of *Serratia* strains associated with insects. *Current Microbiology* 2, 139–142.

Grimont, P.A.D., Grimont, F. and Starr, M.P. (1979b) *Serratia ficaria* sp. nov., a bacterial species associated with Smyrna figs and the fig wasp *Blastophaga psenes*. *Current Microbiology* 2, 277–282.

Grimont, P.A.D., Grimont, F. and Irino, K. (1982) Biochemical characterization of *Serratia liquefaciens sensu stricto*, *Serratia proteamaculans*, and *Serratia grimesii* sp. nov. *Current Microbiology* 7, 69–74.

Hazir, S., Stackebrandt, E., Lang, E., Schumann, P., Ehlers, R.U. and Keskin, N. (2004) Two new subspecies of *Photorhabdus luminescens*, isolated from *Heterorhabditis bacteriophora* (Nematoda: Heterorhabditidae): *Photorhabdus luminescens* subsp. *kayaii* subsp. nov. and *Photorhabdus luminescens* subsp. *thracensis* subsp. nov. *Systematic and Applied Microbiology* 27, 36–42.

Hedegaard, J., Steffensen, S.A., Nørskov-Lauritsen, N., Mortensen, K.K. and Sperling-Petersen, H.U. (1999) Identification of Enterobacteriaceae by partial sequencing of the gene encoding translation initiation factor 2. *International Journal of Systematic Bacteriology* 49, 1531–1538.

Hilario, E., Buckley, T.R. and Young, J.M. (2004) Improved resolution on the phylogenetic relationships among *Pseudomonas* by the combined analysis of *atp*D, *car*A, *rec*A and 16S rDNA. *Antonie van Leeuwenhoek* 86, 51–64.

Hurst, M.R., Glare, T.R., Jackson, T.A. and Ronson, C.W. (2000) Plasmid-located pathogenicity determinants of *Serratia entomophila*, the causal agent of amber disease of grass grub, show similarity to the insecticidal toxins of *Photorhabdus luminescens*. *Journal of Bacteriology* 182, 5127–5138.

Iwabe, N., Kuma, K.I., Hasegawa, M., Osawa, S. and Miyata, T. (1989) Evolutionary relationship of archaebacteria, eubacteria, and eukaryotes inferred from phylogenetic trees of duplicated genes. *Proceedings of the National Academy of Science of the USA* 86, 9355–9359.

Jackson, T.A., Boucias, D.G. and Thaler, J.-O. (2001) Pathobiology of amber disease, caused by *Serratia* spp., in the New Zealand grass grub, *Costelytra zealandica*. *Journal of Invertebrate Pathology* 78, 232–243.

Kersters, K., Ludwig, W., Vancanneyt, M., De Vos, P., Gillis, M. and Schleifer, K.-H. (1996) Recent change in the classification of the pseudomonads: an overview. *Systematic and Applied Microbiology* 19, 465–477.

Kimura, M. (1980) A simple method for estimating evolutionary rates of base substitutions through comparative studies of nucleotide sequences. *Journal of Molecular Evolution* 16, 111–120.

Lerat, E., Daubin, V. and Moran, N.A. (2003) From gene trees to organismal phylogeny in prokaryotes: the case of the γ-proteobacteria. *PLOS Biology* 1, 101–109.

Liu, J., Berry, R., Poinar, G. and Moldenke, A. (1997) Phylogeny of *Photorhabdus* and *Xenorhabdus* species and strains as determined by comparison of partial 16S rRNA gene sequences. *International Journal of Systematic Bacteriology* 47, 948–951.

Margulies, M., Egholm, M., Altman, W.E., Attiya, S., Bader, J.S., Bemben, L.A., Berka, J., Braverman, M.S., Chen, Y.J., Chen, Z., Dewell, S.B., Du, L., Fierro, J.M., Gomes, X.V., Godwin, B.C., He, W., Helgesen, S., Ho, C.H., Irzyk, G.P., Jando, S.C., Alenquer, M.L., Jarvie, T.P., Jirage, K.B., Kim, J.B., Knight, J.R., Lanza, J.R., Leamon, J.H., Lefkowitz, S.M., Lei, M., Li, J., Lohman, K.L., Lu, H., Makhijani, V.B., McDade, K.E., McKenna, M.P., Myers, E.W., Nickerson, E., Nobile, J.R., Plant, R., Puc, B.P., Ronan, M.T., Roth, G.T., Sarkis, G.J., Simons, J.F., Simpson, J.W., Srinivasan, M., Tartaro, K.R., Tomasz, A., Vogt, K.A., Volkmer, G.A., Wang, S.H., Wang, Y., Weiner, M.P., Yu, P., Begley, R.F. and Rothberg, J.M. (2005) Genome sequencing in micro-fabricated high-density picolitre reactors. *Nature* 437, 376–380.

Palleroni, N.J. (1984) Genus I: *Pseudomonas* Migula 1894, 237$^{AL}$. In: Krieg, N.J. and Holt, J.G. (eds) *Bergey's Manual of Systematic Bacteriology*, vol. 1. Williams & Wilkins, Baltimore, Maryland, pp. 141–199.

Palleroni, N.J., Balard, R.W., Ralston, E. and Doudoroff, M. (1972) Deoxyribonucleic acid homologies among some *Pseudomonas* species. *Journal of Bacteriology* 110, 1–11.

Palleroni, N.J., Kunisawa, R., Contopoulou, R. and Doudoroff, M. (1973) Nucleic acid homologies in the genus *Pseudomonas*. *International Journal of Systematic Bacteriology* 23, 333–339.

Priest, F.G., Barker, M., Baillie, L.W.J., Holmes, E.C. and Maiden, M.C.J. (2004) Population structure and evolution of the *Bacillus cereus* group. *Journal of Bacteriology* 186, 7959–7970.

Rainey, F.A., Ehlers, R.U. and Stackebrandt, E. (1995) Inability of the polyphasic approach to systematics to determine the relatedness of the genera *Xenorhabdus* and *Photorhabdus*. *International Journal of Systematic Bacteriology* 45, 379–381.

Rasko, D.A., Altherr, M.R., Han, C.S. and Ravel, J. (2005) Genomics of the *Bacillus cereus* group of organisms. *FEMS Microbiology Reviews* 29, 303–329.

Rokas, A., Williams, B.L., King, N. and Carroll, S.B. (2003) Genome-scale approaches to resolving incongruence in molecular phylogenis. *Nature* 425, 798–804.

Saitou, N. and Nei, M. (1987) The neighbour-joining method: a new method for reconstructing phylogenetic trees. *Molecular and Biological Evolution* 4, 406–425.

Shendure, J., Porreca, G.J., Reppas, N.B., Lin, X., McCutcheon, J.P., Rosenbaum, A.M., Wang, M.D., Zhang, K., Mitra, R.D. and Church, G.M. (2005) Accurate multiplex polony sequencing of an evolved bacterial genome. *Science* 309, 1728–1732.

Spröer, C., Mendrock, U., Swiderski, J., Lang, E. and Stackebrandt, E. (1999) The phylogenetic position of *Serratia*, *Buttiauxella* and some other genera of the family Enterobacteriaceae. *International Journal of Systematic and Evolutionary Microbiology* 49, 1433–1438.

Steinhaus, E.A. (1941) A study of the bacteria associated with thirty species of insects. *Journal of Bacteriology* 42, 757–789.

Steinhaus, E.A. (1959) *Serratia marcescens* Bizio as an insect pathogen. *Hilgardia* 28, 351–380.

Swofford, D.L. (2003) *PAUP*: Phylogenetic Analysis Using Parsimony (*and Other Methods)*, Version 4.0b10. Sinauer Associates, Sunderland, Massachusetts.

Tailliez, P., Pagès, S., Ginibre, N. and Boemare, N. (2006) New insight into diversity in the genus *Xenorhabdus*, including the description of ten new species. *International Journal of Systematic and Evolutionary Microbiology* 56, 2805–2818.

Thompson, C.C., Thompson, F.L., Vandemeulebroecke, K., Hoste, B., Dawyndt, P. and Swings, J. (2004) Use of *recA* as an alternative phylogenetic marker in the family *Vibrionaceae*. *International Journal of Systematic and Evolutionary Microbiology* 54, 919–924.

Vodovar, N., Vallenet, D., Cruveiller, S., Rouy, Z., Barbe, V., Acosta, C., Cattolico, L., Jubin, C., Lajus, A., Segurens, B., Vacherie, B., Wincker, P., Weissenbach, J., Lemaitre, B., Médigue, C. and Boccard, F. (2006) Complete genome sequence of the entomopathogenic and metabolically versatile soil bacterium *Pseudomonas entomophila*. *Nature Biotechnolology* 24, 673–679.

Waterfield, N.R., Bowen, D.J., Fetherston, J.D., Perry, R.D. and ffrench-Constant, R.H. (2001) The tc genes of *Photorhabdus*: a growing family. *Trends in Microbiology* 9, 185–191.

Wertz, J.E., Goldstone, C., Gordon, D.M. and Riley, M.A. (2003) A molecular phylogeny of enteric bacteria and implications for bacterial species concept. *Journal of Evolutionary Biology* 16, 1236–1248.

Woese, C.R., Magrum, L.J. and Fox, G.E. (1978) Archaebacteria. *Journal of Molecular Evolution* 11, 245–251.

Yamada, S., Ohashi, E., Agata, N. and Venkateswaran, K. (1999) Cloning and nucleotide sequence analysis of *gyrB* of *Bacillus cereus*, *B. thuringiensis*, *B. mycoides*, and *B. anthracis* and their application to the detection of *B. cereus* in rice. *Applied and Environmental Microbiology* 65, 1483–1490.

# 7 Molecular Systematics of Entomopathogenic Fungi

S.A. Rehner

*USDA, ARS Systematic Mycology and Microbiology Laboratory, Beltsville, USA*



## 7.1. Introduction

Insect parasitism has multiple and diverse origins within the kingdom fungi, with shifts to trophic specialization on insects having evolved one or more times in each of the four traditionally recognized phyla of fungi, the Ascomycota, Basidiomycota, Chytridiomycota and Zygomycota. The rich legacy of these adaptive shifts to nutritional specialization on insects is evidenced in the more than 750 described species of fungal invertebrate pathogens (Carruthers and Soper, 1987). With the advent of DNA sequencing and phylogenetic analytical methods, from the 1980s onwards, the field of fungal evolutionary biology has developed rapidly, yielding unprecedented insights about the evolutionary history of the fungi. These discoveries have led to significant changes in how we classify, investigate and communicate about Fungi. Because of the disparate origins of species and lineages of entomopathogens within Fungi, a broad knowledge of fungal diversity and evolution, combined with an understanding of the principles by which species are delimited and classified, is fundamental to working with this ecologically and agriculturally important group of organisms.

The goal of this chapter is to review and summarize recent advances in fungal phylogenetics that inform understanding of their evolution and classification, and how molecular data are used to develop modern species concepts and how this relates to the systematic study of entomopathogenic fungi. The chapter is divided into two major sections. The first reviews recent molecular phylogenetic data that have shed new light on the origin and diversification of the fungi and how this information enhances and changes understanding of the evolution and higher-level classification of fungal entomopathogens. The second section reviews species concepts in fungi and the development of phylogenetic species concepts with molecular phylogenetic data, with several examples where this approach has been applied to test and revise species delimitation in entomopathogenic fungi.

## 7.2. Molecular Phylogenies of Fungi and the Origins of Entomopathogens

Over the last two decades, molecular phylogenetic studies have provided an astonishing array of insights bearing on the origin, circumscription and evolutionary relationships among the true Fungi. Molecular phylogenies have demonstrated that Fungi, which traditionally have been grouped with plants, are more closely related to the Metazoa, or animals (Baldauf and Palmer, 1993; Wainright *et al.*, 1993; Baldauf, 1999; Baldauf *et al.*, 2000; Berbee and Taylor, 2001; Lang *et al.*, 2002). Fungi are thus likely to have evolved from a flagellated ancestor (Cavalier-Smith, 2001), which bolsters the long-held assumption that chytrids, which produce flagellated zoospores, represent the earliest surviving lineages of Fungi and point to an aquatic origin for the Fungi. Organisms formerly allied or classified within Fungi, i.e. the plasmodial and cellular slime moulds, and water moulds (Myxomycota, Dictyosteliomycota and Oomycota, respectively) are now known to have originated independently from the Fungi and thus are classified separately. However, the principal scientists who study slime moulds and water moulds are often in fact mycologists. The slime moulds (Mycetozoa) are derived from within the Amoebozoa, which are the presumed sister lineage to the animals and Fungi (Baldauf and Doolittle, 1997; Cavalier-Smith, 1998).

The Oomycota are currently grouped with heterokont algae, including the brown algae and diatoms (Forster *et al.*, 1990; Cavalier-Smith *et al.*, 1994; Leipe *et al.*, 1994; Hausner *et al.*, 2000). Thus, the filamentous mosquito pathogen, *Lagenidium giganteum*, as well as the related filamentous plant pathogens *Phytophthora*, *Pythium* and *Peronospora* share remarkable similarities to the Fungi in morphology and nutritional mode as the result of evolutionary convergence. In contrast, other phylogenetic data suggest that the circumscription of Fungi must be expanded to accommodate the inclusion of the amitochondriate microsporidia. The microsporidia are unicellular, protist-like, obligate intracellular parasites of protozoa, arthropods and metazoa and are characterized by highly reduced morphology and genomes (Fast and Keeling, 2005). The microsporidia are unique in possessing a complex infection mechanism by which they evert their cell contents into the host cytoplasm through a thin polar tube. Although a phylogenetic connection between the microsporidia and fungi is supported by the shared traits of closed mitosis and spores that contain chitin and trehalase (Cavalier-Smith, 2001), the extreme morphological reduction and singular spore structure of microsporidia has previously made their phylogenetic placement and classification a biological enigma.

To date, considerable molecular phylogenetic evidence has now been presented that places the microsporidia at or near the base of the fungal phylogeny (Keeling and McFadden, 1998; Hirt *et al*., 1999; Keeling *et al*., 2000; Cavalier-Smith, 2001; Keeling, 2003; Fast and Keeling, 2005). However, differences in the position and statistical support for the placement of microsporidia have varied according to the genes analysed. The inconsistent positioning of the microsporidia appears to be due to long-branch attraction (Felsenstein, 1978), a phenomenon in phylogenetic reconstruction that artefactually draws divergent sequences together. This is not surprising, as gene sequences in Microsporidia tend to be extremely diverged in comparison to both animals and fungi (Fast and Keeling, 2005).

The 1990s saw an explosion of molecular phylogenetic publications on diverse aspects of fungal evolution and classification. However, taxon and gene sampling in many studies tended to be limited in scope, in part due to resource and technical issues. Inclusive taxon sampling of fungi has been difficult to achieve for lack of material for high-quality DNA extraction stemming variously from difficulties in collecting specimens of rarely encountered taxa, inability to culture particular taxa and challenges in isolating DNA of sufficient quality from herbarium and preserved specimens where fresh or cultured material was unavailable. Additionally, thorough sampling of some taxa has been hampered by the lack of qualified taxonomic specialists capable of providing accurately identified specimens to include in phylogenetic studies.

### 7.2.1. Molecular markers considered

Gene sampling schemes in early molecular investigations of fungi were almost exclusively reliant on one or more nuclear ribosomal regions (rDNA), e.g. the small subunit (SSU rDNA), large subunit (LSU rDNA) or internal transcribed spacer regions (ITS), which includes the ITS 1 and 2 and the intervening 5.8S ribosomal subunit (see Chapter 16, this volume, for further information). Without question, the rDNA primers developed for fungi by White *et al.* (1990) have been integral to the development of fungal molecular phylogenetic systematics and continue to be highly useful molecular markers for systematic and diversity investigations of Fungi. However, molecular phylogenetic biologists have increasingly used data from multiple and functionally different gene regions for phylogenetic inference to obtain independent estimates of phylogeny as well as to increase overall character and statistical support for evolutionary hypotheses.

Until recently, the lack of sequence information for the majority of fungi, coupled with high-sequence divergences among fungal lineages, has been a persistent obstacle to the development of conserved primer sets that enable PCR and sequencing of orthologous genes across the Fungi. However, as whole-genome sequences for a wider diversity of fungi are determined, inter-genomic comparisons of gene sequences with Basic Local Alignment Search Tool (BLAST) (Altschul *et al.*, 1990) and more sophisticated comparative phylogenomic approaches (Robbertse *et al.*, 2006), identification of a common set of gene orthologues for phylogenetic analysis that encompass the Fungi are increasingly available.

## 7.2.2. Fungi and the tree of life

In 2002, the National Science Foundation funded a research initiative, the 'Assembling the Fungal Tree of Life' (AFTOL; http://aftol.org/). The primary goal of the AFTOL project was to synthesize a revised higher-level classification of the Fungi based on a phylogenetic reconstruction using molecular and non-molecular (e.g. morphological) character data representing all major lineages of the Kingdom Fungi. A central goal and activity of the project was the development of a molecular sequence data matrix for 199 species for six genes, including the SSU rDNA, LSU rDNA and ITS and partial sequences of the nuclear-encoded elongation factor-1α (EF1α), RNA polymerase II largest subunit (RPB1) and RNA polymerase II second smallest subunit (RPB2). The results of this study are reported in the focal AFTOL publication of James *et al.* (2006a). Revisions to the classification of Fungi have been proposed in light of this phylogeny and higher-level taxonomic categories proposed for Fungi include six phyla, ten subphyla, 35 classes, 12 subclasses and 129 orders (Hibbett *et al.*, 2007). Also, a special edition of Mycologia edited by Spatafora *et al.* (2006) features publications dealing with molecular dating, structural and biochemical characters for classification of fungi, and 16 studies presenting analyses of selected higher-level taxa, some of which are cited in the following discussion. In Fig. 7.1, the phylogeny of Fungi, based on the study of James *et al.* (2006a), indicates the phylogenetic placement of the major lineages of insect-associated fungi.

### *7.2.2.1. Microsporidia: primitive fungi*

Starting from the base of the fungal tree of life, the Microsporidia, which are now recognized as the phylum Microsporidia (Hibbett *et al.*, 2007), group with the fungal endoparasitic chytrid *Rozella allomycis*, which together may constitute the basal lineage of Fungi (James *et al.*, 2006a,b). However, similar to earlier studies, the placement of Microsporidia remains equivocal. Tests of statistical differences in likelihood scores among alternative topologies (Shimodaira, 2002) performed by James *et al.* (2006a) placing microsporidia either as: (i) the sister to the zygomycete order Entomophthorales (Ustinova *et al.*, 2000); (ii) among the harpellid Trichomycetes (Cavalier-Smith, 2001); (iii) as sister to the blastocladialean chytrids; (iv) as the sister to the zygomycete *Dimargaris*; (v) as the sister to Dikarya (i.e. Basidiomycota + Ascomycota); or (vi) as the sister to the Fungi cannot be differentiated statistically. However, all topologies that place microsporidia outside the Fungi were a statistically unlikely fit to the AFTOL data; thus, the balance of evidence indicates that Microsporidia are Fungi (James *et al.*, 2006). Confident placement of Microsporidia in the fungal tree ranks high among research

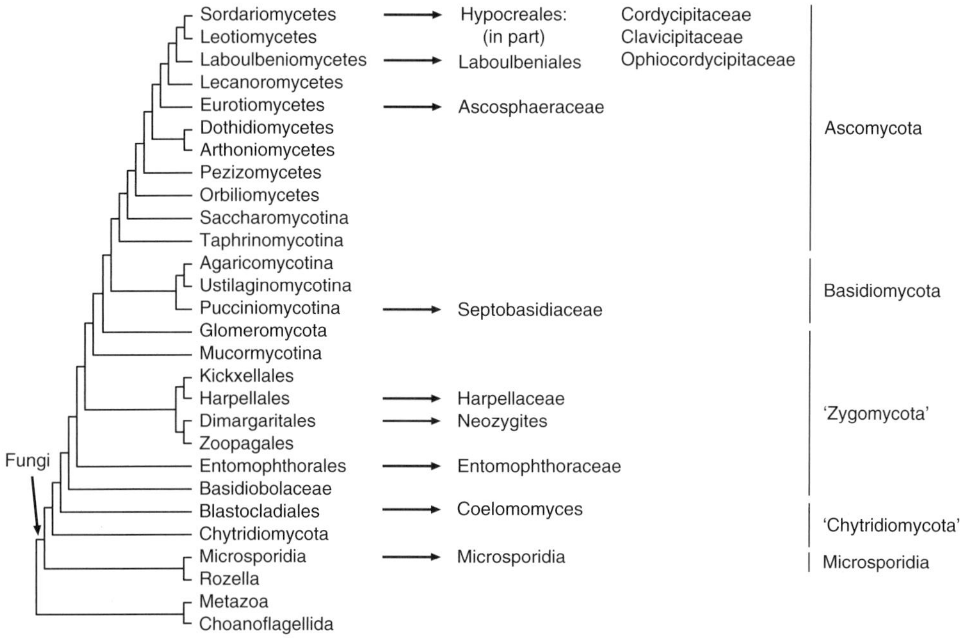


**Fig. 7.1.** A phylogeny of fungi indicating the phylogenetic placement of insect-pathogenic and other obligate insect-associated fungi, based on phylogenies in James *et al.* (2006a) and Spatafora *et al.* (2006). Arrows indicate the least inclusive taxon that encompasses specific lineages of insect-associated fungi. Labels for vertical bars at the right indicate the traditional fungal phyla.

priorities in detailing the phylogeny and evolution of the early-diverging lineages of fungi (James *et al.*, 2006a). It is possible that evolutionary reconstructions with more slowly evolving genes will resolve the evolutionary origin of Microsporidia. Additionally, more data are needed to determine the ancestral hosts and habitats of Microsporidia. Extant species of Microsporidia include pathogens of marine crustaceans and terrestrial arthopods, and species with both host affiliations and ecologies are intermixed within the major lineages that constitute the phylum (Keeling and Fast, 2005). At this time, it remains unclear whether Microsporidia originated in terrestrial or marine habitats and the hosts with which they were associated. Determination of the evolutionary polarity of marine-terrestrial and host affiliations within basal lineages of the phylum may shed light on the ancestral habitats and host associations of early Microsporidia.

#### *7.2.2.2. The Chytrids*

The majority of early-diverging fungi are zoosporic chytrids, which form a paraphyletic grade at the base of the fungal tree (James *et al.*, 2006a,b). Chytrids are primarily terricolous, freshwater or marine saprobes; however, a few species are plant pathogens and, exceptionally, invertebrate or vertebrate pathogens. Previous classifications placed all the zoosporic fungi in a single phylum, the Chytridiomycota. Phylogenetic data by James *et al.* (2006a,b) subdivide the traditional Chytridiomycota into four distinct lineages of which three phyla are now recognized (Hibbett *et al.*, 2007). The Chytridiomycota is retained and includes the majority of species traditionally classified in the phylum, none of which are insect pathogens. The two new phyla recognized (Hibbett *et al.*, 2007) include the Neocallimastigomycota, a group of saprobes, including digestive symbionts of herbivores (Wubah and Fuller, 1993a,b), and the Blastocladiomycetes (James *et al.*, 2006a), which includes plant pathogens, saprobes and the entomopathogens *Myiophagus* and *Coelomomyces*. Based on current information, the Blastocladiomycota is the chytrid lineage that is most closely related to non-zoosporic fungi, and group at the base of a grade of the zygomycete fungi (James *et al.*, 2006a,b; White *et al.*, 2006).

If Microsporidia are found not to represent the basal lineage of Fungi then the Blastocladiomycota would then constitute the earliest extant fungal lineage in which obligate insect parasitism is known. The question of whether insect parasitism in the Blastocladiales has one or more origins and whether it is an ancestral or recently derived trophic specialization will require additional sampling of taxa in the phylum. The fourth chytrid, the plant pathogen *Olpidium brassicae*, groups within the base of the polyphyletic grade of Zygomycota (James *et al.*, 2006a,b) and is near *Basidiobolus*, a non-flagellate taxon previously allied with the Entomophthorales. James *et al.* (2006a) estimated that flagellae have been lost at least four to six times in the early history of the fungi, with *Olpidium* representing the most recently derived fungal lineage diverging from the backbone of the fungal tree known to retain this organelle. This would suggest that the transition from aqueous to aerial spore dispersal has occurred multiple times and that some of the different spore types and dispersal mechanisms found in the basal Zygomycota are independently derived from different chytrid ancestors.

#### *7.2.2.3. Zygomycota and the evolution of fungal adaptations to life on land*

Fungi traditionally classified in the Zygomycota are inferred to be paraphyletic and form a grade between the early-diverging chytrid lineages and the more derived Ascomycota and Basidiomycota (James *et al.*, 2006a,b; White *et al.*, 2006; Redecker and Raab, 2006). The Glomeromycota, a group of obligate terrestrial biotrophs dependent on land plant symbioses, are phylogenetically distinct from the Zygomycota and are recognized as a distinct phylum (Hibbett *et al.*, 2006). Of the six phyla of Fungi presently recognized, the Glomeromycota is the only phylum that is not known to include entomopathogens. The Glomeromycota are sister to the Ascomycota + Basidiomycota and these three phyla together have been informally designated the 'Symbiomycota' (Tehler *et al.*, 2003).

Putting aside the Glomeromycota, the remaining Zygomycota are classified in four subphyla: the Mucormycotina, Zoopagomycotina, Entomophthoromycotina and Kickxellomycotina (Hibbett *et al.*, 2007). Of these four zygomycotan lineages three include insect-associated fungi. First and foremost is the subphylum Entomophthoromycotina in which all species of the family Entomophthoraceae are obligate entomopathogens (Humber, 1989). The Entomophthoromycotina are ontogenetically unique among zygomycete fungi by the production of holoblastic conidia that lack a sporangiospore wall. The conidia are forcibly discharged and some species produce secondary capilloconidia. In addition to entomopathogens, the Entomophthorales also include species that are either saprobes or parasites of arthropod or plants (Humber, 1989).

To date, relatively few entomophthoralean taxa have been incorporated in molecular phylogenetic studies (Jensen *et al.*, 1998); thus, the monophyly of the subphylum and the origin of insect parasitism in this lineage remain to be critically examined. However, the monophyly of the entomogenous Entomophthoraceae is supported by the unique cytology and conidial structure of the group (Humber, 1981). As recently circumscribed, the family is considered to include 12 genera and at least 132 species (Humber, 1989).

A thorough molecular phylogenetic analysis of the family has yet to be conducted. Based on current phylogenetic evidence, the Entomophthoraceae, in addition to the Microsporidia, constitute the deepest originating branches of the fungal tree that have given rise to sustained radiations of obligate entomopathogens.

Two additional examples of trophic specialization on insects have also originated in the Zygomycota. The zygomycete genus *Neozygites*, which was previously been classified in the Entomophthorales, is in fact more closely related to the mycopathogens *Dimargaris*, *Dispira* and *Tieghiomyces*, which together are now classified in the order Dimargaritales in the subphylum Kickxellomycotina and thus represent a second origin of insect parasitism within the Zygomycota (White *et al.*, 2006; Hibbett *et al.*, 2006).

The order Harpellales, a monophyletic group of obligate, lower intestinal endocommensal associates of primarily aquatic insects and isopods, is also derived from within the Kickxellomycotina, but these fungi are not closely related to *Neozygites*. Study of the Harpellales has been complicated by the inability to bring these fungi into axenic culture. Clean dissections of fungal cells free of host and other DNA contaminants are extremely difficult, and remain a distinct impediment to the molecular phylogenetic analysis of these fungi (see Chapter 16, this volume). More information on the phylogeny and trophic biology of the Harpellales is needed for these unique and intimate insect–fungal relationships.

As the first truly terrestrial fungi, the Zygomycota are of pivotal importance in understanding the evolution of fungal adaptations to life on land. Further clarification of the paraphyly of the Zygomycota will provide much needed detail in reconstructing the evolutionary history for the precursors of terrestrial fungi and the evolutionary transitions that made life on land possible.

#### *7.2.2.4. Origin of insect-pathogenic fungi*

Clades of insect pathogens have also originated in both phyla of the subkingdom Dikarya, the Basidiomycota and Ascomycota. The Basidiomycota includes only a single group of insect pathogens, the monophyletic Septobasidiobasidiaceae, which are classified in the subphylum Pucciniomycotina, or rust clade (Aime *et al.*, 2006; Henk and Vilgalys, 2007). The genus *Septobasidium*, which includes the majority of species in the family (~170), and four additional genera are specialized parasites of scale insects (Couch, 1938). *Septobasidium* spp. grow as superficial resupinate mats on branches and leaves of trees within which individuals and colonies of scale insects reside. Depending on the species, Septobasidiaceae overgrow and penetrate one to multiple individuals within a scale-insect colony as they feed on the host plant. Septobasidiaceae form various-shaped haustoria in the haemocoel of the insect through which an indirect transfer of plant nutrients is effected. Although few Septobasidiaceae–scale insect associations have been characterized in detail, the balance between insect parasitism, insect-fungal symbiosis, and plant-parasitism in Septobasidiaceae–scale insect association provide a fertile field for exploring co-evolutionary dynamics in a relatively simple and accessible system.

Three notable radiations of insect pathogens have originated within the Ascomycota (James *et al.*, 2006), once each within the Eurotiomycetes and Sordariomycetes, and also in the singular class Laboulbeniomycetes. The Eurotiomycete genus *Ascosphaera* is an obligate pathogen of bees and *Ascosphaera aphis* in the well-known causal agent of chalkbrood disease of honeybee larvae (*Apis mellifera*). Interestingly, honeybee defences towards *A. apis* appear to have a 'lock-and-key' dynamic, leading to suggestions that this pathogen has pushed bees, over evolutionary time frames, towards lower nest-level relatedness (Tarpy and Seeley, 2006; Seeley and Tarpy, 2007). There are 29 described species of *Ascosphaera* that infect various bee species, each of which generates pathologies similar to that of *A. apis*. A ribosomal phylogeny of selected species of *Ascosphaera* supports the monophyly of the genus and provides support for traditional species concepts (Anderson *et al.*, 1998). A whole genome sequence, the first for any fungal entomopathogen, is available (Qin *et al.*, 2006) and will provide a unique opportunity to analyse the genetic architecture of a specialized entomopathogen.

The Laboulbeniomycetes, which are obligate insect ectoparasites, currently rank as the most speciose radiation of insect-associated fungi with over 2000 described species (Weir and Blackwell, 2001). Due to their reduced and unusual morphology, determinate growth and unculturability, the classification of Laboulbeniomycetes has posed an enigma to fungal systematists since their discovery in the mid-19th century. Relatively few mycologists are familiar with Laboulbeniomycetes and they are frequently neglected or only briefly mentioned in discussions of ascomycete systematics. Indeed, representatives of the Laboulbeniomycetes were not included in the AFTOL taxon sampling scheme (James *et al.*, 2006). However, Blackwell and Malloch (1989), Blackwell (1994) and Weir and Blackwell (2001) have presented definitive morphological and molecular evidence that these fungi form a clade, albeit of uncertain affinity, within Ascomycota that they recognize as a distinct class. Because Laboulbeniomycetes are of little consequence to the health and well-being of their insect hosts, these organisms are largely ignored by insect pathology and biological control scientists. However, the spectacular diversity and ubiquity of these obligate insect associates remains a fertile field for investigating the trophic biology, origin and co-evolution of these fascinating dependent relationships.

The remaining and by far the most complex radiation of insect pathogens in Fungi is that of the clavicipitaceous fungi in the order Hypocreales, class Sordariomycetes. The clavicipitalean fungi are a remarkably diverse assemblage of biotrophs that form an array of pathogenic, parasitic and mutualistic associations with different kingdoms of life, including plants, fungi, insects and other microinvertebrates (Spatafora *et al.*, 2007). Traditionally classified in the family Clavicipitaceae, the taxonomic circumscription of the clavicipitaceous fungi is based on the sexual reproductive stages, or teleomorphs, which include 22 plant-associated genera (Bischoff and White, 2003), including the notorious ergot-producing plant pathogen *Claviceps purpurea*, the well-known grass endophyte *Epichlöe* and also the entomopathogen genera *Cordyceps*, *Torrubiella* and *Hypocrella*. The large genus *Cordyceps*, which contains over 500 described species (www.indexfungorum.org), deserves particular mention because of its wide host range, which includes a plethora of species from ten orders of insects, other microinvertebrates such as nematodes, and also fungi (Spatafora *et al.*, 2007). In addition, nearly all culturable clavicipitaceous fungi reproduce mitotically. The mitiotic forms, termed anamorphs, are generally much more frequently encountered in environmental samples than are the sexual stages. While some anamorphs have been developmentally linked to particular sexual species, a significant number have not, suggesting that species diversity of clavicipitaceous fungi based on named sexual species may be substantially underestimated. In pre-molecular taxonomies, these asexual forms were classified apart from sexual fungi in the now abandoned subdivision Deuteromycotina (Kendrick, 1989). A succession of molecular phylogenetic studies (Spatafora and Blackwell, 1993, 1994; Rehner and Samuels, 1995; Glenn *et al.*, 1996; Sung *et al.*, 2001; Castlebury *et al.*, 2004) has convincingly demonstrated that clavicipitaceous fungi are members of the Hypocreales (Sordariomycetes), including many of the roughly 20 associated entomogenous anamorph genera. Sung *et al.* (2007) presented a multilocus phylogeny of *Cordyceps* and clavicipitaceous fungi, prompting several important revisions in the classification of clavicipitaceous fungi along evolutionary lines.

To briefly summarize, the phylogeny of Sung *et al.* (2007) rejects the monophyly of both *Cordyceps* and Clavicipitaceae, as traditionally circumscribed, but finds support for three clades of clavicipitaceous fungi. These clades, informally designated A, B and C (Sung *et al.*, 2007), are recognized at family rank as Clavicipitaceae, Ophiocordycipitaceae fam. nov. and Cordycipitaceae. Clavicipitaceae consists of two subclades, one consisting of the arthropod pathogens *Metarhizium*, *Nomuraea* and elements of *Paecilomyces*, and the nematode and rotifer pathogens *Pochonia*, *Tolypocladium* and *Rotiferophthora*. The *Metarhizium* teleomorphs formerly classified in *Cordyceps* are transferred to the new genus *Metacordyceps* Sung *et al.* The other subclade of Clavicipitaceae contains the entomopathogens *Hypocrella* and allied *Aschersonia* anamorphs and also the plant pathogens and endophyte genera, including *Claviceps*, *Balansia*, *Epichlöe* and others. The new family Ophiocordycipitaceae contains a diverse mix of arthropod, nematode and fungal pathogens and the teleomorphs are transferred to the new genus *Ophiocordyceps*. The Cordycipitaceae is dominated by arthropod pathogens but also contains parasites of macrofungi and plant pathogenic rust fungi. The genus *Cordyceps* is restricted to teleomorphs of the Cordycipitaceae, and includes anamorphs classified in *Verticillium*, *Lecanicillium*, *Isaria*, *Beauveria* *Engyodontium* and *Simplicillium*. The closest relative to Cordycipitaceae is neither the Ophiocordycipitaceae nor Clavicipitaceae, but Hypocreaceae, a hyperdiverse family that includes fungal parasites, soil saprobes and endophytes. Thus, earlier taxonomic circumscriptions of Clavicipitaceae were non-monophyletic.

Perhaps the most intriguing evolutionary question posed by the clavicipitaceous fungi is the sequence of events leading to the remarkable range of host affiliations and nutritional specializations displayed by different species in this group of fungi. Spatafora *et al.* (2007) used maximum-likelihood character-state reconstruction analysis to develop hypotheses clarifying the numbers and directions of interkingdom host jumps in clavicipitaceous fungi. According to their analysis, Spatafora *et al.* (2007) obtained support for five to eight separate and unidirectional host jumps within the clavicipitaceous fungi, including three to five to fungi, one to two to animals and one to plants. For Clavicipitaceae, significant support was obtained for an ancestral affiliation and nutritional specialization on animals, with a single host jump to plants and then a single jump to fungi. The Ophiocordycipitaceae are also inferred to have an ancestral animal pathogen ecology with inferred host jumps to parasites of the ascomycete genus *Elaphomyces*, and one or more intrakingdom host jumps to nematodes and other animal phyla. The ancestral host affiliation of Cordycipitaceae remains ambiguous in part due to the sister group relationship to Hypocreaceae, which are predominately fungal parasites, and in part due to uncertainties in the relationships between insect-pathogenic and fungal-parasitic species within Cordycipitaceae. Although multiple interkingdom host jumps were inferred to have occurred within Cordycipitaceae, more thorough taxon sampling is needed to reconstruct ancestral hosts and the polarity of subsequent host jumps within this family (Spatafora *et al.*, 2007).

## 7.3. Species Recognition

Species are widely acknowledged as the fundamental units of evolution and biodiversity (Mayr, 1942, 1963; White, 1973; Grant, 1981; Coyne and Orr, 2004; Claridge *et al.*, 1997); thus, it is critical that species recognition criteria accurately encompass these essential attributes. Delineating species of entomopathogenic fungi as meaningful evolutionary units is therefore an essential prerequisite to characterizing other species attributes, including their biogeography, genetic structure and functional ecology. From this starting point, reconstruction of species relationships can shed light on the origins and mechanisms of speciation of entomopathogenic fungi and also for tracing the stepwise evolution of ecologically important traits within groups of related species. Finally, accurate species concepts and precise identification methods can facilitate efforts to identify, select and manage entomopathogenic fungi for the control of pest insects.

Species concepts and speciation have been the focus of extensive debate in biology (Ghiselin, 1987; Mishler and Donoghue, 1982; Ereshefsky, 1992; Hull, 1997; Wheeler and Meier, 2000; Mallet, 2001; Coyne and Orr, 2004). Parallel discussions have also been prominent in the mycological literature (Clémençon, 1977; Burnett, 1983, 2003; Brasier, 1987, 1997; Natvig and May, 1996; Harrington and Rizzo, 1999; Petersen and Hughes, 1999; Taylor *et al.*, 2000; Kohn, 2005). Over the last three decades, empirical investigations of fungal species concepts have evolved from a system in which species are recognized on the basis of morphology (Morphological Species Concept; MSR), to the use of mating tests for Biological Species Recognition (BSR; Mayr, 1942, 1963), and most recently, the use of phylogenetic concordance for Phylogenetic Species Recognition (PSR; Baum, 1992; Avise and Ball, 1990; Baum and Shaw, 1995; Futuyma, 1998; Taylor *et al.*, 2000). This section is intended as a brief summary of the salient features of each conceptual approach to species recognition and their application with selected entomopathogenic fungi.

### 7.3.1. Morphological species concept

The MSR forms the historical basis for existing taxonomic classifications and identification methods for the vast majority of fungi, including entomopathogenic taxa. However, the morphological simplicity and phenotypic plasticity of many fungi render identification of species boundaries extremely difficult, a problem that also holds true for many entomopathogenic fungi. One needs to look no further than the ubiquitous anamorph entomopathogenic genera *Beauveria*, *Metarhizium* and *Paecilomycs*, to appreciate the limitations of morphology for species identification. However, as the historical link to over 200 years of taxonomic mycology, morphology will always remain as a pillar of species taxonomies, although it is now a common practice in taxonomic studies to include molecular analyses to validate species boundaries and infer species relationships.

### 7.3.2. Biological species concept

BSR (Mayr, 1942) defines species as interbreeding populations of individuals that are isolated from other such populations. BSR gained popularity as a species concept because mating, sexual compatibility and progeny viability are expressions of fundamental processes that promote species cohesion and divergence. Investigative mating studies in fungi have revealed numerous cases where traditionally recognized morphological species encompassed two or more reproductively isolated groups, or biological species (Anderson and Ullrich, 1979; Harrington *et al.*, 1989; Vilgalys and Sun, 1994; Hibbett and Donoghue, 2002; Dettman *et al.*, 2003b, 2006). These and other studies provided the first clear indication that fungal species diversity under MSR has frequently been significantly underestimated. However, the use of BSR has been criticized because reproductive compatibility is an ancestral characteristic that may be retained among physically isolated but genetically distinguishable populations (Rosen, 1979). For mycologists, one shortcoming of BSR is that it cannot be applied to the many fungi that are difficult to cultivate or induced to perform their sexual cycle in the laboratory. Also, BSR cannot be applied to homothallic or asexual fungi, which constitute approximately 20% of described species of Fungi. Finally, the application of BSR as an identification tool is cumbersome due to the requirement to maintain collections of reference isolates and the logistics to set up and evaluate extensive pairwise crossing matrices. There are no examples in the literature where BSR has been developed for any fungal entomopathogen, presumably because so many genera are believed to be asexual. However, with the discovery of sexual states for species in genera including *Beauveria* and *Metarhizium* (Liang *et al.*, 1991; Li *et al.*, 2001; Liu *et al.*, 2001), the development of mating compatibility tests in these genera is certainly now a possibility.

### 7.3.3. Phylogenetic species concept

The PSR criterion (Cracraft, 1983, 1989) recognizes as a species any group of organisms that exclusively share one or more unique derived (apomorphic) feature acquired by descent from a common ancestor. Although any type of unique character can be used to define a phylogenetic species, implementation of PSR in fungi has gained wider currency with access to nucleotide sequencing and robust methods for phylogenetic reconstruction. In fungi, PSR with molecular data holds practical advantage over both MSR and BSR because it can be applied to any species for which DNA is available, regardless of their morphology, ability to grow in culture or potential to mate *in vitro*. With relatively modest sequencing effort, nucleotide polymorphisms vastly outnumbering the pool of available informative morphological characters can be readily generated to support robust phylogenetic analyses.

PSR has been validated extensively as an effective strategy for detecting cryptic species in different fungal taxa spanning a wide range of ecological specializations, including plant pathogens (O'Donnell *et al.*, 2000a,b; Burgess *et al.*, 2006), animal and human pathogens (Matute *et al.*, 2006; Pringle *et al.*, 2005), saprotrophs (Vilgalys and Sun, 1994) and mycorrhizal fungi (Aanen *et al.*, 2000), to provide only a few examples. Although early use of PSR with molecular data was often based on data from a single locus, this practice was criticized as there was no basis by which to determine whether a single-gene phylogeny truly reflected the organismal phylogeny (Pamilo and Nei, 1988; Takahata, 1989; Avise and Wollenberg, 1997; Rosenberg, 2002). Taylor *et al.* (2000) described Genealogical Concordance Phylogenetic Species Recognition (GCPSR), which diagnoses species boundaries based on genealogical concordance of multiple gene phylogenies. Because GCPSR is technically straightforward and provides a more objective basis for determining species boundaries than either MSR or BSR, it is rapidly becoming the species recognition criterion of choice in mycology.

Application of GCPSR (Taylor *et al.*, 2000) is now integral in the determination of species boundaries and inference of relationships in many fungi (e.g. Koufopanou *et al.*, 1997; Geiser *et al.*, 1998; Kasuga *et al.*, 1999; O'Donnell *et al.*, 2000a,b; Dettman *et al.*, 2003a,b, 2006; Matute *et al.*, 2006). A frequent outcome of GCPSR studies has been an increase in the numbers of species resolved as compared to MSR and BSR. This result suggests that the rate of molecular divergence frequently exceeds the rate of morphological differentiation among sister lineages (Taylor *et al.*, 2000). Also, GCPSR has revealed instances where mating compatibility was retained between phylogenetically diverged allopatric populations, validating the criticism that BSR may yield overly broad species concepts because the ability to interbreed may be retained between physically separated yet phylogenetically distinct populations (Dettman *et al.*, 2003, 2006).

## 7.4. Species-level Phylogenies of Entomopathogenic Fungi

To date, the majority of species-level molecular phylogenetic investigations of entomopathogenic fungi have focused on genera of clavicipitalean fungi. Chaverri *et al.* (2008) conducted a morphological and molecular phylogenetic monograph of the neotropical species of *Hypocrella* and relatives and aschersonia-like anamorphs. Their multi-gene phylogeny, based on LSU rDNA, EF-1α and RPB1, supported the monophyly of three genera, the previously described *Hypocrella* and *Moelleriella*, and a new genus, *Samuelsia*, described in this study. Also, the monophyly of 28 morphological species was supported, although molecular diversity in some terminal species lineages suggest that these lineages may contain more than one morphologically cryptic species (Chaverri *et al.*, 2008). The *Aschersonia* monograph provides a very useful summary of Neotropical species diversity, their relationships and detailed information on the morphological identification of both the sexual and asexual stages of the life cycle. The phylogeny of *Cordyceps* and clavicipitalean fungi by Sung *et al.* (2007) also supports the monophyly of a *Hypocrella* clade within the Clavicipitaceae and these genera are basal to the clade containing *Claviceps* and *Epichlöe* and other parasitic and plant endophytic taxa.

The genus *Paecilomyces* sect. *Isarioidea* contains about a dozen species of anamorphic entomopathogens, including *Paecilomyces tenuipes*, *Paecilomyces farinosus* and *Paecilomyces fumosoroseus*. However, a SSU rDNA phylogeny of the genus *Paecilomyces sensu* (Samson, 1974) by Luangsa-ard *et al.* (2004) demonstrated that the genus is polyphyletic and includes taxa with affinities to Eurotiomycetes as well to several lineages of clavicipitalean fungi in Hypocreales. In a following phylogenetic study, Luangsa-ard *et al.* (2005) reclassified most species of *Paecilomyces* sect. *Isarioidea* to the genus *Isaria* (Gams *et al.*, 2005; Hodge *et al.*, 2005), family Cordycipitaceae (Sung *et al.*, 2007). Other clavicipitalean *Paecilomyces* spp. have different family affinities with *Paecilomyces lilacinus* derived from within the Ophiocordycipitaceae and *Paecilomyces carneus* and *Paecilomyces marquandii* from within Clavicipitaceae (Sung *et al.*, 2007).

Rehner and Buckley (2005) conducted a global phylogenetic survey of *Beauveria* using ITS and EF-1α sequences. This phylogeny provided congruent support for genus monophyly and for six internal lineages that include all the described morphological species of the genus. A unique finding of this study was the discovery that the morphospecies *Beauveria bassiana* s.l. consists of a pair of phylogenetically divergent, but morphologically indistinguishable species lineages. The more widely distributed and frequently encountered lineage corresponds to *Beauveria bassiana* s.s., whereas the second, less frequently occurring lineage is an undescribed species, informally referred to as clade C (Rehner and Buckley, 2005). Also, EF-1α resolved phylogenetic structure within *B. bassiana* s.s., indicating that this globally distributed species may be a cryptic complex. The species *Beauveria malawiensis* (Rehner *et al.*, 2006) was first detected by phylogenetic screening with EF-1α and ITS, and both morphologically and phylogenetically distinct.

In another study of *B. bassiana* s.s., two nuclear intergenic regions, *Bloc* and *EFutr*, each of which displays greater nucleotide variability than either EF-1α or

ITS in *B. bassiana* s.l., were used in a phylogenetic epidemiological investigation of *B. bassiana* pathogens of the coffee berry borer in the neotropics and Africa (Rehner *et al.*, 2006). A phylogeny based on these loci showed that many groups of isolates from different continents were phylogenetically distinct, demonstrating that *B. bassiana* s.s. consists of multiple lineages, many of which are continental endemics. In both Africa and the neotropics, multiple cryptic species lineages were isolated from coffee berry borer as well as from other insect species, indicating a wide host range even at the level of individual cryptic species. Additionally, *B. bassiana* s.l. was also isolated as an endophyte of coffee, which suggests that members of this complex may have other ecological associations that diversify their position and options in the environment. The dominant cryptic lineage of *B. bassiana* s.l. on coffee berry borer in both the neotropics and Africa were closely related. However, the significantly greater genetic variability observed in the Neotropical population suggests that it may be ancestral to the African population, and dispersed to Africa prior to the global expansion of coffee agriculture.

GCPSR has also been used to critically examine problematic MSRs in the genus *Metarhizium*. Bischoff and Rehner (2006) assessed the status of *Metarhizium anisopliae* var. *frigidum* using a multilocus data set combining partial sequences of EF-1$\alpha$, RPB1 and RPB2. Their analysis supported the recognition of *M. frigidum* as a distinct species, demonstrating that morphological crypsis may occur among distantly related species analogous to that observed in the *B. bassiana* morphospecies. Additionally, the phylogeny supported the monophyly of both *Metarhizium flavoviride* vars. *pemphigi* and *minus* and these are recognized at species rank and are related to *M. frigidum* and *M. flavoviride*, respectively.

GCPSR can also be used as a tool for differentiating species in local or regional ecological surveys where prior knowledge of species diversity is lacking. In a study of *Metarhizium* in forest and agricultural soils in Ontario, Canada, Bidochka *et al.* (2005) used GCPSR on a data set of six nuclear loci to detect two genetically distinct groups within the *M. anisopliae* morphospecies. These groups had previously been described to differ in ecological distribution, growth temperature response, UV-tolerance and by multiple fixed differences in several categories of molecular markers, including isozymes, PCR-RFLP and RAPD markers (Bidochka *et al.*, 2001). Together these studies suggest that phylogenetic species may possess distinctive biological attributes that can provide essential clues as to their ecology and habitat distribution, information that can be extremely useful in the selection and management of these organisms for insect biological control.

## 7.5. Conclusions

Entomopathogenic fungi are not a monophyletic group, but rather a heterogeneous assemblage of fungi independently derived from Ascomycota, Basidiomycota, Entomophthoromycotina, Blastocladiales, Kickxellomycotina, Neocallimastigomycota and Microsporidia. Understanding the origins and diversity of entomopathogenic fungi is thus inextricably linked to progress in understanding the evolution of Fungi. With the increasing efficiency and lower cost of whole-genome sequencing, whole-genome comparisons of entomopathogenic fungi will refine knowledge of their evolutionary history and enable direct insight into the genomic biology of entomopathogenesis.

Nucleotide sequences from EF-1α, RPB1 and RPB2 have proven extremely informative in reconstructing the phylogeny and revising the classification of clavicipitalean entomopathogens in Hypocreales (Sung *et al.*, 2007). Additionally, these same loci have been highly useful in resolving relationships and refining species concepts within several entomopathogenic genera, including *Beauveria*, *Metarhizium and Aschersonia*. The potential to cross-reference molecular phylogenetic data generated in different laboratories by sequencing a common set of loci is perhaps one of the most powerful unifying forces of modern molecular systematics. By working from a common database of data generated from vouchered reference specimens, workers from around the world can more rapidly recognize and characterize novel lineages and species of fungi that contribute to the planet's fungal biodiversity.

## References


Aanen, D.K., Kuyper, T.W., Mes, T.H.M. and Hoekstra, R.F. (2000) The evolution reproductive isolation in the ectomycorrhizal *Hebeloma crustuliniforme* aggregate (Basidiomycetes) in northwestern Europe: a phylogenetic approach. *Evolution* 54, 1192–1206.

Aine, M.C., Matheney, B.P., Henk, D.A., Frieders, E.J., Nilsson, R.H., Piepenbring, M., McLaughlin, D.J., Sizabo, L.J., Begerow, D., Sampaio, J.P., Bauer, R., Weiss, M., Oberwinkler, F. and Hibbett, D.S. (2006) An overview of the higher-level classification of Pucciniomycotina based on combined analyses of nuclear large and small subunit rDNA sequences. *Mycologia* 98, 896–905.

Altschul, S.F., Gish, W., Miller, W., Myers, E.W. and Lipman, D.J. (1990) Basic local alignment search tool. *Journal of Molecular Biology* 215, 403–410.

Anderson, D.L., Gibbs, A.J. and Gibson, N.L. (1998) Identification and phylogeny of spore-cyst fungi (*Ascosphaera* spp.) using ribosomal DNA sequences. *Mycological Research* 102, 541–547.

Anderson, J.B. and Ullrich, R.C. (1979) Biological species of *Armillaria mellea* in North America. *Mycologia* 71, 402–414.

Avise, J.C. and Ball, R.M. (1990) Principles of genealogical concordance in species concepts and biological taxonomy. In: Futuyma, D.J. and Antonovics, J. (eds) *Oxford Surveys in Evolutionary Biology*. Oxford University Press, Oxford, pp. 45–67.

Avise, J.C. and Wollenberg, K. (1997) Phylogenetics and the origin of species. *Proceedings of the National Academy of Science of the USA* 94, 7748–7755.

Baldauf, S.L. (1999) A search for the origins of animals and fungi: comparing and combining molecular data. *American Naturalist* 154(suppl.), S178–S188.

Baldauf, S.L. and Doolittle, W.F. (1997) Origin and evolution of the slime molds (Mycetozoa). *Proceedings of the National Academy of Science of the USA* 94, 12007–12012.

Baldauf, S.L. and Palmer, J.D. (1993) Animals and fungi are each other's closest relatives – congruent evidence from multiple proteins. *Proceedings of the National Academy of Sciences of the USA* 90, 11558–11562.

Baldauf, S.L., Roger, A.J., Wnek-Siefert, I. and Doolittle, W.F. (2000) A kingdom-level phylogeny of eukaryotes based on combined protein data. *Science* 290, 972–977.

Baum, D. (1992) Phylogenetic species concepts. *Trends in Ecology and Evolution* 7, 1–2.

Baum, D. and Shaw, K.L. (1995) Genealogical perspective on the species problem. In: Hoch, P.S. and Stephenson, A.C. (eds) *Experimental and Molecular Approaches to Plant Biosystematics*. Missouri Botanical Garden, St Loius, Montana, pp. 289–303.

Berbee, M.L. and Taylor, J.W. (2001) Fungal molecular evolution: gene trees and geologic time. In: McLaughlin, D.J., McLaughlin, E.G. and Lemke, P.A. (eds) *The Mycota, vol. 7A, Systematics and Evolution*. Springer, New York, pp. 229–245.

Bidochka, M.J., Kamp, A.M., Lavender, T.M., Dekoning, J. and Amritha de Croos, J.N. (2001) Habitat association in two genetic groups of the insect-pathogenic fungus *Metarhizium anisopliae*: uncovering cryptic species? *Applied and Environmental Microbiology* 67, 1335–1342.

Bidochka, M.J., Cherrie-Lee, N.S. and Spironello, M. (2005) Recombination within sympatric cryptic species of the insect pathogenic fungus *Metarhizium anisopliae*. *Environmental Microbiology* 7, 1361–1368.

Bischoff, J.F. and Rehner, S.A. (2006) *Metarhizium frigidum* sp. nov.: a cryptic species of *M. anisopliae* and a member of the *M. flavoviride* complex. *Mycologia* 98, 737–745.

Bischoff, J.F. and White, J. (2003) The plant-infecting Clavicipitaleans. In: White, J.F., Bacon, C.W., Hywel Jones, N.L. and Spatafora, J.W. (eds) *Clavicipitalean Fungi: Evolutionary Biology, Chemistry, Biocontrol, and Cultural Impacts*. Marcel Dekker, Basel, Switzerland, pp. 125–150.

Blackwell, M. (1994) Minute mycological mysteries: the influence of arthropods on the lives of fungi. *Mycologia* 86, 1–17.

Blackwell, M. and Malloch, D. (1989) *Pyxidiophora*: a link between the Laboulbeniales and hyphal ascomycetes. *Memoirs of the New York Botanical Garden* 49, 23–32.

Brasier, C.M. (1987) The dynamics of fungal speciation. In: Rayner, A.D.M., Brasier, C.M. and Moore, D. (eds) *Evolutionary Biology of the Fungi*. Cambridge University Press, Cambridge, pp. 231–260.

Brasier, C.M. (1997) Fungal species in practice: identifying species units in fungi. In: Claridge, M.F., Dawah, H.A. and Wilson, M.R. (eds) *Species: The Units of Biodiversity*. Chapman and Hall, London, pp. 135–170.

Burgess, T.I., Barber, P.A., Mohali, S., Pegg, G., de Beer, W. and Wingfield, J.J. (2006) Three new *Lasiodiplodia* spp. from the tropics, recognized based on DNA sequence comparisons and morphology. *Mycologia* 98, 423–435.

Burnett, J.H. (1983) Speciation in fungi. *Transactions of the British Mycological Society* 81, 1–14.

Burnett, J.H. (2003) *Fungal Populations and Species*. Oxford University Press, Oxford, 348.

Carruthers, R.J. and Soper, D.S. (1987) Fungal diseases. In: Fuxa, J.R. and Tanada, Y. (eds) *Epizootiology of Insect Diseases*. John Wiley & Sons, pp. 357–416.

Castlebury, L.A., Rossman, A.Y., Sung, G.-H., Hyten, A.S. and Spatafora, J.W. (2004) Multigene phylogeny reveals new lineage for *Stachybotrys chartarum*, the indoor air fungus. *Mycological Research* 108, 864–872.

Cavalier-Smith, T. (1998) A revised six-kingdom system of life. *Biological Reviews of the Cambridge Philosophical Society* 73, 203–266.

Cavalier-Smith, T. (2001) What are fungi? In: McLaughlin, D.J., McLaughlin, E.G. and Lemke P.A. (eds) *The Mycota, vol. 7A, Systematics and Evolution*. Springer, New York, pp. 3–37.

Cavalier-Smith, T., Allsop, M.T.E.P. and Chao, E.E. (1994) Thraustochytrids are chromists, not fungi: 18S rRNA signatures of heterokonta. *Philosophical Transactions of the Royal Society of London, Series B* 339, 139–146.

Chaverri, P., Liu, M. and Hodge, K.T. (2008) A monograph of the entomopathogenic genera *Hypocrella*, *Moelleriella*, and *Samuelsia* gen. nov. (Ascomycota, Hypocreales, Clavicipitaceae), and their aschersonia-like anamorphs in the Neotropics. *Studies in Mycology* 60, 1–66.

Claridge, M.F., Dawah, H.A. and Wilson, M.R. (1997) *Species: The Units of Biodiversity*. Chapman and Hall, London, 461.

Clémençon, H. (1977) *The Species Concept in Hymenomycetes*. J. Cramer, Vaduz, 444.

Couch, J.N. (1938) The genus *Septobasidium*. Chapel Hill, UNC Press, North Carolina.

Coyne, J.A. and Orr, H.A. (2004) *Speciation*. Sinauer Associates, Sunderland, Massachusetts, 545.

Cracraft, J. (1983) Species concepts and speciation analysis. *Current Ornithology* 1, 159–187.

Cracraft, J. (1989) Speciation and its ontology. The empirical consequences of alternative species concepts for understanding patterns and processes of differentiation. In: Ott, D. and Endler, J.A. (eds) *Speciation and Its Consequences*. Sinauer Associates, Sunderland, Massachusetts, pp. 28–59.

Dettman, J.R., Jacobson, D.J. and Taylor, J.W. (2003a) A multilocus genealogical approach to phylogenetic species recognition in the model eukaryote *Neurospora*. *Evolution* 57, 2703–2720.

Dettman, J.R., Jacobson, D.J., Turner, E., Pringle, A. and Taylor, J.W. (2003b) Reproductive isolation and phylogenetic divergence in *Neurospora*: comparing methods of species recognition in a model eukaryote. *Evolution* 57, 2721–2741.

Dettman, J.R., Jacobson, D.J. and Taylor, J.W. (2006) Multilocus sequence data reveal extensive phylogenetic species diversity within the *Neurospora discreta* complex. *Mycologia* 98, 436–446.

Ereshefsky, M. (ed.) (1992) *The Units of Evolution: Essays on the Nature of Species*. MIT Press, Cambridge, Massachusetts, 432.

Fast, N.M. and Keeling P.J. (2005) The fungal roots of microsporidian parasites. In: Vega, F.E. and Blackwell, M.E. (eds) *Insect-Fungal Associations: Ecology, and Evolution*. Oxford University Press, New York.

Felsenstein, J. (1978) Cases in which parsimony and compatibility methods will be positively misloading. *Systematic Zoology* 27, 401–410.

Förster, H., Coffey, M.D., Elwood, H. and Sogin, M.L. (1990) Sequence analysis of the small subunit ribosomal RNA of three zoosporic fungi and implications for fungal evolution. *Mycologia* 82, 306–312.

Futuyma, D.J. (1998) *Evolutionary Biology*. Sinauer Associates. Sunderland, Massachusetts, 763.

Gams, W., Hodge, K.T., Samson, R.A., Korf, R.P. and Seifert, K.A. (2005) Proposal to conserve the name Isaria (anamorphic fungi) with a conserved type. *Taxon* 52, 537.

Geiser, D.M., Pitt, J.I. and Taylor, J.W. (1998) Cryptic speciation and recombination in the aflatoxin-producing fungus *Aspergillus flavus*. *Proceedings of the National Academy of Science of the USA* 95, 388–393.

Ghiselin, M.T. (1987) Species concepts, individuality, and objectivity. *Biology and Philosophy* 2, 127–143.

Glenn, A.E., Bacon, C.W., Price, R. and Hanlin, R.T. (1996) Molecular phylogeny of *Acremonium* and its taxonomic implications. *Mycologia* 88, 369–383.

Grant, V. (1981) *Plant Speciation*, 2nd edn. Columbia University Press, New York, 563.

Harrington, T.C. and Rizzo, D.M. (1999) Defining species in the fungi, In: Worrall, J.J. (ed.) *Structure and Dynamics of Fungal Populations*. Kluwer Press, Dordrecht, The Netherlands, pp. 42–71.

Harrington, T.C., Worrall, J.J. and Rizzo, D.M. (1989) Compatibility among host specialized isolates of *Heterobasidion annosum* from western North America. *Phytopathology* 79, 290–296.

Hausner, G., Belkhiri, A. and Klassen, G.R. (2000) Phylogenetic analysis of the small subunit ribosomal RNA gene of the hyphochytrid *Rhizidiomyces apophysatus*. *Canadian Journal of Botany* 78, 124–128.

Henk, D.A. and Vilgalys, R. (2007) Molecular phylogeny suggests a single origin of insect symbiosis in the Urediniomycetidae with support for some relationships within the *Septobasidium*. *American Journal of Botany* 94, 1515–1526.

Hibbett, D.S. and Donoghue, M.J. (2002) Implication of phylogenetic studies for conservation of genetic diversity in shiitake mushrooms. *Conservation Biology* 10, 1321–1327.

Hibbett, D.S., Binder, M., Bischoff, J.F., Blackwell, M., Cannon, P.F., Eriksson, O., Huhndorf, S., James, T., Kirk, P.M., Lücking, R., Lumbsch, T., Lutzoni, F., Matheny, P.B., McLaughlin, D.J., Powell, M.J., Redhead, S., Schoch, C.L., Spatafora, J.W., Stalpers, J.A., Vilgalys, R., Aime, M.C., Aptroot, A., Bauer, R., Begerow, D., Benny, G.L., Castlebury, L.A., Crous, P.W., Dai, Y.-C., Gams, W., Geiser, D.M., Griffith, G.W., Gueidan, C., Hawksworth, D.L., Hestmark, G., Hosaka, K., Humber, R.A., Hyde, K., Koljalg, U., Kurtzman, C.P., Larsson, K.-H., Lichtward, R., Longcore, J., Miadlikowska, J., Miller, A., Monclavo, J.-M., Mozley-Standridge, S., Oberwinkler, F., Parmasto, E., Reeb, V., Rogers, J.D., Roux, C., Ryvarden, L., Sampaio, J.P., Schuessler, A., Sugiyama, J., Thorn, R.G., Tibell, L.,

Untereiner, W.A., Walker, C., Wang, Z., Weir, A., Weiss, M., White, M., Winka, K., Yao, Y.-J. and Zhang, N. (2007) A higher-level phylogenetic classification of the Fungi. *Mycological Research* 111, 509–547.

Hirt, R.P., Logsdon, J.M., Healy, B., Dorey, M.W., Doolittle, W.F. and Embley, T.M. (1999) Microsporidia are related to Fungi: evidence from the largest subunit of RNA polymerase II and other proteins. *Proceedings of the National Academy of Science of the USA* 96, 580–585.

Hodge, K.T., Gams, W., Samson, R.A. and Korf, K.A. (2005) Lectotypification and status of *Isaria* Pers. Fr. *Taxon* 52, 485–489.

Hull, D.L. (1997) The ideal species concept – and why we can't get it. In: Claridge, M.F., Dawah, H.A. and Wilson, M.R. (eds) *Species: The Units of Biodiversity*. Chapman and Hall, London, pp. 357–377.

Humber, R.A. (1981) An alternative view of certain taxonomic criteria used in the Entomopthorales. *Mycotaxon* 13, 191–240.

Humber, R.A. (1989) Synopsis of a revised classification for the Entomopthorales (Zygomycotina). *Mycotaxon* 34, 441–460.

James, T.Y., Kauff, F., Schoch, C., Matheny, B., Hofstetter, V., Cox, C.J., Celio, G., Guiedan, C., Fraker, E., Miadlikowska, J., Lumbsh, T., Rauhut, A., Reeb, V., Arnold, A., Amtoft, A., Stajich, J.E., Hosaka, K., Sung, G., Johnson, D., O'Rourke, B., Crockett, M., Binder, M., Curtis, J.M., Slot, J.C., Wang, Z., Wilson, A.W., Schueller, A., Longcore, J.E., O Donnell, K., Mozley-Standridge, S., Porter, D., Letcher, P.M., Powell, M.J., Taylor, J.W., White, M.M., Griffith, G.W., Davies, D.R., Humber, R.A., Morton, J.B., Sugiyama, J., Rossman, A.Y., Rogers, J.D., Pfister, D.H., Hewitt, D., Hansen, K., Hambleton, S., Shoemaker, R.A., Kohlmeyer, J., Volkmann-Kohlmeyer, B., Spotts, R.A., Serdani, M., Crous, P.W., Hughes, K.W., Matsuura, K., Langer, E., Langer, G., Untereiner, W.A., Lucking, R., Budel, B., Geiser, D.M., Aptroot, D.M., Diederich, P., Schmitt, I., Schultz, M., Yahr, R., Hibbett, D.S., Lutzoni, F., Mclaughlin, D.J., Spatafora, J.W. and Vilgalys, R. (2006a) Reconstructing the early evolution of Fungi using a six-gene phylogeny. *Nature* 443, 818–822.

James, T.Y., Letcher, P.M., Longcore, J.E., Mozley-Standridge, S.E., Porter, D., Powell, M., Griffith, G.W. and Vilgalys, R. (2006b) A molecular phylogeny of the flagellated fungi (Chytridiomycota) and description of a new phylum (Blastocladiomycota). *Mycologia* 98, 860–871.

Jensen, A.G., Gargas, A., Eilenberg, J. and Rosendahl, S. (1998) Relationships of the insect-pathogenic order Entomophthorales (Zygomycota, Fungi) based on phylogenetic analysis of nuclear small subunit ribosomal DNA sequences (SSU rDNA). *Fungal Genetics and Biology* 24, 325–334.

Kasuga, T.J., Taylor, J.W. and White, T.J. (1999) Phylogenetic relationships of varieties and geographical groups of the human pathogenic fungus *Histoplasma capsulatum* Darling. *Journal of Clinical Microbiology* 37, 653–663.

Keeling, P.J. (2003) Congruent evidence from alpha-tubulin and beta-tubulin gene phylogenies for a zygomycete origin of microsporidia. *Fungal Genetics and Biology* 38, 298–309.

Keeling, P.J. and McFadden, G.I. (1998) The origins of microsporidia. *Trends in Microbiology* 6, 19–23.

Keeling, P.J., Luker, M.A. and Palmer, J.D. (2000) Evidence from beta-tubulin phylogeny that microsporidia evolved from within the fungi. *Molecular Biology and Evolution* 17, 23–31.

Kendrick, B. (1989) 'Subdivision Deuteromycotina' – a fungal chimera. *Sydowia* 41, 6–14.

Kohn, L.M. (2005) Mechanisms of fungal speciation. *Annual Review of Phytopathology* 43, 279–308.

Koufopanou, V., Burt, A. and Taylor, J.W. (1997) Condordance of gene genealogies reveals reproductive isolation in the pathogenic fungus *Coccidioides immitis*. *Proceedings of the National Academy of Science of the USA* 94, 5478–5482.

Lang, B.F., O'Kelly, C.O., Nerad, T., Gray, M.W. and Burger, G. (2002) The closest unicellular relative of animals. *Current Biology* 12, 44–68.

Leipe, D.D., Wainright, P., Gunderson, J.H., Porter, D., Patterson, D.J., Valois, F., Himmerich, S. and Sogin, M.L. (1994) The stramenopiles from a molecular perspective: 16S like rRNA sequences from *Labyrinthuloides minuta* and *Cafeteria roenbergensis*. *Phycologia* 33, 369–377.

Li, Z., Li, C., Huang, B. and Fan, M. (2001) Discovery and demonstration of the teleomorph of *Beauveria bassiana* (Bals.) Vuill. an important entomogenous fungus. *Chinese Scientific Bulletin* 46, 751–753.

Liang, Z.-Q., Liu, A.-Y. and Liu, J.-L. (1991) A new species of the genus *Cordyceps* and *Metarhizium* anamorph. *Acta Sinica* 10, 257–262.

Liu, Z.-Y., Liang, Z.-Q., Whalley, A.J.S., Yao, Y.-J. and Liu, A.-y. (2001) *Cordyceps brittlebankisoides*, a new pathogen of grubs and its anamorph, *Metarhiziium anisopliae var. majus*. *Journal of Invertebrate Pathology* 78, 178–182.

Luangsa-ard, J.J., Hywel-Jones, N.L. and Samson, R.A. (2004) The polyphyletic nature of *Paecilomyces* sensu lato based on 18S generated rDNA phylogeny. *Mycological Research* 109, 581–589.

Luangsa-ard, J.J., Hywel-Jones, N.L., Manoch, L. and Samson, R.A. (2005) On the relationships of *Paecilomyces* sect. *Isarioidea* species. *Mycologia* 96, 773–780.

Lutzoni, F., Kauff, F., Cox, C.J., McLaughlin, D., Celio, G., Dentinger, B., Padamsee, M., Hibbett, D., James, T.Y., Baloch, E., Grube, M., Reeb, V., Miadlikowska, J., Spatafora, J., Johnson, D., Sung, G.-H., Lucking, R., Lumbsch, T., O'Donnell, K., Binder, M., Diederich, P., Harris, R.C., Hosaka, K., Lim, Y.W., Matheny, B., Nishida, H., Pfister, D., Rogers, J., Rossman, A., Schmitt, I., Sipman, H., Stone, J., Sugiyama, J., Yahr, R. and Vilgalys, R. (2004) Assembling the fungal tree of life: progress, classification and evolution of subcellular traits. *American Journal of Botany* 91, 1446–1480.

Mallet, J. (2001) Concepts of species. *Encylopedia of Biodiversity* 5, 427–440.

Matute, D.R., McEwen, J.G., Puccia, R., Montes, G.A., San-Blas, G., Bagagli, E., Rauscher, J.T., Restrespo, A., Morais, F., Nino-Vega, G. and Taylor, J.W. (2006) Cryptic speciation and recombination in the fungus *Paracoccidioides brasiliensis* as revealed by gene genealogies. *Molecular Biology and Evolution* 23, 65–73.

Mayr, E. (1942) *Systematics and Origin of Species*. Columbia University Press, New York, 392.

Mayr, E. (1963) *Animal Species and Evolution*. Belknap Press, Cambridge, Massachusetts, 797.

Mishler, B.D. and Donoghue, M.J. (1982) Species concepts: a case for pluralism. *Systematic Zoology* 31, 491–503.

Natvig, D.O. and May, G. (1996) Fungal evolution and speciation. *Journal of Genetics* 75, 441–452.

O'Donnell, K., Kistler, H.C., Tacke, B.K. and Casper, H.H. (2000a) Gene genealogies reveal global phylogeographic structure and reproductive isolation among lineages of *Fusarium graminearum*, the fungus causing wheat scab. *Proceedings of the National Academy of Sciences of the USA* 97, 7905–7910.

O'Donnell, K., Nirenberg, H.I., Aoki, T. and Cigelnik, E. (2000b) A multigene phylogeny of the *Gibberella fujikori* species complex: detection of additional phylogenetically distinct species. *Mycoscience* 41, 61–78.

Pamilo, P. and Nei, M. (1988) Relationship between gene trees and species trees. *Molecular Biology and Evolution* 5, 568–583.

Petersen, R.H. and Hughes, K.W. (1999) Species and speciation in mushrooms. *Bioscience* 49, 440–452.

Pringle, A., Baker, D.M., Platt, J.L., Wares, J.P., Latge, J.P. and Taylor, J.W. (2005) Cryptic speciation in the cosmopolitan and clonal pathogenic fungus *Aspergillus fumigatus*. *Evolution* 59, 1886–1899.

Qin, X., Evans, J.D., Aronstein, K.A., Murray, K.D. and Weinstock, G.M. (2006) Genome sequences of the honeybee pathogens *Paenibacillus larvae* and *Ascosphaera apis*. *Insect Molecular Biology* 15, 715–718.

Redecker, D. and Raab, P. (2006) Phylogeny of the Glomeromycota (arbuscular mycorrhizal fungi): recent developments and new gene markers. *Mycologia* 98, 885–895.

Rehner, S.A. and Buckley, E.P. (2005) A *Beauveria* phylogeny inferred from nuclear ITS and EF1-a sequences: evidence of cryptic diversification and links to *Cordyceps* teleomorphs. *Mycologia* 97, 84–98.

Rehner, S.A. and Samuels, G.J. (1995) Molecular systematics of the Hypocreales: a teleomorph gene phylogeny and the status of their anamorphs. *Canadian Journal of Botany* 73, S816–S823.

Rehner, S.A., Posada, F., Buckley, E.P., Infante, F., Castille, A. and Vega, F.E. (2006) Phylogenetic origins of African and Neotropical *Beauveria bassiana s.l.* pathogens of coffee berry borer, *Hypothenemus hampei*. *Journal of Invertebrate Pathology* 93, 11–21.

Robbertse, B., Reeves, J.B., Schoch, C.L. and Spatafora, J.W. (2006) A phylogenomic analysis of the Ascomycota. *Fungal Genetics and Biology* 43, 715–725.

Rosen, D.E. (1979) Fishes from the uplands and intermontane basins of Guatemala: revisionary studies and comparative geography. *Bulletin of the American Museum of Natural History* 162, 267–375.

Rosenberg, N.A. (2002) The probability of topological concordance of gene trees and species trees. *Theoretical Population Biology* 61, 225–247.

Samson, R.A. (1974) *Paecilomyces* and some allied Hyphomycetes. *Studies in Mycology* 6, 1–119.

Seeley, T.D. and Tarpy, D.R. (2007) Queen promiscuity lowers disease within honeybee colonies. *Proceedings of the Royal Society of London B-Biological Sciences*, 274, 67–72.

Shimodaira, H. (2002) An approximately unbiased test of phylogenetic tree selection. *Systematic Biology* 51, 492–508.

Spatafora, J.W. and Blackwell, M. (1993) Molecular systematics of unitunicate perithecial ascomycetes: the Clavicipitales–Hypocreales connection. *Mycologia* 85, 912–922.

Spatafora, J.W. and Blackwell, M. (1994) The polyphyletic origins of ophiostomatoid fungi. *Mycological Research* 98, 1–9.

Spatafora, J.W., Hughes, K.W. and Blackwell, M. (2006) A phylogeny for kingdom Fungi Deep Hypha Issue. *Mycologia* 98, 1–1105.

Spatafora, J.W., Sung, G.-H., Sung, J.-M., Hywel-Jones, N.L. and White, J.F. (2007) Phylogenetic evidence for an animal pathogen origin of ergot and grass endophytes. *Molecular Ecology* 16, 1701–1711.

Sung, G.-H., Spatafora, J.W., Zare, R., Hodge, K.T. and Gams, W. (2001) A revision of *Verticillium* sect. Prostrata. II. Phylogenetic analyses of SSU and LSU nuclear rDNA sequences from anamorphs and teleomorphs of the Clavicipitaceae. *Nova Hedwigia* 72, 311–328.

Sung, G.H., Hywel-Jones, N., Sung, J.-M., Luangsa-ard, J.J. and Spatafora, J.W. (2007) Phylogenetic classification of *Cordyceps* and the clavicipitaceous fungi. *Studies in Mycology* 57, 1–63.

Takahata, N. (1989) Gene genealogy in three related populations: consistency probability between gene and population trees. *Genetics* 122, 957–966.

Tarpy, D.F. and Seeley, T.D. (2006) Lower disease infection in honeybee (*Apis mellifera*) colonies headed by polyandrous vs. monandrous queens. *Naturwissenschaften* 93, 195–199.

Taylor, J.W., Jacobson, D.J., Kroken, S., Kasuga, T., Geiser, D.M., Hibbett, D.S. and Fisher, M.C. (2000) Phylogenetic species recognition and species concepts in fungi. *Fungal Genetics and Biology* 31, 21–32.

Tehler, A., Little, D.P. and Farres, J.S. (2003) The full-length phylogenetic tree from 1551 ribosomal sequences of chitinous fungi, Fungi. *Mycological Research* 107, 901–916.

Ustinova, I., Kreinitz, L. and Huss, V.A.R. (2000) *Hyaloraphidium curvatum* is not a green alga, but a lower fungus; *Amoebidium parasiticum* is not a fungus, but a member of the DRIPs. *Protist* 151, 253–262.

Vilgalys, R. and Sun, B.L. (1994) Ancient and recent patterns of geographic speciation in the oyster mushroom *Pleurotus* revealed by phylogenetics analysis of ribosomal DNA sequences. *Proceedings of the National Academy of Sciences of the USA* 91, 4599–4603.

Wainright, P.O., Hinkle, G., Sogin, M.L. and Stickel, S.K. (1993) Monophyletic origins of the Metazoa: 'an evolutionary link with fungi.' *Science* 260, 340–342.

Weir, A. and Blackwell, M. (2001) Molecular data support the Laboulbeniales as a separate class of Ascomycota, Laboulbeniomycetes. *Mycological Research* 105, 715–722.
Wheeler, Q.D. and Meier, R. (2000) *Species Concepts and Phylogenetic Theory*. Columbia University Press, New York, 247.
White, M.J.D. (1973) *Animal Cytology and Evolution*. Cambridge University Press, London.
White, M.M., James, T.Y., O'Donnell, K., Cafaro, M., Tanabe, Y. and Sugiyama, J. (2006) Phylogeny of the Zygomycota based on nuclear ribosomal sequence data. *Mycologia* 98, 872–884.
White, T.J., Bruns, T., Lee, S. and Taylor, J. (1990) Amplification and direct sequencing of fungal ribosomal RNA genes for phylogenetics. In: Innis, M.A., Gelfand, D.H., Sninsky, J.J. and White, T.J. (eds) *PCR Protocols: A Guide to Methods and Applications*. Academic Press, San Diego, California, pp. 315–322.
Wubah, D.A. and Fuller, M.S. (1993a) *Neocallomastix*: a comparative morphological study. *Canadian Journal of Botany* 69, 835–843.
Wubah, D.A. and Fuller, M.S. (1993b) *Caecomyces communis*: morphological and development. *Mycologia* 83, 303–310.

# 8 Phylogenetics and Population Genetics of Entomopathogenic and Insect-parasitic Nematodes

S.M. Peat,[1] B.C. Hyman[2] and B.J. Adams[1]

[1]*Department of Microbiology and Molecular Biology, Brigham Young University, Provo, USA;* [2]*Department of Biology, University of California, Riverside, USA*



## 8.1. Introduction

Phylogenetics, and particularly molecular systematics, has played a key role in numerous advances in the study of entomopathogenic, entomophilic and insect-parasitic nematodes. The contribution of molecular systematics and population genetics to both applied and fundamental research on these organisms is most evident in taxonomic endeavours, but has also been integral to expanding knowledge of their biodiversity, geographic distributions, host ranges, ecology, behaviour and co-evolution (Campbell *et al.*, 2003; Adams *et al.*, 2006). In this chapter, we present a brief introduction to phylogenetics, population genetics and DNA bar coding, and discuss the genetic loci and analytical tools (software) that are relevant to applying them to entomogenous nematodes.

## 8.2. Phylogenetics

Phylogenetic systematics, the study and process of recovering the historical relationships among species and taxonomic groups, has greatly aided in the study of insect-parasitic and pathogenic nematode diversity and their evolution. Phylogenies not only reveal the hierarchical relationships among taxa, but they can also be used as a contextual framework to study the evolution of life-history traits, morphological traits, behaviour, pathogenicity and any other characteristic of entomopathogenic nematodes. Phylogenies are also the foundation of research programmes in historical biogeography, phylogeography, historical ecology and co-evolution (Brooks and McLennan, 1991, 2002; Avise, 2000, 2004). A reconstructed past, in the form of a phylogeny, provides the historical context required for inferences of evolutionary change to be tested, shedding light on the origin, mode, tempo and maintenance of entomogenous nematode diversity that we see today.

The development of phylogenetic systematics emerged as biologists began to embrace Darwin's notion that classifications should reflect evolutionary relationships (Andrássy, 1976), with pheneticists and cladists in conflict over which philosophical approach was most appropriate to use for constructing evolutionary relationships. Pheneticists considered overall similarity to be the best indicator of phylogeny, whereas cladists argue that only shared, derived characters appropriately and accurately reflected evolutionary relationships. The cladists prevailed, in terms of their logical and empirical arguments, as well as having their methods widely accepted and adopted by subsequent researchers. Today, a wide variety of methods exist for building phylogenetic trees. These include phenetic methods such as unweighted pair group method with arithmetic mean (UPGMA) or neighbour-joining and parsimony and model-based methods like maximum likelihood and Bayesian analysis. Parsimony and model-based methods have been shown to perform best in simulations (Huelsenbeck and Hillis, 1993; Huelsenbeck, 1995; Siddall, 1998, 2001), but more contemporary methods have been developed for neighbour-joining analyses that improve their performance (Gascuel, 1997; Steel *et al.*, 2000; for further discussion see Hillis *et al.*, 1992, 1994; Huelsenbeck and Hillis, 1993; Siddall, 1998; Swofford *et al.*, 2001).

Prior to the advent of molecular tools, systematists primarily utilized morphological characters to construct phylogenies and infer evolutionary relationships. The relatively conservative morphologies of nematodes have made this exercise difficult. For example, *Caenorhabditis elegans* and *Caenorhabditis briggsae* are virtually indistinguishable based on their morphologies, yet comparative analyses of their genomes suggest that they last shared a common ancestor 80–110 million years ago – comparable to the divergence times of *Anopheles* and *Drosophila* (Stein *et al.*, 2003). Clearly, the benefits of molecular data in nematode phylogenetics are numerous. The advent and refinement of PCR and DNA sequencing techniques have made it possible to produce millions of characters in a matter of days (Hudson, 2007).

Early studies of phylogenetic relationships of insect-parasitic and pathogenic nematodes primarily utilized randomly amplified polymorphic DNA (RAPD) (Liu and Berry, 1996) and morphology and bionomics (Poinar, 1993). Today, phylogenetic relationships of insect-parasitic and pathogenic nematodes can be inferred using sequence data from nuclear and mitochondrial genetic loci, often with a special emphasis on nuclear ribosomal DNA (rDNA) genes, partial messenger RNA (mRNA) copies of protein-coding genes (*E*xpressed *S*equence *T*ags (ESTs)) and more recently, whole genomes (phylogenomics).

One of the more challenging problems facing systematists is finding the optimal solution given the number of possible phylogenetic trees that can be produced. As the number of taxa increases in a phylogenetic systematic study, the number of possible phylogenetic solutions drastically increases (Felsenstein, 1978). For example, in an analysis with four taxa, there are 15 possible rooted phylogenetic reconstructions. In an analysis of only ten taxa, the number of possible rooted phylogenetic reconstructions drastically increases to approximately 34,459,425. For 135 taxa, there are approximately $10^{265}$ possible rooted binary trees, more than the number of electrons in the known universe (Penny *et al.*, 1995).

## 8.3. Population Genetics

Population genetics remains one of the most understudied aspects of this important group of organisms. The roots of population genetics can be traced back to Darwin and Wallace's theory of natural selection as well as Mendel's explanation of the genetic mechanisms of inheritance. From Mendel's ideas, Hardy and Weinberg developed one of the simplest models of population genetics, which has become the null model in describing genetic attributes of a population (Templeton, 2006). Other models/theories exploring changes in allele frequencies within populations, mutation and inbreeding were pioneered by Fisher, Haldane and Wright. Their work provided the framework for a quantitative analysis of Mendelian genetics (Thompson, 1990), and a synthesis of Mendelian heredity and natural selection into the science of population genetics (Provine, 2001). Methods for inferring population processes from genetic patterning have increased tremendously in the past 20 years. In this chapter, we present some of the methods most relevant to entomogenous nematodes.

## 8.4. DNA Bar Coding

Bar coding (see Stock, Chapter 4, this volume) has not been developed to infer deep relationships or group species into kingdoms, phyla or classes, as this is the job of phylogenetic systematics (though Molecular Operational Taxonomic Unit (MOTU) data can be used in phylogenetic analyses (Floyd *et al.*, 2002; Powers, 2004)). However, the information gained from bar coding can be used with pre-existing phylogenies to answer questions that bar coding could not answer by itself. DNA bar coding, taxonomy and systematics should not be thought of as mutually exclusive research pathways. There is an overlap between these three disciplines and together they can work to build a better system for the identification of species, inferring diversity and determining relationships between and among a variety of taxonomic groups.

## 8.5. DNA Markers Considered for Phylogenetic and Population Genetics Studies

### 8.5.1. Ribosomal DNA

Nuclear rDNA has proven extremely useful and has been employed extensively to study nematodes systematics at the molecular level (see Stock, Chapter 4, this volume). The variability of evolutionary rates observed among different genes and spacers within a rDNA transcription unit is useful in that specific segments can be chosen based on the taxonomic or organismal level of study. For example, for a phylogenetic analysis at the species level, small subunit (SSU) ribosomal RNA (rRNA) has been considered suboptimal when attempting to differentiate closely related species (Liu *et al.*, 1997), while the less-conserved regions of the large ribosomal subunit (LSU) (Stock *et al.*, 2001; Nadler *et al.*, 2006b) and the internal transcribed spacer (ITS) that separates rDNA-coding regions (Nguyen *et al.*, 2001; Adams *et al.*, 2006) have proven to be more informative. For studies involving deeper nodes among more distantly related taxa, the more conserved SSU and LSU gene regions are more appropriate.

The single biggest obstacle in using rRNA genes in phylogenetic reconstruction is that the gene product can vary in length without compromising functionality within the ribosome. Whereas the length and composition of protein-coding genes are generally subject to selection by codon usage, rRNA genes are not. For some rDNA regions, insertion and deletion events are as frequent as transitions and transversions. In some cases, insertion and deletion events (indels) can involve blocks of multiples of nucleotides (Adams *et al.*, 1998; Nguyen *et al.*, 2001). Indel events can result in substantial rDNA size differences between sequences (taxa), which complicate the process of generating multiple sequence alignments and reduce confidence in the homology statements for each nucleotide in the multiple sequence alignment. Because phylogenetic reconstruction tests accurate homology statements (i.e. the thymidine at position 123 in the multiple sequence alignment in taxon A is homologous to the thymidine in taxon B at the same position), alignment ambiguity can result in spurious phylogenies.

In fact, this aspect has been explored for *Heterorhabditis* and *Steinernema*, and results suggest there is more variation in tree topology due to differences in the multiple sequence alignment than there is from the different methods used to generate the trees (i.e. parsimony, maximum likelihood and neighbour-joining) (Adams *et al.*, 1998; Nguyen *et al.*, 2001; Spiridonov *et al.*, 2004; Nadler *et al.*, 2006a). Approaches to addressing this problem require thoughtful consideration and include visually inspecting the sequences and removing the alignment-ambiguous regions based on an a priori metric (i.e. remove ambiguous indels that lie between invariant regions), direct optimization (discussed below in the section on phylogenetic methods), comparison of secondary structure based on minimum energy models and minimum posterior probabilities among alternative placements of nucleotides (characters) in the alignment (see discussion of alignment methods below).

Alternatives to rRNA genes include single-copy mitochondrial sequences (discussed in more detail below), nuclear protein-coding genes and intron sequences. The majority of the single-copy nuclear protein-coding genes thus far explored for phylogenetic utility are fairly conserved, and found to be most useful for resolving very deep nodes among distantly related taxa. Some of these include heat shock protein HSP90 (*daf*-21), RNA polymerase II (*ama*-1) and actin (*act*-1/3,2,4; Baldwin *et al.*, 1997; Kovaleva *et al.*, 2004; Skantar and Carta, 2004). Genes encoding ribosomal proteins (rather than rRNA) are promising for the resolution of fairly deep nodes, such as among genera of entomogenous nematodes, but it should be noted that rRNA genes and their highly expressed associated ribosomal proteins appear to evolve in a concerted fashion (Longhorn *et al.*, 2007). Similar to ITS rDNA, introns generally have high rates of nucleotide substitution, and contain numerous indels. Thus, ITS and intron sequences present similar multiple sequence alignment challenges. Intron sequences can be very useful for population genetic studies and phylogenetic analyses among very closely related taxa, although these are only beginning to be explored in entomogenous nematodes (Rolston *et al.*, 2004).

#### *8.5.1.1. Small subunit ribosomal DNA (SSU or 18S)*

The SSU rRNA gene has been the most frequently utilized genetic marker for nematodes, as it has proven useful in studies of deep phylogenetic relationships because of its slow evolutionary rate of change. The conservative nature of SSU rRNA also allows for the development of universal primers which can be used to amplify DNA from groups of nematodes for which little to no molecular data exist (Hillis and Dixon, 1991). Blaxter *et al.* (1998) utilized SSU sequences from 53 nematode taxa to construct a phylogeny from which they could study the evolution of the phylum Nematoda. Though the SSU is highly conserved, Blaxter *et al.* were able to use the SSU to differentiate between major nematode groups, and use their newly constructed evolutionary framework to bring into question the monophyly of previously proposed groups. Holterman *et al.* (2006) conducted a similar study using 339 nematode taxa. The additional taxa resulted in the proposal that the phylum Nematoda be divided into 12 clades rather than the five clades proposed by Blaxter *et al.* (1998). Furthermore, Holterman *et al.* suggested that 18S rRNA may be suitable for differentiation at the species level due to the acceleration of substitution rates in plant- and animal-parasitic clades.

#### *8.5.1.2. Large subunit ribosomal DNA (LSU or 28S)*

While SSU rRNA is primarily used to examine evolutionary events that occurred in the Precambrian time period, LSU rRNA is used primarily to examine evolutionary events which occurred through the Palaeozoic and Mesozoic time periods

(Hillis and Dixon, 1991). For example, Stock *et al.* (2001) utilized 28S rRNA sequence and morphology to investigate phylogenetic relationships among 21 *Steinernema* spp. The use of a combined data set and a larger sampling of taxa within the genus allowed for the construction of a robust evolutionary framework for the genus *Steinernema*, upon which hypotheses of species boundaries and the evolution of morphological features were assessed. A striking example of the explanatory power of comparative methods applied to entomogenous nematodes is that of Campbell *et al.* (2003). Using the relationships inferred from the Stock *et al.* (2001) phylogenetic study of *Steinernema*, Campbell *et al.* (2003) mapped behavioural, ecological and morphological characters onto a *Steinernema* tree to assess the origin, maintenance and evolution of interspecific variation in these traits. Mapping of host-finding strategies supported the inference that the ancestral *Steinernema* sp. was an intermediate forager, and that two other feeding strategies, ambush and cruise foraging, evolved only once.

Though not as extensively studied as their plant-parasitic counterparts, relationships within and among insect-parasitic tylenchid (Hexatylina) genera have been investigated using both rDNA and mitochondrial DNA (mtDNA) loci. To investigate the relationships within the genus *Fergusobia*, Ye *et al.* (2007) utilized SSU data to determine that *Howardula* was the sister taxon to *Fergusobia*. Subsequently, LSU, mitochondrial cytochrome *c* oxidase subunit I (COI) and a combined analysis of both LSU and COI sequences were employed to construct a phylogenetic framework for *Fergusobia* spp., facilitating an investigation of how plant–host associations evolved. Analyses of relationships within *Fergusobia* provide substantial evidence for host-switching within the genus *Fergusobia* with gall types being a labile feature (Ye *et al.*, 2007). While it has been shown that the LSU alone has been unable to adequately resolve relationships between the three subfamilies of Hexatylina first proposed by Chizhov (2004) (Subbotin *et al.*, 2006a), the LSU region may be best suited for resolving relationships at the species level (Ye *et al.*, 2007).

#### *8.5.1.3. Internal transcribed spacer (ITS) region*

Investigations into the utility of ITS regions embedded within the rRNA transcription unit, ITS-1 and ITS-2, indicate that they evolve at a much higher rate than the 18S and 28S genes, making these regions ideal for phylogenetic studies at the species and population levels, population genetic studies, as well as taxonomic identification (Ferris *et al.*, 1993; Chilton *et al.*, 1995; Cherry *et al.*, 1997; Powers *et al.*, 1997). Furthermore, the presence of conserved flanking regions encoding the 18S, 5.8S and 28S rRNA gene products allows for the reliable amplification of both of the ITS regions (Hillis and Dixon, 1991).

Multiple studies have utilized all or a portion of ITS rDNA to investigate phylogenetic relationships of a number of different entomopathogenic and other parasitic nematode genera with varying success. Spiridonov *et al.* (2004) utilized the whole ITS rDNA region to analyse relationships between groups and species of steinernematids using a phylogenetic framework constructed using parsimony. Spiridonov *et al.* (2004) concluded that while providing new information about the composition of five main clades within the genus *Steinernema*, the ITS rDNA region was of little value for resolving relationships between clades. A similar study by Nguyen *et al.* (2001) utilized fewer taxa to evaluate the utility of the ITS rDNA region in identifying species, reconstructing evolutionary histories and delimiting species within the genus *Steinernema*. It was concluded that while suitable for species identification, ITS rDNA is too variable to resolve relationships between all *Steinernema* spp. Adams *et al.* (1998) utilized the ITS1 region to infer phylogenetic relationships among *Heterorhabditis* spp. The study indicated that ITS1 sequences resolved relationships among sister *Heterorhabditis* taxa better than it resolved larger clades within the genus. Finally, the high rate of sequence evolution within *Howardula* data caused significant alignment difficulties for Perlman *et al.* (2003), and as such, 18S and COI data were primarily used in the inference of interspecific relationships for this genus. Each of these examples point to the need for multiple loci exhibiting varying levels of variation to reliably infer relationships among deep and shallow nodes of phylogenetic trees.

### 8.5.2. Mitochondrial DNA

Application of mtDNA sequence analysis to the study of nematode population and evolutionary biology was first reviewed by Hyman (Hyman *et al.*, 1988). Since that time, 27 complete nematode mitochondrial genome sequences have been deposited in GenBank. These circular molecules, ranging in size from 12.6 to 39 kb (Tang and Hyman, 2007), typically encode 12 protein-coding genes (*cox1-cox3, nad1-nad6, nad4L, cob, atp6*), 22 transfer RNAs and two ribosomal RNAs (rrnS and rrnL) (Hu and Gasser, 2006). The vertebrate-parasitic nematode *Trichinella spiralis* mtDNA encodes an additional protein-coding gene, *atp8* (Lavrov and Brown, 2001). The mitochondrial genome maintained within *Globodera pallida* (a plant–plant-parasitic nematode) is not a single circular molecule but instead the mtDNA is multipartite in architecture, with these same mitochondrial genes distributed among several sub-genomic circles (Armstrong *et al.*, 2000).

Favourable aspects of mtDNA analysis for population and evolutionary studies include an accelerated rate of nucleotide substitution at levels measured to be 10–100 times that of nuclear DNA, asexual transmission through maternal lineages and infrequent recombination events. Once thought to be absent, nematodes were among the first systems in which animal mtDNA recombination was demonstrated (Lunt and Hyman 1997; Piganeau *et al.*, 2004). Unlike many taxa, nematode mitochondrial gene orders can vary considerably. While the chromadorean nematodes show some degree of syntenic relationships among mitochondrial genes, no two enoplean nematodes share the same gene order (Tang, 2006). Given an accelerated degree of mtDNA rearrangement coupled with considerable nucleotide substitution among mitochondrial gene orthologues (Powers *et al.*, 1993), it is difficult to design universal primers for amplification across a wide range of taxa.

Mitochondrial genes are evolving at different rates; therefore, it becomes necessary to identify loci that diverge at a rate that provides signal useful to the question being addressed. MtDNA loci that evolve slowly, such as COI, are best suited to deeper lineage phylogeny, such as affinities between genera, as more rapidly evolving genes would obscure ancestral affinities. However, more rapidly evolving mitochondrial genes can often distinguish between congeners.

#### *8.5.2.1. Steinernematidae and Heterorhabditidae*

With respect to entomopathogenic nematodes, Liu *et al.* (Liu *et al.*, 1999) evaluated the mitochondrial ND4 gene as a phylogenetically informative marker for 15 heterorhabditid isolates representing five species. Seven mtDNA sequence haplotypes were identified that could be catalogued into four distinct groupings; the study concluded that ND4 is able to reveal interspecific and intraspecific differences within the genus *Heterorhabditis*, and that the molecular phylogeny constructed using ND4 divergence supports an existing, morphology-based taxonomic framework. Furthermore, the level of ND4 sequence substitution is such that geographic variation within certain species of *Heterorhabditis* (*Heterorhabditis megidis* and *Heterorhabditis indica*) can be resolved. Finally, the 1999 study by Liu *et al.* also suggests that the ND4 region may provide more robust phylogenetic information as compared to the ITS1 region, due to the requirement for gap insertion necessary to align the ITS1 region.

Intraspecific variation with the heterorhabditid ND4 locus has also been exploited to study the genetic structure of *Heterorhabditis marelatus* isolates (Blouin *et al.*, 1999). Four ND4 sequence haplotypes were identified among some 60 total individuals representing six populations distributed along the Oregon and California coasts. ND4 nucleotide sequence diversity was analysed by standard population genetic methodologies to estimate gene flow and effective population sizes. The *H. marelatus* populations were genetically structured; a high proportion of the genetic diversity could be apportioned between populations, suggesting small population sizes and minimal gene flow, characters expected of insect parasites with little opportunity for migration.

The complete mitochondrial genome of the entomopathogenic nematode *Steinernema carpocapsae* has recently been determined (Montiel *et al.*, 2006). *S. carpocapsae* and its congeners parasitize a wide variety of insect pests, and are often used in biological control strategies. The mtDNA sequence was determined for the purpose of identifying genetic markers for use in field applications (Montiel *et al.*, 2006). When placed in a phylogenetic context, the *S. carpocapsae* mtDNA sequence reveals more affinity to that of *Ascaris suum* (a vertebrate parasite) and *C. elegans* (free-living bacteribore nematode) relative to *Strongyloides stercoralis* (a vertebrate parasite), a result that stands in contrast to their phylogenetic position based on nuclear SSU rDNA data. The recent availability of the *S. carpocapsae* mitochondrial genome sequence has not yet enabled its use in population studies.

Interestingly, the mitochondrial protein and rRNA gene order of *S. carpocapsae* is identical to that of its fellow rhabditids *Ancylostoma duodenale*, *C. briggsae*, *C. elegans*, *Cooperia oncophora*, *Necator americanus* and *Haemonchus contortus*, as well as the ascarids *Anisakis simplex* and *A. suum*. However, *S. carpocapsae* is not completely syntenic to the only other entomopathogenic rhabditid mtDNA characterized to date, *Heterorhabditis bacteriophora*. Rather, these two insect-parasitic nematodes diverge at a few gene junctions (Tang, 2006). As such, attempting to map life-history traits onto a phylogeny based on mitochondrial gene order is not always a simple exercise, likely a result of the rapid changes in mitochondrial gene order discussed earlier.

Beyond nucleotide sequence divergence, short repeated sequences, often involving mtDNA non-coding regions, have proven to be markers useful for nematode population genetics. The first such application of *v*ariable *n*umber *t*andem *r*epeat (VNTR) to population genetic analysis (Whipple *et al.*, 1998) involved measuring the copy number of a 63 base pair (bp) tandemly repeated sequence within the *Meloidogyne incognita* mitochondrial genome. The number of 63 bp repeat copies ranged from 1 to 21 within individual mitochondrial genomes, thus defining 21 different alleles. Hierarchical statistical treatment of allele frequencies revealed that most of the genetic diversity resides within individuals, with little differentiation among populations. Diversity of mtDNA molecules within individuals is primarily due to an elevated rate of 'mutation' to different copy numbers as nematodes progress through their life stages. Hypervariation at this level has also been observed within representative genera of the nematode family Mermithidae.

#### *8.5.2.2. Mermithidae*

The Mermithidae is the only taxonomic order within the Enoplea that has evolved obligate invertebrate parasitism. Mermithid nematodes parasitize a wide range of invertebrates, with insects being the most common hosts. Several have been used as biological control agents with a strong emphasis on mosquito management, and as such, are considered entomopathogens.

Within the Mermithidae, mtDNA variation is not a consequence of simple VNTR copy number changes. Rather, lengthy (>1 kb) expanses have become repeated, often incorporating mitochondrial gene-coding sequences. In the absence of selective pressure, loss-of-function mutations can accumulate in all but one gene copy. These alterations are in the form of base substitutions, deletions and inversions to form degenerated pseudogene copies. All mermithid nematode mtDNAs characterized to date contain such large repeat regions.

Such a complex locus resides within the mitochondrial genome of the isopod-parasitic nematode *Thaumamermis cosgrovei* (Tang and Hyman, 2007). Most mitochondrial genes are mapped to a common skeleton shared by all *T. cosgrovei* individuals. The remainder of the mtDNA is occupied by a hypervariable region containing duplicated pseudogene copies of the ATP6 and ND4 genes, four mitochondrial transfer RNA (tRNA) genes, and one or more functional and pseudogene copies of the small ribosomal rRNA (rrnS) gene. These intact or degenerated gene copies are interspersed with a variety of non-coding sequences that themselves have been duplicated and have accumulated substitutions. Deletions and inversions, along with bases substitutions, have resulted in a seemingly endless ensemble of variations, detectable as different banding patterns on electrophoretic gels after restriction enzyme digestion of rolling circle amplified mtDNA (Tang and Hyman, 2005) from individual nematodes.

Haplotype hypervariation can be used to better understand interesting questions in nematode life history and population structure. Typically, isopod hosts are infected by a single *T. cosgrovei* individual. Between 5% and 10% of the hosts are multiply infected with 2–16 nematodes. Little is known of the mechanism by which multiple infections occur. Do these represent independent spatial or temporal parasitism events by genetically unrelated nematodes? Are the parasites genetically related, indicating simultaneous infection of the host?

When isopod hosts are parasitized by multiple nematodes, individuals infecting the same host share the same mtDNA haplotype, indicating that they are derived from the same maternal lineage (Tang and Hyman, 2007). Sharing of mtDNA haplotypes has been observed in 80% of the cohorts dissected from multiply infected hosts. Kaiser (Kaiser, 1991) described two general routes for multiple infection by mermithids that would include the entomopathogens in this nematode family. One proposed mechanism is passive infection that involves host ingestion of an egg clutch containing unhatched J1-stage nematodes. A second possible route to parasitism, termed active infection, suggests that hatched, J2-stage infectious individuals independently infect a single host. That 80% of the parasitized hosts contain individuals with identical mtDNA haplotypes, the only documented occurrence of shared mitochondrial haplotypes within *T. cosgrovei*, indicates that passive infection is the most frequent mode of infection, though active infection can infrequently occur.

It will be exciting to learn of additional examples of mitochondrial genome hypervariation within other entomopathogenic nematodes. It is anticipated that application of mtDNA variation to population structure and life histories of entomopathogens will find an important role in integrated pest management regimes.

## 8.6. Methodology

### 8.6.1. Alignment strategies

When conducting a phylogenetic analysis of molecular data, generating a robust multiple sequence alignment is critical to inferring accurate relationships. The alignment is a statement of positional homology, and tree topology is often more sensitive to alignment methodology than to the method of phylogenetic tree reconstruction that is chosen (Morrison and Ellis, 1997; Phillips *et al.*, 2000). Multiple methods exist for the alignment of sequence data and are briefly described below.

#### *8.6.1.1. Visual inspection*

This was the first and probably most common alignment method. This method consists of using a word processor or other sequence visualization programs (i.e. MacClade, BioEdit, etc.) to view sequences and manually move lines of sequence left or right and/or insert gaps, until the investigator is satisfied with the alignment. The problem with this method is that there are no discernable criteria by which the investigator decides upon a suitable alignment, thus making this method highly subjective. While this method may work well for sequences that are very similar, attempts to align dissimilar sequence data will result in highly subjective and irreproducible alignments.

#### 8.6.1.2. Pairwise alignment

Pairwise alignments can be subdivided into two methods: local and global. Local methods are typically used in a database searching and retrieval function, where the alignment tries to determine if a portion or portions of one sequence is present in another sequence (Phillips *et al.*, 2000). This is the type of method that is used in the Basic Local Alignment Search Tool (BLAST) searches that are conducted in GenBank. Global methods, which are typically conducted for alignments that will be used in phylogenetic analysis, compare the entire sequence of taxon A to the entire sequence of taxon B.

Global pairwise alignment relies on the assignment of costs to changes (transition and transversion) and gaps. The best alignment is the one that minimizes the cost. An extension of the pairwise alignment is multiple sequence alignment (below). When multiple alignment methods are used, the same cost minimization idea is applied to $n$ sequences in $n$ dimensions (Phillips *et al.*, 2000).

#### 8.6.1.3. Multiple alignments and most common software

Numerous multiple sequence alignment programs exist, each with its particular advantages and disadvantages. These algorithms are summarized below.

CLUSTAL. This is one of the most utilized multiple alignment programs (Thompson *et al.*, 1994, 1997). CLUSTAL utilizes a progressive alignment algorithm that first estimates a distance tree (Wheeler, 2001), which is then used to construct pairwise alignments of subtrees within the original guide tree (Edgar, 2004b). Advantages of CLUSTAL include the speed at which alignments are constructed, a friendly graphical user interface (GUI) and an output of only one multiple sequence alignment. Another helpful feature is the profile alignment mode, which allows users to align individual sequences to an already established multiple sequence alignment. Limitations to the method include a lack of guarantee that the minimum cost alignment is found, practicality in that a finite number of sequences can be analysed (typically 500, but varies with computer platform and computational power) and the output represents a single alignment when many equally scoring alternatives may exist. Furthermore, most CLUSTAL alignments often require readjustments by eye.

MALIGN. Similar to CLUSTAL, MALIGN (Wheeler and Gladstein, 1994) also utilizes a guide tree to chaperon the alignment process. MALIGN furthers the idea of using a guide tree by searching multiple guide trees in an attempt to find an optimally minimized cost (parsimonious) alignment. Multiple guide trees are searched in a manner similar to a phylogenetic tree search, where branch swapping and random addition of taxa are used. While MALIGN will generally find more optimal solutions than CLUSTAL, considerably more computational power is employed in constructing the optimal alignment. Occasionally both methods will recover multiple alignments of equal score.

MAFFT. MAFFT was developed to increase efficiency (speed of computation and accuracy) of the multiple alignment process (Katoh *et al.*, 2002). Using a fast

Fourier transform (FFT), MAFFT is able to quickly identify homologous regions of DNA sequence. Along with the FFT, MAFFT also employs a scoring system that is different from earlier alignment programs such as CLUSTAL and T-COFFEE. The MAFFT scoring system is touted as enabling the accurate alignment of sequence data with large insertions or extensions, as well as highly divergent sequence data of similar length (Katoh *et al.*, 2002).

MUSCLE. MUSCLE also utilizes a progressive alignment algorithm, but refines the alignment procedure by applying a horizontal process to its initial progressive alignment (Edgar, 2004a,b). This improves the initial guide tree in an attempt to find the lowest-scoring guide tree for use in directing the alignment process. The improvements made to the progressive alignment process allow MUSCLE to compute alignments of large numbers of taxa (several thousand) in a shorter amount of time, with enhanced biological accuracy relative to CLUSTAL and MAFFT. One drawback of MUSCLE is its use of a command line-driven interface, making it less user-friendly than CLUSTAL.

DIRECT OPTIMIZATION (DO). First proposed by Wheeler (1996), DO strays from the typical multiple alignment procedure where the alignment is conducted in one step and is followed by the construction of a phylogenetic tree in a separate step. Instead, DO constructs the alignment and the phylogenetic tree concurrently, thus applying the same optimality criterion to alignment and tree construction, a feature that is lacking in all other alignment and tree reconstruction methods. DO completely eliminates the multiple alignment procedure, and instead constructs alignments at each node of the phylogenetic tree (Wheeler, 1996). Due to large numbers of possible topologies and possible ways in which sequences can be optimized to those topologies, DO presents an exceedingly computationally complex problem (Terry and Whiting, 2005). The DO method for the construction of phylogenies can be carried out using the software package POY (Wheeler, 2003).

PROALIGN. Another approach with promise for analysis of rDNA sequences is ProAlign (Loytynoja and Milinkovitch, 2003). This approach uses a hidden Markov model, a progressive alignment algorithm (above) and a nucleotide substitution model to identify the minimum posterior probability of each homology statement. Thresholds can be explored and established that allow the user to exclude characters with a low posterior probability from the multiple sequence alignment. This approach reduces the potential for investigator bias regarding the identification and removal of alignment-ambiguous characters, and also allows for clever comparisons among data sets where the effects of indel inclusion/excision are concerned (Nadler *et al.*, 2006a).

SECONDARY STRUCTURE MODELS. Two-dimensional secondary structure modelling of rRNAs can be used to inform homology statements. Ribosomal RNAs have base-paired stem and unpaired loop regions that result in different constraints and substitution rates among bases that comprise these two structural features. There are several methods for inferring secondary structure, including minimum free energy, base-pair probabilities and comparing secondary structures as they vary across a broad range of energies (Hofacker, 2003). Since each of these approaches can yield different optimal secondary structure models, there remains a continuum of multiple sequence alignments that should be evaluated prior to phylogenetic analysis. A library of nematode sequences based on modelled secondary structures for nematodes (ITS and LSU D2, D3 regions) is available at http://www.nemamex.ucr.edu/rna/ (Subbotin *et al.*, 2006b). Phylogenetic analyses of these can be run in Phase, which uses explicit models to account for substitution events that are influenced by secondary structure (Telford *et al.*, 2005).

## 8.6.2. Phylogenetic reconstruction methods

### *8.6.2.1. Parsimony*

Phylogenetic reconstruction methods can be divided into two main categories: parsimony methods and model-based methods. Parsimony, the most widely used method, is based on the principle that the simplest explanation is the explanation best supported by the current data. Accordingly, the optimal phylogenetic tree is the one that minimizes the number of ad hoc hypotheses required to explain the data. Model-based methods, such as maximum likelihood and Bayesian inference, assume a model of DNA sequence evolution and then find the tree that best fits the model.

As mentioned, the first and most important step in the construction of phylogenies is the alignment, as this is the homology statement. Everything that occurs after the alignment is directly dependent on the accuracy of this step. When conducting analyses under the parsimony criterion, the most widely used software package is PAUP* 4.0b10 (Swofford, 2002). PAUP* features a GUI, which allows users to quickly specify parameters and run analyses without the need for in-depth knowledge of command line prompts and keywords. NEXUS files that have been directly outputted from alignment programs or exported from MACCLADE (Maddison and Maddison, 2002) can be loaded directly into PAUP*. Parsimony trees are built via a multi-step process that utilizes Fitch optimization, a method by which the cost of a tree is calculated. The number of character state changes is calculated for each tree to determine which tree has the lowest score. The tree with the lowest score is the most parsimonious (optimal) tree.

Parsimony analyses can be refined through the specification of a number of different tree search parameters, including the type of search algorithm employed, the form of branch swapping that is conducted, and how taxa are added to each reconstruction. PAUP* offers heuristic, branch and bound and exhaustive search algorithms. The exhaustive search guarantees that the globally optimal topology will be discovered by examining every possible topology in the landscape of trees. This is a very computationally expensive method, and thus is only practical for a data set of 20 taxa or less. The branch and bound algorithm also guarantees obtaining the most parsimonious solution, but is much faster because it bypasses known suboptimal topologies. An even less computationally intensive method is the heuristic algorithm. The heuristic search algorithm takes samples (local optima) from the tree landscape, with the idea that if you sample enough local optima, one of them will likely be the global optimum. The heuristic method is useful when working with data sets with more than 20 taxa.

The heuristic search method utilizes multiple methods of branch swapping and tree construction methods to search tree space in an attempt to find the globally optimal topology in a vast forest of trees. To begin a parsimony analysis, an initial tree is constructed using multiple options for the formation of the initial tree provided within PAUP*. Starting tree construction options include neighbour-joining, a distance-based method of tree construction, and stepwise addition. The choice of stepwise addition offers further taxa addition options, including asis, simple, closest and random. Asis adds taxa to the initial tree in the order they are listed in the data matrix. The preferred method of stepwise addition is random addition, which randomly adds taxa to the tree. It is generally suggested that a minimum of 1000 random addition replicates be used in the construction of the initial tree.

Occasionally, a heuristic search will get stuck on a locally optimal solution within the tree space. Branch swapping provides a method by which new reconstructions are proposed, thus enabling new parts of the tree space to be explored and the discovery of trees that are more optimal than previous reconstructions. PAUP* offers three main branch-swapping methods: nearest neighbour interchange (NNI), subtree pruning and re-grafting (SPR) and tree bisection and reconnection (TBR). NNI utilizes subtrees that make up the larger tree. Each subtree has two neighbours, and by swapping a neighbour from one subtree with a neighbour from an adjacent subtree (Felsenstein, 2004), a new arrangement of the taxa is produced. SPR is similar to NNI in that it also utilizes subtrees to form new topologies. However, SPR removes one branch, along with its subtree, and forms new arrangements by reinserting the removed subtree in all possible places within the tree (Felsenstein, 2004). TBR breaks a tree into two separate trees, and subsequently proceeds to assemble new tree rearrangements by attaching a branch from one tree to a branch from the other (Felsenstein, 2004). While TBR is the more computationally intensive of the three methods described above, it generally finds a more optimal solution.

Branch-swapping methods, while useful for smaller data sets (<200 taxa), do not work as well for large data sets, which produce large numbers of suboptimal trees, many of which are similar in topology and length (Nixon, 1999). These large groups of trees are known as 'islands of trees', and branch-swapping methods can get marooned on these large islands. The parsimony ratchet (Nixon, 1999), which is implemented in the software package TNT (Goloboff *et al.*, 2008) and can also be run in PAUP*, reduces the problem of becoming trapped on 'islands', and finds optimal trees for very large data sets (>500 taxa) much faster than other methods. By maximizing starting points, reducing the amount of time spent swapping on each starting point and retaining structure from the existing solution at each point, the parsimony ratchet allows for the majority of the computing time to be spent breaking out of tree islands and improving the current tree (Nixon, 1999).

While a phylogeny alone shows inferred relationships, estimates of support can be generated for each node (monophyletic group) in the tree. Multiple categories of branch support methods exist, including incongruence, support indices and statistical resampling methods. Non-parametric bootstrapping is a method that relies on resampling characters, with replacement, to estimate confidence limits of internal branches (Hillis and Bull, 1993) and is incorporated in many phylogeny reconstruction software packages. Bremer support, also called the decay index, is a measure of the number of extra steps beyond the most parsimonious tree that must be allowed before trees are found which do not include the monophyletic group of interest (DeBry, 2001). Bremer support can also be defined as the number of steps required to dissolve a node. While Bremer support values can be informative, their interpretation can be misleading. The number of informative characters present at each node will have a large effect on the interpretation of a Bremer support value (DeBry, 2001), and thus should always be taken into account when evaluating the utility or measure of support that is indicated by a specific Bremer support value. In analyses that incorporate multiple data sets (i.e. sequence data for multiple loci), partitioned Bremer support values, or Bremer support values for each data set, can be calculated. Bremer support values can be calculated using PAUP*, AutoDecay, TreeRot and many other programs. Partitioned Bremer support values, calculated using the program TreeRot (Sorenson, 1999), allow for the relative amount of support each data set contributes to each node to be assessed. Thus, one can tell if a particular gene gives more support to terminal taxa or to internal nodes of the tree. This information may be useful when examining conflicts among partitions.

#### *8.6.2.2. Maximum likelihood*

Maximum likelihood is a model-based method of phylogenetic reconstruction. That is, it relies on a specific model of sequence evolution to infer the probability of the data given by a particular phylogenetic tree. Critics of likelihood methods assert that even the most general and parameter-rich models cannot possibly capture all of the processes that generated a particular sequence or sequences (Sullivan and Swofford, 2001), thus hampering the ability of the method to recover the 'true' tree. Likelihood analysis of molecular data begins by constructing a suitable alignment of the sequence data. The alignment is then input into MODELTEST (Posada and Crandall, 1998), or other model selection programs that determine the best-fit model of evolution for the specified group of sequences. This step is important, as the selection of an incorrect model may lead to the recovery of an incorrect tree (Posada and Buckley, 2004). MODELTEST selects models using two different criteria: the likelihood-ratio test (LRT) and the Akaike information criterion (AIC). The LRT, while possessing unfavourable qualities such as a propensity to always select the most complex model, has nonetheless been the most widely used model selection criteria. Recently, the AIC has gained popularity, likely due to its ability to calculate an AIC for each model in isolation and the inclusion of penalties for over-parameterization (Sullivan and Joyce, 2005). Model parameters output from MODELTEST are inserted at the end of a PAUP* block of the corresponding aligned sequence data, and executed in PAUP*. Likelihood analysis proceeds by simply switching the optimality criterion to likelihood, and selecting a tree-searching method. Likelihood analyses conducted in PAUP* are typically very time-intensive, with analyses running for weeks or even months before completion. Other, less time-intensive software for conducting likelihood analyses include GARLI (http://www.zo.utexas.edu/faculty/antisense/Download.html), a program that utilizes a genetic algorithm approach to efficiently and accurately find the most likely tree, TREEFINDER (Jobb, 2008), a fast method that allows the use of partitioned data and bootstrap calculations, and PHYML (Guindon and Gascuel, 2003) which builds an initial tree using a simple hill-climbing algorithm, then modifies the topology and branch lengths simultaneously and progressively to find the optimal tree.

#### *8.6.2.3. Bayesian analysis*

Bayesian phylogenetic inference is another model-based tree reconstruction approach. Bayesian inference utilizes the Metropolis Coupled Markov Chain Monte Carlo (MCMCMC) method to randomly explore the tree space so as to settle down into an equilibrium distribution of trees with the desired distribution (Felsenstein, 2004). Bayesian phylogenetic inference has been lauded for its ability to incorporate priors, to handle large numbers of taxa and to efficiently sample tree space.

MRBAYES (Huelsenbeck and Ronquist, 2001) is a command line-driven program that can be used to conduct Bayesian analysis of phylogenies. MRBAYES reads aligned DNA or amino acid sequences in a NEXUS formatted file. As with a likelihood analysis, the specific parameters for the best-fit model of evolution should be included at the end of the alignment file, along with the number of chains to be used, total generations, number of runs and frequency with which trees are saved. Upon completion of a Bayesian analysis, the burn-in value (defined below) must be calculated and all trees falling within the burn-in phase of the analysis must be removed prior to calculating a Bayesian tree.

The burn-in refers to all of the tree topologies prior to the point at which the equilibrium distribution of trees is reached. To determine when the analysis reaches equilibrium, the likelihood values file (file will end in .p) is opened in EXCEL or any other spreadsheet program. The numbers in the first two columns are plotted as a scatter plot to identify the point on the plot where the distribution levels out. The $x$ value of this point is the burn-in value. Since the burn-in contains trees that are not in the desired distribution, these trees will be eliminated from the tree file (file will end in .t) following the completion of a full Bayesian run. This process can also be conducted in TRACER v1.3 (Rambaut and Drummond, 2003). TRACER allows the user to input multiple '.p' files (from different independent runs) at the same time, allowing visualization of the point where all runs reach stationarity as well as enabling the user to ensure that all runs converged to a similar likelihood value. While one could build a Bayesian tree from a single run, a more thorough method would be to conduct multiple runs, ensure all runs converge on a similar likelihood score and calculate the burn-in for each run, eliminate the burn-in from the tree files (.t file), combine the tree files from each independent run into a single-tree file, and then compute the final Bayesian tree (majority rule consensus tree).

The number of generations that are considered suitable for an optimal run in a Bayesian analysis varies based on individual data sets, though no less than 10–20 million should probably be run. This depth allows for sufficient search of the tree space, and detects whether a second set of desired tree distributions exists. Similarly, the number of runs and the number of chains used in an analysis will also vary based on knowledge of the data set in question as well as available computing power, though more runs and more chains (>4) will only improve the analyses.

MRBAYES 3 (Ronquist and Huelsenbeck, 2003), a rewritten and restructured version of MRBAYES, enables the incorporation of mixed models into a Bayesian analysis. This strategy is beneficial for studies that contain heterogeneous sequence data. The use of mixed models in a simultaneous analysis allows for the partitioning of a gene or genes based on the best-fit model of evolution of each gene or gene region, thus allowing each data set to be analysed independently according to its specific best-fit model.

To illustrate the benefit of a mixed models analysis in MRBAYES 3 over a single model analysis in MRBAYES, consider a hypothetical situation where a phylogeny of *Heterorhabditis* is constructed using 18S rRNA, ITS1 rRNA and ND4 data. If this analysis were run in MRBAYES, MODELTEST would first be conducted on the combined data set of the three genes to identify a single best-fit model. To conduct a mixed models Bayesian analysis, the data set for each individual gene is run through MODELTEST, resulting in a specific model for each gene. Since both nuclear ribosomal gene and a mitochondrial protein-coding locus are being used in the hypothetical study, the chance that a single model sufficiently accounts for the rates of nucleotide evolution in all three genes is low. Thus, an analysis that applies a best-fit model to each gene has a higher probability of recovering the 'true' tree than an analysis of one that utilizes one model for all three genes.

Bayesian analysis produces a unique form of branch support known as posterior probabilities. A posterior probability represents the number of trees (represented as a percentage in decimal form) that supported the grouping of the clade of interest. Care should be taken when interpreting Bayesian posterior probability values, as posterior probabilities can be potentially inflated, especially relative to bootstrap support values (Pérez-Losada *et al.*, 2007).

## 8.7. Co-phylogenesis and Cospeciation

### 8.7.1. General concepts

Cospeciation, the joint speciation of two organisms living in close association with one another, has been detected in numerous systems, including parasitic (Hafner and Nadler, 1988; Perlman *et al.*, 2003) and mutualistic (Clark *et al.*, 2000) relationships. Assessment of cospeciation is carried out through the use of phylogenetics, and more specifically co-phylogenetic analyses. Through the process of a co-phylogenetic analysis, phylogenies of both the host and associate are compared and the amount of congruence between the two phylogenies is assessed. The null hypothesis in cospeciation studies is one of strict congruence or strict cospeciation. As such, the phylogeny of the associate mirrors the phylogeny of the host, though this is a rare occurrence (Johnson *et al.*, 2003; Downie and Gullan, 2005). A lack of correspondence between host and associate phylogenies can be explained by numerous other events. These events include host-switching, duplications and sorting events. Sorting events are events that account for the absence of a parasite lineage on a host. These can include instances of 'missing the boat' or extinction events. Duplications refer to an occasion where the parasite speciates in the absence of host speciation.

### 8.7.2. Methodology

Numerous methods exist to assess the amount of phylogenetic congruence between host and associate phylogenies. One of the earliest methods is Brooks parsimony analysis (BPA) (Brooks, 1981), a method that utilizes parsimony and the Wagner algorithm, as well as additive binary coding of the parasite phylogeny, to detect cospeciation (a strategy also used in studies of historical biogeography) (Dowling, 2002). BPA analysis of co-evolution begins by converting all terminal taxa and nodes of the parasite phylogenetic tree into a binary character matrix (Brooks, 1981). The parasite names in the character matrix are then replaced with the name of the host known to associate with each parasite. The end result is a phylogeny that minimizes the number of host-switching events and extinctions (Dowling, 2002). This earliest version of BPA has been modified several times since its inception to further the development of this method for conducting cospeciation and historical biogeographical analyses (Wiley, 1988a,b; Brooks, 1990).

#### *8.7.2.1. Software*

COMPONENT (PAGE, 1993). This algorithm is utilized for the comparison of phylogenetic trees. Thus, this software is not specific for cospeciation studies, and like BPA, can also be used to conduct studies of historical biogeography. COMPONENT employs the 'tree reconciliation' method developed by Page (1990), which uses duplication and loss events to fit the parasite tree to the host tree (Slowinski, 1993). One criticism of COMPONENT is its inability to allow for host switches (Charleston, 1998).

TREEMAP (PAGE, 1994). Similar to COMPONENT, this software program is also a reconciliation-based method of co-phylogenetic analysis (Page, 1994). As with all other co-phylogenetic analysis methods, the goal of TREEMAP is to maximize the number of cospeciation events, while minimizing the number of ad hoc hypotheses in the form of duplications and sorting events. The reconciliation process used in TREEMAP is executed by labelling all internal nodes and terminal tips of both the host and associate tree. Each node on the parasite tree is mapped on to a corresponding node in the host tree, and paths are drawn to trace the path between the corresponding host tree and the parasite tree (Dowling, 2002). From this process, the number of cospeciation, duplication and sorting events are determined. Included in the TREEMAP v1.0 program is an option that allows for testing of the least costly reconciliation against a number of randomly generated trees, to determine if the result obtained is statistically significant from chance alone.

Some common criticisms of TreeMap include its prohibition of host-switching, the requirement of input trees that are fully resolved, and its overestimation of duplications and sorting events (Ronquist, 1995; Dowling, 2002). A later release of TreeMap, version 2.0 (Charleston and Page, 2002), made major improvements to the tree reconciliation process through the use of a method called 'jungles'. One of the improvements is the inclusion of more event types, including host-switching, lineage sorting, duplications, and of course, cospeciation events (Charleston, 1998).

CASE STUDIES. Perlman *et al.* (2003) investigated associations between *Drosophila* and its nematode parasite *Howardula*. Following construction of both host (COI, II and III) and parasite (18S, ITS and COI) phylogenies, tree reconciliation analysis performed using TreeMap v1.0 demonstrated that *Drosophila* and *Howardula* phylogenies were not congruent, as indicated by a lack of statistically significant cospeciation events. Perlman *et al.* (2003) conclude that incongruence between host and parasite phylogenies is most likely attributed to a relatively high degree of host-switching and infection of novel hosts other than *Drosophila*. Conversely, preliminary analyses of cospeciation between *Heterorhabditis* and its bacterial endosymbiont *Photorhabdus* (Peat *et al.*, 2006) indicate the presence of statistically significant cospeciation within this highly specialized symbiotic system.

The close associations of *Heterorhabditis* and *Steinernema* with their respective bacterial endosymbionts (*Photorhabdus* and *Xenorhabdus*) also provide research avenues that could uncover clues to the origin and maintenance of these nematode/bacterium associations. Numerous resources and opportunities exist for studying the co-evolution of entomopathogenic nematodes and their associates (insects and/or bacteria), driving this nascent but rapidly growing area of research.

## 8.8. Population Genetics Methods

### 8.8.1. Software and analysis of data

The following section describes population genetics software that can be used to address questions focusing on population structure, demography, genetic diversity, gene flow, linkage disequilibrium and selection on genomes. When using any of these programs, it is imperative that the assumptions used in each algorithm are fully understood to enable accurate and meaningful interpretations of data outputs. As a complete review of all contemporary, computational approaches to population genetics is beyond the scope of the present discussion, we suggest Labate (2000) and Excoffier and Heckel (2006) for a more exhaustive review of population genetics software applications.

Studies investigating properties of populations often begin by using descriptive statistics. Common approaches include measures of genetic diversity, linkage disequilibrium (LD), and tests of Hardy-Weinburg equilibrium (HWE). Genetic diversity, a measure of variation within populations, measures the numbers of polymorphic loci, catalogues distinct haplotypes and allele frequencies, and estimates proportions of heterozygotes. Several software programs, including Arlequin (Excoffier and Schneider, 2005), DnaSP (Rozas and Rozas, 1995, 1997, 1999),

GDA (Lewis and Zaykin, 2001) and GENEPOP (Raymond and Rousset, 1995), offer calculations of these and other measures of diversity. ARLEQUIN and GENEPOP provide tests of HWE and LD, and DNASP and GDA compute LD indices (Excoffier and Heckel, 2006). Detection of population structure and measures of population subdivision (using *F*-statistics) are common in a number of broad-spectrum and specific population genetics software programs.

#### *8.8.1.1. STRUCTURE (Falush et al., 2003; Pritchard et al., 2000)*

This program was designed to detect genetic structure present in a set of individuals in the absence of user-defined population information, and can be utilized to determine the number of unique populations that exist from a data set of individuals. It can assign individuals of unknown origin to a predefined population, or to identify cryptic population structure (Pritchard *et al.*, 2000). Other programs available to detect population structure and/or population subdivision include GDA, ARLEQUIN and DNASP.

#### *8.8.1.2. LAMARC v 2.0.2 (Kuhner et al., 2005)*

This program couples the primary abilities of four different programs (MIGRATE, FLUCTUATE, COALESCE and RECOMBINE) into one interface. LAMARC estimates effective population sizes, population exponential growth rates, divergence times, recombination rates and past migration rates for one to *n* populations using single nucleotide polymorphism (SNP), microsatellite, DNA or RNA data (Kuhner *et al.*, 2005). A useful feature provided by LAMARC is the estimation of population size or migration rates between more than two populations (Excoffier and Heckel, 2006), a feature not offered in most population genetics programs. Estimation of migration rates requires the separation of each population into its own separate data file using either PHYLIP or MIGRATE file formatting.

#### *8.8.1.3. IM (Nielsen and Wakeley, 2001)*

IM is another program well suited for inference of population size, divergence times and migration rates. IM furthers the estimation of migration rates by jointly estimating divergence times and migration rates from DNA sequence data (Nielsen and Wakeley, 2001). Other programs that can be used to estimate migration rates and divergence times include ARLEQUIN, GENEPOP and DNASP.

User interface, computation time, input file format, accepted data types and model assumptions vary across all population genetics software. It is important to understand what population-related questions will be addressed and what programs can be used to answer proposed research questions, prior to selecting a particular population genetics program. If only descriptive statistics are required, ARLEQUIN is the most appropriate, leaving STRUCTURE for more sophisticated analyses. Defining hypotheses prior to beginning population genetic analyses will facilitate proper selection of appropriate tests, narrowing the choice of software, and allowing for a more refined selection of appropriate tests/software based on data type (i.e. microsatellite, SNP, etc.) and fit the assumptions of the system being analysed.

We have discussed only a few of the phylogenetic and population genetic analysis programs available; all computer programs mentioned in this paper, and many more, are available online from the following web sites:

http://evolution.genetics.washington.edu/phylip/software.html
http://www.biology.lsu.edu/general/software.html
http://evonet.sdsc.edu/ROADS/subject-listing/softwrpopgen.html
http://pritch.bsd.uchicago.edu/software/structure2_1.html

## References

Adams, B.J., Burnell, A.M. and Powers, T.O. (1998) A phylogenetic analysis of *Heterorhabditis* (Nemata: Rhabditidae) based on internal transcribed spacer 1 DNA sequence data. *Journal of Nematology* 30, 22–39.

Adams, B.J., Fodor, A., Koppenhöfer, H.S., Stackebrandt, E., Stock, S.P. and Klein, M.G. (2006) Biodiversity and systematics of nematode-bacterium entomopathogens. *Biological Control* 37, 32–49.

Andrassy, I. (1976) *Evolution as a Basis for the Systematization of Nematodes.* Pitman Publishing, London.

Armstrong, M.R., Blok, V.C. and Phillips, M.S. (2000) A multipartite mitochondrial genome in the potato cyst nematode *Globodera pallida. Genetics* 154, 181–192.

Avise, J.C. (2000) *Phylogeography: The History and Formation of Species.* Harvard University Press, Cambridge, Massachusetts.

Avise, J.C. (2004) *Molecular Markers, Natural History, and Evolution.* Sinauer Associates, Sunderland, Massachusetts.

Baldwin, J.G., Frisse, L.M., Vida, J.T., Eddleman, C.D. and Thomas, W.K. (1997) An evolutionary framework for the study of developmental evolution in a set of nematodes related to *Caenorhabditis elegans. Molecular Phylogenetics and Evolution* 8, 249–259.

Blaxter, M.L., De Ley, P., Garey, J.R., Liu, L.X., Scheldeman, P., Vierstraete, A., Vanfleteren, J.R., Mackey, L.Y., Dorris, M., Frisse, L.M., Vida, J.T. and Thomas, W.K. (1998) A molecular evolutionary framework for the phylum Nematoda. *Nature* 392, 71–75.

Blouin, M.S., Liu, J. and Berry, R.E. (1999) Life cycle variation and the genetic structure of nematode populations. *Heredity* 83, 253–259.

Brooks, D.R. (1981) Hennig's parasitological method: a proposed solution. *Systematic Zoology* 30, 229–249.

Brooks, D.R. (1990) Parsimony analysis in historical biogeography and coevolution: methodological and theoretical update. *Systematic Zoology* 39, 14–30.

Brooks, D.R. and McLennan, D.A. (1991) *Phylogeny, Ecology, and Behavior.* The University of Chicago Press, Chicago, Illinois.

Brooks, D.R. and McLennan, D.A. (2002) *The Nature of Diversity: An Evolutionary Voyage of Discovery.* University of Chicago Press, Chicago, Illinois.

Campbell, J.F., Lewis, E.E., Stock, S.P., Nadler, S. and Kaya, H.K. (2003) Evolution of host search strategies in entomopathogenic nematodes. *Journal of Nematology* 35, 142–145.

Charleston, M.A. (1998) Jungles: a new solution to the host/parasite phylogeny reconciliation problem. *Mathematical Biosciences* 149, 191–223.

Charleston, M.A. and Page, R.D.M. (2002) TreeMap v2.0 Application for Apple Macintosh, University of Oxford, Oxford.

Cherry, T., Szalanski, A.L., Todd, T.C. and Powers, T.O. (1997) The internal transcribed spacer region of *Belonolaimus* (Nemata: Belonolaimidae). *Journal of Nematology* 29, 23–29.

Chilton, N.B., Gasser, R.B. and Beveridge, I. (1995) Differences in a ribosomal DNA sequence of morphologically indistinguishable species within the Hypodontus macropi complex (Nematoda, Strongyloidea). *International Journal for Parasitology* 25, 647–651.

Chizhov, V.N. (2004) Entomogenous nematodes from the order Tylenchida (Nematoda). In: Sonin, E.D. (ed.) *Parasitic Nematodes of Plants and Insects*. Nauka, Moscow, pp. 277–293.

Clark, M.A., Moran, N.A., Baumann, P. and Wernegreen, J.J. (2000) Cospeciation between bacterial endosymbionts (*Buchnera*) and a recent radiation of aphids (*Uroleucon*) and pitfalls of testing for phylogenetic congruence. *Evolution* 54, 517–525.

Debry, R.W. (2001) Improving interpretation of the decay index for DNA sequence data. *Systematic Biology* 50, 742–752.

Dowling, A.P.G. (2002) Testing the accuracy of TreeMap and Brooks parsimony analyses of coevolutionary patterns using artificial associations. *Cladistics* 18, 416–435.

Downie, D.A. and Gullan, P.J. (2005) Phylogenetic congruence of mealybugs and their primary endosymbionts. *Journal of Evolutionary Biology* 18, 315–324.

Edgar, R.C. (2004a) MUSCLE: a multiple sequence alignment method with reduced time and space complexity. *BMC Bioinformatics* 5, 1–19.

Edgar, R.C. (2004b) MUSCLE: multiple sequence alignment with high accuracy and high throughput. *Nucleic Acids Research* 32, 1792–1797.

Excoffier, L. and Heckel, G. (2006) Computer programs for population genetics data analysis: a survival guide. *Nature Reviews Genetics* 7, 745–758.

Excoffier, L.G.L. and Schneider, S. (2005) Arlequin ver. 3.0: an integrated software package for population genetics data analysis. *Evolutionary Bioinformatics Online* 1, 47–50.

Falush, D., Stephens, M. and Pritchard, J.K. (2003) Inference of population structure using multilocus genotype data: linked loci and correlated allele frequencies. *Genetics* 164, 1567–1587.

Felsenstein, J. (1978) The number of evolutionary trees. *Systematic Zoology* 27, 27–33.

Felsenstein, J. (2004) *Inferring Phylogenies*. Sinauer Associates, Sunderland, Massachusetts.

Ferris, V.R., Ferris, J.M. and Faghihi, J. (1993) Variation in spacer ribosomal DNA in some cyst-forming species of plant parasitic nematodes. *Fundamental and Applied Nematology* 16, 177–184.

Floyd, R., Abebe, E., Papert, A. and Blaxter, M. (2002) Molecular barcodes for soil nematode identification. *Molecular Ecology* 11, 839–850.

Gascuel, O. (1997) BIONJ: an improved version of the NJ algorithm based on a simple model of sequence data. *Molecular Biology and Evolution* 14, 685–695.

Goloboff, P.A., Farris, J.S. and Nixon, K.C. (2008) TNT, a free program for phylogenetic analysis. *Cladistics* 24, 774–786.

Guindon, S. and Gascuel, O. (2003) A simple, fast, and accurate algorithm to estimate large phylogenies by maximum likelihood. *Systematic Biology* 52, 696–704.

Hafner, M.S. and Nadler, S.A. (1988) Phylogenetic trees support the coevolution of parasites and their hosts. *Nature* 332, 258–259.

Hillis, D.M. and Bull, J.J. (1993) An empirical test of bootstrapping as a method for assessing confidence in phylogenetic analysis. *Systematic Biology* 42, 182–192.

Hillis, D.M. and Dixon, M.T. (1991) Ribosomal DNA: molecular evolution and phylogenetic inference. *The Quarterly Review of Biology* 66, 411–453.

Hillis, D.M., Bull, J.J., White, M.E., Badgett, M.R. and Molineux, I.J. (1992) Experimental phylogenetics – generation of a known phylogeny. *Science* 255, 589–592.

Hillis, D.M., Huelsenbeck, J.P. and Cunningham, C.W. (1994) Application and accuracy of molecular phylogenies. *Science* 264, 671–677.

Hofacker, I.L. (2003) Vienna RNA secondary structure server. *Nucleic Acids Research* 31, 3429–3431.

Holterman, M., Van Der Wurff, A., Van Den Elsen, S., Van Megen, H., Bongers, T., Holovachov, O., Bakker, J. and Helder, J. (2006) Phylum-wide analysis of SSU rDNA reveals deep phylogenetic

relationships among nematodes and accelerated evolution toward crown clades. *Molecular Biology and Evolution* 23, 1792–1800.
Hu, M. and Gasser, R.B. (2006) Mitochondrial genomes of parasitic nematodes – progress and perspectives. *Trends in Parasitology* 22, 78–84.
Hudson, M.E. (2007) Sequencing breakthroughs for genomic ecology and evolutionary biology. *Molecular Ecology Notes*, doi: 10.1111/j.1471-8286.2007.02019.
Huelsenbeck, J.P. (1995) Performance of phylogenetic methods in simulation. *Systematic Biology* 44, 17–48.
Huelsenbeck, J.P. and Hillis, D.M. (1993) Success of phylogenetic methods in the four-taxon case. *Systematic Biology* 42, 247–264.
Huelsenbeck, J.P. and Ronquist, F. (2001) MRBAYES: Bayesian inference of phylogenetic trees. *Bioinformatics* 17, 754–755.
Hyman, B.C., Beck, J.L. and Weiss, K.C. (1988) Sequence amplification and gene rearrangement in parasitic nematode mitochondrial DNA. *Genetics* 120, 707–712.
Jobb, G. (2008) TREEFINDER version of March 2008. Munich, Germany. Distributed by the author at www.treefinder.de
Johnson, K.P., Adams, R.J., Page, R.D.M. and Clayton, D.H. (2003) When do parasites fail to speciate in response to host speciation? *Systematic Biology* 52, 37–47.
Kaiser, H. (1991) Terrestrial and semi-terrestrial Mermithidae. In: Nickle, W.R. (ed.) *Manual of Agricultural Nematology*. Marcel Dekker, New York.
Katoh, K., Misawa, K., Kuma, K. and Miyata, T. (2002) MAFFT: a novel method for rapid multiple sequence alignment based on fast Fourier transform. *Nucleic Acids Research* 30, 3059–3066.
Kovaleva, E.S., Masler, E.P., Skantar, A.M. and Chitwood, D.J. (2004) Novel matrix metalloproteinase from the cyst nematodes *Heterodera glycines* and *Globodera rostochiensis*. *Molecular and Biochemical Parasitology* 136, 109–112.
Kuhner, M.K., Yamato, J., Beerli, P. and Felsenstein, J. (2005) LAMARC. 2.0.2. University of Washington, Washington.
Labate, J.A. (2000) Software for population genetic analyses of molecular marker data. *Crop Science* 40, 1521–1528.
Lavrov, D.V. and Brown, W.M. (2001) *Trichinella spiralis* mtDNA: a nematode mitochondrial genome that encodes a putative ATP8 and normally structured tRNAs and has a gene arrangement relatable to those of coelomate metazoans. *Genetics* 157, 621–637.
Lewis, P.O. and Zaykin, D. (2001) Genetic data analysis: computer program for the analysis of allelic data. Version 1.0 (d12). Distributed by the authors. Available at: http://lewis.eeb.ucon.n.edu/lewishome/software.html.
Liu, J. and Berry, R.E. (1996) Phylogenetic analysis of the genus *Steinernema* by morphological characters and randomly amplified polymorphic DNA fragments. *Fundamental and Applied Nematology* 19, 463–469.
Liu, J., Berry, R.E. and Moldenke, A.F. (1997) Phylogenetic relationships of entomopathogenic nematodes (Heterorhabditidae and Steinernematidae) inferred from partial 18S rRNA gene sequences. *Journal of Invertebrate Pathology* 69, 246–252.
Liu, J., Berry, R.E. and Blouin, M.S. (1999) Molecular differentiation and phylogeny of entomopathogenic nematodes (Rhabditida: Heterorhabditidae) based on ND4 gene sequences of mitochondrial DNA. *Journal of Parasitology* 85, 709–715.
Longhorn, S.J., Foster, P.G. and Vogler, A.P. (2007) The nematode-arthropod clade revisited: phylogenomic analyses from ribosomal protein genes misled by shared evolutionary biases. *Cladistics* 23, 130–144.
Loytynoja, A. and Milinkovitch, M.C. (2003) A hidden Markov model for progressive multiple alignment. *Bioinformatics* 19, 1505–1513.
Lunt, D.H. and Hyman, B.C. (1997) Animal mitochondrial DNA recombination. *Nature* 387, 247.

Maddison, W.P. and Maddison, D.R. (2002) *MacClade version 4.0.5*. Sinauer Associates, Sunderland, Massachusetts.

Montiel, R., Lucena, M.A., Medeiros, J. and Simoes, N. (2006) The complete mitochondrial genome of the entomopathogenic nematode *Steinernema carpocapsae*: insights into nematode mitochondrial DNA evolution and phylogeny. *Journal of Molecular Evolution* 62, 211-215.

Morrison, D.A. and Ellis, J.T. (1997) Effects of nucleotide sequence alignment on phylogeny estimation: a case study of 18S rDNAs of Apicomplexa. *Molecular Biology and Evolution* 14, 428–441.

Nadler, S.A., Bolotin, E. and Stock, S.P. (2006a) Phylogenetic relationships of *Steinernema* Travassos, 1927 (Nematoda: Cephalobina: Steinernematidae) based on nuclear, mitochondrial and morphological data. *Systematic Parasitology* 63, 161–181.

Nadler, S.A., De Ley, P., Mundo-Ocampo, M., Smythe, A.B., Stock, S.P., Bumbarger, D., Adams, B.J., De Ley, I.T., Holovachov, O. and Baldwin, J.G. (2006b) Phylogeny of Cephalobina (Nematoda): molecular evidence for recurrent evolution of probolae and incongruence with traditional classifications. *Molecular Phylogenetics and Evolution* 40, 696–711.

Nguyen, K.B., Maruniak, J. and Adams, B.J. (2001) Diagnostic and phylogenetic utility of the rDNA internal transcribed spacer sequences of *Steinernema. Journal of Nematology* 33, 73–82.

Nielsen, R. and Wakeley, J. (2001) Distinguishing migration from isolation: a Markov Chain Monte Carlo Approach. *Genetics* 158, 885–896.

Nixon, K.C. (1999) The Parsimony Ratchet, a new method for rapid parsimony analysis. *Cladistics* 15, 407–414.

Page, R.D.M. (1990) Component analysis: a variant failure? *Cladistics* 6, 119–136.

Page, R.D.M. (1993) COMPONENT: tree comparison software for Microsoft Windows, version 2.0. The Natural History Museum, London.

Page, R.D.M. (1994) Parallel phylogenies: reconstructing the history of host–parasite assemblages. *Cladistics* 10, 155–173.

Peat, S., ffrench-constant, R., Waterfield, N. and Adams, B.J. (2006) The effect of rooting and outgroup choice in recovering significant cospeciation between *Heterorhabditis and Photorhabdus. Journal of Nematology* 38, 286.

Penny, D., Steel, M., Waddell, P.J. and Hendy, M.D. (1995) Improved analyses of human mtDNA sequences support a recent African origin for *Homo sapiens. Molecular Biology and Evolution* 12, 863–882.

Pérez-Losada, M., Porter, M.L., Tazi, L. and Crandall, K.A. (2007) New methods for inferring population dynamics from microbial sequences. *Infection, Genetics, and Evolution* 7, 24–43.

Perlman, S.J., Spicer, G.S., Shoemaker, D.D. and Jaenike, J. (2003) Associations between mycophagous *Drosophila* and their *Howardula* nematode parasites: a worldwide phylogenetic shuffle. *Molecular Ecology* 12, 237–249.

Phillips, A., Janies, D. and Wheeler, W. (2000) Multiple sequence alignment in phylogenetic analysis. *Molecular Phylogenetics and Evolution* 16, 317–330.

Piganeau, G., Gardner, M. and Eyre-Walker, A. (2004) A broad survey of recombination in animal mitochondria. *Molecular Biology and Evolution* 21, 2319–2325.

Poinar, G.O., Jr (1993) Origins and phylogenetic relationships of the entomophilic rhabditids, *Heterorhabditis* and *Steinernema. Fundamental and Applied Nematology* 16, 333–338.

Posada, D. and Buckley, T.R. (2004) Model selection and model averaging in phylogenetics: advantages of Akaike information criterion and Bayesian approaches over likelihood ratio tests. *Systematic Biology* 53, 793–808.

Posada, D. and Crandall, K.A. (1998) MODELTEST: testing the model of DNA substitution. *Bioinformatics* 14, 817–818.

Powers, T.O. (2004) Nematode molecular diagnostics: from bands to barcodes. *Annual Review of Phytopathology* 42, 367–383.

Powers, T.O., Harris, T.S. and Hyman, B.C. (1993) Mitochondrial DNA sequence divergence among *Meloidogyne incognita, Romanomermis culicivorax, Ascaris suum,* and *Caenorhabditis elegans. Journal of Nematology* 25, 564–572.

Powers, T.O., Todd, T.C., Burnell, A.M., Murray, P.C.B., Fleming, C.C., Szalanski, A.L., Adams, B.A. and Harris, T.S. (1997) The rDNA internal transcribed spacer region as a taxonomic marker for nematodes. *Journal of Nematology* 29, 441–450.

Pritchard, J.K., Stephens, M. and Donnelly, P. (2000) Inference of population structure using multi-locus genotype data. *Genetics* 155, 945–959.

Provine, W.B. (2001) *The Origins of Theoretical Population Genetics.* University of Chicago Press, Chicago, Illinois.

Rambaut, A. and Drummond, A.J. (2003) Tracer v1.3. Available at: http://evolve.zoo.ox.ac.uk/software.html?id = tracer

Raymond, M. and Rousset, F. (1995) Genepop (version 1.2): population genetics software for exact tests and ecumenicism. *Journal of Heredity* 86, 248–249.

Rolston, A.N., Boyle, S., Kakouli-Duarte, T., Griffin, C.T. and Downes, M.J. (2004) Distribution of, and intraspecific variation between, populations of entomopathogenic nematodes from Bull Island, Republic of Ireland. *Journal of Nematology* 36, 344.

Ronquist, F. (1995) Reconstructing the history of host–parasite associations using generalised parsimony. *Cladistics* 11, 73–89.

Ronquist, F. and Huelsenbeck, J.P. (2003) MRBAYES 3: bayesian phylogenetic inference under mixed models. *Bioinformatics* 19, 1572–1574.

Rozas, J. and Rozas, R. (1995) DnaSP, DNA sequence polymorphism: an interactive program for estimating population genetics parameters from DNA sequence data. *Computer Applications in the Biosciences* 11, 621–625.

Rozas, J. and Rozas, R. (1997) DnaSP version 2.0: a novel software package for extensive molecular population genetics analysis. *Computer Applications in the Biosciences* 13, 307–311.

Rozas, J. and Rozas, R. (1999) DnaSP version 3: an integrated program for molecular population genetics and molecular evolution analysis. *Bioinformatics* 15, 174–175.

Siddall, M.E. (1998) Success of parsimony in the four-taxon case: long-branch repulsion by likelihood in the Farris Zone. *Cladistics* 14, 209–220.

Siddall, M.E. (2001) Philosophy and phylogenetic inference: a comparison of likelihood and parsimony methods in the context of Karl Popper's writings on corroboration. *Cladistics* 17, 395–399.

Skantar, A.M. and Carta, L.K. (2004) Molecular characterization and phylogenetic evaluation of the Hsp90 gene from selected nematodes. *Journal of Nematology* 34, 466–480.

Slowinski, J.B. (1993) COMPONENT, version 2.0. *Cladistics* 9, 351–353.

Sorenson, M. (1999) *TreeRot Version 2.* MA Boston University, Boston, Massachusetts.

Spiridonov, S.E., Reid, A.P., Podrucka, K., Subbotin, S.A. and Moens, M. (2004) Phylogenetic relationships within the genus *Steinernema* (Nematoda: Rhabditida) as inferred from analyses of sequences of the ITSI-5.8S-ITS2 region of rDNA and morphological features. *Nematology* 6, 547–566.

Steel, M., Huson, D. and Lockhart, P.J. (2000) Invariable sites models and their use in phylogeny reconstruction. *Systematic Biology* 49, 225–232.

Stein, L.D., Bao, Z., Blasiar, D., Blumenthal, T., Brent, M.R., Chen, N., Chinwalla, A., Clarke, L., Clee, C., Coghlan, A., Coulson, A., D'eustachio, P., Fitch, D.H.A., Fulton, L.A., Fulton, R.E., Griffiths-Jones, S., Harris, T.W., Hillier, L.W., Kamath, R., Kuwabara, P.E., Mardis, E.R., Marra, M.A., Miner, T.L., Minx, P., Mullikin, J.C., Plumb, R.W., Rogers, J., Schein, J.E., Sohrmann, M., Spieth, J., Stajich, J.E., Wei, C., Willey, D., Wilson, R.K., Durbin, R. and Waterston, R.H. (2003) The genome sequence of *Caenorhabditis briggsae*: a platform for comparative genomics. *PLOS Biology* 1, 166–192.

Stock, S.P., Campbell, J.F. and Nadler, S.A. (2001) Phylogeny of *Steinernema* Travassos, 1927 (Cephalobina: Steinernematidae) inferred from ribosomal DNA sequences and morphological characters. *Journal of Parasitology* 87, 877–889.

Subbotin, S.A., Sturhan, D., Chizhov, V.N., Vovlas, N. and Baldwin, J.G. (2006a) Phylogenetic analysis of Tylenchida Thorne, 1949 as inferred from D2 and D3 expansion fragment of the 28S rRNA gene sequences. *Nematology* 8, 455–474.

Subbotin, S.A., Sturhan, D., Vovlas, N., Castillo, P., Tanyi Tambe, L., Moens, M. and Baldwin, J.G. (2006b) Application of secondary structure model of rRNA for phylogeny: D2–D3 expansion segments of the LSU gene of plant-parasitic nematodes from the family Hoplolaimidae Filipjev, 1934. *Molecular Phylogenetics and Evolution* 43, 881–890.

Sullivan, J. and Joyce, P. (2005) Model selection in phylogenetics. *Annual Review of Ecology Evolution and Systematics* 36, 445–466.

Sullivan, J. and Swofford, D.L. (2001) Should we use model-based methods for phylogenetic inference when we know that assumptions about among-site rate variation and nucleotide substitution pattern are violated? *Systematic Biology* 50, 723–729.

Swofford, D.L. (2002) *PAUP*. Phylogenetic Analysis Using Parsimony (*and Other Methods)*. Sinauer Associates, Sunderland, Massachusetts.

Swofford, D.L., Waddell, P.J., Huelsenbeck, J.P., Foster, P.G., Lewis, P.O. and Rogers, J.S. (2001) Bias in phylogenetic estimation and its relevance to the choice between parsimony and likelihood methods. *Systematic Biology* 50, 525–539.

Tang, S. (2006) Comparative mitochondrial genomics of the Nematoda: the family Mermithidae represents rapid enoplean mitochondrial DNA evolution. PhD Dissertation, University of California-Riverside, California, 252.

Tang, S. and Hyman, B.C. (2005) Rolling circle amplification of complete nematode mitochondrial genomes. *Journal of Nematology* 37, 236–241.

Tang, S. and Hyman, B.C. (2007) Mitochondrial genome haplotype hypervariation within the isopod parasitic nematode *Thaumamermis cosgrovei. Genetics* 176, 1139–1150.

Telford, M.J., Wise, M.J. and Gowri-Shankar, V. (2005) Consideration of RNA secondary structure significantly improves likelihood-based estimates of phylogeny: examples from the bilateria. *Molecular Biology and Evolution* 22, 1129–1136.

Templeton, A.R. (2006) *Population Genetics and Microevolutionary Theory*. Wiley-Liss, Hoboken, New Jersey.

Terry, M.D. and Whiting, M.F. (2005) Comparison of two alignment techniques within a single complex data set: POY versus CLUSTAL. *Cladistics* 21, 272–281.

Thompson, E.A. (1990) Fisher, R.A.'s Contributions to Genetic Statistics. *Biometrics* 46, 905–914.

Thompson, J.D., Higgins, D.G. and Gibson, T.J. (1994) CLUSTAL W: improving the sensitivity of progressive multiple sequence alignment through sequence weighting, position-specific gap penalties and weight matrix choice. *Nucleic Acids Research* 22, 4673–4680.

Thompson, J.D., Gibson, T.J., Plewniak, F., Jeanmougin, F. and Higgins, D.G. (1997) The CLUSTAL_X windows interface: flexible strategies for multiple sequence alignment aided by quality analysis tools. *Nucleic Acids Research* 25, 4876–4882.

Wheeler, W. (1996) Optimization alignment: the end of multiple sequence alignment in phylogenetics? *Cladistics* 12, 1–9.

Wheeler, W. (2001) Homology and the optimization of DNA sequence data. *Cladistics* 17, S3–S11.

Wheeler, W. and Gladstein, D. (1994) MALIGN. 1.99 ed. American Museum of Natural History, New York.

Wheeler, W.C. (2003) Implied alignment: a synapomorphy-based multiple-sequence alignment method and its use in cladogram search. *Cladistics* 19, 261–268.

Whipple, L.E., Lunt, D.H. and Hyman, B.C. (1998) Mitochondrial DNA length variation in *Meloidogyne incognita* isolates of established genetic relationships: utility for nematode population studies. *Fundamental and Applied Nematology* 21, 265–271.

Wiley, E.O. (1988a) Parsimony analysis and vicariance biogeography. *Systematic Zoology* 37, 271–290.

Wiley, E.O. (1988b) Vicariance biogeography. *Annual Review of Ecology and Systematics* 19, 513–542.

Ye, W., Giblin-Davis, R.M., Davies, K.A., Purcell, M.F., Scheffer, S.J., Taylor, G.S., Center, T.D., Morris, K. and Thomas, W.K. (2007) Molecular phylogenetics and the evolution of host plant associations in the nematode genus *Fergusobia* (Tylenchida: Fergusobiinae). *Molecular Phylogenetics and Evolution* 45, 123–141.

# III Host–Pathogen Interactions

# 9 Host–Virus Interactions

J.P. Burand,[1] M. Nakai[2] and I. Smith[3]

[1]*Division of Entomology, Department of Plant Soil and Insect Science, University of Massachusetts, Amherst, USA;* [2]*Graduate School of Agriculture, Tokyo University of Agriculture and Technology, Fuchu, Tokyo, Japan;* [3]*Nara Institute of Science and Technology, Nara, Japan*



## 9.1. Introduction

Viruses are now recognized as the causative agents of naturally occurring infections that have been observed for centuries in insect populations, including the jaundice disease in the silkworm described by Nysten in 1808 and Maestri in 1856, the 'wilting disease' of the nun moth described by von Tubeuf in 1892 (reviewed

 *Insect Pathogens: Molecular Approaches and Techniques* (eds S.P. Stock *et al.*)

in Benz, 1986; see also Miller, 1997) and, more recently, virus epizootics in the gypsy moth (Stiles *et al.*, 1983). Their ability to devastate insect populations has stimulated interest in deploying pathogenic viruses as insecticidal agents, and has been the driving force behind research into their biology and molecular biology. This in turn has led to our steadily improving knowledge of insect viruses, which includes a growing body of research into the molecular basis of a remarkable range of virus–host interactions that are involved in disease pathology and virus transmission. It has also led to the development of molecular tools, particularly recombinant baculoviruses carrying additional genes or gene deletions, which are yielding important insights into virus–host interactions as well as advancing both basic and applied research in many different disciplines of biological sciences beyond invertebrate pathology.

Insect virology is an immensely broad and diverse field, encompassing many kinds of viruses that infect many kinds of insects in many different ways. Rather than attempting to cover the entire spectrum of interactions that are known to occur between these viruses and their hosts, we have chosen to focus primarily on insect-pathogenic viruses, and particularly on those which we believe have the greatest potential for practical control of insect pests. The purpose of this chapter, therefore, is to describe some examples of how viruses have already been used to control insect pests, and then to illustrate the range of interactions that occur between selected insect viruses and their hosts, highlighting the molecular techniques used to study these interactions. In doing so, we hope the reader will gain insights into how molecular approaches have contributed to our evolving understanding of insect virus–host interactions, and an appreciation of the central role played by such interactions in the overall progression of viral disease.

## 9.2. Use of Viruses as Insect Pest Control Agents

Of the several different kinds of viruses that have been used to control insect pests, the rod-shaped DNA viruses, particularly members of the family *Baculoviridae*, have received most attention. Viruses of this type have only been reported to infect arthropods, principally insects, and their absence as pathogens of plants or higher animals implies that they are safe for large-scale environmental use to control insect pests, posing no threat to non-target organisms. The baculoviruses have also been of especial interest because of the broad range of insect species that they are known to infect. In addition, baculovirus particles are embedded within a paracrystalline proteinaceous occlusion body (OB), which confers environmental stability and allows baculoviruses to be formulated and applied in much the same way as traditional insect pest control agents.

The earliest example of pest suppression by an introduced insect virus was recorded in the 1930s. The European spruce sawfly, *Gilpinia hercyniae*, was brought accidentally from Scandinavia to Canada, and classical biological control using parasitoids was undertaken to manage this insect. Several parasitoid species were introduced and a few became established; however, some of them carried a virus disease of the host, described as a polyhedrosis, which spread effectively throughout a severely infested area of about 3 million ha until, by 1940,

*G. hercyniae* had ceased to be a pest in Canada (Balch and Bird, 1944). Although this was an impressively successful example of 'permanent introduction' using a viral disease of insect pests, most viruses used subsequently as control agents have been applied by 'inundative release' or 'inoculative release' approaches.

To date more than 40 viral pesticides have been registered worldwide for insect control. *Helicoverpa zea* nucleopolyhedrovirus (NPV) was the first registered baculovirus for commercial use in the USA, and was sold from 1975 to 1982 (Ignoffo and Couch, 1981). A cypovirus (CPV) of *Dendrolimus spectabilis*, registered in Japan in 1974, was used to suppress populations of the pine moth, but is not currently being produced. Aerial applications of the gypsy moth *Lymantria dispar* NPV (LdMNPV) have been used successfully in forests in the USA to control this insect (Webb *et al.*, 1999). In Brazil, *Anticarsia gemmatalis* NPV (AgNPV) is currently being applied annually to over 1 million ha of soybean fields to control the noctuid pest *A. gemmatalis* (Moscardi, 1999). Until now, AgNPV has been formulated crudely from cadavers of virus-infected *A. gemmatalis* larvae collected in the field, but production may soon move into the laboratory due to increased demand (Szewczyk *et al.*, 2006). China also produces viral pesticides for commercial use: *Helicoverpa armigera* NPV is most widely used, with 200–300 t produced annually for application on approximately 100,000 ha of cotton. Eight NPVs, three granuloviruses (GVs), a CPV and a densovirus of *Periplaneta fuliginosare* have been commercially available in China for several years (X. Sun, Wuhan, 2006, personal communication; see http://www.chinapesticide.gov.cn).

A GV of the codling moth *Cydia pomonella* (CpGV) has been used inundatively to control codling moth larvae in apple and pear orchards. CpGV is highly virulent against codling moth larvae, with a median lethal dose ($LD_{50}$) of approximately one OB in neonates, the stage at which larvae are targeted before they enter the fruit (Huber, 1986). CpGV has been applied to approximately 80,000 ha in Europe and 4000 ha in the USA (Lacey *et al.*, 2004). A viral agent containing the NPV of *Spodoptera exigua* (SeMNPV), a pest of vegetables and flowers, is commercially available in the USA, Europe and Thailand, while *Spodptera littoralis* NPV is also registered in Europe to control pests of cotton, maize and other vegetables.

A non-occluded DNA virus, the *Oryctes* virus, has also been used very successfully to control the rhinoceros beetle, *Oryctes rhinoceros*, in oil palm plantations in Malaysia (Bedford, 1980); the newly discovered Hz-2V, another non-occluded DNA virus, also has considerable promise as an insect-sterilizing agent (Raina and Adams, 1995; Hamm *et al.*, 1996). These two viruses are discussed further in Section 9.7.

## 9.3. Development of Baculoviruses for Foreign Gene Expression

### 9.3.1. Genetic manipulation strategies

The major protein component of almost all baculovirus OBs, polyhedrin in the NPVs and granulin in the GVs, is a ~30-kDa polypeptide which is synthesized in copious quantities during the terminal stages of virus replication. This phase is preceded by the assembly and release from cells of one of the two distinct baculovirus particle phenotypes, budded virus (BV), which is responsible for transmission of virus from one infected cell to the next. The virus particle phenotype that becomes embedded in, and later liberated from, the OB is referred to as occlusion-derived virus (ODV). In cell culture, therefore, the NPV *polyhedrin* gene – *polh* – is dispensable for the production of progeny virus in the form of BV; in nature, the principal role of polyhedrin is to protect virus during the prolonged periods in which it may have to survive outside the host.

Following the first demonstrations, in the laboratories of Summers (Smith *et al.*, 1983) and Miller (Pennock *et al.*, 1984), that the *polh* coding region could be replaced with that of a foreign gene, which would then be expressed in a similar fashion to *polh* itself, the use of *Autographa californica* NPV (AcMNPV) as a baculovirus expression vector (BEV) rapidly became widespread. It remains popular more than 20 years later: the BEV system has yielded thousands of publications and several protocols books, and is integral to the success of a number of commercial enterprises.

The classical procedure for generating recombinant AcMNPV has three essential steps: (i) insertion of the foreign gene into a plasmid containing a portion of the AcMNPV genome, typically at a site adjacent to the *polh* promoter, so that the foreign gene becomes sandwiched between two flanking stretches of viral DNA; (ii) incorporation of this modified AcMNPV sequence into a whole AcMNPV genome; and (iii) cloning of this recombinant AcMNPV genome to isolate it from wild-type virus genomes. Step (i) is accomplished in *Escherichia coli*, where routine cloning procedures permit sophisticated manipulations of both viral and foreign gene sequences to be made quickly and confirmed by DNA sequencing. In step (ii), purified plasmid DNA and genomic AcMNPV DNA are transfected together into cultured insect cells, where homologous recombination between collinear viral sequences on the plasmid and the genome yields recombinant AcMNPV DNA molecules, which then replicate and become packaged into BV and are finally released into the culture medium. The resultant BV population, however, also includes an extremely high proportion of wild-type particles, and in step (iii) recombinant virus is separated, on the basis of some visible phenotypic difference, from the unwanted wild type by isolating virus from 'plaques' of infected cells, each the result of infection by a single BV particle within the mixed-genotype BV suspension from step (ii).

This protocol, although effective, was slow and laborious: the whole procedure took several weeks, and was further hampered by the difficulty (especially to non-virologists interested only in producing quantities of the foreign protein sufficient for further studies) of identifying *polh*-negative plaques containing recombinant virus, whose appearance was not readily distinguishable from that of surrounding uninfected cells. Kost *et al.* (2005) have reviewed the major technical improvements that have been made to simplify BEV protocols, beginning with the use of linearized genomic AcMNPV DNA (Kitts *et al.*, 1990) to reduce and ultimately almost eliminate the occurrence of wild-type BVs at the end of step (ii). A second breakthrough was achieved by developing the bacmid vector (Luckow *et al.*, 1993), which eliminated the co-transfection at the beginning of step (ii) and enabled all of the manipulations required for generating recombinant AcMNPV DNA to be undertaken in *E. coli*. Bacmid-based BEV systems are now commercially available, and can yield pure stocks of recombinant BVs in 1 week. In relation to improving our understanding of host–virus interactions, bacmids offer the exciting prospect that they can be used to construct recombinant viruses bearing deletions in genes that may be essential for viral replication, a procedure which is impossible to conduct in permissive insect cells. Stewart *et al.* (2005), for example, have recently taken advantage of the bacmid system to investigate the role of an essential and highly conserved 'immediate-early' gene, *ie0-ie1*, in AcMNPV replication.

The above brief historical overview masks an extraordinary range of modifications and refinements that have been made to the BEV system, and the interested reader is urged to study the excellent survey by Kost *et al.* (2005) to gain an impression of the variety of applications currently being pursued. Notable topics covered are glycosylation of foreign proteins, a particular concern for medical applications; selection of promoters and other viral DNA elements to improve expression of the protein of interest, or indeed the use of several promoters to express multiple proteins simultaneously; insertion of sequences encoding foreign proteins or peptides into the *gp64* gene so that, for example, immunogens can be displayed on the surface of virus particles; and the use of so-called BacMam viruses to deliver genes into mammalian cells.

### 9.3.2. Genetically modified baculovirus insecticides

Interest in the use of genetically modified baculoviruses (GMBs) as enhanced biopesticides appears to be waning, after a period of intense commercial activity during the 1990s. Whether or not there is a future for GMB products will depend both on commercial perceptions about their likely profitability and on political considerations regarding the social acceptability of such viruses (and, indeed, of genetically modified (GM) products of any description). Efforts to produce GMBs have usually been motivated by a desire to increase 'speed of kill' – essentially, to minimize the lag between ingestion of virus and cessation of feeding by an infected host larva – a parameter for which baculoviruses compare unfavourably with faster-acting agents such as chemical pesticides and *Bacillus thuringiensis* toxins. Genes encoding a variety of arthropod-specific toxins from scorpions, mites and ants have been inserted into AcMNPV, where they are usually expressed under the strong *polh* or *p10* viral promoters, and have often led to substantially increased virulence. Recent examples include the expression of a lepidopteran-specific scorpion toxin (*Buthus tamulus* ButaIT) under the *p10* promoter, which reduced the median survival time ($ST_{50}$) of infected *Heliothis virescens* larvae by 42% compared with that of larvae infected with wild-type AcMNPV (Rajendra *et al.*, 2006); of an insect-specific predatory ant toxin (*Paraponera clavata* poneratoxin) under the *polh* promoter, which reduced the $ST_{50}$ of infected *Spodoptera frugiperda* larvae by 15% (Szolajska *et al.*, 2004); and of another insect-specific scorpion toxin (*Leiurus quinquestriatus* LqhIT2) under early and late viral promoters, which reduced the $ST_{50}$ of *H. virescens* larvae by up to 55% (van Beek *et al.*, 2003).

The design of vectors used to express these toxins, and of bioassays used to assess their effectiveness in GMBs, includes a number of variables that merit brief consideration. Although the choice of promoter might be expected to affect both the timing and the intensity of toxin gene expression, a review by van Beek *et al.* (2003) of earlier studies and of their own data suggests that there is usually little difference in speed of kill for particular GMBs expressing toxins under early (e.g. *ie1*) or late (e.g. *p10*) promoters. In contrast, the choice of signal peptide sequence inserted to ensure that the toxin is secreted from infected cells may have substantial effects: for viruses encoding LqhIT2 expressed under the *ie1* promoter, van Beek *et al.* (2003) reported that the response time of third-instar *H. virescens* larvae varied dramatically, ranging from ~64 h with the *Bombyx mori* bombyxin signal sequence to ~110 h with the *Androctonus australis* scorpion toxin AaIT signal sequence. With respect to bioassays, it is often difficult to make comparisons between data presented by different authors. The dose of inoculum used affects the time taken for the host to succumb to infection, and even when this is standardized discrepancies may remain. For neonate *H. virescens* larvae infected *per os* by the droplet-feeding method, using the same concentration of wild-type AcMNPV OBs, van Beek *et al.* (2003) and Rajendra *et al.* (2006) reported median response times of 85 and 112 h, respectively. Such differences may be due partly to varied insect-rearing conditions, or to the host strain or AcMNPV genotype used, but they are comparable in magnitude to differences between the response times of wild-type and certain toxin-expressing viruses.

Following several unsuccessful attempts by others to produce more virulent AcMNPV genotypes expressing *B. thuringiensis* toxin genes under the *polh* promoter, Chang *et al.* (2003) reported a different strategy, based on a technique first described by McLinden *et al.* (1992) and crucially modified by Je *et al.* (2003), in which sequences encoding the N-terminal region of a *B. thuringiensis* Cry1Ac polypeptide were placed downstream of the *polh*-coding region and upstream of a marker open reading frame (ORF) encoding green fluorescent protein (GFP). The transfer vector used to express this chimaeric protein also carried a second copy of *polh*, transcribed independently from an ectopic *p10* promoter. Remarkably, Chang *et al.* (2003) found that phenotypically normal OBs formed in *S. frugiperda* Sf9 cells infected with the resultant recombinant virus, and gold-labelled antibodies against Cry1Ac and GFP bound to sections of the OBs. Bioassays in second- and third-instar *Plutella xylostella* larvae showed that the activity of these OBs was very similar to that of comparable amounts of purified Cry1Ac protein: the $ST_{50}$ was reduced by 63% relative to wild-type AcMNPV, and the $LD_{50}$ by 100-fold. An important extension of this work has been described by Seo *et al.* (2005), who fused the same region of *cry1Ac* to *polh* and expressed the chimaeric protein in *E. coli*, where it formed insoluble inclusion bodies that were also active against *P. xylostella* larvae. The authors argue that such protein-based insecticides can be mass-produced in bacteria at low cost, and, furthermore, could overcome some of the safety concerns associated with the use of GMBs.

### 9.3.3. Insect genome manipulation

Other than AcMNPV, GMB-related activity has so far been restricted to *B. mori* NPV (BmNPV). This virus is of particular interest in eastern Asia, for two reasons.

First, pioneering work by Maeda *et al.* (1985) showed that biologically active interferon could be recovered from the haemolymph of silkworms infected with recombinant BmNPV and suggested that *B. mori* – whose larvae are unusually large and docile, and were first domesticated in China more than 2000 years ago – could be used as a 'factory' for producing BmNPV-expressed foreign proteins at much lower cost than would be achievable in the cell-culture systems used with AcMNPV. Second, unlike most other lepidopteran species, *B. mori* is a beneficial insect of considerable economic importance in the region; rather than attempting to increase viral potency, transgenic work in this area is geared towards making the host more resistant to viral pathogens including BmNPV. A study by Isobe *et al.* (2004), for example, probed the use of RNA interference (RNAi) to generate silkworms having reduced susceptibility to BmNPV infection. RNAi comprises a group of techniques that enable genes to be silenced by targeting their messenger RNA (mRNA) for degradation (reviewed by Hannon, 2002; for a recent review of antiviral RNAi, see Morris and Rossi, 2006). By incorporating into the *B. mori* genome sequences expected to suppress mRNA of the essential BmNPV *lef-1* gene, Isobe *et al.* (2004) reported a 50% reduction in the titre of BV released from transgenic *B. mori* BmN cells infected with BmNPV, as well as partial resistance to the virus in transgenic larvae.

Transformation of the host genome in these experiments was achieved by means of the *piggyBac* vector, which is based on a transposable element from the *Trichoplusia ni* genome that was first discovered in spontaneous AcMNPV mutants arising during infection of *T. ni* TN368 cells (Fraser *et al.*, 1983). Although various approaches to transduction and transformation of insect cells have been described (for a recent review of germline transformation and its applications in the silkworm, see Goldsmith *et al.*, 2005), *piggyBac* appears to be particularly promising: as well as transforming numerous invertebrate species, including *Drosophila melanogaster* and the malaria parasite *Plasmodium falciparum* (see Balu *et al.*, 2005), it is also capable of transforming mice (Ding *et al.*, 2005).

## 9.4. Insect Defences Against Viruses

Insects are among the most successful of all animal groups. In occupying a wide variety of habitats, insects constantly encounter microorganisms and parasites, which are attempting to exploit them as a food and energy source for their own survival. Despite this continuous exposure, relatively few microbes – including viruses – are successful at invading insects. A major reason for the paucity of insect pathogens is that insects have evolved elaborate defences to protect themselves from invaders.

There are three major routes of entry known for insect viruses. These are the pathways used by baculoviruses, entomopoxviruses, CPVs and the *Oryctes* virus, namely ingestion or *per os* infection; transmission directly into the haemocoel, used by ascoviruses, which are injected by parasitic insects during oviposition; and sexual transmission during mating, used by the newly described insect virus Hz-2V. The route of infection taken by particular viruses reflects biological attributes that each has acquired to overcome specific host barriers to infection.

Like most other complex eukaryotes, insects have evolved both physical barriers and cellular defence mechanisms to protect themselves from invasion by viruses (reviewed by Narayanan, 2004). The primary physical defence is the chitinous cuticular lining, which covers most of the external surface of an insect and extends through the foregut, the hindgut, the tracheal tubes and most of the reproductive tissues. The external cuticle is virtually impenetrable to viruses except for those, like the ascoviruses, that recruit parasitic insects to inject them into the host insect, along with the parasite's eggs, during penetration of the cuticle by a needle-like ovipositor (Hamm *et al.*, 1985; Tillman *et al.*, 2004).

Since phytophagous insects routinely encounter viruses and other microbes on the plant material on which they feed, the gut is the primary site of entry for numerous insect-pathogenic viruses. The physiological conditions of the insect gut include a variety of digestive enzymes, providing considerable protection from invasion by viruses. Lining the midgut, protecting the epithelial cells from damage and serving as a barrier against virus infections, is the peritrophic matrix or membrane (PM), which consists of chitin along with glucosaminoglycans, glycoproteins and a variety of other proteins (reviewed by Tellam *et al.*, 1999). Even if a virus manages to establish itself in the midgut epithelium, the basal lamina, a protective layer of protein, collagen and elastin between the haemocoel and many of the insect's organs, constitutes a further physical barrier to spread of the virus within the host (Narayanan, 2004).

The primary cellular defence mechanism in insects to viral infection is programmed cell death, or apoptosis, of infected cells (Clem, 2005), which is discussed in more detail in Section 9.5.4. While other cellular responses, including haemocyte encapsulation and melanization, have also been shown to play a role in protection from viral infection (Trudeau *et al.*, 2001), these are not specific to viruses but components of a general, innate host response to microbial infection.

## 9.5. Baculovirus Pathogenesis

Baculoviruses are unusual among viruses in that their virus particles, as noted in Section 9.3, have two physically different infectious forms. One form, the ODV, is assembled and enveloped in the cell nucleus during the terminal stages of replication, where it becomes embedded within the OB. In the other form, the BV, nucleocapsids become enveloped as they bud through the plasma membrane into the extracellular environment. BVs are thus involved in transmission from cell to cell, and ODVs in transmission from insect to insect. However, it should be noted that neither of the hymenopteran NPVs whose genomes have been sequenced encode homologues of the BV envelope proteins that are known to mediate cell-to-cell transmission of lepidopteran NPVs (Lauzon *et al.*, 2006); NPVs isolated from Hymenoptera and Diptera, as well as a GV from the western grapeleaf skeletonizer *Harrisina brillians*, appear to replicate only in the midgut epithelium (Federici, 1997). The description that follows in this section should therefore be considered as applying principally, albeit not exclusively, to the lepidopteran NPVs.

### 9.5.1. Infection of the midgut epithelium

After ingestion, baculovirus OBs dissolve rapidly in the alkaline midgut, releasing the ODVs. The lepidopteran midgut is a harsh environment for the ODV to survive in, having a pH generally above 9 and abundant proteases and other enzymes capable of inactivating the virus (Nakazawa *et al.*, 2004). Baculoviruses have somehow adapted to these surroundings, and the large amount of protein solubilized upon dissolution of OBs may afford the liberated ODVs some protection and prolong their viability.

Midgut epithelial cells are the principal targets of the ODV, acting as the site for primary infection. To overcome the physical barrier presented by the PM (see Section 9.4), a number of baculoviruses have acquired genes encoding enhancins, OB-associated metalloproteinases that degrade components of the PM and allow the ODV to gain access to the epithelium (Lepore *et al.*, 1996). Interestingly, LdMNPV possesses two *enhancin* genes, and both enhancins were found by immunogold staining to be associated specifically with ODVs (Slavicek and Popham, 2005). Popham *et al.* (2001) had previously demonstrated, in bioassays using recombinant LdMNPVs lacking one or both of the *enhancin* genes, that one enhancin could compensate for the absence of the other, but that viruses lacking both *enhancin* genes were 12-fold less potent than wild-type LdMNPV. A recombinant AcMNPV harbouring the *enhancin* gene from another virus, *Mamestra configurata* NPV, is about four times more potent than wild-type AcMNPV (Li *et al.*, 2002).

Once the ODV has crossed the PM, the infection process begins with virus replication in columnar midgut epithelial cells. The virus enters these cells by a two-step process, of attachment to specific receptors on microvilli and fusion of the viral envelope with the cell membrane, which involves at least four different genes coding for *per os* infectivity factors (PIFs): *p74*, *pif1*, *pif2* and *pif3*. It is not known precisely how each of the proteins encoded by these genes functions in the process of viral entry into midgut cells. However, *in vivo* fluorescence dequenching assays, using recombinant viruses in which individual genes were deleted, indicate that all of them except PIF3 are required for ODV binding to microvilli (Haas-Stapleton *et al.*, 2004; Ohkawa *et al.*, 2005). The mechanism by which the ODV envelope fuses with the microvillar membrane is not yet known; nor is it clear how the ODV nucleocapsid, which is 30–60 nm wide and 200–300 nm long, is transported through microvilli measuring 100 nm by 1 μm to the nucleus where virus replication takes place.

As part of normal insect development, epithelial cells lining the midgut are routinely renewed, with new cells arising from regenerative cells and old cells being sloughed into the midgut by apoptosis. This disposal and regeneration of midgut cells present a further challenge to virus infection, since cells infected by virus may undergo apoptosis prior to the production and release of progeny virus required for systemic spread of the infection throughout the insect. However, the anti-apoptotic genes encoded by baculoviruses (see below) are likely to thwart this process, and the non-lepidopteran NPVs, whose replication is restricted to the midgut, presumably possess robust mechanisms to ensure that infected gut cells remain *in situ* for prolonged periods. LdMNPV, atypically for an NPV, appears unable to replicate in midgut cells of the gypsy moth (Shields, 1985), but when OBs are administered along with an optical brightener (e.g. Calcoflour M2R/Tinopal LPW) normal virus replication occurs in these cells (Adams *et al.*, 1994). Optical brighteners are known to increase *per os* infectivity of other baculoviruses including AcMNPV, and are thought to do so both by altering the integrity of the PM (Wang and Granados, 2000) and by suppressing the sloughing of midgut epithelial cells (Washburn *et al.*, 1998) by inhibiting apoptosis (Dougherty *et al.*, 2006).

Baculoviruses have evolved at least one other way to overcome the barrier to infection presented by the rapid turnover of midgut cells. Granados and Lawler (1981), using electron microscopy and *in vitro* plaque assays of haemolymph collected from infected larvae, argued that AcMNPV nucleocapsids could circumvent replication in midgut cells altogether by traversing infected cells, exiting by budding at their base at the plasmalemma. This scenario seems plausible since the gene encoding GP64, the major BV envelope protein of AcMNPV, has an early promoter, which is responsible for rapid production of large amounts of GP64 during replication (Blissard and Rohrmann, 1989). Early transcription of *gp64* could thus provide sufficient membrane-bound GP64 for ODV-derived nucleocapsids, having traversed the cytoplasm of an infected cell, to acquire a BV-like envelope and bud from the cell. Support for this model has been provided by Zhang *et al.* (2004), who showed that AcMNPV gene expression in secondarily infected tracheal cells of *T. ni* and *S. exigua* began within 4 h of the onset of expression in midgut cells, an interval too short for replication of the incoming ODV to have occurred. Upon leaving the primary site of infection, BVs must negotiate the basal lamina, a fibrous extracellular matrix, before systemic infection of the insect can proceed. Engelhard *et al.* (1994) constructed an AcMNPV mutant containing the *lacZ* reporter gene and visualized the early infection process using light microscopy. Their results showed that the insect tracheal system provides a major conduit for BV to bypass the basal lamina and thereafter spread through the entire insect body.

### 9.5.2. Perturbation of the host endocrine system

One of the most interesting host–baculovirus interactions is the pathogen's ability to interfere with host development by suppressing the moulting hormone ecdysone. In baculovirus-infected insects, the activity of ecdysone has been shown to be affected by the viral ecdysteroid UDP-glucosyltransferase (*egt*) gene, whose product blocks moulting and pupation, thereby increasing the amount of progeny virus produced by an infected insect and enhancing viral fitness. Baculovirus regulation of this aspect of host development occurs through the conjugation of sugars to ecdysone by EGT, which effectively renders the hormone inactive (O'Reilly and Miller, 1989; Kelly *et al.*, 1995; Clarke *et al.*, 1996). O'Reilly and Miller (1991) and Slavicek *et al.* (1999) showed that deletion of *egt* in AcMNPV and LdMNPV resulted in genotypes that killed insects significantly faster than the wild-type virus, and larvae infected with *egt*$^-$ LdMNPV yielded fewer OBs than those infected with wild-type virus. The faster killing speed of *egt*$^-$ AcMNPV in infected *S. exigua* larvae may be due to accelerated degeneration of Malpighian tubules (Flipsen *et al.*, 1995). Comparing an *egt*$^-$ mutant with wild-type LdMNPV, Burand *et al.* (1996) also demonstrated that while the titre of ecdysteroids released from prothoracic glands of wild-type LdMNPV-infected insects was actually elevated, the titre of ecdysteroids released in *egt*$^-$-infected insects remained the same as that produced by glands in healthy *L. dispar* larvae. This suggests that sugar-conjugated ecdysteroids do not act as negative feedback regulators to inhibit ecdysteroid production from prothoracic glands.

Fifth-instar *Adoxophyes honmai* leaf-roller larvae infected with a GV (AdhoGV) also show elevated levels of ecdysteroids compared with healthy last-instar larvae (Nakai *et al.*, 2002). Although it is not known what proportions of these ecdysteroids are free and conjugated, sugar-conjugated forms predominate in LdMNPV-infected gypsy moth larvae (Kelly *et al.*, 1995). Like the NPVs described above, AdhoGV prevents pupation of *A. honmai* larvae and prolongs their larval stage. However, because OB yield has already reached a plateau by this time, the greatly extended larval period characteristic of AdhoGV infection may benefit this virus in another way: its OBs remain protected within the host, harboured in leaf nests, until they can be transmitted efficiently to the larval progeny of neighbouring healthy insects. Interestingly, *A. honmai* larvae infected with an entomopoxvirus (AdhoEPV; another large, double-stranded DNA virus) also fail to pupate and exhibit a prolonged larval stage similar to that shown by AdhoGV-infected larvae, but they lack detectable EGT activity (Nakai *et al.*, 2004). Ecdysteroids are known to have various functions in insects, and the acquisition of a gene to regulate ecdysteroid activity may have been crucial to the evolution of certain insect viruses. However, the *egt* gene is not ubiquitous and, as suggested by Burand *et al.* (1996), insect viruses probably encode additional functions that allow them to disrupt other hormonally controlled aspects of host development.

### 9.5.3. Liquefaction of the moribund host

The final event in baculovirus replication in most insects is liquefaction of the infected host preceding the release of billions of viral OBs. Liquefaction, which is thought to enhance dissemination of virus, is due to the breakdown of the host epidermis, and is associated with expression of two viral genes encoding chitinase (*chiA*) and cathepsin (*v-cath*). In bioassay experiments using AcMNPV *chiA* and *v-cath* deletion mutants, Hawtin *et al.* (1997) showed that deletion of either gene alone inhibited liquefaction of infected *T. ni* larvae, but that insects infected with a mixture of the two deletion mutants liquefied normally. AcMNPV chitinase possesses a C-terminal KDEL motif that acts as an endoplasmic reticulum retrieval signal. Secretion of chitinase from cells infected with a *chiA* KDEL deletion mutant was detected by Western blotting at times prior to the onset of cell lysis caused by wild-type AcMNPV infection (Saville *et al.*, 2004). Furthermore, the KDEL deletion mutant increased the lethal time and lethal dose for AcMNPV in *T. ni* larvae, suggesting that localization of chitinase can affect killing speed and virulence of NPVs. Cathepsin, on the other hand, is synthesized as an inactive proenzyme (proV-CATH) and requires chitinase for its activation, which occurs only after cell death (Hom and Volkman, 2000; Hom *et al.*, 2002). Daimon *et al.* (2006) also showed that deletion of BmNPV *chiA* inhibits cathepsin activity and liquefaction in infected *B. mori* larvae. Liquefaction of BmNPV-infected silkworms is, furthermore, prevented by deletion of the *25K FP* gene (Katsuma *et al.*, 1999). Katsuma *et al.* (2004) suggested that 25K FP is associated with secretion of cathepsin from infected cells.

Hymenopteran and dipteran NPVs do not encode either *chiA* or *v-cath*, and replicate only in midgut tissue (Afonso *et al.*, 2001; Lauzon *et al.*, 2006). Unlike baculovirus-infected lepidopteran larvae, insects infected with these viruses do not liquefy. Among those baculoviruses whose full genome sequences are currently known, most lepidopteran baculoviruses encode both *chiA* and *v-cath* genes, except for *A. honmai* NPV (AdhoNPV), which lacks *chiA* (Nakai *et al.*, 2003), and three GVs – *P. xylostella* GV (PxGV; Hashimoto *et al.*, 2000), *Phthorimaea operculella* GV (accession NC_004062) and *Adoxophyes orana* GV (AdorGV; Wormleaton *et al.*, 2003) – in which both genes are absent. *A. orana* larvae infected with AdorGV do not liquefy at the end of infection (Wormleaton *et al.*, 2003). To account for the observed liquefaction of PxGV-infected *P. xylostella*, Hashimoto *et al.* (2000) proposed a role for a metalloproteinase encoded by this GV.

### 9.5.4. Intracellular host–virus interactions

Within a baculovirus-infected cell, numerous interactions occur between the replicative programme of the pathogen and the intrinsic components of the cell in which it finds itself. Initial expression of viral genes requires only that the viral genome reach the cell nucleus: purified baculovirus DNA is infectious (Burand *et al.*, 1980), and expression of 'early' viral genes by the host's transcription machinery occurs both in susceptible cells and indeed in a wide range of cells, including mammalian cells (see Kost *et al.*, 2005), which do not support complete replication. Although a detailed description of the orchestrated expression of baculovirus genes is beyond the scope of this chapter, a critical role is played by *ie1*, whose multifunctional product coordinates the expression of other early genes and the replication of viral DNA (reviewed by Friesen, 1997). However, expression of *ie1* and other early genes appears to be related to the triggering of apoptosis, and in AcMNPV IE1 may itself play an essential part in preventing apoptosis by stimulating the expression of *p35* (Friesen, 1997; Stewart *et al.*, 2005), one of several anti-apoptotic genes encoded by the lepidopteran baculoviruses.

Two distinct methods of countering apoptosis are deployed by baculoviruses. In the more common, the cell's initiator caspases are blocked by inhibitor of apoptosis proteins (IAPs), while the second method relies on p35 to inhibit effector caspases (reviewed in Clarke and Clem, 2003b). Means *et al.* (2003), for example, used RNAi to show that the *iap* gene in *Orgyia pseudotsugata* NPV (OpMNPV) was responsible for preventing apoptosis during infection of Ld652Y cells, while Clarke and Clem (2003a) constructed an AcMNPV p35 deletion mutant expressing GFP to monitor infection in permissive and semi-permissive insect species. Although suppression of apoptosis *in vivo* by IAP and p35 proteins is likely to play a decisive role in allowing the lepidopteran baculoviruses to complete their replication in midgut cells and other tissues, the likelihood of an anti-apoptotic response during hymenopteran and dipteran baculovirus infections is less clear: *Culex nigripalpus* NPV, in particular, encodes no IAPs and its *p35*-like gene lacks a C-terminal region that is considered essential for anti-apoptotic activity (Afonso *et al.*, 2001). The genetic distance between the latter NPVs and those infecting Lepidoptera is underscored by the finding that homologues of genes encoding IE1, which, as noted above, is essential for AcMNPV replication, are not evident in the sequenced genomes of viruses that infect Hymenoptera and Diptera.

Other than the apoptotic pathway, little is currently known about intracellular responses to baculovirus infection. In *B. mori* Bm5 cells, Lu *et al.* (1996) reported that BmNPV IE1, expressed from a transfected plasmid, could *trans*-activate the expression of reporter genes linked to a host cell housekeeping gene promoter. While this experiment showed that direct stimulation of host gene expression by a viral gene product could occur, its significance *in vivo* is unclear because the authors found that no such stimulation ensued when the same cells were infected with BmNPV. Two studies have investigated global host gene expression patterns in response to NPV infection. Both found an overall decline in host gene transcription during the later stages of infection, which correlates with the well-established shutdown of host protein synthesis that occurs as infection progresses. However, Okano *et al.* (2001) sequenced several thousand complementary DNA (cDNA) clones and found that a few BmN cell nuclear genes were upregulated late in BmNPV infection; conversely, Nobiron *et al.* (2003) used differential display to identify a small number of Sf9 genes that were upregulated at early times during AcMNPV infection. These large-scale analytical approaches may eventually uncover interesting details of the host response to both permissive and non-permissive baculovirus infections.

## 9.6. Baculovirus Host Range

As bioinsecticides, baculoviruses offer the apparent advantage over most chemical insecticides of a relatively high degree of host specificity. Productive replication (culminating in host mortality) is restricted exclusively to invertebrate species, predominantly within the Lepidoptera, and baculoviruses have no direct detrimental effects on beneficial invertebrates such as the hymenopteran parasitoids. In economic terms, the advantage of high host specificity pertains when plant damage is due primarily to a single insect species, as in the case of *L. dispar* in North American and Russian forest systems; but it disappears if, as is common in vegetable crops, the plant is attacked simultaneously by larvae of several pest species, each susceptible to perhaps only one virus (Black *et al.*, 1997). In such cases, a single chemical product would give sufficient control, whereas a cocktail of multiple viruses might be required to achieve a comparable effect.

Very few baculoviruses have been tested thoroughly to assess their host range. In an early review, Gröner (1986) identified AcMNPV as the most promiscuous baculovirus, capable of killing larvae of more than 30 species; Bishop *et al.* (1995), in a study conducted for the purpose of safety testing prior to a field release of GM virus, raised this number to 45 lepidopteran species that were classified as either 'permissive' (seven species) or 'semi-permissive' (38 species) for AcMNPV. Nominally susceptible host species may differ by several orders of magnitude in the dose of virus that is required to kill larvae of a given developmental stage, and such differences are clearly important economic factors for commercialization of baculovirus insecticides (Black *et al.*, 1997).

It is hardly surprising that our current understanding of the molecular determinants of baculovirus host range is patchy, given the complexity of infection in terms of both the physical progression of virus through the host and the number of viral genes that are expressed. Genome-wide transcription profiling experiments have revealed that, during AcMNPV infection of Sf9 cells, almost all of the ~150 ORFs identified by sequencing the genome are transcribed; the same is true of the ~140 ORFs of BmNPV in BmN cells (Iwanaga *et al.*, 2004). The gene manipulations that are required to modify host specificity also impose technical constraints on the range of viruses that can be studied. Nevertheless, for NPVs that can grow well in lepidopteran cell lines, several genes have been identified that confer on a recipient genotype the ability to replicate in cells that are refractory to the wild-type parent.

The first host-range expansion of this kind was achieved by replacing a portion of an AcMNPV gene encoding helicase with the corresponding sequence from BmNPV. These two viruses are genetically similar, with a highly collinear arrangement of genes sharing an overall nucleotide sequence identity of about 90% (Gomi *et al.*, 1999), sufficient to permit homologous recombination. Both viruses grow well in established cell lines, but neither is able to replicate in cells that are permissive for the other. Concomitant work in the laboratories of Croizier and Maeda yielded a series of papers in which increasingly subtle replacements of AcMNPV nucleotides by their counterparts in BmNPV enabled a modified AcMNPV genotype to replicate in *B. mori* cells. Both groups initially identified a 572-base pair (bp) restriction fragment in the central part of the helicase coding region, which, when co-transfected with AcMNPV genomic DNA into Sf9 cells, yielded BVs in the supernatant that could replicate in BmN cells. Further experiments involving *in vitro* mutagenesis or progressive shortening of this fragment revealed that minimal replacements of one (Kamita and Maeda, 1997) or two (Argaud *et al.*, 1998) AcMNPV amino acids by those occurring in BmNPV helicase enabled AcMNPV to infect BmN cells or *B. mori* larvae, respectively.

For reasons that are unclear, however, Argaud *et al.* (1998) found that AcMNPV BVs containing the single substitution identified by Kamita and Maeda (1997) failed to kill any of 53 injected larvae. Viruses bearing the two-amino acid substitution caused 17% mortality, while those encoding the majority of the BmNPV helicase polypeptide killed 87% of the injected insects. Although minimal substitutions enable a modified AcMNPV to replicate in *B. mori* cells, it appears that they confer only a moderate degree of pathogenicity towards the whole insect. Using the 572-bp BmNPV fragment mentioned above, in which there are 14 amino acid differences between AcMNPV and BmNPV, Deo *et al.* (2006) have recently constructed a modified AcMNPV bacmid that grows well enough in silkworm larvae to enable good expression of foreign proteins *in vivo*.

Another important species that wild-type AcMNPV cannot infect is the forest defoliator *L. dispar*. In the *L. dispar* cell line Ld652Y, AcMNPV replication does not occur and the virus causes the host cell to shut down synthesis of both cellular and viral proteins. By transfecting AcMNPV genomic DNA together with cosmid and plasmid clones containing progressively shorter regions of DNA from LdMNPV, Thiem *et al.* (1996) succeeded in defining a single LdMNPV gene, named *hrf-1*, that allowed AcMNPV to overcome the block in protein synthesis. The resultant AcMNPV virus particles were released into the culture medium and detected by their ability to infect Sf21 cells. Of particular note is that AcMNPV genomes harbouring *hrf-1* can replicate in and kill *L. dispar* larvae. In neonate bioassays, the median lethal concentration ($LC_{50}$) of the modified virus was about tenfold higher than that of LdMNPV itself, and against second-instar *S. exigua*, a species that is highly permissive for AcMNPV, the $LC_{50}$s of wild-type and hrf-1-expressing viruses were indistinguishable (Chen *et al.*, 1998). The authors observed, however, that the modified virus took much longer to kill *L. dispar* larvae than did LdMNPV.

Genotypic differences more subtle than those between AcMNPV and BmNPV can also have profound effects on host spectrum. A group of GV genotypes sharing ~99% sequence identity, for example, vary dramatically in their pathogenicity for two *Pieris* species (Smith and Crook, 1988). Likewise, strain differences between conspecific hosts can modulate susceptibility to a particular virus: a striking example is the recent report that 14 of 31 tested strains of *B. mori* are permissive for AcMNPV (Guo *et al.*, 2005). Inter-strain genetic crosses suggest that a single *B. mori* gene dictates whether or not AcMNPV can kill inoculated silkworm larvae, and identification of this gene will be a major addition to our knowledge about the interactions between baculoviruses and their hosts.

While attempts to modify baculovirus host range by manipulating individual genes are currently restricted to those viruses that grow in cultured cells, it is now possible to study gene-specific aspects of host resistance in whole insects. In a fascinating polymerase chain reaction (PCR)-based study of infection by three NPVs in three *Spodoptera* species, Simón *et al.* (2004) showed that SeMNPV, a virus that fails to kill *S. frugiperda* or *S. littoralis* larvae, establishes a partial infection in both species, which is subsequently cleared by the host. Using reverse transcriptase polymerase chain reaction (RT-PCR) to monitor the expression of several SeMNPV genes in midgut and haemocoel tissues, the authors found that transcripts of *polh*, a very late viral gene whose expression requires prior DNA replication and the expression of numerous earlier genes, were present in both tissues for up to 5 days following oral administration of SeMNPV OBs, but thereafter virtually disappeared. SeMPNV therefore appears to be capable of replicating successfully in midgut cells of non-permissive species, indicating that infection is blocked at a relatively later stage of infection.

The fate of AcMNPV in a semi-permissive host, *H. zea*, has been monitored using a recombinant virus which carries *lacZ*, under the constitutively active *Drosophila hsp70* promoter, in an otherwise wild-type AcMNPV genome (Trudeau *et al.*, 2001). Compared with the progression of this virus through *H. virescens*, a highly susceptible host, foci of β-galactosidase-expressing tracheal cells are much smaller in *H. zea*, and the haemocytes of the latter species appear to play a decisive role in blocking infection: in *H. virescens*, infected haemocytes are an important source of BV that ensures systemic infection, whereas those in *H. zea* are resistant to infection but can sequester BV from the haemolymph. Furthermore, they aggregate in regions of infected tracheae and participate in an encapsulation response. Together, these observations indicate that *H. zea* haemocytes are crucial to the insect's ability to overcome and survive AcMNPV infection.

Experiments intended to elucidate or modify baculovirus host range frequently involve co-infection of cells with two different viruses. In such situations, a pathogenic genotype may facilitate replication of a virus which, by itself, is incapable of productive infection. Simón *et al.* (2004) found that SeMNPV DNA replication was enhanced in non-permissive species when the virus was inoculated together with a pathogenic NPV genotype, while Yanase *et al.* (1998) reported that AcMNPV could enable SeMNPV to replicate in the normally refractory Sf21 cell line. Conversely, in the latter case, SeMNPV appeared to inhibit AcMNPV replication, a phenomenon also noted by Kamita and Maeda (1993) for BmNPV replication in BmN cells co-infected with AcMNPV and attributed to interference by the latter's helicase. The fates of two viruses co-infecting a single host may thus depend, at the molecular level, on whether a *trans*-acting gene product(s) of one virus complements or competes with its homologue in the other virus (Bideshi and Federici, 2000).

## 9.7. Unclassified DNA Viruses

As noted in Section 9.1, insects are hosts for a diverse range of viral pathogens. While most of these viruses belong to established taxonomic groups (see Fauquet *et al.*, 2005), several are not accommodated within the current system of virus classification but are none the less of interest because of their unusual host interaction strategies. Three such viruses, whose DNA genomes are large and only distantly related to those of other insect viruses, are described briefly here.

### 9.7.1. *Oryctes* virus

The rod-shaped *Oryctes* virus has been used successfully to control the rhinoceros beetle, *O. rhinoceros*, on palms in numerous Pacific islands (Bedford, 1980). As with baculoviruses, the route of infection of the *Oryctes* virus is *per os*, with the midgut being the primary site of virus replication; however, the *Oryctes* virus lacks the protective OB common to baculoviruses (Huger and Krieg, 1991). Until the sequence of the ~130-kbp *Oryctes* virus genome is determined, information about how viral gene products function during infection and virus replication is scarce. Nevertheless, since its infection process resembles those of the dipteran and hymenopteran NPVs, the *Oryctes* virus is likely to encode proteins that carry out functions analogous to those of the PIFs and IAPs, and perhaps metalloproteinases, found ubiquitously in baculoviruses.

Infections of *Oryctes* virus in larvae, pupae and adults are often fatal, although in adult beetles infection is often chronic with infected adults often showing no overt disease symptoms. Virus replication in the midgut epithelium results in the proliferation of midgut cells to the point where they may actually fill the midgut lumen (reviewed by Huger and Krieg, 1991; Burand, 1998; Huger, 2005).

Extensive virus replication in the nuclei of midgut cells leads to release of progeny into the gut, and eventual excretion in the faeces of infected beetles. The chronic nature of *Oryctes* virus infection in adult beetles contributes to the transmission of the virus: infected adults may actively disseminate virus for several weeks while they travel to breeding areas and feeding burrows in palms, shedding large quantities of virus at these sites. The major means of auto-dissemination of the *Oryctes* virus is through transmission to other adults by oral faecal contact, during mating or co-occupation of the same habitat by healthy and infected insects (Zelazny, 1976).

### 9.7.2. Hz-1V and Hz-2V

Hz-1V and the closely related Hz-2V (Burand *et al.*, 2005) are rod-shaped, enveloped DNA viruses that resemble baculoviruses in size and structure although, like the *Oryctes* virus, both lack OBs (reviewed by Burand, 1998). In the case of Hz-2V, this is a direct reflection of the close association that has evolved between the virus and its host *H. zea*, in which direct contact between insects is required for virus transmission. Unlike baculoviruses, Hz-2V is not highly infectious *per os* (Hamm *et al.*, 1996) and does not remain viable for long periods outside the host. The major routes of transmission of Hz-2V in nature appear to be transovarial, inside eggs laid by infected females, and through direct contact between moths during mating attempts between healthy and infected insects. Although the exact pathway used by Hz-2V to infect adult reproductive tissues is not known, the Hz-1V genome contains genes that share homology with the baculovirus *pif*, *iap* and *metalloproteinase* genes (Cheng *et al.*, 2002), suggesting that the route used by Hz-2V to infect epithelial cells in the adult reproductive tract is similar to that used by baculoviruses to infect epithelial cells in the larval midgut.

Unlike the *Oryctes* virus and baculoviruses, Hz-2V infections do not result in insect mortality but rather in sterility of infected moths. The Hz-2V replication cycle includes both a persistent phase and a productive phase. In the persistent phase, the virus replicates in the infected insect without any disease symptoms and is transmitted transovarially to progeny by asymptomatic carrier females (Hamm *et al.*, 1996). Larvae in which productive replication occurs appear normal, but emerge as sterile adults with abnormal reproductive systems, resulting in a condition referred to as 'agonadal' (AG) (Raina and Adams, 1995; Hamm *et al.*, 1996; Rallis and Burand, 2002a,b). It is not known what determines if an Hz-2V-infected insect will be AG, but this may be linked to the amount of virus an individual insect receives from the infected female parent.

Infected AG males do not have accessory glands and therefore cannot produce or transfer the anti-calling factor pheromonostatic peptide (PSP) to females during contacts made while attempting to mate (Burand and Tan, 2006). As a result, healthy females contacted by infected, sterile males can become infected with Hz-2V and subsequently transfer virus to healthy males which mate with and fertilize them. Virus replication in females typically results in the accumulation of a large number of virus-filled vesicles that make up a 'virus plug' covering the tip of the abdomen of an infected moth. The virus plug serves both to block the transfer of PSP from healthy males and as a source of contaminating virus for males that attempt to mate with infected females (Burand *et al.*, 2004). Hz-2V-infected females produce more sex pheromones and are more attractive to healthy males than are healthy females (Burand *et al.*, 2005). These infected females call more frequently and are occupied by males for shorter periods than their healthy counterparts, making them a continual source of virus for males they attract and contaminate during mating attempts. The ability of Hz-2V to manipulate the physiology and behaviour of infected females so that they attract and contaminate a succession of healthy males, which in turn disseminate the virus to other healthy females during subsequent matings, makes Hz-2V an auto-disseminating virus reminiscent of the *Oryctes* virus.

## 9.8. Mechanisms of Insect Virus Persistence

To ensure a continued association with their hosts from one generation to the next, insect viruses may have to survive for prolonged periods when their hosts are inactive. Insects inhabiting cooler latitudes, for example, typically undergo one reproductive cycle per year, breeding during the summer and then overwintering as larvae or pupae. How do viruses cope with these quiescent periods?

There are two possible states in which viral genomes can persist between sporadic bursts of replication in their hosts. Viruses possessing an OB (*Baculoviridae* and *Cypoviridae*) can survive within this structure in the external environment for prolonged periods: Thompson *et al.* (1981) reported that OBs of OpMNPV remained active in Canadian forest soil for more than 40 years, and LdMNPV can persist in tree bark for more than 1 year (Podgwaite *et al.*, 1979). Such niches afford some degree of protection from the destructive effects of sunlight, which is known to inactivate exposed OBs (for example, those on a leaf surface) very rapidly (Petrik *et al.*, 2003). Alternatively, the virus may persist in some inert state within the host. Although this type of association is well documented for other large DNA viruses such as herpes simplex virus, which can be propagated episomally in symptomless vertebrate hosts (reviewed by Efstathiou and Preston, 2005), the idea that baculoviruses can remain within their hosts in a similar 'latent' form has long been contentious (for discussions of this issue, see Burand *et al.*, 1986; Kukan, 1999; Cory and Myers, 2003; Ilyinykh *et al.*, 2004; and references therein). However, there is now reasonably strong evidence that a latent NPV occurs in both wild and domesticated populations of *Mamestra brassicae* in the UK (Hughes *et al.*, 1997; Burden *et al.*, 2006).

Powerful techniques such as PCR, and the use of marker genes to track the ability of viruses to persist vertically within the host from one generation to the next, now render the study of this intriguing type of virus–host interaction amenable to rigorous analysis. Mori *et al.* (1995) incorporated the firefly *luc* gene into AcMNPV and reported that the modified virus could be transmitted to the progeny of infected *B. mori*, a species normally refractory to AcMNPV (see Section 9.6), with luciferase activity detected in the haemocytes of a small proportion of next-generation larvae. Evidence for vertical transmission of single-stranded RNA viruses in honeybees (*Apis mellifera*) has also been presented by Chen *et al.*

(2006), in a study of ten laboratory colonies in the USA, using genotype-specific primers for RT-PCR to detect RNA of up to five discrete viruses in single queens as well as their larval and adult progeny. Although the mechanisms of persistence and pathogenicity of honeybee viruses are not well understood, RT-PCR provides a promising tool to delineate the molecular ecology of viral pathogens in this vital beneficial insect.

Hz-2V can exist in a persistent or latent state within the host, becoming activated into productive replication either during a later stage in the life cycle of the insect or in progeny insects. In mating experiments, Hamm *et al.* (1996) established that some Hz-2V-infected female moths were asymptomatic carriers (AS) of the virus. When mated with healthy males, these AS females produced some sterile AG progeny. Interestingly, the virus persists in some latent form, since virus particles are virtually undetectable in the early larval instars of these healthy and apparently normal AG insects and only in the later stages of their development does the virus become activated into productive replication. Activation appears to coincide with differentiation of adult reproductive tissues, whereupon productive virus replication causes them to become malformed and results in the sterility of these infected insects.

It is unclear precisely what determines if an Hz-2V-infected larva will become an AS or AG adult; however, using PCR, Burand *et al.* (2004) were able to determine that female moths receiving a high virus dose yielded a higher percentage of AG progeny than those receiving a low dose, and produced a higher percentage of AS progeny. This suggests that the fate of an Hz-2V-infected insect is determined by the amount of virus transmitted to it through the egg by the female.

Hz-1V, which was originally isolated from an ovarian cell line (Granados *et al.*, 1978) was shown by Burand *et al.* (1983, 1986), using restriction enzyme analysis and Southern blotting, to be able to persist *in vitro* through the formation of defective interfering particles (DIPs). DIPs possess smaller viral genomes than standard virus particles, and are apparently able to outcompete intact genomes for gene products that are required for productive viral replication (Huang and Baltimore, 1970; Bangham and Kirkwood, 1990; Holland, 1990). In NPVs, DIPs arise during serial passage and long-term propagation *in vitro* (reviewed by Krell, 1996), resulting in a reduction of OB production and BV titre. Pijlman *et al.* (2002) identified a region of the SeMNPV genome containing a non-homologous region origin of DNA replication (non-*hr ori*), which was preferentially retained in defective interfering genomes. Deletion of this region by recombination in *E. coli* resulted in a bacmid which was used to produce a non-*hr ori*-negative mutant. This mutant virus was able to sustain high levels of OB production and BV titre on passage in culture, implicating the non-*hr ori* in baculovirus DIP formation.

Northern analysis of RNAs from insect cell lines persistently infected with Hz-1V conducted by Chao *et al.* (1992, 1998) revealed that a single viral transcript (*pat1*) was constitutively expressed in all cell lines, suggesting that PAT1 expression is important in the persistence of Hz-1V *in vitro*. Using pulse-field gel electrophoresis and chromosomal fluorescence *in situ* hybridization, Lin *et al.* (1999) demonstrated that cells persistently infected with Hz-1V contained viral genomes that were both episomal and integrated into the host genome. Since Hz-2V can also establish persistent infections *in vitro* and contains a *pat1* gene homologue (Burand, unpublished data), it is possible that this virus persists in insects in the same manner as Hz-1V.

With respect to the *Baculoviridae*, regardless of whether viral DNA is integrated into the genome of its host or remains independent, co-localization of host and virus genomes in the host cell nucleus during persistent or productive infection provides opportunities for transfer of genetic information between the two. Transposable elements have been witnessed moving from the *T. ni* to the AcMNPV genome (see Section 9.3), and sequence analysis of baculovirus genes has suggested that during evolution at least two, encoding IAP and EGT, were acquired from insect genomes; a third, encoding chitinase, appears to have been derived from a bacterium (Hughes and Friedman, 2003). Evidence for evolutionary movements of envelope fusion protein genes into and out of the *Baculoviridae*, involving exchanges with other viruses, has also been reviewed recently by Okano *et al.* (2006).

## 9.9. Conclusions and Future Prospects

Insect pathogenic viruses display a wide variety of interactions with their hosts that facilitate their replication and transmission, including strategies for evading the host's defences against invasion by microbes and for manipulating host physiology and behaviour. By applying a wide range of molecular techniques and approaches, we now have a better understanding of how at least some of these interactions occur and of the roles played by both host and viral genes. Although the majority of this work – and therefore also of this review – has focused on viruses that have been demonstrated to have potential for controlling insect pests, successful use of viruses in insect control strategies will require a deeper knowledge about the interactions between far more of these viruses and their hosts. Future research on different types of insect-pathogenic viruses, targeting the molecular mechanisms by which they elude host defences and spread within the insect, will provide additional molecular tools for biologically based insect pest management systems. Understanding how Hz-2V infection leads to the malformation of host reproductive tissues may, for example, allow us to identify molecular targets in the host that could be exploited in control strategies involving sterile insects.

Genetic manipulation of baculoviruses, including arming them with genes encoding insect-specific scorpion toxins, has demonstrated that their speed of action can be increased dramatically, and there is much room for further refinement to enhance the levels of toxin expression, activity, and delivery to targets within the insect. Deletion of viral genes such as *egt* and *chiA*, which are not required for infectivity or pathogenesis but increase OB production, has also been shown to improve potency and productivity. Rational approaches to expanding virus host range, already being exploited in the BmNPV–AcMNPV system to improve BEVs, will become increasingly realistic as information accumulates about the range of molecules affecting host specificity. Despite uncertainties about the prospects for public acceptance of GMBs and of genetic modification in general, it is clear from a technical perspective that substantial improvements to baculoviruses as insect control agents can be achieved through genetic manipulation.

Although *in vivo* production of baculoviruses such as AgNPV has proven to be successful, the commercialization of these viruses, particularly for GMBs, will most likely require *in vitro* virus production. Successful *in vitro* commercial production will in turn depend on the development of inexpensive, large-scale production methods and cheaper media for growing insect cells and GMBs *in vitro*, and perhaps on the development of engineered cell lines, to maximize OB production.

## References

Adams, J.R., Sheppard, C.A., Shapiro, M. and Tompkins, G.J. (1994) Light and electron microscopic investigations on the histopathology of the midgut of gypsy moth larvae infected with LdMNPV plus a fluorescent brightener. *Journal of Invertebrate Pathology* 64, 156–159.

Afonso, C.L., Tulman, E.R., Lu, Z., Balinsky, C.A., Moser, B.A., Becnel, J.J., Rock, D.L. and Kutish, G.F. (2001) Genome sequence of a baculovirus pathogenic for *Culex nigripalpus*. *Journal of Virology* 75, 11157–11165.

Argaud, O., Croizier, L., López-Ferber, M. and Croizier, G. (1998) Two key mutations in the host-range specificity domain of the *p143* gene of *Autographa californica* nucleopolyhedrovirus are required to kill *Bombyx mori* larvae. *Journal of General Virology* 79, 931–935.

Balch, R.E. and Bird, F.T. (1944) A disease of the European spruce sawfly, *Gilpinia hercyniae* (Htg) and its place in natural control. *Scientific Agriculture* 25, 65–80.

Balu, B., Shoue, D.A., Fraser, M.J. and Adams, J.H. (2005) High-efficiency transformation of *Plasmodium falciparum* by the lepidopteran transposable element *piggyBac*. *Proceedings of the National Academy of Sciences of the United States of America* 102, 16391–16396.

Bangham, C.R.M. and Kirkwood, T.B.L. (1990) Defective interfering particles: effects in modulating virus growth and persistence. *Virology* 179, 821–826.

Barbehenn, R.V. and Martin, M.M. (1995) Peritrophic envelope permeability in herbivorous insects. *Journal of Insect Physiology* 41, 303–311.

Bedford, G.O. (1980) Biology, ecology, and control of palm rhinoceros beetles. *Annual Review of Entomology* 25, 309–339.

Bel Haj Rhouma, R., Cérutti-Duonor, M., Benkhadir, K., Goudey-Perrière, F., El Ayeb, M., Lopez-Ferber, M. and Karoui, H. (2005) Insecticidal effects of *Buthus occitanus tunetanus* BotIT6 toxin expressed in *Escherichia coli* and baculovirus/insect cells. *Journal of Insect Physiology* 51, 1376–1383.

Benz, A.G. (1986) Introduction: historical perspectives. In: Granados, R.R. and Federici, B.A. (eds) *The Biology of Baculoviruses*, Vol. 1. CRC Press, Boca Raton, Florida, pp. 1–35.

Bideshi, D.K. and Federici, B.A. (2000) The *Trichoplusia ni* granulovirus helicase is unable to support replication of *Autographa californica* multicapsid nucleopolyhedrovirus in cells and larvae of *T. ni*. *Journal of General Virology* 81, 1593–1599.

Bishop, D.H.L., Hirst, M.L., Possee, R.D. and Cory, J.S. (1995) Genetic engineering of microbes: virus insecticides – a case study. In: Hunter, P.A., Darby, G.K. and Russell, N.J. (eds) *Fifty Years of Microbials: Past Perspectives and Future Trends*. Cambridge University Press, Cambridge, pp. 249–277.

Black, B.C., Brennan, L.A., Dierks, P.M. and Gard, I.E. (1997) Commercialization of baculoviral insecticides. In: Miller, L.K. (ed.) *The Baculoviruses*. Plenum Press, New York, pp. 341–387.

Blissard, G.W. and Rohrmann, G.F. (1989) Location, sequence, transcriptional mapping, and temporal expression of the gp64 envelope glycoprotein gene of the *Orgyia pseudotsugata* multicapsid nuclear polyhedrosis virus. *Virology* 170, 537–555.

Burand, J.P. (1998) Nudiviruses. In: Miller, L.K. and Ball, L.A. (eds) *The Insect Viruses*. Plenum Press, New York, pp. 69–90.
Burand, J.P. and Rallis, C.P. (2004) *In vivo* dose-response of insects to Hz-2V infection. *Virology Journal* 1: 15 (http://www.virologyj.com/content/1/1/15).
Burand, J.P. and Tan, W. (2006) Mate preference and mating behavior of male *Helicoverpa zea* (Lepidoptera: Noctuidae) moths infected with the sexually transmitted insect virus Hz-2V. *Annals of the Entomological Society of America* (in press).
Burand, J.P., Summers, M.D. and Smith, G.E. (1980) Transfection with baculovirus DNA. *Virology* 101, 286–290.
Burand, J.P., Wood, H.A. and Summers, M.D. (1983) Defective particles from a persistent baculovirus infection in *Trichoplusia ni* tissue culture cells. *Journal of General Virology* 64, 391–398.
Burand, J.P., Kawanishi, C.H. and Huang, Y.-S. (1986) Persistent baculovirus infections. In: Granados, R.R. and Federici, B.A. (eds), *The Biology of Baculoviruses*, Vol. 1. CRC Press, Boca Raton, Florida, pp. 159–175.
Burand, J.P., Park, E.J. and Kelly, T.J. (1996) Dependence of ecdysteroid metabolism and development in host larvae on the time of baculovirus infection and the activity of the UDP-glucosyltransferase gene. *Insect Biochemistry and Molecular Biology* 26, 845–852.
Burand, J.P., Rallis, C.P. and Tan, W. (2004) Horizontal transmission of Hz-2V by virus infected *Helicoverpa zea* moths. *Journal of Invertebrate Pathology* 85, 128–131.
Burand, J.P., Tan, W.J., Kim, W.J., Nojima, S. and Roelofs, W. (2005) Infection with the insect virus Hz-2v alters mating behavior and pheromone production in female *Helicoverpa zea* moths. *Journal of Insect Science* 5.06 (http://www.insectscience.org).
Burden, J.P., Possee, R.D., Sait, S.M., King, L.A. and Hails, R.S. (2006) Phenotypic and genotypic characterisation of persistent baculovirus infections in populations of the cabbage moth (*Mamestra brassicae*) within the British Isles. *Archives of Virology* 151, 635–649.
Chang, J.H., Choi, J.Y., Jin, B.R., Roh, J.Y., Olszewski, J.A., Seo, S.J., O'Reilly, D.R. and Je, Y.H. (2003) An improved baculovirus insecticide producing occlusion bodies that contain *Bacillus thuringiensis* insect toxin. *Journal of Invertebrate Pathology* 84, 30–37.
Chao, Y.-C., Wood, H.A., Chang, C.Y., Lee, H.J., Shen, W.C. and Lee, H.T. (1992) Differential expression of Hz-1 baculovirus genes during productive and persistent viral infections. *Journal of Virology* 66, 1442–1448.
Chao,Y.-C., Lee, S.-T., Chang, M.-C., Chen, H.-H., Chen, S.-S., Wu, T.-Y., Liu, F.-H., Hsu, E.-L. and Hou, R.F. (1998) A 2.9-kilobase noncoding nuclear RNA functions in the establishment of persistent Hz-1 viral infection. *Journal of Virology* 72, 2233–2245.
Chen, C.J., Quentin, M.E., Brennan, L.A., Kukel, C. and Thiem, S.M. (1998) *Lymantria dispar* nucleopolyhedrovirus hrf-1 expands the larval host range of *Autographa californica* nucleopolyhedrovirus. *Journal of Virology* 72, 2526–2531.
Chen, Y.P., Pettis, J.S., Collins, A. and Feldlaufer, M.F. (2006) Prevalence and transmission of honeybee viruses. *Applied and Environmental Microbiology* 72, 606–611.
Cheng, C.-H., Liu, S.-M., Chow, T.-Y., Hsiao, Y.-Y., Wang, D.-P., Huang, J.-J. and Chen, H.-H. (2002) Analysis of the complete genome sequence of the Hz-1 virus suggests that it is related to members of the *Baculoviridae*. *Journal of Virology* 76, 9024–9034.
Clarke, E.E., Tristem, M., Cory, J.S. and O'Reilly, D.R. (1996) Characterization of the ecdysteroid UDP-glucosyltransferase gene from *Mamestra brassicae* nucleopolyhedrovirus. *Journal of General Virology* 77, 2865–2871.
Clarke, T.E. and Clem, R.J. (2003a) *In vivo* induction of apoptosis correlating with reduced infectivity during baculovirus Infection. *Journal of Virology* 77, 2227–2232.
Clarke, T.E. and Clem, R.J. (2003b) Insect defenses against virus infection: the role of apoptosis. *International Reviews of Immunology* 22, 401–424.
Clem, R.J. (2005) The role of apoptosis in defense against baculovirus infection in insects. *Current Topics in Microbiology and Immunology* 289, 113–130.

Cory, J.S. and Myers, J.H. (2003) The ecology and evolution of insect baculoviruses. *Annual Review of Ecology, Evolution and Systematics* 34, 239–272.
Daimon, T., Katusma, S., Kang, W.-K. and Shimada, T. (2006) Comparative studies of *Bombyx mori* nucleopolyhedrovirus chitinase and its host ortholog, BmChi-h. *Biochemical and Biophysical Research Communications* 345, 825–833.
Deo, V.K., Hiyoshi, M. and Park, E.Y. (2006) Construction of hybrid *Autographa californica* nuclear polyhedrosis bacmid by modification of p143 helicase. *Journal of Virological Methods* 134, 212–216.
Ding, S., Wu, X., Li, G., Han, M., Zhuang, Y. and Xu, T. (2005) Efficient transposition of the piggyBac (PB) transposon in mammalian cells and mice. *Cell* 122, 473–483.
Dougherty, E.M., Narang, N., Loeb, M., Lynn, D.E. and Shapiro, M. (2006) Fluorescent brightener inhibits apoptosis in baculovirus-infected gypsy moth larval midgut cells *in vitro*. *Biocontrol Science and Technology* 16, 157–168.
Efstathiou, S. and Preston, C.M. (2005) Towards an understanding of the molecular basis of herpes simplex virus latency. *Virus Research* 111, 108–119.
Engelhard, K.K., Kam-Morgan, L.N.W., Washburn, J.O. and Volkman, L.E. (1994) The insect tracheal system: a conduit for the systemic spread of *Autographa californica* nuclear polyhedrosis virus. *Proceedings of the National Academy of Sciences of the United States of America* 91, 3224–3227.
Fauquet, C.M., Mayo, M.A., Maniloff, J., Desselberger, U. and Ball, L.A. (eds) (2005) *Virus Taxonomy, VIIIth Report of the ICTV*. Elsevier/Academic Press, London.
Federici, B.A. (1997) Baculovirus pathogenesis. In: Miller, L.K. (ed.) *The Baculoviruses*. Plenum Press, New York, pp. 33–59.
Flipsen, J.T.M., Mans, R.M.W., Kleefsman, A.W.F., Knebel-Mörsdorf, D. and Vlak, J.M. (1995) Deletion of the baculovirus ecdysteroid UDP-glucosyltransferase gene induces early degeneration of Malpighian tubules in infected insects. *Journal of Virology* 69, 4529–4532.
Fraser, M.J., Smith, G.E. and Summers, M.D. (1983) The acquisition of host cell DNA sequences by baculoviruses: relation between host DNA insertions and FP mutants of *Autographa californica* and *Galleria mellonella* NPVs. *Journal of Virology* 47, 287–300.
Friesen, P.D. (1997) Regulation of baculovirus early gene expression. In: Miller, L.K. (ed.) *The Baculoviruses*. Plenum Press, New York, pp. 141–170.
Goldsmith, M.R., Shimada, T. and Abe, H. (2005) The genetics and genomics of the silkworm, *Bombyx mori*. *Annual Review of Entomology* 50, 71–100.
Gomi, S., Majima, K. and Maeda, S. (1999) Sequence analysis of the genome of *Bombyx mori* nucleopolyhedrovirus. *Journal of General Virology* 80, 1323–1337.
Granados, R.R. and Lawler, K.A. (1981) *In vivo* pathway of *Autographa californica* baculovirus invasion and infection. *Virology* 108, 297–308.
Granados, R.R., Nguyen, T. and Cato, B. (1978) An insect cell line persistently infected with a baculovirus-like particle. *Intervirology* 10, 309–317.
Gröner, A. (1986) Specificity and safety of baculoviruses. In: Granados, R.R. and Federici, B.A. (eds) *The Biology of Baculoviruses*, Vol. 1. CRC Press, Boca Raton, Florida, pp. 177–202.
Guo, T., Wang, S., Guo, X. and Lu, C. (2005) Productive infection of *Autographa californica* nucleopolyhedrovirus in silkworm *Bombyx mori* strain Haoyue due to the absence of a host antiviral factor. *Virology* 341, 231–237.
Haas-Stapleton, E.J., Washburn, J.O. and Volkman, L.E. (2004) P74 mediates specific binding of *Autographa californica* M nucleopolyhedrovirus occlusion-derived virus to primary cellular targets in the midgut epithelia of *Heliothis virescens* larvae. *Journal of Virology* 78, 6786–6791.
Hamm, J.J., Nordlung, D.A. and Marti, O.G. (1985) Effects of a nonoccluded virus of *Spodoptera frugiperda* (Lepidoptera, Noctuidae) on the development of a parasitoid, *Cotesia narginiventris* (Hymenoptera, Braconidae). *Environmental Entomology* 14, 258–261.

Hamm, J.J., Carpenter, J.E. and Styer, E.L. (1996) Oviposition day effect on incidence of agonadal progeny of *Helicoverpa zea* (Lepidoptera: Noctuidae) infected with a virus. *Annals of the Entomological Society of America* 89, 266–275.

Hannon, G.J. (2002) RNA interference. *Nature* 418, 244–251.

Hashimoto, Y., Hayakawa, T., Ueno, Y., Fujita, T., Sano, Y. and Matsumoto, T. (2000) Sequence analysis of the *Plutella xylostella* granulovirus genome. *Virology* 275, 358–372.

Hawtin, R.E., Zarkowska, T., Arnold, K., Thomas, C.J., Gooday, G.W., King, L.A., Kuzio, J.A. and Possee, R.D. (1997) Liquefaction of *Autographa californica* nucleopolyhedrovirus-infected insects is dependent on the integrity of virus-encoded chitinase and cathepsin genes. *Virology* 238, 243–253.

Holland, J.J. (1990) Defective viral genomes. In: Fields, B.N., Knipe, D.M., Chanock, R.M., Hirsch, M.S., Melnick, J.L., Monath, T.P. and Roizman, B. (eds) *Virology*, 2nd edn. Raven Press, New York, pp. 151–165.

Hom, L.G. and Volkman, L.E. (2000) *Autographa californica* M nucleopolyhedrovirus chiA is required for processing of V-CATH. *Virology* 277, 178–183.

Hom, L.G., Ohkawa, T., Trudeau, D. and Volkman, L.E. (2002) *Autographa californica* M nucleopolyhedrovirus proV-CATH is activated during infected cell death. *Virology* 296, 212–218.

Huang, A.S. and Baltimore, D. (1970) Defective viral particles and viral disease processes. *Nature* 226, 325–227.

Huber, J. (1986) Use of baculoviruses in pest management programs. In: Granados, R.R. and Federici, B.A. (eds) *The Biology of Baculoviruses*, Vol. 2. CRC Press, Boca Raton, Florida, pp. 181–202.

Huger, A.M. (2005) The *Oryctes* virus: its detection, identification, and implementation in biological control of the coconut palm rhinoceros beetle, *Oryctes rhinoceros* (Coleoptera: Scarabaeidae). *Journal of Invertebrate Pathology* 89, 78–84.

Huger, A.M. and Krieg, A. (1991) *Baculoviridae*. Nonoccluded baculoviruses. In: Adams, J.R. and Bonami, J.R. (eds) *Atlas of Invertebrate Viruses*. CRC Press, Boca Raton, Florida, pp. 287–316.

Hughes, A.L. and Friedman, R. (2003) Genome-wide survey for genes horizontally transferred from cellular organisms to baculoviruses. *Molecular Biology and Evolution* 20, 979–987.

Hughes, D.S., Possee, R.D. and King, L.A. (1997) Evidence for the presence of a low-level, persistent baculovirus infection of *Mamestra brassicae* insects. *Journal of General Virology* 78, 1801–1805.

Ignoffo, C.M. and Couch, T.L. (1981) The nucleopolyhedrosis virus of *Heliothis* species as a microbial insecticide. In: Burgess, H.D. (ed.) *Microbial Control of Pests and Plant Diseases*. Academic Press, New York, pp. 1970–1980.

Ilyinykh, A.V., Shternshis, M.V. and Kuzminov, S.V. (2004) Exploration into a mechanism of transgenerational transmission of nucleopolyhedrovirus in *Lymantria dispar* L. in Western Siberia. *BioControl* 49, 441–454.

Isobe, R., Kojima, K., Matsuyama, T., Quan, G.X., Kanda, T., Tamura, T., Sahara, K., Asano, S.I. and Bando, H. (2004) Use of RNAi technology to confer enhanced resistance to BmNPV on transgenic silkworms. *Archives of Virology* 149, 1931–1940.

Iwanaga, M., Takaya, K., Katsuma, S., Ote, M., Tanaka, S., Kamita, S.G., Kang, W., Shimada, T. and Kobayashi, M. (2004) Expression profiling of baculovirus genes in permissive and nonpermissive cell lines. *Biochemical and Biophysical Research Communications* 323, 599–614.

Je, Y.H., Jin, B.R., Park, H.W., Roh, J.Y., Chang, J.H., Seo, S.J., Olszewski, J.A., O'Reilly, D.R. and Kang, S.K. (2003) Baculovirus expression vectors that incorporate the foreign protein into viral occlusion bodies. *BioTechniques* 34, 81–87.

Kamita, S.G. and Maeda, S. (1993) Inhibition of *Bombyx mori* nuclear polyhedrosis virus (NPV) replication by the putative DNA helicase gene of *Autographa californica* NPV. *Journal of Virology* 67, 6239–6245.

Kamita, S.G. and Maeda, S. (1997) Sequencing of the putative DNA helicase-encoding gene of the *Bombyx mori* nuclear polyhedrosis virus and fine-mapping of a region involved in host range expansion. *Gene* 190, 173–179.

Katsuma, S., Noguchi, Y., Zhou, C.E., Kobayashi, M. and Maeda, S. (1999) Characterization of the 25K FP gene of the baculovirus *Bombyx mori* nucleopolyhedrovirus: implications for postmortem host degradation. *Journal of General Virology* 80, 783–791.

Katsuma, S., Tanaka, S., Shimada, T. and Kobayashi, M. (2004) Reduced cysteine protease activity of the hemolymph of *Bombyx mori* larvae infected with *fp25K*-inactivated *Bombyx mori* nucleopolyhedrovirus results in the reduced postmortem host degradation. *Archives of Virology* 149, 1773–1782.

Kelly, T.J., Park, E.J., Masler, C.A. and Burand, J.P. (1995) Characterization of the glycosylated ecdysteroids in the hemolymph of baculovirus-infected gypsy moth larvae and cells in culture. *European Journal of Entomology* 92, 51–61.

Kitts, P.A., Ayres, M.D. and Possee, R.D. (1990) Linearization of baculovirus DNA enhances the recovery of recombinant virus expression vectors. *Nucleic Acids Research* 18, 5667–5672.

Kost, T.A., Condreay, J.P. and Jarvis, D.L. (2005) Baculoviruses as versatile vectors for protein expression in insect and mammalian cells. *Nature Biotechnology* 23, 567–575.

Kukan, B. (1999) Vertical transmission of nucleopolyhedrovirus in insects. *Journal of Invertebrate Pathology* 74, 103–111.

Lacey, L.A., Arthurs, S., Thomson, D., Fritts, R. and Granatstein, D. (2004) Codling moth granulovirus and insect specific nematodes for control of codling moth in the Pacific Northwest. *Tilth Producers Quarterly* 13, 10–12.

Lauzon, H.A.M., Garcia-Maruniak, A., Zanotto, P.M. de A., Clemente, J.C., Herniou, E.A., Lucarotti, C.J., Arif, B.M. and Maruniak, J.E. (2006) Genomic comparison of *Neodiprion sertifer* and *Neodiprion lecontei* nucleopolyhedroviruses and identification of potential hymenopteran baculovirus-specific open reading frames. *Journal of General Virology* 87, 1477–1489.

Lepore, L.S., Roelvink, P.R. and Granados, R.R. (1996) Enhancin, the granulosis virus protein that facilitates nucleopolyhedrovirus (NPV) infections, is a metalloprotease. *Journal of Invertebrate Pathology* 68, 131–140.

Li, Q.J., Li, L.L., Moore, K., Donly, C., Theilmann, D.A. and Erlandson, M. (2005) Characterization of *Mamestra configurata* nucleopolyhedrovirus enhancin and its functional analysis via expression in an *Autographa californica* M nucleopolyhedrovirus recombinant. *Journal of General Virology* 84, 123–132.

Lin, C.-L., Lee, J.-C., Chen, S.-S., Wood, H.A., Li, M.-L., Li, C.-F. and Chao, Y.-C. (1999) Persistent Hz-1 virus infection in insect cells: evidence for insertion of viral DNA into host chromosomes and viral infection in a latent status. *Journal of Virology* 73, 128–139.

Lu, M., Johnson, R.R. and Iatrou, K. (1996) *Trans*-activation of a cell housekeeping gene promoter by the IE1 gene product of baculoviruses. *Virology* 218, 103–113.

Luckow, V.A., Lee, S.C., Barry, G.F. and Olins, P.O. (1993) Efficient generation of infectious recombinant baculoviruses by site-specific transposon-mediated insertion of foreign genes into a baculovirus genome propagated in *Escherichia coli*. *Journal of Virology* 67, 4566–4579.

Maeda, S., Kawai, T., Obinata, M., Fujiwara, H., Horiuchi, T., Saeki, Y., Sato, Y. and Furusawa, M. (1985) Production of human α-interferon in silkworm using a baculovirus vector. *Nature* 315, 592–594.

McLinden, J., Ploplis, V.A., Fraser, M.J. and Rosen, E.D. (1992) Expression of foreign epitopes on recombinant occlusion bodies of baculoviruses. *Vaccine* 10, 231–238.

Means, J.C., Muro, I. and Clem, R.J. (2003) Silencing of the baculovirus Op-*iap3* gene by RNA interference reveals that it is required for prevention of apoptosis during *Orgyia pseudotsugata* M nucleopolyhedrovirus infection of Ld652Y cells. *Journal of Virology* 77, 4481–4488.

Miller, L.K. (1997) Introduction to the baculoviruses. In: Miller, L.K. (ed.) *The Baculoviruses*. Plenum Press, New York, pp. 1–6.

Mori, H., Yamao, M., Nakazawa, H., Sugahara, Y., Shirai, N., Matsubara, F., Sumida, M. and Imamura, T. (1995) Transovarian transmission of a foreign gene in the silkworm, *Bombyx mori*, by *Autographa californica* nuclear polyhedrosis virus. *Bio/Technology* 13, 1005–1007.

Morris, K.V. and Rossi, J.J. (2006) Antiviral applications of RNAi. *Current Opinion in Molecular Therapy* 8, 115–121.

Moscardi, F. (1999) Assessment of the application of baculoviruses for control of Lepidoptera. *Annual Review of Entomology* 44, 257–289.

Nakai, M., Goto, C., Shiotsuki, T. and Kunimi, Y. (2002) Granulovirus prevents pupation and retards development of *Adoxophyes honmai* larvae. *Physiological Entomology* 27, 157–164.

Nakai, M., Goto, C., Kang, W., Shikata, M., Luque, T. and Kunimi, Y. (2003) Genome sequence and organization of a nucleopolyhedrovirus isolated from the smaller tea tortrix *Adoxophyes honmai*. *Virology* 314, 171–183.

Nakai, M., Shiotsuki, T. and Kunimi, Y. (2004) An entomopoxvirus and a granulovirus use different mechanisms to prevent pupation of *Adoxophyes honmai*. *Virus Research* 101, 185–191.

Nakazawa, H., Tsuneishi, E., Ponnuvel, K.M., Furukawa, S., Asaoka, A., Tanaka, H., Ishibashi, J. and Yamakawa, M. (2004) Antiviral activity of a serine protease from the digestive juice of *Bombyx mori* larvae against nucleopolyhedrovirus. *Virology* 321, 154–162.

Narayanan, K. (2004) Insect defence: its impact on microbial control of insect pests. *Current Science* 86, 800–814.

Nobiron, I., O'Reilly, D.R. and Olszewski, J.A. (2003) *Autographa californica* nucleopolyhedrovirus infection of *Spodoptera frugiperda* cells: a global analysis of host gene regulation during infection, using a differential display approach. *Journal of General Virology* 84, 3029–3039.

O'Reilly, D.R. and Miller, L.K. (1989) A baculovirus blocks insect molting by producing ecdysteroid UDP-glucosyl transferase. *Science* 245, 1110–1112.

O'Reilly, D.R. and Miller, L.K. (1991) Improvement of a baculovirus pesticide by deletion of the *egt* gene. *Bio/Technology* 9, 1086–1089.

Ohkawa, T., Washburn, J.O., Sitapara, R., Sid, E. and Volkman, L.E. (2005) Specific binding of *Autographa californica* M nucleopolyhedrovirus occlusion-derived virus to midgut cells of *Heliothis virescens* larvae is mediated by products of *pif* genes *Ac119* and *Ac022* but not by *Ac115*. *Journal of Virology* 79, 15258–15264.

Okano, K., Shimada, T., Mita, K. and Maeda, S. (2001) Comparative expressed-sequence-tag analysis of differential gene expression profiles in BmNPV-infected BmN cells. *Virology* 282, 348–356.

Okano, K., Vanarsdall, A.L., Mikhailov, V.S. and Rohrmann, G.F. (2006) Conserved molecular systems of the *Baculoviridae*. *Virology* 344, 77–87.

Pennock, G.D., Shoemaker, C. and Miller, L.K. (1984) Strong and regulated expression of *Escherichia coli* ß-galactosidase in insect cells with a baculovirus vector. *Molecular and Cellular Biology* 4, 399–406.

Petrik, D.T., Iseli, A., Montelone, B.A., Van Etten, J.L. and Clem, R.J. (2003) Improving baculovirus resistance to UV inactivation: increased virulence resulting from expression of a DNA repair enzyme. *Journal of Invertebrate Pathology* 82, 50–56.

Pijlman, G.P., Dortmans, J.C.F.M., Vermeesch, A.M.G., Yang, K., Martens, D.E., Goldbach, R.W. and Vlak, J.M. (2002) Pivotal role of the non-*hr* origin of DNA replication in the genesis of defective interfering baculoviruses. *Journal of Virology* 76, 5605–5611.

Podgwaite, J.D., Shields, K.S., Zerillo, R.T. and Bruen, R.B. (1979) Environmental persistence of the nucleopolyhedrosis virus of the gypsy moth, *Lymantria dispar*. *Environmental Entomology* 8, 528–536.

Popham, H.J.R., Bischoff, D.S. and Slavicek, J.M. (2001) Both *Lymantria dispar* nucleopolyhedrovirus *enhancin* genes contribute to viral potency. *Journal of Virology* 75, 8639–8648.

Raina, A.K. and Adams, J.R. (1995) Gonad-specific virus of corn earworm. *Nature* 374, 770.

Rajendra, W., Hackett, K.J., Buckley, E. and Hammock, B.D. (2006) Functional expression of lepidopteran-selective neurotoxin in baculovirus: potential for effective pest management. *Biochimica et Biophysica Acta* 1760, 158–163.

Rallis, C.P. and Burand, J.P. (2002a) Pathology and ultrastructure of Hz-2V infection in the agonadal female corn earworm, *Helicoverpa zea*. *Journal of Invertebrate Pathology* 81, 33–44.

Rallis, C.P. and Burand, J.P. (2002b) Pathology and ultrastructure of the insect virus, Hz-2V, infecting agonadal male corn earworms, *Helicoverpa zea*. *Journal of Invertebrate Pathology* 80, 81–89.

Saville, G.P., Patmanidi, A.L., Possee, R.D. and King, L.A. (2004) Deletion of the *Autographa californica* nucleopolyhedrovirus chitinase KDEL motif and *in vitro* and *in vivo* analysis of the modified virus. *Journal of General Virology* 85, 821–831.

Seo, J.H., Yeo, J.S. and Cha, H.J. (2005) Baculoviral polyhedrin-*Bacillus thuringiensis* toxin fusion protein: a protein-based bio-insecticide expressed in *Escherichia coli*. *Biotechnology and Bioengineering* 92, 166–172.

Shields, K.S. (1985) Pathways of nucleopolyhedrosis virus infection of gypsy moth, *Lymantria dispar*. In: Grimble, D.G. and Lewis, F.B. (eds) Microbial control of spruce budworms and gypsy moths. United States Forest Service, GTR-NE-100, pp. 123–124.

Simón, O., Williams, T., López-Ferber, M. and Caballero, P. (2004) Virus entry or the primary infection cycle are not the principal determinants of host specificity of *Spodoptera* spp. nucleopolyhedroviruses. *Journal of General Virology* 85, 2845–2855.

Slavicek, J.M. and Popham, H.J.R. (2005) The *Lymantria dispar* nucleopolyhedrovirus enhancins are components of occlusion-derived virus. *Journal of Virology* 79, 10578–10588.

Slavicek, J.M., Popham, H.J.R. and Riegel, C.I. (1999) Deletion of the *Lymantria dispar* multicapsid nucleopolyhedrovirus ecdysteroid UDP-glucosyl transferase gene enhances viral killing speed in the last instar of the gypsy moth. *Biological Control* 16, 91–103.

Smith, G.E., Summers, M.D. and Fraser, M.J. (1983) Production of human beta interferon in insect cells infected with a baculovirus expression vector. *Molecular and Cellular Biology* 3, 2156–2165.

Smith, I.R.L. and Crook, N.E. (1988) *In vivo* isolation of baculovirus genotypes. *Virology* 166, 240–244.

Stewart, T.M., Huijskens, I., Willis, L.G. and Theilmann, D.A. (2005) The *Autographa californica* multiple nucleopolyhedrovirus *ie0-ie1* gene complex is essential for wild-type virus replication, but either IE0 or IE1 can support virus growth. *Journal of Virology* 79, 4619–4629.

Stiles, B., Burand, J.P., Meda, M. and Wood, H.A. (1983) Characterization of gypsy moth *Lymantria dispar* nuclear polyhedrosis virus. *Applied and Environmental Microbiology* 46, 297–303.

Szewczyk, B., Hoyos-Carvajal, L., Paluszek, M., Skrzecz, I. and de Souza, M.L. (2006) Baculoviruses – re-emerging biopesticides. *Biotechnology Advances* 24, 143–160.

Szolajska, E., Poznanski, J., López Ferber, M., Michalik, J., Gout, E., Fender, P., Bailly, I., Dublet, B. and Chroboczek, J. (2004) Poneratoxin, a neurotoxin from ant venom. Structure and expression in insect cells and construction of a bio-insecticide. *European Journal of Biochemistry* 271, 2127–2136.

Tellam, R.L., Wijffels, G. and Willadsen, P. (1999) Peritrophic matrix proteins. *Insect Biochemistry and Molecular Biology* 29, 87–101.

Thiem, S.M., Du, X., Quentin, M.E. and Berner, M.M. (1996) Identification of a baculovirus gene that promotes *Autographa californica* nuclear polyhedrosis virus replication in a nonpermissive insect cell line. *Journal of Virology* 70, 2221–2229.

Thompson, C.G., Scott, D.W. and Wickman, B.E. (1981) Long-term persistence of the nuclear polyhedrosis virus of the Douglas-fir tussock moth, *Orgyia pseudotsugata* (Lepidoptera: Lymantriidae), in forest soil. *Environmental Entomology* 10, 254–256.

Tillman, P.G., Styer, E.L. and Hamm, J.J. (2004) Transmission of Ascovirus from *Heliothis virescens* (Lepidoptera: Noctuidae) by three parasitoids and effects of virus on survival of parasitoid *Cardiochiles nigriceps* (Hymenoptera: Braconidae). *Environmental Entomology* 33, 633–643.

Trudeau, D., Washburn, J.O. and Volkman, L.E. (2001) Central role of hemocytes in *Autographa californica* M nucleopolyhedrovirus pathogenesis in *Heliothis virescens* and *Helicoverpa zea*. *Journal of Virology* 75, 996–1003.

van Beek, N., Lu, A., Presnail, J., Davis, D., Greenamoyer, C., Joraski, K., Moore, L., Pierson, M., Herrmann, R., Flexner, L., Foster, J., Van, A., Wong, J., Jarvis, D., Hollingshaus, G. and

McCutchen, B. (2003) Effect of signal sequence and promoter on the speed of action of a genetically modified *Autographa californica* nucleopolyhedrovirus expressing the scorpion toxin LqhIT2. *Biological Control* 27, 53–64.

Wang, P. and Granados, R.R. (2000) Calcofluor disrupts the midgut defense system in insects. *Insect Biochemistry and Molecular Biology* 30, 135–143.

Washburn, J.O., Kirkpatrick, B.A., Haas-Stapleton, E., and Volkman, L.E. (1998) Evidence that the stilbene-derived optical brightener M2R enhances *Autographa californica* M nucleopolyhedrovirus infection of *Trichoplusia ni* and *Heliothis virescens* by preventing sloughing of infected midgut epithelial cells. *Biological Control* 11, 58–69.

Webb, R.E., Thorpe, K.W., Podgwaite, J.D., Reardon, R.C., White, G.B. and Talley, S.E. (1999) Field evaluation of an improved formulation of Gypchek (a nuclear polyhedrosis virus product) against the gypsy moth (Lepidoptera: Lymantriidae). *Journal of Entomological Science* 34, 72–83.

Wormleaton, S., Kuzio, J. and Winstanley, D. (2003) The complete sequence of the *Adoxophyes orana* granulovirus genome. *Virology* 311, 350–365.

Yanase, T., Yasunaga, C., Hara, T. and Kawarabata, T. (1998) Coinfection of *Spodoptera exigua* and *Spodoptera frugiperda* cell lines with the nuclear polyhedrosis viruses of *Autographa californica* and *Spodoptera exigua*. *Intervirology* 41, 244–252.

Zelazny, B. (1976) Transmission of a baculovirus in populations of *Oryctes rhinoceros*. *Journal of Invertebrate Pathology* 27, 221–227.

Zhang, J.H., Washburn, J.O., Jarvis, D.L. and Volkman, L.E. (2004) *Autographa californica* M nucleopolyhedrovirus early GP64 synthesis mitigates developmental resistance in orally infected noctuid hosts. *Journal of General Virology* 85, 833–842.

# 10 Insect–Protozoa–Bacteria Associations: a Model System for Investigating Host–Parasite Interactions

B.L. Weiss, G.M. Attardo and S. Aksoy

*Department of Epidemiology and Public Health, Section of Vector Biology, Yale University School of Medicine, New Haven, USA*



## 10.1. Introduction

Despite a significant effort by public health officials, many insect-borne infectious diseases continue to negatively impact humans and animals on a global scale. Malaria continues to kill approximately 2 million people annually, and the incidence of leishmania, chagas disease, filariasis, trypanosomiasis and numerous abroviruses is increasing (Waterfield *et al.*, 2004). Current methods for controlling these insect-borne diseases involve the use of traps and chemical pesticides. While these approaches are effective in the short term, they have significant limitations including environmental toxicity, adverse effects on human health, high costs associated with repeated applications and the emergence of insect resistance (Hemingway, 1999). Furthermore, setting traps and applying insecticides requires continuous community participation, often across international borders. This degree of cooperation is difficult to achieve in developing and often war-torn countries most affected by these diseases.

In an effort to develop new approaches to controlling vector insects and the pathogens they transmit, significant research is under way to acquire a better understanding of the interactions between these two groups of organisms. This chapter focuses on insect immune response to infection, pathogen development in the host and interactions between insect host microbial flora and pathogens. Knowledge acquired from this work can then be applied to the development of technologies that will reduce the incidence of insect-borne diseases in endemic areas. Research of this nature is being performed on other similar model systems including mosquitoes/malaria (Abraham *et al.*, 2005; Billingsley *et al.*, 2006), sandlfies/leishmania (Sacks and Kamhawi, 2001), triatomine bugs/chagas disease (Beard *et al.*, 2002; Lopez *et al.*, 2003) and ticks/spirochetes (Schwan and Piesman, 2002). This review summarizes the results of research conducted on one insect host-parasite model system – the tsetse fly and its associated microorganisms.

## 10.2. Trypanosomatid Protozoa and Tsetse Flies

Trypanosomiases are a group of debilitating zoonotic diseases caused by pathogenic trypanosomes (see Obornik, Chapter 5, this volume). Within this group of diseases are human African trypanosomiasis (HAT) and nagana, both of which are restricted to sub-Saharan Africa, vectored by tsetse flies of the genus *Glossina* (Diptera: Glossinidae), and invariably fatal if left untreated. HAT is the human form of the disease, while nagana is manifested in domesticated animals that lack immunity to the disease (unlike many wild animals that serve as reservoirs). HAT, which was largely under control in the mid-20th century, has made a voracious comeback. The most recently available epidemiological data (World Health Organization, 2006) indicate 17,500 reported cases. However, many times that number likely exist, as the disease often effects people living in rural, hard-to-reach locations.

Chemotherapeutic treatment of individuals afflicted with HAT has been met with little past success. Toxicity in the patient, pathogen resistance and a lack of guaranteed drug supply are all associated problems (Barrett, 2000). Furthermore, the phenomenon called 'antigenic variation', during which trypanosomes alter their surface protein coat, has inhibited the development of anti-trypanosomal vaccines (Borst *et al.*, 1996). In light of these issues, the most practical means of controlling HAT may be to disrupt the disease transmission cycle by managing tsetse vector populations. Current strategies involve the use of traps and/or insecticides. Unfortunately, these options have not been explored to a significant degree for the control of HAT because they require extensive community participation (Grant, 2001). Recent advances in molecular biology, and their potential application to insects, provide the opportunity to develop new approaches for use in reducing vector populations (reviewed by Aksoy, 2003).

In this chapter we review tsetse's digestive system in the context of trypanosome development. Also discussed are the interactions between trypanosome infection, tsetse's immune system and symbiotic bacteria. In conclusion, we discuss tsetse control strategies currently being investigated and outline some future experiments that we speculate will enhance our ability to control this devastating disease.

### 10.2.1. Tsetse's digestive system

Pathogenic African trypanosomes are transmitted from the tsetse vector to their mammalian host via the saliva during feeding. However, before reaching this point of transmission, trypanosomes undergo a complex series of developmental changes that occur within distinct portions of the fly's alimentary canal. Thus, we provide below a brief description of tsetse's digestive system (for more detailed reviews, see Wigglesworthia, 1929; Buxton, 1955; Aksoy *et al.*, 2003) as a prelude to describing trypanosome development in the fly.

The *Glossina* alimentary canal is a continuous tube divided into three parts: the proboscis, midgut and hindgut. The proboscis (also commonly referred to as the 'mouthparts') is divided into the labrum, hypopharynx and labium. During feeding these structures penetrate the host's cuticle, delivering saliva to the wound and blood to the fly's food canal. The food canal continues into the cuticle-lined foregut, which connects to the anterior midgut at a junction defined by an organ called the 'proventriculus'. This portion of tsetse's digestive tract is responsible for concentrating the blood meal. Tsetse's primary endosymbiont *Wigglesworthia glossinidia* is housed in an organ located midway along the anterior midgut called the 'bacteriome' (described later in more detail). The midgut continues distally, where it becomes the hindgut. Tsetse's Malpighian tubules are located at this intersection. Finally, waste exits the hindgut via the rectum.

The proventriculus is of particular interest with regard to tsetse–trypanosome interactions due to its production of a structure called the 'peritrophic matrix' (PM). Tsetse's PM forms a continuous sheath (or series of concentric sheathes) that lines the entire midgut (Lehane, 1997). The most common functions attributed to this structure are protection of the midgut epithelium from mechanical injury as well as providing a physical barrier against the establishment of trypanosome infections in tsetse's midgut (Sudha and Muthu, 1988; Lehane, 1997). The PM is composed of glycosaminoglycans and glycoproteins embedded in a chitinous backbone. The latter structures include the oligosaccharides N-acetylglucosamine (GlcNAc) and α-linked N-acetlygalactosamine (GalNAc), both of which bind tsetse midgut lectins that are in turn capable of binding trypanosomes (Ibrahim *et al.*, 1984; Lehane *et al.*, 1996). The roles these substances play in trypanosome differentiation and tsetse refractoriness to infection remain to be elucidated.

### 10.2.2. Trypanosome development in tsetse

A dramatic change in environment occurs when trypanosomes are taken up by a tsetse fly while feeding on an infected host. The parasites are immediately subjected to a dramatic change in temperature, and are exposed to foreign digestive enzymes and potent immune molecules. The trypanosomes must respond to these changes by undergoing a remarkable physiological transformation (Aksoy *et al.*, 2003). These events, summarized below, are in reference to the causative agent of HAT, *Trypanosoma brucei* spp. (for more detailed reviews, see Vickerman, 1985; Vickerman *et al.*, 1988; Van den Abbeele *et al.*, 1999; Aksoy *et al.*, 2003).

Prior to entry into tsetse, some mammalian 'bloodstream forms' transition from a long and slender shape to a short and stumpy one. In the vertebrate host, this morphological change is marked by the development of functional mitochondria, thus eliminating the trypanosome's dependence on mammalian glycolysis as an energy source (Vickerman, 1965). Once the trypanosomes are taken up during feeding, they begin the process of differentiating into the insect 'procyclic form'. Within 8 h of acquisition, transcription of the variant surface glycoprotein halts and procyclin expression begins. These procyclic cells are actively dividing within 24 h of ingestion by the fly (Matthews and Gull, 1998), and are no longer able to infect mammals (Ghiotto *et al.*, 1979). Several days post-ingestion, actively dividing procyclics cross the PM into the ectoperitrophic space and within a week migrate forward and establish infections at the anterior portion of the midgut. Here the cells cease division, undergo another round of differentiation (into highly motile mesocyclic trypomastigotes) and travel to the fly's salivary glands by passing back across the PM and migrating up the oesophagus. Finally, salivary gland infections undergo more differentiation, multiply rapidly and are ready to infect a new host as mature infective metacyclics. This entire process takes about 3–5 weeks (Vickerman *et al.*, 1988).

## 10.3. Molecular Approaches and Their Application to Study Insect Host Immune Responses

Determining the infection status of natural tsetse populations has until now been accomplished largely by trapping flies and microscopically examining their midguts and salivary glands for the presence of trypanosomes. Current estimates derived using this method indicate that approximately 1–5% of flies contain salivary gland infections (Msangi *et al.*, 1998). However, higher infection rates are often encountered when PCR-based technologies are applied. Furthermore, microscopic visual inspection can miss early-stage infections, and those that are detected can be misidentified (Lehane *et al.*, 2000). In the laboratory, where all flies are given an infectious blood meal, parasite infection prevalence is similarly restricted.

The refractory nature of natural tsetse populations to trypanosome infection can be the result of the insect's efficient immune system. This system is triggered by proteolytic cascades that originate in the fat body or gut, and can result in the sequestering and subsequent death of invading pathogens via melanization, phagocytosis-mediated clearance of foreign organisms following opsonization and the synthesis of haemolymph-borne molecules that have antimicrobial properties (Lehane *et al.*, 2004). Haemolymph-borne molecules, midgut-synthesized agglutinins and reactive oxygen intermediates have received the most attention to date. These substances include nitric oxide and the phenoloxidase cascade (Lehane *et al.*, 1997; Luckhart *et al.*, 1998), agglutinis (Ibrahim *et al.*, 1984), antimicrobial peptides (AMPs; Kaaya *et al.*, 1987; Hao *et al.*, 2001; Boulanger *et al.*, 2002) and midgut lectins (Welburn *et al.*, 1989, 1994).

In tsetse lectins may be associated with the PM and are capable of adhering to trypanosomes (Lehane and Msangi, 1991; Grubhoffer *et al.*, 1997). Feeding tsetse flies the lectin-specific sugar GlcNAc increases dramatically the incidence of trypanosome infection in these flies. This phenomenon thus implicates lectin in tsetse–trypanosome interactions (Maudlin and Welburn, 1987). An interesting facet of this hypothesis involves tsetse's secondary symbiont, *Sodalis glossinidius*. This organism exhibits chitinolytic activity, and one of the by-products of this reaction is GlcNAc, which can bind and reduce the availability of trypanocidal lectins (Welburn *et al.*, 1993). Thus, this symbiont may play a role in influencing the level of trypanosome refractoriness exhibited by individual tsetse species (Welburn and Maudlin, 1999).

Like other insects such as fruit flies and mosquitoes, tsetse's immune reaction is most likely activated by receptors that recognize specific molecular markers on the pathogen's surface (Barillas-Mury *et al.*, 2000; Dimopoulos *et al.*, 2001). Stimulation of these receptors results in the activation of two signal transduction pathways, Toll and immunodeficiency (IMD). Recent analysis of tsetse's fat body transcriptome identified homologues of *Drosophila* pattern recognition receptors, including gram-negative bacteria binding protein GNBP1 and the peptidoglycan recognition proteins PGRP-LB, PGRP-LC, PGRP-LCx and PGRP-SA (Attardo *et al.*, 2006). These molecules may recognize and interact with invading microbes (Werner *et al.*, 2003), and activate Toll or IMD pathway cascades resulting in the synthesis of transcription factors such as Spätzle, Relish and nuclear factor-κβ. These in turn lead to the synthesis of a battery of AMPs (Ghosh *et al.*, 1998; Choe *et al.*, 2002; Gobert *et al.*, 2003). Hao *et al.* (2001) evaluated the expression pattern of three AMPs (*attacin*, defensin and diptericin) in tsetse flies challenged with different immune stimuli. The authors found that the genes encoding these peptides are differentially regulated, and that tsetse's immune system is specific to pathogenic stimuli. For example, *Escherichia coli* (both injected into or fed to tsetse) induced a rapid and robust expression of *attacin* and *defensin*, while trypanosomes induced a similar response only when provided in the blood meal. Furthermore, the immune response evoked following trypanosome acquisition could not be detected until parasite titres had increased for 3–6 days. The expression of AMPs continued in infected flies, while it dropped significantly in those that were able to clear their infection. The latter group of self-cured flies were subsequently able to mount an immune response when challenged with different microorganisms, suggesting they are not immuno-compromised (Aksoy *et al.*, 2003).

Recently, experiments were performed to acquire a more complete understanding of tsetse's IMD pathway (Hu and Aksoy, 2006). Upon infection with gram-negative microbes, this pathway results in the expression of several AMPs, including attacin and cecropin. RNA interference (RNAi) was used to silence the expression of attacin. Subsequently, these flies exhibited a significant increase in their ability to transmit parasites. The same RNAi-mediated knockdown of *relish*, the transcriptional activator at the downstream end of the IMD pathway, inhibited the expression of both *attacin* and *cecropin*. The incidences of midgut and salivary gland trypanosome infections, as well as the intensity of midgut parasite infections, were significantly higher in flies subjected to *relish* RNAi treatment (Chang and Aksoy, 2006). These experiments provide the first concrete evidence that AMPs are directly involved in regulating the trypanosome transmission ability of tsetse.

## 10.4. Tsetse Endosymbionts

Most insects that feed on a single food source, such as blood, sap or wood, harbour endosymbiotic bacteria that provide nutrients missing from the diet. Tsetse (which feeds exclusively on blood) harbours three such microorganisms that are of great interest from a physiological perspective, mainly because in their absence flies are rendered sterile (Fig. 10.1). Two of these bacteria, the obligate mutualist

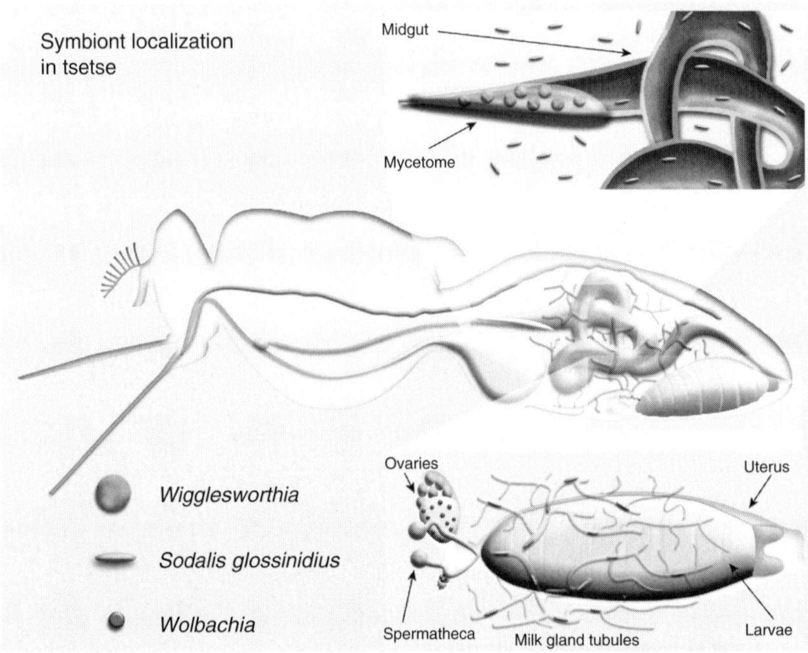


**Fig. 10.1.** Localization of symbiotic bacteria in tsetse flies. Tsetse flies harbour three distinct symbiotic bacteria: *Wigglesworthia glossinidia*, *Sodalis glossinidius* and members of the genus *Wolbachia. Wigglesworthia* resides exclusively within specialized 'bacteriocyte' cells that together comprise an organ called the 'mycetome'. Tsetse's mycetome is a component of its anterior midgut. *Wolbachia*, a parasitic bacterium whose function is unknown in tsetse, is also intracellular and can be found within the fly's reproductive tract. *Sodalis* has a broad tissue distribution and can be found both intracellularly and extracellularly in tsetse's midgut, fat body, muscle, haemolymph, milk gland and salivary glands of certain species.

*W. glossinidia* and the secondary symbiont *S. glossinidius*, are both members of the family Enterobacteriaceae. The third symbiont, members of the genus *Wolbachia*, is rickettsia-like parasitic bacteria (family Rickettsiaceae) found only in certain populations of tsetse (Cheng *et al.*, 2000). The role of *Wolbachia* in tsetse is unknown, although in other organisms it does cause a variety of reproductive abnormalities, the most common of which is called 'cytoplasmic incompatibility' (Stouthammer *et al.*, 1999).

### 10.4.1. *Wigglesworthia*

The concordant evolution exhibited between *Wigglesworthia* and its tsetse host indicates that these two organisms have coexisted together for approximately 50–80 million years (Chen *et al.*, 1999). This bacterium is an intracellular bacterium that resides exclusively in the cells (bacteriocytes) of a specialized organ called the 'bacteriome', which is located within the fly's anterior midgut. Analysis of *Wigglesworthia*'s highly streamlined 700 kb genome (which encodes 629 putative protein products) indicates that this bacterium likely supplements tsetse's vertebrate blood-specific diet. This hypothesis is based on the presence of several vitamin biosynthesis pathways, including those that produce biotin, thiazole, lipoic acid, FAD (riboflavin, B2), folate, pantothenate, thiamine (B1), pyridoxine (B6), protoheme and nicotinamide. In further support of this hypothesis is the fact that female tsetse become sterile when cleared of their *Wigglesworthia* via treatment with antibiotics. This phenomenon can be partially reversed in these 'aposymbiotic' flies by supplementing their diet with a complex of B vitamins (Nogge, 1976, 1982).

Several unique characteristics of *Wigglesworthia*'s genome are noteworthy. First, like other intracellular obligates, *Wigglesworthia*'s chromosome has an exceptionally high adenosine-thymidine (A + T) content of 78% (Akman *et al.*, 2002). Second, *Wigglesworthia*'s DNA replication machinery appears to lack the robustness of that found in closely related free-living eubacteria. In fact, the replication initiation protein, DnaA, is absent from *Wigglesworthia*'s chromosome. Another unique feature of *Wigglesworthia*'s genome is the presence of genes encoding a complete flagellar structure (Akman *et al.*, 2002). While this suggests a functional role, *Wigglesworthia* analysed from adult bacteriomes appears to lack a flagella and is immobile. However, from a hypothetical perspective, *Wigglesworthia* could express a flagellar structure to facilitate invasion through the female's milk gland during its transmission to the intrauterine progeny. Alternatively, once inside the offspring, a flagellum could be necessary for subsequent invasion of larval bacteriocytes (Aksoy *et al.*, 2003).

### 10.4.2. *Sodalis*

Tsetse's third symbiont, *Sodalis*, can be found both intercellularly and intracellularly in the midgut, muscle, fat body, milk gland and salivary glands of certain species (Cheng and Aksoy, 1999). Unlike *Wigglesworthia*'s ancient association with tsetse, *Sodalis* is apparently more recently acquired. In fact, phylogenetic analysis of *Sodalis* from distantly related host species implies that these organisms recently descended from a common ancestor, and that distinct tsetse species then acquired *Sodalis* independently via multiple horizontal transfer events (Aksoy *et al.*, 1997; Weiss *et al.*, 2006). *Sodalis*' recently sequenced and annotated 4.2 Mb genome indicates that this bacterium is closely related to several free-living enterics (including *Samonella* and *Yersinia*) both in regards to chromosomal size and synteny (Toh *et al.*, 2006). However, the high number of pseudogenes present on the chromosome (972), together with an exceptionally low protein-coding capacity (49%), indicates that *Sodalis* is evolving towards a true symbiotic lifestyle and away from a free-living one. Furthermore, the large numbers of the above-mentioned pseudogenes are clustered in groups homologous to known proteins that function in defence as well as transport and metabolism of carbohydrates and inorganic ions. This genomic characteristic implies that *Sodalis* is degeneratively adapting in light of tsetse's innate immune capabilities and restricted nutritional ecology.

Several theories are available regarding the function of *Sodalis* in tsetse, including the earlier-mentioned implication that it enhances tsetse's susceptibility to trypanosome infection (Welburn *et al.*, 1993, 1994; Welburn and Maudlin, 1999). In support of this theory is the fact that Maudlin *et al.* (1990) found a positive correlation between high symbiont densities and trypanosome infection in natural Liberian tsetse populations. In another study, flies that had their *Sodalis* selectively eliminated by treatment with the antibiotic streptozotocin exhibited a significant decrease in longevity compared to their wild-type counterparts (however, this treatment had no effect on reproductive capability; Dale and Welburn, 2001). In contrast to this finding, a recent study from our laboratory demonstrated that *Sodalis*-negative females exhibited no reduction in longevity or fecundity (Weiss *et al.*, 2006). The availability of *Sodalis*' full genome sequence, coupled with efficient genetic transformation and tsetse reconstitution systems, will facilitate future studies to determine the functional role of this symbiosis.

## 10.5. Control of Insect-borne Diseases

Due to the ineffectiveness of current management strategies, novel methods are now under development that will more effectively control insect disease vectors. In this section, we briefly summarize the most recent studies.

### 10.5.1. Insect transgenics

Modern molecular biological techniques have opened a promising avenue to genetically manipulate insect vectors with the intent of increasing their refractoriness to infection (Beaty, 2000). At the core of these efforts is a procedure called 'transgenesis', a process whereby circular plasmid DNA encoding a transposable element (TE) is injected into syncytial embryos. The TE subsequently inserts into the recipient insects germline, at which point all subsequent offspring inherit it. Several types of TEs, derived from a wide variety of eukaryotic organisms, are currently available for germline transformation experiments. TEs can carry a wide variety of exogenous DNA, including marker (i.e. green fluorescent protein (GFP) or luciferase) or effector (i.e. *attacin*) genes. Furthermore, prior to injection, donor DNA can be engineered to express temporally or spatially by incorporating specific promoters into the sequence (Atkinson *et al.*, 2001).

### 10.5.2. Paratransgenesis

Transgenesis has been used to ectopically express foreign genes in several important insect vectors, including a mosquito that vectors malaria in Asia (*Anopheles stephensi*; Catteruccia *et al.*, 2000) as well as the yellow fever mosquito, *Aedes aegypti* (Kakoza *et al.*, 2000). However, the physiology of certain insect disease vectors is such that current transgenic technologies are not the best option for foreign gene expression. Thus, a novel approach called 'paratransgenesis' has been developed with the intent of decreasing the vectorial capacity of these insects. Paratransgenesis involves isolating symbiotic bacteria from the insect and genetically modifying it *in vitro* so that it expresses and exports a molecule that interferes with disease transmission. The recombinant symbionts are then introduced back into their host, where they subsequently increase host refractoriness (Beard *et al.*, 2002).

Paratransgenesis has been successfully implemented in triatome bugs and tsetse, both of which exploit unique means of transmitting symbionts to their offspring. The former group serves as a vector for *Trypanosomma cruzi* (the causative agent of Chagas disease), and harbours a symbiont called *Rhodococcus rhodnii* that lives in the bug's midgut. This bacterium is acquired by naive, early nymphal *Rhodnius* via coprophagy, or the ingestion of faeces from other individuals. The reproductive biology of female tsetse flies is also different than most other insects, as they produce a single egg per gonotrophic cycle (viviparous reproduction). This egg hatches *in utero* and the larva matures through three instars, all the while being nourished with milk gland secretions containing symbionts. Upon completion of larval development the mother deposits the non-feeding larva, which immediately pupates.

This distinctive reproductive characteristic means germline transformation via embryo injections would be difficult with tsetse; thus, a paratransgenic strategy has been developed whereby a foreign gene product is expressed in *Sodalis* (Beard *et al.*, 1993). With this approach, genes are not inserted into tsetse's chromosome, but instead into the chromosome of this secondary symbiont. Subsequently, the transgenic bacteria are introduced back into the haemocoel of fertile female flies (Fig. 10.2).

*Sodalis* is well suited to express foreign gene products for many reasons:

**1.** *Sodalis* resides in tsetse's gut in close proximity to pathogenic trypanosomes. Thus, trypanocidal substances produced by recombinant cells will have a higher likelihood of detrimentally effecting the pathogen.

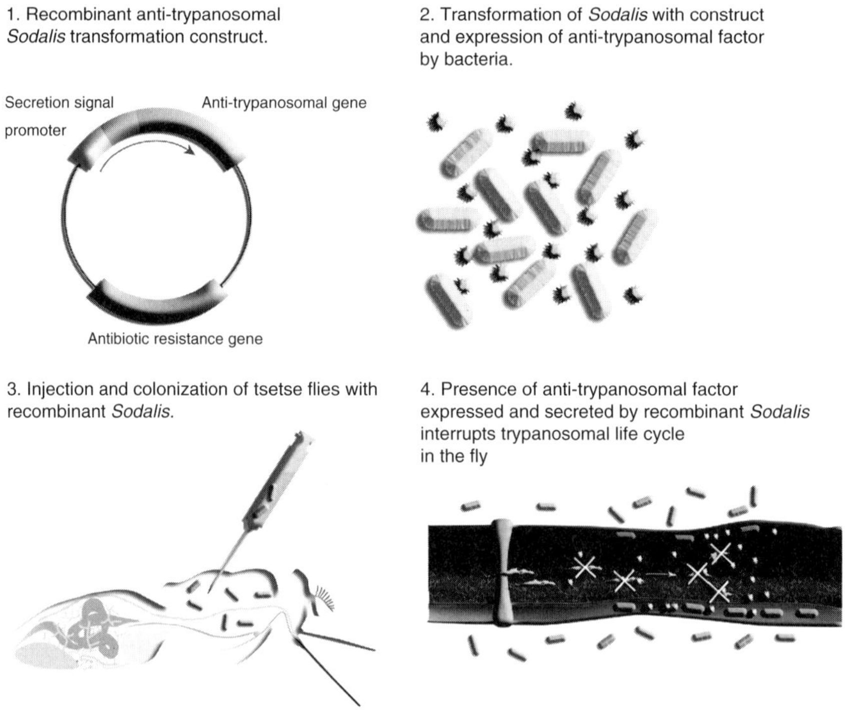


**Fig. 10.2.** Diagrammatic representation of the steps involved in using paratransgenesis to reduce tsetse's ability to spread African trypanosomes. Step 1: A construct is engineered with a promoter and secretion signal upstream of a gene encoding a trypanocidal peptide. The arrow indicates direction of transcription. Step 2: Wild-type *Sodalis* are transformed with the construct, and recombinant clones are selected by plating the transformation reaction on to an antibiotic-supplemented medium. Expression of the effector gene is confirmed by functional assay and Western blot analysis. Step 3: Newly-emerged tsetse flies are injected with a physiological dose of recombinant *Sodalis*. The reconstituted flies are subsequently maintained on a diet containing antibiotics to eliminate wild-type bacteria while enriching the recombinant population. Step 4: Recombinant *Sodalis* become established in tsetse's midgut. When a fly consumes a trypanosome-infected blood meal, *Sodalis* begin to express and secrete the recombinant trypanocidal peptide. This molecule kills trypanosomes in the fly's midgut.

**2.** A system for culturing *Sodalis in vitro* (both in liquid media and on agar plates) has been developed and can be used in conjunction with standard molecular biology techniques to insert and express foreign genes of interest in this bacterium (Beard *et al.*, 1993; Dale and Maudlin, 1999; Matthew *et al.*, 2005; Pontes and Dale, 2006).
**3.** *Sodalis* is highly resistant to many trypanocidal peptides (Hu and Aksoy, 2005; Haines *et al.*, 2003).
**4.** *Sodalis* can be reintroduced into tsetse by thoracic microinjection and passed on to future progeny (Cheng and Aksoy, 1999; Rio *et al.*, 2004).

**5.** Isolates from one tsetse species can be transinfected into different tsetse species to streamline the paratransgenesis process (Weiss *et al.*, 2006).
**6.** *Sodalis*' genome is completely sequenced and annotated, and this information will serve as a valuable resource that can be exploited to improve the efficiency of our expression system (Toh *et al.*, 2006).
**7.** This symbiont has highly restricted metabolic capabilities, and a specific, anchored relationship with its host genera, that would severely hinder (or most likely completely eliminate) survival outside of its normal host (Rio *et al.*, 2003; Toh *et al.*, 2006).

Previous work in our laboratory developed a heat-shock-based transformation procedure to introduce the shuttle plasmid pSUP204 (with a *Pseudomonas* origin of replication) into *Sodalis*. Transformants were selected based on their plasmid-encoded resistance to multiple antibiotics (Beard *et al.*, 1993). In subsequent experiments, *Sodalis* cultures were transformed with a plasmid that expresses the GFP marker gene. When the recombinant symbionts were microinjected into the haemocoel of fertile female flies, they were detected in first and second generation adults by PCR-amplification of gut tissue with GFP-specific primers. Finally, cultures of recombinant *Sodalis* from first generation adult females were established and GFP expression was confirmed by fluorescent microscopy (Cheng and Aksoy, 1999).

### 10.5.3. Gene-driving mechanisms

The ability to spread refractory laboratory-generated phenotypes into susceptible field populations is a crucial component of using transgenics to control vector-borne diseases. *Wolbachia*, tsetse's third symbiont, may provide one potential mechanism for driving a genetically modified *Sodalis* into natural tsetse populations (this approach could also be used with other transgenic insect vectors). Infections with this bacterium often induce a pathology called 'cytoplasmic incompatibility' (CI), which results in zygotic death during embryogenesis. In this instance, the sperm enters the egg but does not successfully transfer its genetic material (Zabalou *et al.*, 2004). CI results when uninfected females mate with infected males or with males that carry a different *Wolbachia* strain (this includes 'super-infections', which means the effected individual carries two different *Wolbachia* strains). Thus, CI confers a reproductive advantage to infected females over their uninfected counterparts because they can mate with both types of males. Novel combinations of paratransgenic tsetse and *Wolbachia* could be developed (and released) that could move through and outcompete natural, wild-type populations since both microorganisms are maternally transmitted into the developing larva (Sinkins and Gould, 2006). *Wolbachia*-mediated CI has also been used successfully to spread other maternally transmitted genetic markers such as mitochondrial DNA (Turelli *et al.*, 2002).

## 10.6. Current and Future Work

### 10.6.1. Identify novel effector molecules

The importance of identifying trypanocidal molecules has intensified with the availability of a tsetse paratransgenesis system. Two types of effector molecules may be expressed to kill trypanosomes: (i) transmission-blocking agents; and (ii) small peptides.

#### *10.6.1.1. Transmission-blocking agents*

Transmission-blocking agents are capable of disrupting parasite development and pathogenicity by binding proteins necessary for these events to occur. For practical purposes, these molecules can be expressed as target-specific single-chain antibody fragments from one gene. A molecule of this nature was successfully expressed by symbiotic bacteria (*R. rhodnii*) living in the gut of *Rhodnius prolixus*. In this experiment functional single-chain antibody rDB3 (which encodes murine V(H)/K that binds progesterone) was exported into the host insect's gut lumen (Durvasula *et al.*, 1999). While single-chain antibodies have yet to be expressed in *Sodalis*, several antibodies that target major procyclic *T. brucei* surface proteins have been reported (Nantulya and Moloo, 1988).

#### *10.6.1.2. Small peptides*

Small peptides are one component of the innate immune system of many higher multicellular organisms. As mentioned in the section on host immunity, three such molecules (attacin, defensin and diptericin) have been identified in tsetse (Hao *et al.*, 2001). These well-studied substances are produced by tsetse's immune tissues when challenged by a pathogen(s), and they may be also responsible for the pathogen-refractory phenotypes of many vector species.

Co-evolution over millions of years has allowed parasites to evolve mechanisms that permit them to partially evade their specific insect host's immune response. This same phenomenon may make that parasite extremely vulnerable when exposed to an antimicrobial peptide from another organism (Zambrano-Villa *et al.*, 2002). Thus, the choice of which antimicrobial peptide gene(s) will most efficiently destroy trypanosomes when expressed in *Sodalis* may best be made by observing the endogenous immune response of other afflicted animals. One antimicrobial peptide of interest, BMAP-27, is produced by bovine neutrophils. Recent experiments by Haines *et al.* (2003) revealed that BMAP-27 is highly lethal to both bloodstream-form and procyclic-form trypanosomes. Further experiments revealed that *Sodalis* is resistant to 65-fold higher concentrations of *BMAP-27* than bloodstream-form trypanosomes. These results indicate that this peptide will be very useful in paratransgenesis experiments. *Drosophila melanogaster* has also served as an excellent model for the study of these peptides in insects. The immune system of *Drosophila* produces AMPs from seven different families. These peptides have a broad spectrum of activity, although targeting of specific pathogen types has been demonstrated (Hetru *et al.*, 2003). Research on the use of both anti-trypanosomal single-chain antibody fragments and trypanocidal peptides in tsetse's paratransgenic system is currently under way.

Another promising candidate for use in our paratransgenic system is the human trypanocide apolipoprotein L-1 (apoL-1). This protein, which is a component of normal human serum, lyses trypanosomes that do not cause HAT (Vanhamme and Pays, 2004). Resistant forms, such as *Trypanosoma brucei rhodesiense*, express a surface protein called SRA, which interacts with the C-terminus of apoL-1 and thus inactivates it (Xong *et al.*, 1998). However, *T. b. rhodesiense* is susceptible when incubated with a truncated version of apoL-1 (Tr-apoL-1) that lacks a C-terminal SRA-interacting domain (Vanhamme *et al.*, 2003). As a means of developing this protein for potential HAT therapy, Baral *et al.* (2006) fused Tr-apoL-1 to the single-domain 'nanobody' (NbAn33) that specifically targets conserved epitopes of the trypanosome variant surface glycoprotein, thus allowing the conjugate to outcompete endogenous apoL-1. Treatment of HAT-infected mice with NbAn33-Tr-apoL-1 caused no adverse physiological effects, and definitively cured animals with both acute and chronic infections. This type of nanobody-conjugated trypanocide, combined with a *Sodalis*-specific secretion signal, has promising potential to increase tsetse's refractoriness to infection.

### 10.6.2. Identify novel promoters and secretion signals

Efficient expression of foreign gene products by *Sodalis*, and the subsequent secretion of recombinant proteins into the midgut environment, is crucial to the success of tsetse paratransgenesis. Thus, the identification of novel promoters and secretion signals becomes of paramount importance. The ideal promoter for this type of system would be endogenous to *Sodalis* and function temporally and spatially so that expression of the transgene occurs only at a specific time and position. The most suitable time for expression of trypanocidal compounds would be immediately following a blood meal, or better yet, immediately following the acquisition of an *infected* blood meal. From a spatial perspective, expression of trypanocides specifically within tsetse's adult midgut would likely reduce any host fitness costs potentially associated with this procedure. With this in mind, techniques such as reverse transcription PCR could be used to determine if any *Sodalis* genes become up-regulated under the desired circumstances. Regulatory elements cloned from these genes could then be placed into constructs upstream of sequences that encode trypanocides. In theory, these promoters would then only drive transgene expression in the presence of one of the above-mentioned stimuli. This type of system would limit the time that *Sodalis* (and tsetse for that matter) is exposed to antimicrobial substances, and would facilitate the cloning of constructs that encode toxic gene products in susceptible bacteria such as *E. coli*.

Translocation of recombinant effector molecules across *Sodalis*' outer membrane and into the tsetse's gut lumen, where newly acquired trypanosomes begin the differentiation process, is also required for paratransgenesis to be successful. Several candidate systems that may accomplish this goal are currently under consideration. For example, signal sequences from secreted proteins can be used to produce an effector molecule in an expression construct. Some potential signals are the pectate lysate N-terminal pel-B leader sequence from *Erwinia carotovora* (Lei *et al.*, 1987), as well as signals on *Sodalis*-specific genes such as *spaR* and *insulinase* (Toh *et al.*, 2006). Several studies have also elegantly demonstrated the use of *E. coli*'s α-haemolysin secretion system (a type I secretion system) to translocate recombinant peptides across the outer membrane of different bacterial species (Tzschaschel *et al.*, 1996; Gentschev *et al.*, 2002). *Sodalis*' chromosome encodes a full-length haemolysin gene and homologues of some of the necessary *E. coli* type I secretion system apparatus genes (Toh *et al.*, 2006). A more detailed analysis is necessary to evaluate the usefulness of this system as a mechanism for secreting recombinant proteins from *Sodalis*.

## 10.7. Conclusions

Central to effectively managing insect disease vectors and their associated pathogens is acquiring a better understanding of the interactions between these groups of organisms. This chapter provides an overview of the organisms associated with the spread of HAT: tsetse flies, their endosymbionts and pathogenic trypanosomes. Further unanswered questions regarding the biology of these organisms, as well as the use of this information to develop more effective control mechanisms, must be addressed before this devastating disease can be controlled.

## References

Abraham, E.G., Donnelly-Doman, M., Fujioka, H., Ghosh, A., Moreira, L. and Jacobs-Lorena, M. (2005) Driving midgut-specific expression and secretion of a foreign protein in transgenic mosquitoes with AgAper1 regulatory elements. *Insect Molecular Biology* 14, 271–279.

Akman, L., Yamashita, A., Watanabe, H., Oshima, K., Shiba, T., Hattori, M. and Aksoy, S. (2002) Genome sequence of the endocellular obligate symbiont of tsetse flies, *Wigglseworthia glossinidia. Nature Genetics* 32, 402–407.

Aksoy, S. (2003) Control of tsetse flies and trypanosomes using molecular genetics. *Veterinary Parasitology* 105, 125–145.

Aksoy, S., Chen, X. and Hypsa, V. (1997) Phylogeny and potential transmission routes of midgut-associated endosymbionts of tsetse (Diptera: Glossinidae). *Insect Molecular Biology* 6, 183–190.

Aksoy, S., Gibson, W.C. and Lehane, M.J. (2003) Interactions between tsetse and trypanosomes with implications for the control of trypanosomiasis. *Advances in Parasitology* 53, 1–83.

Atkinson, P.W., Pinkerton, A.C. and O'Brochta, D.O. (2001) Genetic transformation systems in insects. *Annual Review of Entomology* 46, 317–346.

Attardo, G.M., Strickler-Dinglasan, P., Perkin, S.A.H., Caler, E., Bonaldo, M.F., Soares, M.B., El-Sayeed, N. and Alsoy, S. (2006) Analysis of fat body transcriptome from adult tsetse fly, *Glossina morsitans. Insect Molecular Biology* 15, 410–424.

Baral, T.N., Magez, S., Stijlemans, B., Conrath, K., Vanhollebeke, B., Pays, E., Muyldermans, S. and De Baetselier, P. (2006) Experimental therapy of African trypanosomiasis with a nanobody-conjugated human trypanolytic factor. *Nature Medicine* 12, 580–584.

Barillas-Mury, C., Wizel, B. and Han, Y.S. (2000) Mosquito immune responses and malaria transmission: lessons from insect model systems and implications for vertebrate innate immunity and vaccine development. *Insect Biochemistry and Molecular Biology* 30, 429–442.

Barrett, M.P. (2000) Problems for the chemotherapy of human African trypanosomiasis. *Current Opinion in Infectious Diseases* 13, 647–651.

Beard, C.B., O'Neill, S.L., Mason, P., Mandelco, L., Woese, C.R., Tesh, R.B., Richards, F.F. and Aksoy, S. (1993) Genetic transformation and phylogeny of bacterial symbionts from tsetse. *Insect Molecular Biology* 1, 123–131.

Beard, C.B., Cordon-Rosales, C. and Durvasula, R.V. (2002) Bacterial symbionts of the triatominae and their potential use in control of Chagas disease transmission. *Annual Review of Entomology* 47, 123–141.

Beaty, B. (2000) Genetic manipulation of vectors: a potential novel approach for control of vector-borne diseases. *Proceedings of the National Academy of Sciences of the USA* 97, 10295–10297.

Billingsley, P.F., Baird, J., Mitchell, J.A. and Drakeley, C. (2006) Immune interactions between mosquitoes and their hosts. *Parasite Immunology* 28, 143–153.

Borst, P., Rudenko, G., Taylor, M.C., Blundell, P.A., Van Leeuwen, F., Bitter, W., Cross, M. and McCulloch, R. (1996) Antigenic variation in trypanosomes. *Archives of Medical Research* 27, 379–388.

Boulanger, N., Brun, R., Ehret-Sabatier, L., Kunz, C. and Bulet, P. (2002) Immunopeptides in the defense reactions of *Glossina morsitans* to bacterial and *Trypanosoma brucei brucei* infections. *Insect Biochemistry and Molecular Biology* 32, 369–375.

Buxton, P.A. (1955) *The Natural History of Tsetse Flies*. London School of Hygeine and Tropical Medicine, Memoir no. 10. H.K. Lewis, London.

Catteruccia, F., Nolan, T., Loukeris, T.G., Blass, C., Savakis, C., Kafatos, F.C. and Crisanti, A. (2000) Stable germline transformation of the malaria mosquito *Anopheles stephensi*. *Nature* 405, 959–962.

Chen, X., Li, S. and Aksoy, S. (1999) Concordant evolution of a symbiont with its host species: molecular phylogeny of genus *Glossina* and its bacteriome-associated endosymbiont, *Wigglesworthia glossinidia*. *Journal of Molecular Evolution* 48, 49–58.

Cheng, Q. and Aksoy, S. (1999) Tissue tropism, transmission and expression of foreign genes *in vivo* in midgut symbionts of tsetse flies. *Insect Molecular Biology* 8, 125–132.

Cheng, Q., Ruel, T.D., Zhou, W., Moloo, S.K., Majiwa, P., O'Neill, S.L. and Aksoy, S. (2000) Tissue distribution and prevalence of *Wolbachia* infections in tsetse flies, *Glossina spp. Medical and Veterinary Entomology* 14, 44–50.

Choe, K.M., Werner, T., Stoven, S., Hultmark, D. and Anderson, K.V. (2002) Requirement for a peptidoglycan protein (PGRP) in relish activation and antibacterial immune responses in *Drosophila*. *Science* 296, 359–362.

Dale, C. and Maudlin, I. (1999) *Sodalis* gen. nov. and *Sodalis glossinidius* sp. nov., a microaerophilic secondary endosymbiont of the tsetse fly *Glossina morsitans morsitans*. *International Journal of Systematic Bacteriology* 49, 267–275.

Dale, C. and Welburn, S.C. (2001) The endosymbionts of tsetse flies: manipulating host–parasite interactions. *International Journal for Parasitology* 31, 628–631.

Dimopoulos, G., Richman, A., Muller, H.M. and Kafatos, F.C. (2001) Innate immune defense against malaria infection in the mosquito. *Current Opinion in Immunology* 13, 79–88.

Durvasula, R.V., Gumbs, A., Panackal, A., Kruglov, O., Taneja, J., Kang, A.S., Cordon- Rosales, C., Richards, F.F., Whitham, R.G. and Beard, C.B. (1999) Expression of a functional antibody fragment in the gut of *Rhodnius prolixus* via transgenic bacterial symbiont *Rhodococcus rhodnii*. *Medical and Veterinary Entomology* 13, 105–109.

Gentschev, I., Dietrich, G. and Goebel, W. (2002) The *E. coli* alpha-hemolysin secretion system and its use in vaccine development. *Trends in Microbiology* 10, 39–45.

Ghiotto, V., Brun, R., Jenni, L. and Hecker, H. (1979) *Trypanosoma brucei*: morphometric changes and loss of infectivity during transformation of bloodstream forms to procyclic culture forms *in vitro*. *Experimental Parasitology* 48, 447–456.

Ghosh, S., May, M.J. and Kopp, E.B. (1998) NF-kappa B and Rel proteins: evolutionarily conserved mediators of immune responses. *Annual Review of Immunology* 16, 225–260.

Gobert, V., Gottar, M., Matskevitch, A.A., Rutschmann, S., Royet, J., Belvin, M. *et al.* (2003) Dual activation of the *Drosophila* toll pathway by two pattern recognition receptors. *Science* 302, 2126–2130.

Grant, I. (2001) Insecticides for tsetse and trypanosomiasis control: is the environmental risk acceptable? *Trends in Parasitology* 17, 10–14.

Grubhoffer, L., Hypsa, V. and Volf, P. (1997) Lectins (hemagglutinins) in the gut of important disease vectors. *Parasite* 4, 203–216.

Haines, R.L., Hancock, R.W. and Pearson, T.W. (2003) Cationic antimicrobial peptide killing of African trypanosomes and *Sodalis glossinidius*, a bacterial symbiont of the insect vector of sleeping sickness. *Vector-borne and Zoonotic Diseases* 3, 175–186.

Hao, Z., Kasumba, I., Lehane, M.J., Gibson, W.C., Kwon, J. and Aksoy, S. (2001) Tsetse immune responses and trypanosome transmission: implications for the development of tsetse-based strategies to reduce trypanosomiasis. *Proceedings of the National Academy of Sciences of the USA* 98, 12648–12653.

Hemingway, J. (1999) Insecticide resistance in malaria vectors: a new approach to an old subject. *Parasitologia* 41, 315–318.

Hetru, C., Troxler, L. and Hoffmann, J.A. (2003) *Drosophila melanogaster* antimicrobial defense. *Journal of Infectious Disease* 187(Suppl. 2), S327–S334.

Hu, C. and Aksoy, S. (2006) Innate immune responses regulate trypanosome parasite infection of the tsetse fly, *Glossina morsitans morsitans*. *Molecular Microbiology* 60, 1094–1204.

Hu, Y. and Aksoy, S. (2005) An antimicrobial peptide with trypanocidal activity characterized from *Glossina morsitans morsitans*. *Insect Biochemistry and Molecular Biology* 35, 105–115.

Ibrahim, E.A.R., Ingram, G.A. and Molyneux, D.H. (1984) Haemagglutinins and parasite agglutinins in haemolymph and gut of *Glossina*. *Tropenmedical Parasitolgy* 35, 151–156.

Kaaya, G.P., Flyg, C. and Boman, H.G. (1987) Induction of cecropin and attacin-like antibacterial factors in the haemolymph of *Glossina morsitans morsitans*. *Insect Biochemistry* 17, 309–315.

Kakoza, V., Ahmed, A., Cho, W.L., Jasinskiene, N., James, A.A. and Raikhel, A. (2000) Engineering bloodmeal-activated system immunity in the yellow fever mosquito, *Aedes aegypti*. *Proceedings of the National Academy of Sciences of the USA* 97, 9144–9149.

Lehane, M.J. (1997) Peritrophic matrix structure and function. *Annual Review of Entomology* 42, 525–550.

Lehane, M.J. and Msangi, A.R. (1991) Lectin and peritrophic membrane development in the gut of *Glossina m.morsitans* and a discussion of their role in protecting the fly against trypanosome infection. *Medical and Veterinary Entomology* 5, 495–501.

Lehane, M.J., Allingham, P.G. and Weglicki, P. (1996) Composition of the peritrophic matrix of the tsetse fly, *Glossina morsitans morsitans*. *Cell and Tissue Research* 283, 375–384.

Lehane, M.J., Wu, D. and Lehane, S.M. (1997) Midgut-specific immune molecules are produced by the blood-sucking insect *Stomoxys calcitrans*. *Proceedings of the National Academy of Sciences of the USA* 94, 10502–10507.

Lehane, M.J., Msangi, A.R., Whitaker, C.J. and Lehane, S.M. (2000) Grouping of trypanosome species in mixed infections in *Glossina pallidipes*. *Parasitology* 120, 583–592.

Lehane, M.J., Aksoy, S. and Levahina, E. (2004) Immune responses and parasite transmission in blood-feeding insects. *Trends in Parasitology* 20, 433–439.

Lei, S.P., Lin, H.C., Wang, J., Callaway, G. and Wilcox, G. (1987) Characterization of the *Erwinia carotovora* pelB gene and its product, pectate lysate. *Journal of Bacteriology* 169, 4379–4383.

Lopez, L., Morales, G., Ursic, R., Wolff, M. and Lowenberger, C. (2003) Isolation and characterization of a novel insect defensin from *Rhodnius prolixus*, a vector of Chagas disease. *Insect Biochemistry and Molecular Biology* 33, 439–447.

Luckhart, S., Vodovotz, Y., Cui, L. and Rosenberg, R. (1998) The mosquito *Anopheles stephensi* limits malaria parasite development with inducible synthesis of nitric oxide. *Proceedings of the National Academy of Sciences of the USA* 95, 5700–5705.

Matthew, C.Z., Darby, A.C., Young, S.A., Hume, L.H. and Welburn, S.C. (2005) The rapid isolation and growth dynamics of the tsetse symbiont *Sodalis glossinidius*. *FEMMS Microbiology Letters* 248, 69–74.

Matthews, K.R. and Gull, K. (1998) Identification of stage-regulated and differentiation-enriched transcripts during transformation of the African trypanosome from its bloodstream to procyclic form. *Molecular and Biochemical Parasitology* 95, 81–95.

Maudlin, I. and Welburn, S.C. (1987) Lectin mediated establishment of midgut infections of *Trypanosoma congolense* and *Trypanosoma brucei* in *Glossina morsitans*. *Tropical Medicine and Parasitology* 38, 167–170.

Maudlin, I., Welburn, S.C. and Mehlitz, D. (1990) The relationship between rickettsia-like-organisms and trypanosome infections in natural populations of tsetse in Liberia. *Tropical Medicine and Parasitology* 41, 265–267.

Msangi, A.R., Whitaker, C.J. and Lehane, M.J. (1998) Factors influencing the prevalence of trypanosome infection of *Glossina pallidipes* on the Ruvu flood plain in eastern Tanzania. *Acta Tropica* 70, 143–155.

Nantulya, V.M. and Moloo, S.K. (1988) Suppression of cyclical development of *Trypanosoma brucei brucei* in *Glossina morsitans centralis* by an anti-procyclics monoclonal antibody. *Acta Tropica* 45, 137–144.

Nogge, G. (1976) Sterility in tsetse flies (*Glossina morsitans* Westwood) caused by loss of symbionts. *Experientia* 32, 995–996.

Nogge, G. (1982) Significance of symbionts for the maintenance of an optimal nutritional state for successful reproduction in hematophagous arthropods. *Parasitology* 82, 299–304.

Pontes, M.H. and Dale, C. (2006) Culture and manipulation of insect facultative symbionts. *Trends in Microbiology* 14, 406–412.

Rio, R.V., Lefevre, C., Heddi, A. and Aksoy, S. (2003) Comparative genomics of insect-symbiotic bacteria: influence of host environment on microbial genome composition. *Applied and Environmental Microbiology* 69, 6825–6832.

Rio, R.V., Hu, Y. and Aksoy, S. (2004) Strategies of the home-team: symbioses exploited for vector-borne disease control. *Trends in Microbiology* 12, 325–336.

Sacks, D. and Kamhawi, S. (2001) Molecular aspects of parasite–vector and vector–host interactions in leishmaniasis. *Annual Review of Microbiology* 55, 453–483.

Schwan, T.G. and Piesman, J. (2002) Vector interactions and molecular adaptations of lyme disease and relapsing fever spirochetes associated with transmission by ticks. *Emerging and Infectious Disease* 8, 105–121.

Sinkins, S.P. and Gould, F. (2006) Gene drive systems for insect disease vectors. *Nature Reviews Genetics* 7, 427–435.

Stouthammer, R., Breeuwer, J.A.J. and Hurst, G.D.D. (1999) *Wolbachia pipientis*: microbial manipulator of arthropod reproduction. *Annual Review of Microbiology* 53, 71–102.

Sudha, P.M. and Muthu, S.P. (1988) Damage to the midgut epithelium caused by food in the absence of peritrophic membrane. *Current Science* 57, 624–625.

Toh, H., Weiss, B.L., Perkin, S.A., Yamashita, A., Oshima, K., Hattori, M. and Aksoy, S. (2006) Massive genome erosion and functional adaptations provide insights into the symbiotic lifestyle of *Sodalis glossinidius* in the tsetse host. *Genome Research* 16, 149–156.

Turelli, M., Hoffmann, A.A. and McKechnie, S.W. (2002) Dynamics of cytoplasmic incompatibility and mtDNA variation in natural *Drosophila simulans* populations. *Genetics* 132, 713–723.

Tzschaschel, B.D., Guzman, C.A., Timmis, K.N. and de Lorenzo, V. (1996) An *Escherichia coli* hemolysin transport system-based vector for the export of polypeptides: export of Shiga-like toxin IIeB subunit by *Salmonella typhimurium* aroA. *Nature Biotechnology* 14, 765–769.

Van den Abbeele, J., Claes, Y., Bockstaele, D.V., le Ray, D. and Coosemans, M. (1999) *Trypanosoma brucei sspp.* development in the tsetse fly: characterization of the post-mesocyclic stages in the foregut and proboscis. *Parasitology* 108, 469–478.

Vanhamme, L. and Pays, E. (2004) The trypanosome lytic factor of human serum and the molecular basis of sleeping sickness. *International Journal of Parasitology* 34, 887–898.

Vanhamme, L., Paturiaux-Hanocq, F., Poelvoorde, P., Nolan, D.P., Lins, L., Van Den Abbeele, J., Pays, A., Tebabi, P., Van Xong, H., Jacquet, A., Moguilevsky, N., Dieu, M., Kane, J.P., De Baetselier, P., Brasseur, R. and Pays, E. (2003) Apolipoprotein L-I is the trypanosome lytic factor of human serum. *Nature* 422, 83–87.

Vickerman, K. (1965) Polymorphism and mitochondrial activity in sleeping sickness trypanosomes. *Nature* 208, 762–766.

Vickerman, K. (1985) Developmental cycles and biology of pathogenic trypanosomes. *British Medical Journal* 41, 105–114.

Vickerman, K., Tetley, L., Hendry, A. and Turner, M. (1988) Biology of African trypanosomes in tsetse fly. *Biology of the Cell* 64, 109–119.

Waterfield, N.R., Wren, B.W. and French-Constant, R.H. (2004) Invertebrates as a source of emerging human pathogens. *Nature Reviews Microbiology* 2, 833–841.

Weiss, B.L., Mouchotte, R., Rio, R.V.M., Wu, Y., Wu, Z., Heddi, A. and Aksoy, S. (2006) Inter-specific transfer of bacterial endosymbionts between tsetse species: infection establishment and effect on host fitness. *Applied and Environmental Microbiology* 72, 7013–7021.

Welburn, S.C. and Maudlin, I. (1999) Tsetse–trypanosome interactions: rites of passage. *Parasitology Today* 15, 399–403.

Welburn, S.C., Maudlin, I. and Ellis, D.S. (1989) Rate of trypanosome killing by lectins in midguts of different species and strains of *Glossina. Medical and Veterinary Entomology* 3, 77–82.

Welburn, S.C., Arnold, K., Maudlin, I. and Gooday, G.W. (1993) *Rickettsia*-like organisms and chitinase production in relation to transmission of trypanosomes by tsetse flies. *Parasitology* 107, 141–145.

Welburn, S.C., Maudlin, I. and Molyneux, D.H. (1994) Midgut lectin activity and sugar specificity in teneral and fed tsetse. *Medical and Veterinary Entomology* 8, 81–87.

Werner, T., Borge-Renberg, K., Mellroth, K., Steiner, P. and Hultmark, D. (2003) Functional diversity of the *Drosophila* PGRP-LC gene cluster in the response to lipopolysaccharide and peptidoglycan. *Journal of Biological Chemistry* 278, 26319–26322.

Wigglesworthia, V.B. (1929) Digestion in the tsetse-fly: a study of structure and function. *Parasitology* 21, 288–321.

World Health Organization (2006) Human African trypanosomiasis (sleeping sickness): epidemiological update. *Weekly Epidemiological Record* 81, 71–80.

Xong, H.V., Vanhamm, L., Chamekh, M., Chimfwembe, C.E., Van Den Abbeele, J., Pays, A., Van Meirvenne, N., Hamers, R., De Baetselier, P. and Pays, E. (1998) A VSG expression site-associated gene confers resistance to human serum in *Trypanosoma rhodesiense. Cell* 95, 839–846.

Zabalou, S., Riegler, M., Theodorakopoulou, M., Stauffer, C., Savakis, C. and Bourtzis, K. (2004) *Wolbachia*-induced cytoplasmic incompatibility as a means for insect pest population control. *Proceedings of the National Academy of Sciences of the USA* 101, 15042–15045.

Zambrano-Villa, S., Rosales-Borjas, D., Carrero, J.C. and Ortiz-Ortiz, L. (2002) How protozoan parasites evade the immune response. *Trends in Parasitology* 18, 272–278.

# 11 Methods in Investigating Nematode–Bacterium–Insect Symbiosis

H. Goodrich-Blair,[1] D.J. Clarke,[2] P.S. Grewal[3] and T.A. Ciche[4]

[1]*University of Wisconsin, Madison, USA;* [2]*University College Cork, Ireland;* [3]*Ohio State University, Wooster, USA;* [4]*Michigan State University, East Lansing, USA*

## 11.1. Introduction

In biology, new functions and behaviours can emerge from interspecies alliances, such as when two or more species cooperate to promote disease. For example, filarid nematodes are associated with *Wolbachia* endosymbionts, and both organisms contribute to disease in mammals (Taylor *et al.*, 2005). In such systems, to fully understand the host–pathogen interaction, in addition to probing the interface between the pathogens and the host, it is crucial to investigate the mechanisms that underlie and stabilize the relationship between the allied organisms. While many nematode parasites are a significant concern for human health and agriculture, the entomopathogenic (or insect-parasitic) nematodes (EPNs) are excellent biological control agents of insect pests (Grewal *et al.*, 2005). The two types of EPNs, Heterorhabditidae and Steinernematidae (DeLay and Blaxter, 2002), are symbiotically associated with entomopathogenic bacteria (EPB) *Photorhabdus* and *Xenorhabdus*, respectively; a monophyletic group in the Enterobacteriaceae (Boemare, 2002). The infective juvenile (IJ) or dauer larve stage of the nematode transmits these bacterial symbionts and persists in soil in search of a susceptible insect host (see Erlandson and Theilmann, Chapter 1, this volume). Following entry through the cuticle or natural body openings, the IJs release the symbiotic bacteria into the insect haemocoel (Poinar and Thomas, 1966; Ciche and Ensign, 2003; Martens *et al.*, 2004; Sicard *et al.*, 2004; Snyder *et al.*, 2007) where the bacteria grow and kill the insect host within 24–48 h (Eleftherianos *et al.*, 2006a; Cowles *et al.*, 2007). Nematodes feed on symbiotic bacteria, complete one to three generations in the host cadaver, and as food resources are depleted new IJs are produced which disperse in search of new hosts (Poinar, 1990). The symbiotic bacteria interact with EPNs in at least two niches or states (Forst and Clarke, 2002): the phoretic state in which the bacteria are retained in, and interact with, the gut epithelium of the non-feeding IJ nematode (Boemare *et al.*, 1996) and the vegetative state in which the bacteria produce an arsenal of virulence factors ensuring rapid insect mortality (ffrench-Constant and Waterfield, 2006; Goodrich-Blair and Clarke, 2007). Bioconversion of the insect cadaver by bacterial exoenzymes allows the bacteria to multiply and nematodes to reproduce. During this phase the bacteria also produce secondary metabolites to inhibit invasion of the insect cadaver by competing soil microbes (Webster *et al.*, 2002), enabling the nematodes and bacteria to re-associate in a protected niche. Major events in the infection process of the EPNB symbiotic complex are listed in Table 11.1.

EPNs and their bacterial partners are technically tractable. Each of them can be readily cultivated, and their individual contributions to pathogenesis can be assessed separately from the complex. Furthermore, recent developments now set the stage for an even more detailed examination of the genetic and molecular mechanisms of interaction between entomopathogeic nematodes, symbiotic bacteria and the host. First, the genomes of three EPN bacterial symbionts have been fully sequenced: *Photorhabdus luminescens* ssp. *luminescens* TT01, *Xenorhabdus*

**Table 11.1.** Chronological events in the infection process of entomopathogenic nematode and bacteria complexes and attack and defence strategies used by the parasite/pathogen and the host, respectively.

| Infection event | Parasite/pathogen attack strategy | Host response |
|---|---|---|
| Host finding | Sensing host habitat, host volatiles or contact cues | Detection evasion |
| Host recognition | Sensing host contact or volatile cues | Detection evasion |
| Host penetration | Entry through natural openings or cuticle | Defensive and aggressive behaviour |
| Haemocoel penetration | Mechanical or enzymatic entry through gut wall or tracheae | Immune response activation against the nematode |
| Infective juvenile recovery | Evasion, tolerance or suppression of host immune response | Nematode encapsulation |
| Bacterial release | Production of toxins by the nematode<br>Inhibition of host's antibacterial proteins by the nematode | Immune response activation against the bacteria<br>Bacterial nodulation |
| Bacterial proliferation | Production of toxins and other virulence factors by the bacteria. Bacterial suppression of host cellular and humoral immune response<br>Nematodes feed on bacteria and host tissues and reproduce | |
| Bacteria–nematode re-association | Upon depletion of food resources nematodes produce next generation IJs which re-associate with symbiotic bacteria and exit the host cadaver in search of new hosts | |

*nematophila* (ATCC 19061) and *Xenorhabdus bovienii*, the bacterial symbionts of *Heterorhabditis bacteriophora* TTO1, *Steinernema carpocapsae* and *Steinernema jolli-eti*, respectively (Duchaud *et al.*, 2003; Goodrich-Blair *et al.*, in preparation) Second, the availability of the genome sequences of two related free-living nematodes, *Caenorhabditis elegans* and *Caenorhabditis briggsae*, together with a full complement of genetic and molecular tools (*C. elegans* genome sequencing consortium, 1998; Stein *et al.*, 2003) will facilitate identification of nematode genes involved in pathogenesis towards insects and symbiosis with bacteria. Gene orthology between *C. elegans* and *H. bacteriophora* has been confirmed, and data from the first complementary DNA (cDNA)-sequencing project of *H. bacteriophora* GPS11 have revealed the presence of several genes of interest in the cDNA library of the IJ stage (Sandhu *et al.*, 2006; Bai *et al.*, 2007). Forward genetics by mutagenesis using ethyl methane sulfonate (EMS) was successfully applied to obtain dumpy mutants (Zioni *et al.*, 1992) and a desiccation tolerant mutant in *H. bacteriophora* (O'Leary and Burnell, 1997). Moreover, techniques for genetic diversity assessment (Hashmi and Gaugler, 1998; Jagdale *et al.*, 2006), genetic selection (Gaugler *et al.*, 1989; Glazer *et al.*, 1991; Grewal *et al.*, 1996a,b; Segal and Glazer, 1998), hybridization (Shapiro *et al.*, 1997), subtractive amplification (Gal *et al.*, 2003; Bai and Grewal, 2007), proteomics (Gal *et al.*, 2005; Chen *et al.*, 2006) and DNA transformation

(Hashmi *et al.*, 1995) have been previously developed and proven. Transformation of the *H. bacteriophora* germ line with the *C. elegans* heat-shock promoter transcriptionally fused to beta-galactosidase (Hashmi *et al.*, 1995) and mechanosensitive (*mec-4*) promoter transcriptionally fused to green fluorescent protein (GFP) (Hashmi *et al.*, 1997) suggests that functional analysis of *H. bacteriophora* genes is possible. Reverse genetics, by gene silencing using RNA interference (RNAi), has also been shown in *H. bacteriophora* by Ciche and Sternberg (2007).

Several biochemical approaches have been used to analyse nematode or nematode–bacterium complex mechanisms of pathogenesis and the engineering of *Photorhabdus* and *Xenorhabdus* symbionts to express GFP has facilitated the monitoring of these bacteria both within the nematode and during infection (Ciche and Ensign, 2003; Martens *et al.*, 2003b; Sicard *et al.*, 2004; Ciche *et al.*, 2008). The virulence of the bacterial symbionts has also been investigated by directly injecting them (without their nematode host) into the insect. Insect immunity has been monitored and manipulated using dissection and microscopy to analyse cellular responses such as haemocyte aggregation and nodulation (Park and Kim, 2000; Cowles *et al.*, 2007; Park *et al.*, 2007), Northern blots and quantitative PCR to measure the abundance of individual immune gene transcripts (Ji and Kim, 2004; Park *et al.*, 2007), application of pharmaceuticals to suppress or activate specific branches of immunity (Park *et al.*, 2003) and RNAi to suppress expression of individual genes (Eleftherianos *et al.*, 2006a,b). The use of such techniques has led to the understanding that both the nematode and the bacterium harbour mechanisms to evade, tolerate and suppress both cellular and humoral aspects of insect immunity. However, it should be emphasized that specific details of the pathogen–immune interface vary depending on genetic and environmental parameters, and that there can be significant differences in the infection process depending on the species, and even strains of nematodes, bacteria and insects used in the experimental design (Li *et al.*, 2007).

## 11.2. Molecular Tools to Study Entomopathogenic Nematodes

### 11.2.1. Identification of genes based on homology

The availability of the *C. elegans* genome and the broad knowledge of its developmental and behavioural biology provide an outstanding resource for directed genetic analyses in EPNs (Grewal *et al.*, 2006). For example, knowledge of chemosensory pathways in *C. elegans* is being used to understand host finding by EPNs. Chemoreception is the primary host-finding cue for the infective stages of plant- and animal-parasitic nematodes which can sense aliphatic and aromatic compounds with diverse functional groups (Bargmann and Horvitz, 1991). Olfactory signal transduction is mediated by G-protein-coupled transmembrane receptors (Prasad and Reed, 1999) with seven-transmembrane (7-TM) spanning topology. The *C. elegans* genome may encode ~550 functional chemoreceptor genes and ~250 pseudogenes, which together represent ~6% of the genome (Robertson, 2001). The downstream effectors of the 7-TM chemoreceptors are heterotrimeric G-proteins, comprised of $\alpha$, $\beta$ and $\gamma$ subunits, each subunit encoded by a different gene. There are *21G$\alpha$, 2G$\beta$* and *2G$\gamma$* genes in *C. elegans* (Jansen *et al.*, 1999). The activated 7-TM receptor catalyses the exchange of guanosine diphosphate (GDP) for guanosine triphosphate (GTP) at a specific binding site on the Gα, protein. The Gα protein (with its bound GTP) diffuses from the receptor G-protein complex and interacts with ODR-1, a guanylyl cyclase which catalyses the conversion of GTP to cyclic guanosine monophosphate (cGMP). cGMP is released from the Gα protein and activates a cation channel consisting of two subunits: TAX-2 and TAX-4 (O'Halloran and Burnell, 2003). Thus, the coupling of chemoreceptor activation to electrical activity in sensory neurons of *C. elegans* is mediated via cGMP.

Using degenerate PCR primers, nine candidate G-protein α subunit gene fragments have been cloned from *H. bacteriophora* (O'Halloran *et al.*, 2006). Degenerate primers were designed based on conserved motifs identified from multiple alignments of Gα amino acid sequences from a variety of organisms. Some of the more divergent G-protein α subunit genes from *H. bacteriophora* were cloned by designing PCR primers based on alignments of homologues of these genes from *C. elegans* and *C. briggsae*. Analyses of these sequences and extensive comparisons with G-protein α subunit sequences in all available databases indicate that nematodes have evolved a unique cluster of orthologous G-protein α subunit genes soon after their divergence from other metazoans (O'Halloran *et al.*, 2006). The very large number of G-protein-coupled receptors in the nematode genome, coupled with a nematode-specific expansion of the G-protein α subunit genes, indicates that there exists in nematodes an increased capacity for integrating chemosensory information at single neurons. By contrast, in mammals this level of integration occurs not in neurons but in higher brain centres. This information allows the use of homology-based approaches to identify genes involved in chemosensory responses in EPNs. Indeed, degenerate primers designed from alignments of candidate G-protein-coupled receptors from *C. elegans* have also been used to isolate a putative G-protein-coupled gene from *H. bacteriophora* and the *H. bacteriophora* homologue of the *odr-3* gene. The 5' and 3' rapid amplification of cDNA ends (RACE) has been employed to obtain the full-length coding sequences from these genes (D.M. O'Halloran and A.M. Burnell, Maynooth, 2007, personal communication).

### 11.2.2. Identification of differentially expressed genes

The suppression subtractive hybridization (SSH) procedure originally described by Diatchenko *et al.* (1996) can be used to identify genes differentially expressed by IJs during different phases of the infection process or in different states of interaction with the bacterial symbiont. In this technique, transcripts that are enriched in one condition relative to another are selectively amplified. The SSH technique can be applied to any two conditions of interest to the investigator. For example, it was used to identify *Steinernema feltiae* genes differentially expressed in response to desiccation (Gal *et al.*, 2003) and *H. bacteriophora* genes down-regulated in IJs during contact with host haemolymph (Bai and Grewal, 2007).

Nematode stock populations can be obtained by propagation in last instar *Galleria mellonella* larvae or other suitable target insects following the methods described by Kaya and Stock (1997). The infected host larvae are transferred to White traps and the emerged IJs are collected, washed and stored at 10°C until used (Kaya and Stock, 1997). To ensure viability, populations should be used within ~1 month.

Total RNA can be isolated from each of the two conditions to be tested, for example, uncolonized versus colonized IJs or IJs exposed to water versus haemolymph. Total RNA can be isolated using Trizol (Invitrogen, Carlsbad, California) or the ToTALLY RNA kit (Ambion, Austin, Texas) and messenger RNA (mRNA) can be selectively extracted from total RNA using Qiagen Oligotex. Any genomic DNA contamination is removed by LiCl precipitation or by deoxyribonuclease (DNase) treatment.

- Measure the A260/A280 ratio to assess RNA purity and use agarose electrophoresis to evaluate the integrity of the RNA samples.
- Perform cDNA generation and subtractive amplification method by using commercially available kits that use *s*witching *m*echanism *a*t 5' end of *R*NA *t*emplate (SMART) (Zhu *et al.*, 2001) and SSH (Diatchenko *et al.*, 1996) approaches (e.g. BD Biosciences-Clontech; Super SMART kit and PCR-Select cDNA subtraction kit).
- Apply a mirror orientation selection (MOS) procedure (Rebrikov *et al.*, 2000) to eliminate background noise and enhance the certainty of capturing differentially expressed target cDNA.

The reader is also referred to a recent chapter describing the SSH method (Lukyanov *et al.*, 2007).

- Ligate the resulting subtracted cDNA into an appropriate vector and transform the ligation products into *Escherichia coli* cells, resulting in two subtracted libraries (one per direction).
- Plate the clones in each library in 96-well plates, and screen to determine the percentage of differentially expressed clones in each subtracted library.
- Subsequently, isolate the plasmid DNA and purify from the differential clones and sequence.
- Confirm differential expression by Northern hybridization or quantitative PCR (see below).
- Annotate and analyse the resulting sequences using standard programs that are publicly available (http://www.ncbi.nlm.nih.gov/Tools/).

### 11.2.3. Examination of tissue-specific gene expression using fluorescent *in situ* hybridization (FISH)

SSH techniques can identify transcripts that are more abundant in one condition relative to another, which suggests these transcripts may serve a specific function under that condition. Such a hypothesis would be further supported by demonstrating that the gene of interest is expressed in relevant tissues, as detected by fluorescent *in situ* hybridization (FISH) (Kimbell *et al.*, 2006). In FISH, sense and antisense digoxigenin (DIG)-labelled transcript probes are generated *in vitro* and hybridized to fixed nematode tissue (B. Wimpee and M.J. McFall-Ngai, Wisconsin, 2007, personal communication).

Preparing probes for hybridization:

- To prepare the probes, polyacrylamide gel electrophoresis-purified primers specific to the gene target are generated such that the T3 (5'-AATTAACCCT CACTAAAGGG-3') and T7 (5'-TAATACGACTCACTATAGGG-3') promoter sequences are included at the 5' end of each respectively.

- Use the primers in PCR amplification with the cloned cDNA of interest as the template and Accuprime Pfx (Invitrogen) with a three-stage cycling protocol: after the initial 2 min melting step at 94°C, is five cycles of 94°C for 30 s, 50°C for 30 s and 68°C for 1 min, followed by five then 20 cycles with the annealing temperature at 55°C and 60°C, respectively.
- Verify the product by agarose gel electrophoresis, quantify and purify (e.g. using Qiagen's PCR clean up kit).
- Use ~100–200 ng of purified product as template for *in vitro* transcriptions with the MEGAscript High Yield Transcription kit (Ambion #1334 T7; #1338 T3). Split the template into two tubes, one for T7 transcription and one for T3 transcription, to generate the sense and antisense probes.
- Carry out reactions according to manufacturers' instructions, except reduce the volume of ribonucleotide uridine triphosphate (UTP) to 71% of the other ribonucleoside triphosphates (NTPs), and add DIG-labelled UTP (Boehringer-Mannheim) to 1.46 mM final concentration.
- Hybridize the transcription products to fixed nematode tissue.

Fixing nematode tissues:

- Where appropriate exsheath nematodes, for example by incubation in 5% sodium hypochlorite for 8 min, from which they are removed and washed with sterile buffer as soon as the sheath is removed. Alternatively, for steinernematid IJs, the intestine (and bacterial colonization site) can be extruded by chopping with a razor blade (Vivas and Goodrich-Blair, 2001).
- Fix exsheathed nematodes or extruded intestines in 4% paraformaldehyde overnight.
- Wash tissue (5 × 10 min) in phosphate-buffered saline with 0.1% Tween PBS-T.
- Dehydrate tissue for 5 min in 75%, 50% and 25% methanol (diluted with PBS-T).
- Incubate the samples twice in 100% methanol, 5 min each.
- Rehydrate the sample by incubating in 60% and 30% methanol in PBS-T.
- Rinse 4 × 5 min rinses in PBS-T.

Hybridizing probes:

- Prehybridize fixed nematode tissue overnight at 60°C in hybridization buffer (50% formamide; 5X saline sodium citrate (SSC); 50 μg heparin (Sigma); 0.1% Tween; 1% sodium dodecyl sulfate (SDS); 50 μg salmon sperm DNA (Sigma)).
- Add the denatured probe to the tissue in new hybridization buffer and incubate without agitation.
- Remove probe after 1–2 days.
- Wash the sample in hybridization buffer for 20 min at 60°C.
- Rinse the samples 20 min, 60°C in 2X SSC/hybridization buffer at ratios of 25:75, 50:50, 75:25 and 100:0 in order.
- Wash in 0.05 × SSC at 60°C.
- Rinse the samples 10 min at room temperature (with agitation) in 0.05 × SSC/PBS-T at ratios of 75:25; 50:50, 25:75 and 0:100 in order.

Detection of DIG-labelled probe in the tissues:

- Treat samples with 10 μg/ml RNase for 1 h at 37°C.
- Wash samples with phosphate buffer with Tween (PBT) (0.1% PBS; 0.2% Triton-X100; 0.1% bovine serum albumin (BSA)) 5 × 5 min.
- Incubate 1 h at room temperature in blocking agent (Boehringer-Mannheim).
- Add the primary antibody (mouse anti-DIG; Roche) and incubate overnight at 4°C.
- Wash the tissue 5 × 10 min in PBT at 4°C.
- Block again for 1 h.
- Add the second antibody (goat anti-mouse fluorescein isothiocyanate (FITC) ). To stain actin, alexaflour 546 phalloidin can also be added at this stage.
- Wash again in PBT (4 × 30 min) and 1 × PBS (2 × 15 min).
- Mount hybridized samples on a glass slide, and visualize using fluorescence (or confocal) microscopy fitted for FITC and Cy3 detection.

### 11.2.4. Reverse genetics in EPNs using RNAi

Targeted gene silencing by RNAi is a powerful reverse genetic tool to elucidate gene function in metazoans. The discovery of RNAi in 1998 using the nematode *C. elegans* led to the award of the Nobel Prize in Medicine to Fire and Mellow in 2006 (Fire *et al.*, 1998). RNAi was originally performed by injecting double-stranded RNA (dsRNA) into the body of L4 animals (Guo and Kemphues, 1995; Fire *et al.*, 1998) and has also been shown to be effective by soaking (Tabara *et al.*, 1998), feeding nematodes *Escherichia Coli* bacteria expressing dsRNA (Timmons *et al.*, 2001) or by expressing dsRNA in *C. elegans* cells (Tavernarakis *et al.*, 2000). Genome-wide screens using RNAi have been performed, primarily by feeding *E. coli* expressing gene-specific dsRNAs to *C. elegans* (Kamath and Ahringer, 2003), although these have also been performed using soaking (Maeda and Sugimoto, 2001) and injection (Sonnichsen *et al.*, 2005) methodologies. The dependence of *H. bacteriophora* for symbiotic bacteria for growth and reproduction makes RNAi by feeding problematic. Both RNAi by feeding and soaking depend on the efficient uptake of environmental dsRNA (Winston *et al.*, 2002). However, many nematodes closely related to *C. elegans* are not amenable to RNAi using environmental dsRNA, but are when dsRNA is delivered by microinjection (Winston *et al.*, 2002). Fortunately, in *H. bacteriophora* TT01 RNAi by soaking resulted in potent and specific gene silencing (Ciche and Sternberg, 2007).

Critical for successful RNAi in *H. bacteriophora* is the preparation and incubation of eggs and L1 larvae. Using monoxenic or axenic (symbiont-free) IJs are strongly recommended since contaminants can cause adult or L1 mortality.

Preparation of *H. bacteriophora* nematodes for RNAi:

- Remove contaminants from eggs by surface sterilization using a 5 min treatment in 1% Chlorox bleach followed by three washes in Ringer's solution (100 mM NaCl, 1.8 mM KCl, 2 mM $CaCl_2$, 1 mM $MgCl_2$, 5 mM 4-(2-hydroxyethyl)-1-piperazineethanesulfonic acid (HEPES) pH 6.9).

- Inoculate liquid media (e.g. 2% Proteose Peptone #3 (Difico, Detoit, Michigan) + 0.5% NaCl) with symbiotic bacteria and incubate for 18–24 h at 28°C while shaking at 150 rpm.
- Spread *c.* 50 µl of broth culture on to Nutrient Agar (Difco) supplemented with 10 µg/ml cholesterol (from a 5 mg/ml cholesterol/ethanol) and incubate for 18–24 h at 28°C.
- To improve IJ recovery (i.e. exit from diapause and resumption of development), wash ~10,000 IJs 3× in 15 ml Ringer's in conical tubes, centrifuging for 1 min using a clinical centrifuge set at speed 4 or 5.
- Resuspend the material in 500–1000 µl of Ringer's and add 100–150 IJs to each symbiont lawn.
- Harvest eggs 82–86 h after the addition of IJs to the symbiont lawns where most of the eggs should be at the pretzel (i.e. where the head crosses the tail) stage of development and some J1s are present.
- Wash eggs and hermaphrodites off the lawns with Ringer's by repeated pipetting or by scraping off the nematodes and the bacterial lawn using a flame-sterilized scraper (e.g. a sealed and bent Pasteur pipette).
- Remove bacteria by vacuum filtration using a 10 µm mesh (application of too strong vacuum causes the eggs/L1s to become stuck to the membrane) followed by three washes with Ringer's.
- Separate eggs/L1s from the hermaphrodites by applying the nematodes and eggs to a 36 µm mesh nylon filter suspended in Ringer's where eggs/J1s pass through the filter.
- Concentrate the eggs/L1s by centrifugation in a 15 ml conical tube for 1 min in a clinical centrifuge set at speed 4 or 5 and resuspend to a concentration of 5–25 eggs/L1 per µl.
- Add 5 µl of 5–7.5 mg/ml of dsRNA to 10 µl of nematodes.
- Soak the nematodes for at least 24 h at 28°C without shaking after which the nematodes are added to a lawn of symbiotic bacteria and screened.

Genes to be silenced can be identified by a variety of means (see above) and recently this has been facilitated by the deposition of more than 13,000 *H. bacteriophora* TT01 expressed sequence tags (ESTs) present on National Center for Biotechnology Information (NCBI)'s ENTREZ EST database (http://www.ncbi.nlm.nih.gov/sites/entrez). These ESTs were generated as part of the *H. bacteriophora* TT01 genome-sequencing project, currently targeted for production at Washington University Genome Sciences Center in January 2008 (J. Spieth, St Louis, 2007, personal communication). This strain was chosen because the species is widely distributed and is the host of *P. luminescens* ssp. *laumondii* strain TT01 whose genome has been completely sequenced (Duchaud *et al.*, 2003). More specifically an inbred strain, M31e, of this nematode, which was self-fertilized for 13 generations (Hallem *et al.*, 2007) is being sequenced.

Generating dsRNA for soaking:

- PCR amplify 300–500 bp (smaller fragments such as 150 work sometimes with less efficiency) of the gene to be silenced. Use primer sequences (usually 18–20 bp) specific to the gene of interest with the T7 promoter DNA (5'-taatacgactcactatagggaga-3') added to the 5' ends of each and genomic DNA (100 ng) or cDNA as template. cDNA is preferred as it typically lacks introns.

- After PCR amplification, analyse the products for appropriate size by agarose gel electrophoresis.
- Remove primers by ExoSAP-IT treatment according to the manufacturer's instructions (USB, Cleveland, Ohio).
- Use 5 μl of the PCR reaction directly for *in vitro* transcription using the Megascribe T7 kit (Ambion, Austin, Texas) or T7 RiboMax (Promega) according to the instructions provided, except incubate the transcription reactions for >6 h at 37°C.
- Remove the DNA templates by DNase treatment.
- Precipitate the dsRNA by adding 1/10 volume of 5 M ammonium acetate and 2.5 volume of 100% ethanol and incubating for >1 h at 4°C.
- Centrifuge the precipitated dsRNA for 30 min at 16,000 × g.
- Wash the sample with 70% ethanol prepared in RNase-free water.
- Air-dry the sample for 5 min and dissolve the pellet in 25 μl of RNase-free water.
- Determine the quality of the transcribed RNA by running 1 μl on a 1.2% agarose gel and quantifying ($A_{260}$) using a NanoDrop (Nanodrop Technologies, Wilmington, Delaware).

A variety of postembryonic phenotypes can be scored by direct comparison with an unsilenced control (e.g. a dsRNA not resulting in a phenotype, like dsGFP). In the proof of principle experiments, genes were silenced that led to sterility of the adult nematodes (Ciche and Sternberg, 2007). Silencing was evident by the very transparent nature of the adult nematodes due to the absence of a gonad that normally occupies the majority of the maternal body cavity. Quantitative real-time PCR (RT-PCR) was also used to demonstrate severe and specific decreases in the mRNA corresponding to the gene being silenced. The penetrance of the mutant phenotypes was comparable to RNAi by feeding of orthologues in *C. elegans*. These results suggest that this strain of EPN is amenable to RNAi using environmental DNA, which should greatly facilitate the study of gene function, especially related to symbiosis and parasitism.

## 11.3. Basic Molecular Tools for the Study of Entomopathogenic Bacteria

*Photorhabdus* and *Xenorhabdus* are amenable to genetic manipulation and currently *X. nematophila* and *P. luminescens* are the best studied at the genetic level. Site-directed genomic mutations can be constructed using Campbell integration or insertion deletion methods (Vivas and Goodrich-Blair, 2001; Orchard and Goodrich-Blair, 2004; Cowles and Goodrich-Blair, 2005; Park and Forst, 2006; Brachmann *et al.*, 2007) and various transposons have been effectively used for random mutagenesis screens (Heungens *et al.*, 2002; Joyce and Clarke, 2003; Martens *et al.*, 2003a; Williams *et al.*, 2005). Once constructed, mutants have been tested for virulence and mutualism phenotypes. *X. nematophila* gene expression has been monitored and quantified using the classic method of fusing the gene or promoter of interest to a promoter-less *lacZ* gene, and measuring beta-galactosidase activity (Cowles and Goodrich-Blair, 2004, 2006). Similar fusions to the GFP, or to genes encoding luciferase, may allow monitoring of gene expression during host interactions, although such an approach has not been reported. Northern blots and quantitative RT-PCR (Q-RT-PCR) have also been valuable techniques for monitoring expression of both bacterial and insect genes (Joyce and Clarke, 2003; Martens *et al.*, 2003b; Cowles and Goodrich-Blair, 2005; Eleftherianos *et al.*, 2006a,b; 2007; Park *et al.*, 2007). While such techniques will continue to be useful, particularly as they are expanded for use in other species of *Xenorhabdus* and *Photorhabdus*, the sequenced genomes of *P. luminescens* and *X. nematophila* pave the way for microarray analyses that will dramatically expand our knowledge of effectors involved in mutualism and pathogenesis. The following sections describe microbiological and molecular techniques used in the study of *Photorhabdus* and *Xenorhabdus* with their hosts. In many cases, distinct techniques have been developed for each genus of bacterium, but wherever possible common techniques are presented.

### 11.3.1. General considerations when working with *P. luminescens* and *X. nematophila*

Growth considerations:

- For all microbiological manipulations of *Photorhabdus* and *Xenorhabdus* care must be taken to store growth media in the dark or supplement it with 0.1% (w/v) final concentration of pyruvate (Xu and Hurlbert, 1990).
- Generally antibiotic concentrations used for plasmid selection are: ampicillin (Amp), 100–150 μg/ml; chloramphenicol (Cm), 15 μg/ml; erythromycin (Erm), 200 μg/ml; streptomycin (Str), 12.5 μg/ml; kanamycin (kan), 20–25 μg/ml.
- Although most work with *Photorhabdus* and *Xenorhabdus* has been conducted on the rich, undefined medium Luria-Bertani broth (LB) or similar (Miller, 1972), a defined solid medium that does support the growth of *X. nematophila* (22 mM $KH_2PO_4$, 40.2 mM $K_2HPO_4$, 15.1 mM $(NH_4)_2SO_4$, 0.41 mM nicotinic acid, 9.1 mM Na-pyruvate, 50 mM glucose, 10 ml/l SL4 salts (Atlas, 1997), 50 mg/l each amino acids and 15 g/l agar) has been reported (Orchard and Goodrich-Blair, 2004).

Long-term storage of bacterial strains at −80°C:

- Grow the bacteria overnight at 30°C in LB broth.
- Transfer a sample of the overnight culture to a sterile NUNC cryovial.
- Gently mix with sterile glycerol to 20% (v/v).
- Place vials in a −80°C freezer.
- Retrieve bacteria from the frozen stock, by scraping a visible chunk of frozen material from the tube using sterile sticks or a flamed hot loop and transferring this material immediately to liquid or solid medium. Although dependent on the inoculum size, such cultures should grow to turbidity within 24 h. However, if antibiotics are used for selection culture growth can take longer.

### 11.3.2. Genomic DNA Extraction

- Grow bacteria overnight in 3 ml LB.
- Centrifuge the culture at 4000 rpm for 10 min to collect the cells.
- Resuspend the cell pellet in 567 µl transposable element (TE) buffer by repeated pipetting.
- Transfer suspension to an Eppendorf tube.
- Add 30 µl of 10% SDS, 3 µl of 20 mg/ml proteinase K and 3 µl RNase.
- Mix the solution and incubate for 1 h in a water bath at 37°C.
- Add 100 µl of 5 M NaCl and 80 µl of preheated (65°C) cetyl trimethylammonium bromide (CTAB)/NaCl and mix thoroughly.
- Incubate at 65°C for 5 min, shake and incubate for a further 5 min at 65°C.
- Add an approximately equal volume (0.7–0.8 ml) of chloroform/isoamylalcohol and mix.
- Separate aqueous and organic phases by centrifugation in a microcentrifuge at 10,000 rpm for 5 min.
- Transfer the aqueous, viscous supernatant (containing the DNA) to a fresh microcentrifuge tube, leaving the white interface behind.
- Add an equal volume of phenol/chloroform/isoamyl alcohol to the aqueous phase and mix the contents thoroughly.
- Spin the tube in a microcentrifuge at 10,000 rpm for 5 min.
- Transfer the aqueous supernatant to a fresh tube and add 0.6 volume isopropanol.
- Incubate the sample at −80°C for at least 30 min to precipitate the nucleic acids.
- Spin the tube in a microcentrifuge at 13,200 rpm for 10 min to pellet the DNA.
- Discard the supernatant and wash the pellet in ice-cold 70% ethanol.
- Spin sample as above and discard the supernatant.
- Air-dry the pellet.
- Add 100 µl sterile nuclease-free water.

### 11.3.3. Production of allele-specific deletions in *Photorhabdus* and *Xenorhabdus*

Unmarked deletion mutants can be constructed using a protocol involving three PCR steps followed by cloning into a suicide vector (e.g. pDS132), conjugation into *Photorhabdus* or *Xenorhabdus* and screening (colony PCR) for the mutation (Brachmann *et al.*, 2007). The advantages of this system over traditional methods of knockout mutagenesis include: (i) the mutation constructed will be non-polar; (ii) there will be no fitness cost associated with the mutation as the expression of an antibiotic resistance gene is not required; and (iii) multiple deletions can be constructed in the same strain (especially useful in determining the functions of genes present in several copies). This technique has been used in *Photorhabdus* to delete chromosomal fragments ranging in size from 60 bp to 10 kb as well as the construction of a triple gene deletion strain (D.J. Clarke, unpublished data).

Construction of a mutant, from initial PCR to final confirmation, takes 2–3 weeks and it is possible to construct a number of mutants in parallel. Each open reading frame (ORF) requires six oligonucleotide primers in order to construct and verify the mutation (Fig. 11.1). Two pairs of primers (UPFwd and UPRev; DNFwd and DNRev (average length = 50-mer)) are used to amplify a region of genomic DNA upstream (UP) and downstream (DN) from the ORF to be deleted (Fig. 11.1). In *Photorhabdus* these fragments can range from 300 to 600 bp, but in *Xenorhabdus* at least 1 kb is optimal.

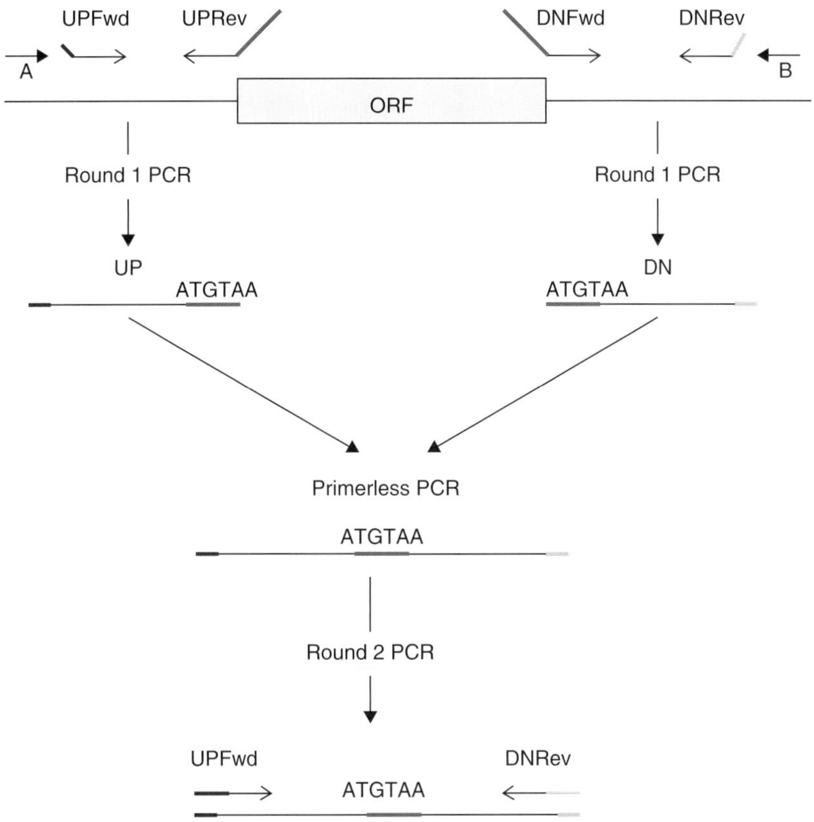


**Fig. 11.1.** Strategy for the construction of a marker-less deletion mutants in *Photorhabdus*; as outlined in the text, six oligonucleotide primers are required for the construction and verification of a marker-less deletion mutant in *Photorhabdus*. The primers UPFwd and DNRev contain restriction enzyme sites at their 5'ends (indicated by blue lines) to facilitate cloning into the suicide vector, pDS132. The primers UPRev and DNFwd overlap the start and stop codon of the ORF to be deleted and the 3'-end of UPRev is identical to the 5'-end of DNFwd. Therefore, the UP and DN PCR products will have complementary 3' and 5'-ends, respectively (indicated by red lines), and this region of complementarity must extend for at least 15 nucleotides to facilitate self-priming during the primerless PCR step. Not shown are the outlying primers used to screen for mutants.

Primers UPFwd and DNRev contain sites for *Sph*I, *Sac*I or *Pst*I at their 5' ends to facilitate cloning into the suicide vector. The primers UPRev and DNFwd overlap the start and stop codon, respectively, of the ORF to be deleted. In addition, the 3' end of UPRev is identical to the 5' end of DNFwd so that the UP and DN PCR products will have complementary 3' and 5' ends, respectively, and this region of complementarity should extend for at least 15 nt. Additional primers, designed to flank UPFwd and DNRev, are used to amplify the locus after mutagenesis to verify the presence of the deletion. It is essential that these primers are upstream and downstream, respectively, from UPFwd and DNRev in order to prevent false positives from, for example, illegitimately recombined plasmid during the screening of potential mutant colonies. All primers should be selected to have a guanine cytosine (GC) content of approximately 50% (the average GC content of the *Photorhabdus* genome in 42.8%) with at least a single G or C at the 3' end.

- Set up PCR reactions using KOD polymerase and standard amplification conditions; 0.2 mM deoxyribonucleotide triphosphates (dNTPs), 1 mM $MgCl_2$, 20 pmoles each primer and 1 U KOD DNA polymerase in a 50 µl reaction volume.
- Cycling conditions are also standard: 95°C for 5 min, 30 cycles of 72°C for 1 min; 55°C for 1 min; 72°C for1 min followed by a 10 min run at 72°C. (Note: In our experience >95% of PCR reactions have given a positive result using these standard reaction and cycling conditions.)
- Clean PCR products to remove primers and excess dNTPs and 1 µl of the UP and DN PCR products are mixed and used as a template for primerless PCR. (Note: In this reaction do not add new primers since UP and DN PCR products, through their small regions of complementarity, serve to prime the synthesis of each other resulting in the production of a chimeric DNA fragment (Fig. 11.1). It is important that the same amount of DNA is added from the UP and DN reactions and this can be confirmed by gel electrophoresis.)
- After ten rounds of amplification (see reaction and cycling conditions above) use 1 µl of the primerless PCR reaction as the template in PCR2 to which the UPFwd and DNRev primers are added to amplify the full-length product (using the same reaction and cycling conditions as described above).
- Following PCR2, clean and analyse the quality of the PCR products by agarose gel electrophoresis.
- Digest full-length PCR product with the appropriate enzymes and ligate into pDS132 (or other suicide vector) digested with the same enzymes. If these restriction sites are present in the PCR product, other sites are available for subcloning, or other vectors are available, and this should be assessed during primer design. (Note: Plasmid pDS132 is recommended because it has been optimized for allelic replacement work in Gram-negative bacteria and contains the gene for chloramphenicol resistance (Phillipe *et al.*, 2004).)
- Transform the ligated DNA into electrocompetent *E. coli* S17–1 (λ*pir*), a strain containing the λ*pir* gene (required for replication of suicide vectors such as pDS132 and pKNG101) and the *mob* genes (required for mobilization of pDS132).
- Identify the bacteria carrying the correct construct by colony PCR.

- Verify the correct construct by preparing a mini-prep of the plasmid DNA and digesting with restriction enzymes.
- Store the bacteria containing the plasmid by freezing at −80°C. This bank of plasmid clones is a valuable resource as they can be used to construct multiple deletions in the same strain (by sequential conjugation) thus facilitating the functional analysis of genes with redundant functions.

In some cases, it may be useful to use antibiotic resistance cassettes to generate mutants:

- Similar to the process described above, amplify separate fragments of the flanking DNA upstream and downstream of the region to be deleted. The primers used for amplification of the upstream and downstream fragments are engineered to include restriction sites. For example, the 5' primer of the upstream fragment incorporates a *Kpn*I site and the 3' primer of the upstream fragment contains a *Bam*HI site, while the 5' primer of the downstream fragment has a *Bam*HI site and the 3' primer has a *Sac*I site.
- After amplification, digest the two fragments with *Bam*HI, ligate together and use as template in another PCR amplification using the outermost primers. The resulting product has a *Bam*HI site at the site of the deletion, an upstream *Kpn*I site and a downstream *Sac*I site.
- Clone fragments into a general cloning vector such as pBC using the restriction sites of the outermost primers.
- Insert an antibiotic cassette into the *Bam*HI site formed by the ligation of the two fragments.
- Subclone the entire fragment using the *Kpn*I and *Sac*I sites into a suicide vector (e.g. pKR100 K. Visick, Loyola University, Chicago) and use the antibiotic cassette for selection.

Another method used to create targeted insertion mutations is the GeneJumper™ kit (Invitrogen, Carlsbad, California) (Orchard and Goodrich-Blair, 2004). This method allows isolation of plasmids with randomly inserted transposon cassettes within a cloned gene. The site of insertion is sequenced, and selected constructs are subcloned into suicide vectors based on the antibiotic cassette carried by the transposon.

### 11.3.4. Insertion of DNA in single copy

Verification of causality of the mutation to observed phenotypic differences is accomplished by providing the mutated gene(s) *in trans*, either on a multi-copy plasmid or in a single copy in the chromosome. The latter is ideal, since it is more likely to mimic the wild-type expression levels of the genes being tested. A particularly useful tool that has facilitated introduction of DNA into the chromosome is the Tn*7* transposon, which inserts into the genome at a single locus, *att*Tn*7* (Waddell and Craig, 1989). Control strains, in which an 'empty' Tn*7*, lacking bacterial genetic material, are used for comparison, although insertions at *att*Tn*7* appear to have negligible effects on *X. nematophila* host interactions

(Martens *et al.*, 2003b; Cowles *et al.*, 2007). Care should be taken to incorporate the endogenous promoter region of any gene cloned into the Tn*7* site. The Tn*7* plasmid vector used thus far for *X. nematophila* is pEVS107 (Stabb and Ruby, 2002). The Tn*7* system is functional in *Photorhabdus* and has been used to insert a copy of the *gfp* gene on to the TT01 chromosome and for the *in trans* complementation of various mutant alleles (Hallem *et al.*, 2007; R.J. Watson and D.J. Clarke, unpublished data).

### 11.3.5. Electroporation of *Photorhabdus luminescens* TT01

- Inoculate 100 ml of LB with 200 µl overnight *Photorhabdus* culture.
- Grow cells to early exponential phase ($OD_{600}$ 0.2–0.3).
- Place the culture on ice for 90 min.
- Harvest cells by centrifugation at 4°C.
- Discard the supernatant and resuspend the pellet in 100 ml ice-cold 5% sucrose, 1 mM HEPES.
- Re-harvest the cells by centrifugation and sequentially resuspend pellet in 50 ml then 1.6 ml and finally in 160 µl ice-cold 5% sucrose, 1 mM HEPES (Bennett and Clarke, 2005).

Note: We have noticed that electrocompetent *Photorhabdus* does not store well at −80°C and therefore cells must be prepared fresh. To date there are no reports of successful electroporation of *X. nematophila*.

### 11.3.6. Conjugation of *Photorhabdus*

- Grow rifampicin-resistant *P. luminescens* TT01 (the recipient) and *E. coli* carrying the donor plasmid (pUT-Km2 for Tn5 mutagenesis or pDS132 for specific gene deletion) in LB broth overnight.
- Inoculate 1 ml of overnight cultures into 50 ml LB in a 250 ml conical flask. For the recipient *Photorhabdus* culture supplement the LB with 1 mM $MgCl_2$.
- Grow *Photorhabdus* culture at 28°C to an $OD_{600}$ of 0.5. (This should take 4–5 h.) Grow the *E. coli* culture at either 30°C or 37°C to an $OD_{600}$ of 0.5. (This should take 2–3 h.)
- Spin 4 ml of *Photorhabdus* culture, wash twice and resuspend in 200 µl LB ($MgCl_2$). Spin 1 ml of *E. coli* culture, wash twice in LB ($MgCl_2$) and resuspend in 100 µl of LB ($MgCl_2$).
- Mix *P. luminescens* and *E. coli* cells in a sterile Eppendorf and pipette the resulting 300 µl on to an LB ($MgCl_2$) agar plate containing no antibiotics.
- Incubate undisturbed overnight at 25°C.

Note: All solid media is supplemented with 0.1% (w/v) pyruvate.

- Recover the exconjugants by pipetting 1 ml of LB on to surface of the agar plate and carefully resuspending cells with a sterile spreader.

- Transfer resuspended cells to a sterile 14 ml tube. Repeat three times.
- Plate 100 μl of the final cell suspension on to LB ($MgCl_2$) plates containing rifampicin and chloramphenicol (for pDS132) or rifampicin and kanamycin (for pUT-Km2) to determine the conjugation efficiency.
- Store the mutant library at −80°C by adding glycerol to a final concentration of 20% (v/v).

Note: We have observed a significant reduction in the number of exconjugants recovered from frozen libraries compared to non-frozen libraries.

### 11.3.7. Conjugation of *Xenorhabdus*

- Grow overnight cultures of a donor strain of *E. coli* (e.g. S17–1 λpir or SM10;) carrying the vector to be conjugated and the recipient strain of *X. nematophila*.
- Dilute cultures 100-fold into 2 ml of fresh media with the appropriate antibiotics and grow for 3 h at 30°C.
- Pellet the entire culture for each strain in a 2 ml Eppendorf tube.
- Wash once with 2 ml LB medium and resuspend in 1 ml LB.
- In a fresh tube, mix equal volumes of each culture and plate 100 μl of mixture on to LB plates containing 0.1% pyruvate.
- Incubate plates for 1 day at 30°C then resuspend and scrape cells into 5 ml LB.
- From this suspension spread a 100 μl sample on to selective medium: selecting for the presence of the marker being transferred, while counter-selecting against the *E. coli* donor strain. Although a rifampicin-resistant strain of *X. nematophila* AN6/1 (isolated by S. Forst, available as HGB081 from H. Goodrich-Blair) has been used for conjugations, counter selection against the *E. coli* donor can also be achieved using 150 μg/ml ampicillin, to which *X. nematophila* is naturally resistant, or by growing the selection plates at room temperature, a condition under which *X. nematophila* colonies grow faster than *E. coli*. Ampicillin selection is only successful if the *E. coli* donor strain does not carry an ampicillin-resistance cassette (*bla* gene).
- Verify exconjugant cells as *X. nematophila* by their negative catalase reaction when hydrogen peroxide is applied and by their characteristic smell (which you just have to experience for yourself).

### 11.3.8. Triparental matings

For delivery of pEVS107 (Tn*7* vector) to both *Photorhabdus* and *Xenorhabdus*, a triparental mating is necessary. In this process an equal volume of an additional ‘helper’ strain carrying pUX-BF13 (Bao *et al.*, 1991) (necessary for Tn*7* transposition) is mixed with the donor and recipient (after processing as described above).

### 11.3.9. Screening for mutant strains

Suicide vectors such as pKR100, pKNG101, pEVS107 and pDS132 will not replicate in *Photorhabdus* or *Xenorhabdus*. Therefore, selection for the antibiotic cassette of the plasmid selects for those cells that have recombined the plasmid into their genome by homologous recombination.

- When using pKNG101 or pDS132 (which encode the *sacB* gene) select some number (~5) of exconjugant colonies that have recombined the plasmid into the chromosome and grow them overnight in LB broth (without selection to allow excision of the plasmid in some cells).
- Spread dilutions on to LB agar containing 0.2% (w/v) sucrose. The *sacB* gene encodes a protein (levansucrase) that is toxic to Gram-negative bacteria in the presence of sucrose. Therefore, this 'negative selection' will select for bacterial cells that have undergone a second recombination event that removes the plasmid DNA from the genome.

As excision can occur in a variety of ways it is necessary to screen (through PCR) the resulting $Suc^R$ colonies for those that have deleted the selected ORF. Identify the strain carrying the correct mutant allele by colony PCR using the A and B primer pairs designed specifically for each ORF (Fig. 11.1). These primers flank UPFwd and DNRev and therefore mutant strains will be detected based on the size of the PCR product amplified using these primers. (A wild-type allele will give a band of the predicted wild-type size, a deletion will result in a PCR product of a predictably smaller size.)

- For colony PCR pick individual colonies with the correct resistance profile (sensitive to the antibiotic to which the donor plasmid confers resistance, but resistant to sucrose) into sterile 0.2 ml Eppendorf tubes containing 50 µl of $H_2O$.
- Incubate the cell suspension at 95°C for 5 min (to facilitate cell lysis).
- Remove debris by centrifugation at maximum speed in a microfuge for 1 min.
- Use the supernatant as template in a PCR reaction; typically 1–5 µl of supernatant is used in a 10–50 µl reaction.

Note: In our experience mutant colonies are present at a high frequency (10–30%) and are therefore readily detected by screening.

- Sequence the PCR product to confirm the integrity of the mutation.

## 11.4. Techniques to Investigate Bacteria–Nematode Mutualism

### 11.4.1. Co-cultivation of nematodes and bacteria

The ability to culture nematodes and bacteria outside of insects allows visualization of the process of colonization initiation and increases the number of bacterial strains that can be monitored for their ability to colonize (Fig. 11.2). Cultivation

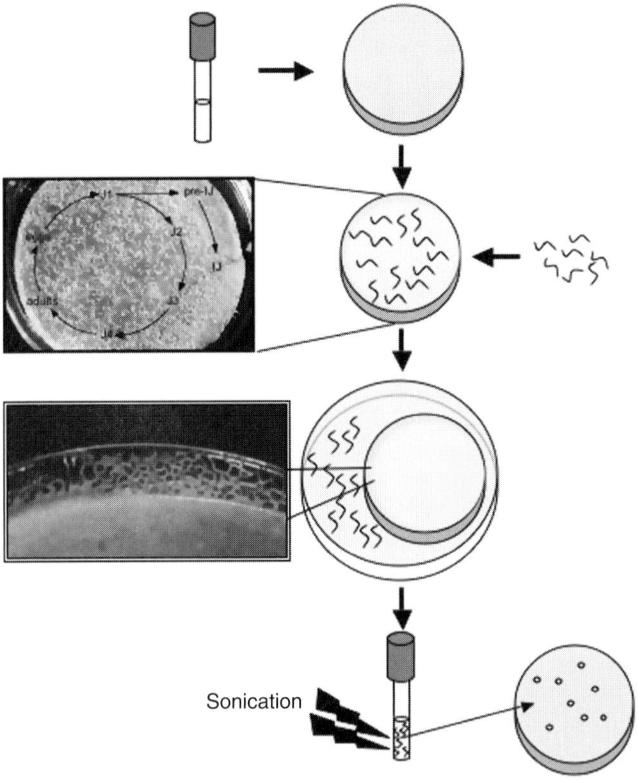


**Fig. 11.2.** *In vitro* cultivation of nematodes and bacteria; the nematode–bacterium symbiosis can be monitored outside of insects. Laboratory-culture-grown bacteria are combined with sterile nematodes (either eggs or axenic infective juveniles (IJs)) on agar plates containing lipids. Nematodes ingest the bacteria and reproduce and develop as they do in insects. Over the course of a week, IJs develop and migrate off the plate into water traps. To monitor colonization, approximately 10,000 IJs are collected from the water, bleach-sterilized and sonicated briefly to release colonizing bacteria. The sonicate is dilution plated to determine the average number of colony-forming units per IJ. Alternatively, the bacterial load of an individual nematode can be assessed after its disruption with a tissue grinder. Finally, the percentage of colonized nematodes in a population can be determined using *Xenorhabdus nematophila* strains expressing green fluorescent protein, which allows visualization by fluorescence microscopy.

on lawns of bacteria requires addition of a lipid source, typically a combination of maize oil and maize syrup or cholesterol. Generally, lipid agar (LA) plates are used.

- Make LA plates by autoclaving in a 2 l flask with a stir bar: 8 g nutrient broth, 5 g yeast extract, 15 g agar and 10 ml of a 0.2 g/ml $MgCl_2 \cdot 6H_2O$ solution in 890 ml $H_2O$. After autoclaving, 96 ml of sterile maize syrup mix (7 ml of maize syrup in 89 ml $H_2O$) and 4 ml maize oil are added to the molten agar while the flask is mixed on a stir plate. To ensure an even distribution of oil to each plate, continually stir the mixture while dispensing a measured volume of media into each Petri dish or well. (Variation in agar thickness or oil distribution leads to inconsistent nematode development between plates.)

- Inoculate fresh LA plates with bacteria (from overnight cultures) and incubate for 1–2 (*X. nematophila*) or 3–4 days (in the case of *Photorhabdus*) before the nematodes are added.
- Approximately 40 IJs, or several thousand J1 juveniles hatched from sterile eggs (see below) should be added directly to the bacteria lawn to initiate the symbiosis assay.

Although IJs carry symbiotic bacterial cells, these cells have not been found to affect our assays. However, if axenic nematodes are necessary, they can be isolated by either culturing the nematodes on bacterial mutants that are unable to colonize the IJ (e.g. *X. nematophila rpoS*; Vivas and Goodrich-Blair, 2001; Heungens *et al.*, 2002) or *P. luminescens pbgPE* (Bennett and Clarke, 2005) mutants or by culturing the nematodes on a non-cognate bacterium (e.g. cultivate *S. carpocapsae* on *Xenorhabdus bovienii* or *H. bacteriophora* TT01 nematodes on *Photorhabdus temperata* K122 bacteria). Indeed we have observed that *P. temperata* K122 will support the growth of all *Heterorhabditis* spp. tested but this bacterium will only colonize its cognate nematode partner, *Heterorhabditis downesi* (S.A. Joyce and D.J. Clarke, unpublished data).

Axenic nematodes also can be obtained by isolating nematode eggs from gravid females. For this the following procedure should be followed:

- Inoculate five 10 cm lipid agar plates with 800 µl each of an overnight LB culture of bacteria and swirl the plates until the surfaces are covered.
- Incubate the plates overnight at 30°C then add approximately 5000 IJ nematodes to the lawns. Incubate the plates in the dark at room temperature for 4–5 days until adults develop.
- Add sterile distilled $H_2O$ to the plates to dislodge the nematodes then using a Pasteur pipette transfer the nematodes to a sterile 50 ml capped conical tube. Let the nematodes settle down to the bottom of the tube.
- Discard the water and resuspend the nematodes in 45 ml sterile $H_2O$. Let them settle and repeat washes 2–3 additional times until water is clear.
- Add 50 ml of axenizing solution (2.4% (v/v) NaOCl, 0.25 N KOH) and incubate for 10 min with shaking.
- Centrifuge the tubes in a tabletop centrifuge 7–10 min at ~2000 × g. Discard the supernatant and resuspend the pellet in axenizing solution as above.
- Repeat procedure.
- Resuspend the pellet in LB and wash two to three times. Finally, resuspend the eggs in 5–10 ml of LB and split in two small Petri dishes.
- Eggs are good for 2–3 days at room temperature and can be stored at 4°C to extend life up to 4–5 days. Eggs hatch into J1 juveniles overnight at room temperature.
- For inoculation of bacterial lawns add ~1500 J1 larvae to each 6 cm LA plate.

### 11.4.2. Monitoring colonization

Progeny IJ nematodes can be harvested by placing the insect or Petri dish in a water trap into which the IJs migrate. When harvesting IJs from LA lawns, the agar slab is transferred to the lid of the 6 cm plate that is then floated in sterile water within a 10 cm Petri plate. Alternatively, up to four 6 cm LA agar slabs (in the Petri dish bottom) can be simultaneously floated in a large (15 cm) Petri dish. The nematodes will migrate off the agar slab and collect in the water. Such nematode populations can be stored in tissue culture flasks either at room temperature or at 4°C. During *in vitro* symbiosis assays, *Heterorhabditis* IJs tend to migrate to the lid of the Petri dish where they can be readily collected.

Several approaches have been used to monitor bacterial colonization of IJ progeny. Microscopic methods allow an examination of distribution among the population and therefore allow quantitative measurement of colonization efficiency (Martens *et al.*, 2003b). Bacteria can be visualized by extruding the nematode intestines by chopping the tip of the head with a razor blade, drying to a glass slide (either by quickly passing the slide through a flame, or by simple air drying) and staining with 0.1% (w/v) crystal violet: cover the surface of the glass slide with the crystal violet using a Pasteur pipette, then rinse with water. The purple rod shapes are readily visible at 40X and 100X magnification (Vivas and Goodrich-Blair, 2001). Alternatively, the bacteria can be engineered to express a fluorescent protein using the Tn*7* transposition method described above, or by campbell integration at a neutral site on the chromosome (Martens *et al.*, 2003b, 2005). Colonization is then monitored in intact nematodes using fluorescence microscopy.

Colonization levels can also be monitored by directly determining the colony-forming units (cfu) present in a lysate following the release of the bacteria from the nematode. Bacterial release is best achieved by physical disruption of the nematode. This can be done on individual or multiple nematodes using a Tenbroeck tissue grinder or a motorized polypropylene pestle (Heungens *et al.*, 2002; Goetsch *et al.*, 2006) or on populations in a sonicating water bath (Heungens *et al.*, 2002). Regardless of release method, it is important to sterilize the surface of the nematodes with bleach, hyamine or merthiolate prior to disruption.

Cultivation of nematodes on bacterial lawns generally leads to a lower colonization level than that which occurs in nematodes cultivated within an insect host (Goetsch *et al.*, 2006; Flores-Lara *et al.*, 2007).

## 11.5. Techniques in Studying EPB Virulence

The infection of insects by EPNs and bacteria is naturally occurring and technically facile to investigate, yet reflects the complicated, multidimensional aspects of disease. Insects that have been used to investigate molecular aspects of the host–EPNB interface include the lepidopteran *G. mellonella*, *Manduca sexta* and *Spodoptera* spp. and the dipteran *Drosophila melanogaster*. While the genetic manipulation of leptidopteran insects cannot rival that of *D. melanogaster*, there is a growing body of knowledge regarding the biochemistry, cell and molecular biology of lepidopteran immunity (Gillespie *et al.*, 1997; Kanost *et al.*, 2004). Furthermore, comparative invertebrate immunology has the power to provide broad insights into the evolution and function of innate immunity (Loker *et al.*, 2004). Several laboratories have begun to focus on *M. sexta* to monitor EPNB pathogenesis due to the significant knowledge regarding the innate immune defences of this insect in response to a variety of pathogens (Kanost *et al.*, 2004). This knowledge has facilitated the experimental modulation of *M. sexta* immunity through RNAi (Levin *et al.*, 2005; Eleftherianos *et al.*, 2006a,b). Another powerful technique that can be used to identify bacterial genes upregulated *in vivo* during host infection is the selective capture of transcribed sequences (Graham and Clark-Curtiss, 1999). This technique has been applied to identify upregulated and virulence genes in *Moraxella osloensis*, a bacterial symbiont of the slug-parasitic nematode *Phasmarhabditis hermaphrodita* (An and Grewal, 2007; An *et al.*, 2007).

### 11.5.1. Insect husbandry

*G. mellonella* larvae are available commercially from wholesale bait shops (e.g. Vanderhorst Wholesale Inc. http://www.ridertown.com/shop/address/vndrhrst.html and Livefood UK at http://www.livefoods.co.uk) and can be kept at 4°C for approximately 1 month before use as a host for propagation of the nematodes. The Goodrich-Blair laboratory has used *M. sexta* eggs obtained from NC State Insectary (http://www.cals.ncsu.edu/entomology/INSECTARY/homepg.html). Preparation of insects should be performed according to the guidelines provided below:

- Soak eggs upon arrival in 10% NaOCl for 3 min.
- Rinse three times in sterile water, each time collecting the eggs on membrane using a filter apparatus.
- Place the cleaned eggs on filter paper in a Petri dish and allow to dry 1 h or overnight at 26°C.
- Prepare a hatching chamber by tacking a rubber bottle stopper into the bottom of each of several sundae cups (Sweetheart plastic food cups and lids, Cat. # F5DB and LMC45).
- For insect diet, autoclave 15 g agar (USP Grade, MP Biomedicals, Cat. No. 100262) in 1 l water.
- Immediately after autoclaving blend in 166 g Gypsy moth wheat-germ diet (MP Biomedicals, Cat No. 960293). Blend thoroughly.
- Pour liquid food into glass dishes that have been cleaned with 75% EtOH. Solidified food should be stored at 4°C.
- Cover each stopper of the hatching chamber with a large amount of food.
- After eggs are dried, distribute them among the hatching chambers (~30/cup). Use a small tissue to create a buffer between the cup and the lid and cut a hole in the centre of the tissue so that it does not touch the food. This helps

prevent the insects being crushed between the cup and the lid. The lid should have several holes in it for air circulation.

- Incubate the eggs at 26°C or until they hatch, at which point they will crawl away from the egg casings on to the food podium. Once most of the eggs have hatched, separate them (~25 insects/cup) into new cups with one large square of food on the bottom of the cup (Fig. 11.3).
- By day 5 the insects will be in late second instar (indicated by their small size, undefined features and dark tails that break off easily). At this stage the insects produce a sticky substance and will stick together.
- Pull insects apart carefully and separate each insect into an individual inverted plastic condiment cup, on to which a small square of food (~0.5 $cm^3$) has been placed. Incubate in a 26°C insect incubator with a 16 h:8 h light/dark cycle.
- Replace food as needed. Typically the insects will be in their fourth instar by ~7 days post-hatch. At this stage they have well-defined features with black-padded feet, red spiracle spots and striped sides. A white head cap is apparent on third instar larvae and this is lost by the fourth instar.

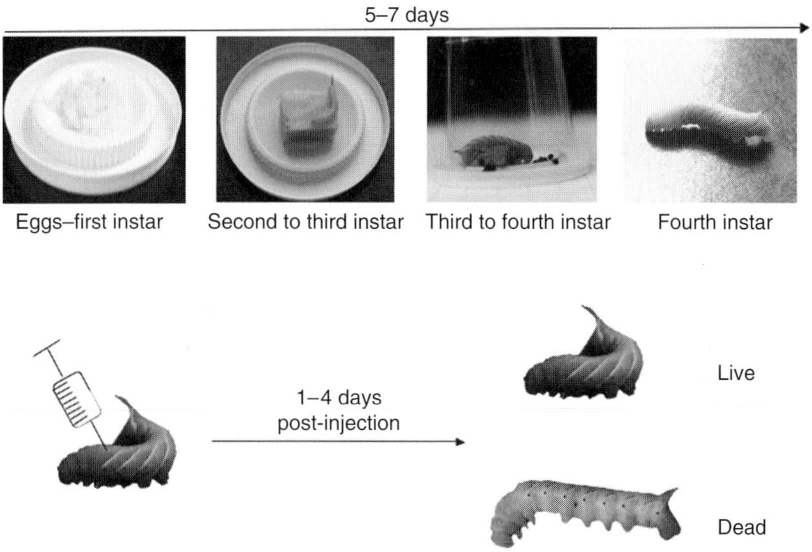


**Fig. 11.3.** Monitoring EPB virulence in *Manduca sexta* insects; virulence can be assayed by direct injection into a susceptible insect host. Commercially available *Manduca sexta* insect eggs are bleached and kept at 26°C. Once hatched, larvae are reared in small individual cages on a sterile wheat-germ-based diet for 5–7 days, progressing through a series of moults until the fourth or fifth instar. Bacterial cells grown to log or stationary phase in liquid culture are injected by syringe into the first proleg of insects. Typically, ten insects are injected per strain. After injection, insects are periodically monitored for up to 4 days for death or signs of disease, including vomiting, diarrhoea, limpness and change in colour from blue-green to green. Virulence data can be expressed in terms of per cent mortality of insects or in terms of the time required for 50% of the injected insects to die ($LT_{50}$).

## 11.5.2. Injections and assessment of bacterial virulence

Although *Photorhabdus* and *Xenorhabdus* are naturally vectored into insect hosts by their respective nematode hosts, they are also virulent when experimentally injected into insect haemolymph.

- For injections, inoculate bacterial cultures from −80°C stocks (see above) into 2 ml liquid LB medium and grow for 16 h at 30°C.
- Subculture 1–100 into fresh LB medium and grow to the desired growth phase.

Note: It has been observed that logarithmic phase *X. nematophila* are more virulent than stationary phase cells (Cowles *et al.*, 2007).

- Set up dilutions so that the desired number of bacteria is injected in a volume of 10 µl.
- Pellet 500 µl of culture (microfuge for 1–2 min at maximum speed), wash in 1 ml PBS, pellet again and resuspend in 500 µl PBS.
- Keep bacterial suspensions on ice.
- For one strain at a time prepare serial tenfold dilutions of the bacterial suspension in a microtitre dish. In each well, dispense 270 µl of PBS. Then add 30 µl of the bacterial suspension to the top well in a column, mix several times, change tips, transfer 30 µl from well A to B, mix and continue (changing tips each time) down to the bottom of the row.
- Determine average cfu per millilitre by plating 10 µl of each dilution on to an LB pyruvate plate both before and after injections.

Virulence assays can be conducted in either *G. mellonella* or *M. sexta* larvae. The latter are more resistant to *X. nematophila* and *P. luminescens* infections, which can be useful for revealing subtle virulence phenotypes of bacterial mutants, while the former are easier to obtain and use. The final instar, non-feeding larvae of *G. mellonella* obtained from a commercial source (e.g. Van der Horst Wholesale or Livefood UK) are used for injections. Typically the Goodrich-Blair laboratory uses feeding fourth instar *M. sexta*, which weigh 0.2 or more grams, and are not moulting (i.e. they lack a bulb on the tip of the head).

Injection protocol:

- Place 1–2 ml 95% EtOH and sterile distilled water in separate yellow cap tubes.
- Group the insects by strain and inoculum and place those insects to be injected first on ice (in their containers – remove food and faeces from *M. sexta* containers). *G. mellonella* larvae are smaller than *M. sexta* and, therefore, do not need to be chilled.
- Rinse a Hamilton 30-gauge syringe in the ethanol three times, then rinse in the water three times.
- From the appropriate dilution in the microtitre plate, load 10 µl into the syringe (normal doses range from 10 to 100 cfu/µl).
- Inject an insect behind the first proleg, being careful to slide the needle just under the skin (avoiding the gut) once only.

- Return insect to container (remove from ice) and continue injections, rinsing the syringe after a given level of inoculum.
- To monitor virulence, typically ten insects are used for each treatment (inoculum).
- When each bacterial strain is completed, the next strain is prepared. A PBS-only control should be included, to ensure insects are not being killed by the injection procedure.

*G. mellonella* insects can be placed on filter paper in a Petri dish, while *M. sexta* are provided with new food and returned to the incubator. Monitor periodically over 96 h for death (failure to respond to stimuli and change in colour) and, in *M. sexta*, disease also can be noted by the onset of diarrhoea. In the case of *G. mellonella*, insects that turn black within 1–2 h of injection are discarded as this is an indicator of significant internal damage.

### 11.5.3. Oral toxicity

Both *Photorhabdus* and *Xenorhabdus* secrete orally active insect toxins (ffrench-Constant and Waterfield, 2006). To monitor the presence of such toxins an oral-feeding assay can be performed.

- Grow bacterial culture in liquid medium for ~24 h and pellet cells.
- Concentrate the supernatant (e.g. using a 10 kDa molecular weight cut-off Centricon filter (Millipore)) to 500 μl.

Suitable controls include concentrated growth medium without bacteria, or concentrated supernatants of *E. coli*.

- Dip a 1 $cm^3$ portion of wheat-germ diet (see above) into the concentrated treatments and provide to second instar larvae that have been weighed.
- Weigh the insects daily for 2 days.

## 11.6. Conclusions and Future Prospects

In addition to the potential for EPNs as biocontrol agents the tripartite association between EPN, EPB and their insect hosts provides researchers with a unique opportunity to study both pathogenicity and mutualism. The recent development of a range of molecular techniques in both *Photorhabdus* and *Xenorhabdus* will now facilitate analysis of these interactions at a molecular level. It is clear that many of the effectors used by EPB to infect and kill their insect hosts have homologues in closely related mammalian pathogens such as *Yersinia* and *E. coli*. Therefore, further studies on EPB virulence will shed more light on the fundamental mechanisms underlying virulence in other bacteria. Moreover, the very recent development of genomic and post-genomic technologies for both the insect and nematode host of EPB suggests that the in-depth analysis of specific gene–gene interactions and their role in controlling bacteria–host interactions is imminent, e.g. in determining host susceptibility to infection.

## Acknowledgements

The authors wish to thank members of their laboratories for their contributions to the development of protocols and images used in figures, Dr. E.E. Herbert Tran for creating figures 11.2 and 11.3, and Professor Ann Burnell (National University of Ireland, Maynooth) for communicating unpublished data. Research in the H. Goodrich-Blair laboratory is supported by awards from the Burroughs Welcome Foundation (1003707), The National Institutes of Health (GM059776–08) and the National Science Foundation (IBN-0416783). Research in the Parwinder S. Grewal (PSG) laboratory is supported by awards from the US Department of Agriculture, Cooperative State Research Education and Extension Service and the Ohio Agricultural Research and Development Center. Research in the D.J. Clarke laboratory has been supported by grants from the Biotechnology and Biological Science Research Council (BBSRC), the Leverhulme Trust and Science Foundation Ireland. The work of T.A. Ciche has been supported by a fellowship from the Gordon Ross Foundation and the Howard Hughes Medical Institute (awarded to P. Sternberg) while he was a postdoctoral fellow in the laboratory of P. Sternberg. The Ciche laboratory is partially supported by the Center for Microbial Pathogenesis at Michigan State University.

## References

An, R. and Grewal, P.S. (2007) Differences in virulence of *Heterorhabditis bacteriophora* and *Steinernema scarabaei* to three white grub species: the relative contribution of the nematodes and their symbiotic bacteria. *Biological Control* 43, 12–23.

An, R., Sreevatsan, S. and Grewal, P.S. (2007) *Moraxella osloensis* gene expression in the primitive mollusk host *Deroceras reticulatum. BMC Microbiology* 8, 19–29.

Atlas, R.M. (1997) *Handbook of Microbiological Media*, CRC, Boca Raton, Florida.

Bai, X. and Grewal, P.S. (2007) Identification of two down-regulated genes in the entomopathogenic nematode *Heterorhabditis bacteriophora* infective juveniles upon contact with insect hemolymph. *Molecular Biochemistry and Parasitology* 156, 162–166.

Bai, X., Grewal, P.S., Hogenhout, S.A., Adams, B.A., Ciche, T.A., Gaugler, R. and Sternberg, P.W. (2007) Expressed sequence tag analysis of gene representation in the insect parasitic nematode *Heterorhabditis bacteriophora. Journal of Parasitology* 93, 1343–1349.

Bao, Y., Lies, D.P., Fu, H. and Roberts, G.P. (1991) An improved Tn*7*-based system for the single-copy insertion of cloned genes into chromosomes of Gram-negative bacteria. *Gene* 109, 167–168.

Bargmann, C.I. and Horvitz, H.R. (1991) Control of larval development by chemosensory neurons in *Caenorhabditis elegans. Science* 251, 1243–1246.

Bennett, H.P.J. and Clarke, D.J. (2005) The *pbgPE* operon in *Photorhabdus luminescens* is required for pathogenicity and symbiosis. *Journal of Bacteriology* 187, 77–84.

Boemare, N. (2002) Biology, taxonomy and systematics of *Photorhabdus* and *Xenorhabdus*. In: Gaugler, R. (ed.) *Entomopathogenic Nematology*. CAB International, Wallingford, UK, pp. 57–78.

Boemare, N., Laumond, C. and Mauleon, H. (1996) The entomopathogenic nematode–bacterium complex: biology, life cycle and vertebrate safety. *Biocontrol Science and Technology* 6, 333–345.

Brachmann, A.O., Joyce, S.A., Jenke-Kodama, Schwar, G., Clarke, D.J. and Bode, H.B. (2007) A type II polyketide synthase is responsible for anthraquinone biosynthesis in *Photorhabdus luminescens. Chem BioChem* 8, 1721–1728 (in press).

*C. elegans* genome sequencing consortium (1998) Genome sequence of the nematode *C. elegans*: a platform for investigating biology. *Science* 282, 2012–2018.

Chen, S., Glazer, I., Gollop, N., Cash, P., Argo, E., Innes, A., Steward, E., Davidson, I. and Wilson, M.J. (2006) Proteomic analysis of the entomopathogenic nematode *Steinernema feltiae* IS-6 IJs under evaporative and osmotic stresses. *Molecular and Biochemical Parasitology* 145, 195–204.

Ciche, T.A. and Ensign, J.C. (2003) For the insect pathogen, *Photorhabdus luminescens*, which end of a nematode is out? *Applied and Environmental Microbiology* 69, 1890–1897.

Ciche, T.A. and Sternberg, P.W. (2007) Postembryonic RNAi in *Heterorhabditis bacteriophora*: a nematode insect parasite and host for insect pathogenic symbionts. *BMC Developmental Biology* 7, 101.

Ciche, T.A., Kim, K.S., Kaufmann-Daszczuk, B., Nguyen, K.C. and Hall, D.H. (2008) Cell invasion and matricide during *Photorhabdus luminescens* transmission by Heterorhabditis bacteriophora Nematodes. *Applied and Environmental Microbiology* 74, 2275–2287.

Cowles, C.E. and Goodrich-Blair, H. (2004) Characterization of a lipoprotein, NilC, required by *Xenorhabdus nematophila* for mutualism with its nematode host. *Molecular Microbiology* 54, 464–477.

Cowles, C.E. and Goodrich-Blair, H. (2006) *nilR* is necessary for co-ordinate repression of *Xenorhabdus nematophila* mutualism genes. *Molecular Microbiology* 62, 760–771.

Cowles, K.N. and Goodrich-Blair, H. (2005) Expression and activity of a *Xenorhabdus nematophila* haemolysin required for full virulence towards *Manduca sexta* insects. *Cellular Microbiology* 2, 209–219.

Cowles, K.N., Cowles, C.E., Richards, G.R., Martens, E.C. and Goodrich-Blair, H. (2007) The global regulator Lrp contributes to mutualism, pathogenesis and phenotypic variation in the bacterium *Xenorhabdus nematophila*. *Cellular Microbiology* 9, 1311–1323.

De Ley, P. and Blaxter, M. (2002) Systematic position and phylogeny. In: Lee, D.L. (ed.) *The Biology of Nematodes*. Taylor & Francis, London, pp. 1–30.

Diatchenko, L., Lau, Y.F., Campbell, A.P., Chenchik, A., Moqadam, F., Huang, B., Lukyanov, S., Lukyanov, K., Gurskaya, N., Sverdlov, E.D. and Siebert, P.D. (1996) Suppression subtractive hybridization: a method for generating differentially regulated or tissue-specific cDNA probes and libraries. *Proceedings of the National Academy of Science of the USA* 93, 6025–6030.

Duchaud, E., Rusniok, C., Frangeul, L., Buchrieser, C., Givaudan, A., Taourit, S., Bocs, S., Boursaux-Eude, C., Chandler, M., Charles, J.F., Dassa, E., DeRose, R., Derzelle, S., Freyssinet, G., Gaudriault, S., Medigue, C., Lanois, A., Powell, K., Siguier, P., Vincent, R., Wingate, V., Zouine, M., Glaser, P., Boemare, N., Danchin, A. and Kunst, F. (2003) The genome sequence of the entomopathogenic bacterium *Photorhabdus luminescens*. *Nature* 21, 1307–1313.

Eleftherianos, I., Marokhazi, J., Millichap, P.J., Hodgkinson, A.J., Sribbonlert, A., ffrench-Constant, R. and Reynolds, S.E. (2006a) Prior infection of *Manduca sexta* with non-pathogenic *Escherichia coli* elicits immunity to pathogenic *Photorhabdus luminscens*: roles of immune-related proteins shown by RNA interference. *Insect Biochemistry and Molecular Biology* 36, 527–525.

Eleftherianos, I., Millichap, P.J., ffrench-Constant, R.H. and Reynolds, S.E. (2006b) RNAi suppression of recognition protein mediated immune responses in the tobacco hornworm *Manduca sexta* causes increased susceptibility to the insect pathogen *Photorhabdus*. *Developmental and Comparative Immunology* 30, 1099–1107.

Eleftherianos, I., Boundy, S., Joyce, S.A., Aslman, S., Marshall, J.W., Cox, R., Simpson, T.J., Clarke, D.J., ffrench-Constant, R. and Reynolds, S.E. (2007) An antibiotic produced by an insect-pathogenic bacterium suppresses host defenses through phenoloxidase inhibition. *Proceedings of the National Academy of Science of the USA* 104, 2419–2424.

ffrench-Constant, R. and Waterfield, N. (2006) An ABC guide to the bacterial toxin complexes. *Advances in Applied Microbiology* 58, 169–183.

Fire, A., Xu, S., Montgomery, M.K., Kostas, S.A., Driver, S.E. and Mello, C.C. (1998) Potent and specific genetic interference by double-stranded RNA in Caenorhabditis elegans. *Nature* 391, 806–811.

Flores-Lara, Y., Renneckar, D., Forst, S., Goodrich-Blair, H. and Stock, P. (2007) Influence of nematode age and culture conditions on morphological and physiological parameters in the bacterial vesicle of *Steinernema carpocapsae* (Nematoda: Steinernematidae). *Journal of Invertebrate Pathology* 95, 110–118.

Forst, S. and Clarke, D. (2002) Bacteria–nematode symbioses. In: Gaugler, R. (ed.) *Entomopathogenic Nematology*. CAB International, Wallingford, UK, pp. 57–77.

Gal, T.Z., Glazer, I. and Koltai, H. (2003) Differential gene expression during desiccation stress in the insect-killing nematode *Steinernema feltiae* IS-6. *Journal of Parasitology* 89, 761–766.

Gal, T.Z., Glazer, I., Sherman, A. and Koltai, H. (2005) Protein interaction of nucleosome assembly protein 1 and casein kinase 2 during desiccation response in the insect-killing nematode *Steinernema feltiae* IS-6. *Journal of Parasitology* 91, 691–693.

Gaugler, R., Campbell, J.F. and McGuire, T.R. (1989) Selection for host-finding in *Steinernema feltiae. Journal of Invertebrate Pathology* 54, 363–372.

Gillespie, J.P., Kanost, M.R. and Trenczek, T. (1997) Biological mediators of insect immunity. *Annual Review of Entomology* 42, 611–643.

Glazer, I., Gaugler, R. and Segal, D. (1991) Genetics of the nematode *Heterorhabditis bacteriophora* Strain HP88: the diversity of beneficial traits. *Journal of Nematology* 23, 324–333.

Goetsch, M., Owen, H., Goldman, B. and Forst, S. (2006) Analysis of the PixA inclusion body protein of *Xenorhabdus nematophila. Journal of Bacteriology* 188, 2706–2710.

Goodrich-Blair, H. and Clarke, D.J. (2007) Mutualism and pathogenesis in *Xenorhabdus* and *Photorhabdus*: two roads to the same destination. *Molecular Microbiology* 64, 260–268.

Graham, J.E. and Clark-Curtiss, J.E. (1999) Identification of *Mycobacterium tuberculosis* RNAs synthesized in response to phagocytosis by human macrophages by selective capture of transcribed sequences (SCOTS). *Proceedings of the National Academy of Science of the USA* 96, 11554–11559.

Grewal, P.S., Gaugler, R. and Shupe, C. (1996a) Rapid changes in thermal sensitivity of entomopathogenic nematodes in response to selection at temperature extremes. *Journal of Invertebrate Pathology* 68, 65–73.

Grewal, P.S., Gaugler, R. and Wang, Y. (1996b) Enhanced cold tolerance of the entomopathogenic nematode *Steinernema feltiae* through genetic selection. *Annals of Applied Biology* 129, 335–341.

Grewal, P.S., Ehlers, R.U. and Shapiro-Ilan, D.I. (2005) *Nematodes as Biocontrol Agents*. CAB International, Wallingford, UK.

Grewal, P.S., Bornstein-Forst, S., Burnell, A.M., Glazer, I. and Jagdale, G.B. (2006) Physiological, genetic, and molecular mechanisms of chemoreception, thermobiosis, and anhydrobiosis in entomopathogenic nematodes. *Biological Control* 38, 54–65.

Guo, S. and Kemphues, K.J. (1995) par-1, a gene required for establishing polarity in *C. elegans* embryos, encodes a putative Ser/Thr kinase that is asymmetrically distributed. *Cell* 81, 611–620.

Hallem, E.A., Rengarajan, M., Ciche, T.A. and Sternberg, P.W. (2007) Nematodes, bacteria, and flies: a tripartite model for nematode parasitism. *Current Biology* 17, 898–904.

Hashmi, G. and Gaugler, R. (1998) Genetic diversity in insect-parasitic nematodes (Rhabditida: Heterorhabditidae). *Journal of Invertebrate Pathology* 72, 185–189.

Hashmi, S., Hashmi, G. and Gaugler, R. (1995) Genetic transformation of an entomopathogenic nematode by microinjection. *Journal of Invertebrate Pathology* 66, 293–296.

Hashmi, S., Hatab, M.A.A. and Gaugler, R.R. (1997) GFP: green fluorescent protein a versatile gene marker for entomopathogenic nematodes. *Fundamental and Applied Nematology* 20, 323–327.

Heungens, K., Cowles, C.E. and Goodrich-Blair, H. (2002) Identification of *Xenorhabdus nematophila* genes required for mutualistic colonization of *Steinernema carpocapsae* nematodes. *Molecular Microbiology* 45, 1337–1353.

Jagdale, G.B., Saeb, A.T., Somasekhar, N. and Grewal, P.S. (2006) Genetic variation and relationships between isolates and species of the entomopathogenic nematode genus *Heterorhabditis* deciphered through isozyme profiles. *Journal of Parasitology* 92, 509–516.

Jansen, G., Thijssen, K.L., Werner, P., van der Horst, M., Hazendonk, E. and Plasterk, R.H.A. (1999) The complete family of genes encoding G proteins of *Caenorhabditis elegans*. *Nature Genetics* 21, 414–419.

Ji, D. and Kim, Y. (2004) An entomopathogenic bacterium, *Xenorhabdus nematophila*, inhibits the expression of an antibacterial peptide, cecropin, of the beet armyworm, *Spodoptera exigua*. *Journal of Insect Physiology* 50, 489–496.

Joyce, S. and Clarke, D. (2003) A *hexA* homolog of *Photorhabdus* regulates pathogenicity, symbiosis, and phenotypic variation. *Molecular Microbiology* 47, 1445–1457.

Kamath, R.S. and Ahringer, J. (2003) Genome-wide RNAi screening in *Caenorhabditis elegans*. *Methods* 30, 313–321.

Kanost, M.R., Jiang, H. and Yu, X.Q. (2004) Innate immune responses of a lepidopteran insect, *Manduca sexta*. *Immunological Reviews* 198, 97–105.

Kaya, H.K. and Stock, S.P. (1997) Techniques in insect nematology. In: Lacey, L.A. (ed.) *Manual of Techniques in Insect Pathology*. Academic Press, London, pp. 281–324.

Kimbell, J.L., Koropatnick, T.A. and McFall-Ngai, M.J. (2006) Evidence for the participation of the proteasome in symbiont-induced tissue morphogenesis. *Biological Bulletin* 21, 1–6.

Levin, D.M., Breuer, L.N., Zhuang, S., Anderson, S.A., Nardi, J.B. and Kanost, M.R. (2005) A hemocyte-specific integrin required for hemocytic encapsulation in the tobacco hornworm, *Manduca sexta*. *Insect Biochemistry and Molecular Biology* 35, 369–380.

Li, X.Y., Cowles, R.S., Cowles, E.A., Gaugler, R. and Cox-Foster, D.L. (2007) Relationship between the successful infection by entomopathogenic nematodes and the host immune response. *International Journal of Parasitology* 37, 365–374.

Loker, E.S., Adema, C.M., Zhang, S.-M. and Kepler, T.B. (2004) Invertebrate immune systems – not homogenous, not simple, not well understood. *Immunological Reviews* 198, 10–24.

Lukyanov, S.A., Rebrikov, D. and Buzdin, A.A. (2007) Suppression subtractive hybridization. In: Buzdin, A.A. and Lukyanov, S.A. (eds) *Nucleic Acids Hybridization Modern Applications*. Springer, The Netherlands.

Maeda, I. and Sugimoto, A. (2001) Systematic functional analysis of the *C. elegans* genome. *Tanpakushitsu Kakusan Koso* 46, 2432–2435.

Martens, E.C., Gawronski-Salerno, J., Vokal, D.L., Pellitteri, M.C., Menard, M.L. and Goodrich-Blair, H. (2003a) *Xenorhabdus nematophila* requires an intact *isc-hsc-fdx* locus to colonize *Steinernema carpocapsae* nematodes. *Journal of Bacteriology* 185, 3678–3682.

Martens, E.C., Heungens, K. and Goodrich-Blair, H. (2003b) Early colonization events in the mutualistic association between *Steinernema carpocapsae* nematodes and *Xenorhabdus nematophila* bacteria. *Journal of Bacteriology* 185, 3147–3154.

Martens, E.C., Vivas, E.I., Heungens, K., Cowles, C.E. and Goodrich-Blair, H. (2004) Investigating mutualism between entomopathogenic bacteria and nematodes. *Nematology Monographs and Perspectives (2): Proceedings of the Fourth International Congress on Nematology* 2, 447–462.

Martens, E.C., Russell, F.M. and Goodrich-Blair, H. (2005) Analysis of *Xenorhabdus nematophila* metabolic mutants yields insight into stages of *Steinernema carpocapsae* nematode intestinal colonization. *Molecular Microbiology* 51, 28–45.

Miller, J.H. (1972) *Experiments in Molecular Genetics*. Cold Spring Harbor Laboratory Press, Cold Spring Harbor, New York.

O'Halloran, D.M. and Burnell, A.M. (2003) An investigation of chemotaxis in the insect parasitic nematode *Heterorhabditis bacteriophora*. *Parasitology* 127, 375–385.

O'Halloran, D.M., Fitzpatrick, D.A., McCormack, G.P., McInerney, J.O. and Burnell, A.M. (2006) The molecular phylogeny of a nematode-specific clade of heterotrimeric G-protein alpha-subunit genes. *Journal of Molecular Evolution* 63, 87–94.

O'Leary, S.A. and Burnell, A.M. (1997) The isolation of mutants of Heterorhabditis megidis (strain UK211) with increased desiccation tolerance. *Fundamental and Applied Nematology* 20, 197–205.

Orchard, S.S. and Goodrich-Blair, H. (2004) Identification and functional characterization of the *Xenorhabdus nematophila* oligopeptide permease. *Applied and Environmental Microbiology* 70, 5621–5627.

Park, D. and Forst, S. (2006) Co-regulation of motility, exoenzyme and antibiotic production by the EnvZ-OmpR-FlhDC-FliA pathway in *Xenorahbdus nematophila. Molecular Microbiology* 61, 1397–1412.

Park, Y. and Kim, Y. (2000) Eicosanoids rescue *Spodoptera exigua* infected with *Xenorhabdus nematophilus*, the symbiotic bacteria to the entomopathogenic nematode *Steinernema carpocapsae. Journal of Insect Physiology* 46, 1469–1476.

Park, Y., Kim, Y., Putnam, S.M. and Stanley, D.W. (2003) The bacterium *Xenorhabdus nematophilus* depresses nodulation reaction to infection by inhibiting eicosanoid biosynthesis in tobacco hornworms, *Manduca sexta. Archives of Insect Biochemistry and Physiology* 52, 71–80.

Park, Y., Herbert, E.E., Cowles, K.N., Cowles, C.E., Menard, M.L., Orchard, S.S. and Goodrich-Blair, H. (2007) Clonal variation in *Xenorhabdus nematophila* virulence and suppression of *Manduca sexta* immunity. *Cellular Microbiology* 9, 645–656.

Phillipe, N., Alcaraz, J.P., Coursange, E., Geiselmann, J. and Schneider, D. (2004) Improvement of pCVD442, a suicide plasmid for gene allelic exchange in bacteria. *Plasmid* 51, 246–255.

Poinar, G.O. Jr (1990) Taxonomy and biology of Steinernematidae and Heterorhabditidae. In: Gaugler, R. and Kaya, H.K. (eds) *Entomopathogenic Nemtaodes in Biological Control*. CRC Press, Boca Raton, Florida, pp. 23–61.

Poinar, G.O.J. and Thomas, G.M. (1966) Significance of *Achromobacter nematophilus* Poinar and Thomas (Achromobacteraceae: Eubacteriales) in the development of the nematode, DD-136 (*Neoaplectana* sp. Steinernematidae). *Parasitology* 56, 385–390.

Prasad, B.C. and Reed, R.R. (1999) Chemosensation: molecular mechanisms in worms and mammals. *Trends in Genetics* 15, 150–153.

Rebrikov, D.V., Britanova, O.V., Gurskaya, N.G., Lukyanov, K.A., Tarabykin, V.S. and Lukyanov, S.A. (2000) Mirror orientation selection (MOS): a method for eliminating false positive clones from libraries generated by suppression subtractive hybridization. *Nucleic Acids Research* 28, E90.

Robertson, H.M. (2001) Updating the str and srj (stl) families of chemoreceptors in *Caenorhabditis* nematodes reveals frequent gene movement within and between chromosomes. *Chemical Senses* 26, 151–159.

Sandhu, S.K., Jagdale, G.B., Hogenhout, S.A. and Grewal, P.S. (2006) Comparative analysis of the expressed genome of the infective juvenile entomopathogenic nematode, *Heterorhabditis bacteriophora. Molecular and Biochemical Parasitology* 145, 239–244.

Segal, D. and Glazer, I. (1998) Genetic approaches for enhancing beneficial traits in entomopathogenic nematodes. Recent development of biological control by beneficial nematodes. *Japanese Journal of Nematology* 28, 61–67.

Shapiro, D.I., Glazer, I. and Segal, D. (1997) Genetic diversity in wild and laboratory populations of Heterorhabditis bacteriophora as determined by RAPD-PCR analysis. *Fundamental and Applied Nematology* 20, 581–585.

Sicard, M., Brugirard-Ricaud, K., Pages, S., Lanois, A., Boemare, N.E., Brehelin, M. and Givaudan, A. (2004) Stages of infection during the tripartite interaction between *Xenorhabdus nematophila*, its nematode vector, and insect hosts. *Applied and Environmental Microbiology* 70, 6473–6480.

Snyder, H.A., Stock, S.P., Flores-Lara, Y., Kim, S.K. and Forst, S. (2007) New insights into the colonization and release process of *Xenorhabdus nematophila* and the morphology and ultrastructure of the bacterial receptacle of its nematode host, *Steinernema carpocapsae. Applied and Environmental Microbiology* 73, 5338–5346.

Sonnichsen, B., Koski, L.B., Walsh, A., Marschall, P., Neumann, B., Brehm, M., Alleaume, A.M., Artelt, J., Bettencourt, P., Cassin, E., Hewitson, M., Holz, C., Khan, M., Lazik, S., Martin, C., Nitzsche, B., Ruer, M., Stamford, J., Winzi, M., Heinkel, R., Roder, M., Finell, J., Hantsch, H., Jones, S.J., Jones, M., Piano, F., Gunsalus, K.C., Oegema, K., Gonczy, P., Coulson, A., Hyman, A.A. and Echeverri, C.J. (2005) Full-genome RNAi profiling of early embryogenesis in *Caenorhabditis elegans. Nature* 434, 462–469.

Spiridonov, S.E., Krasomil-Osterfeld, K. and Moens, M. (2004) "*Steinerne ma jollieti* sp. n. (Rhabditida: Steinernematidae), a new entomopathogenic nematode from the American Midwest." *Russian Journal of Nematology* 12(1), 85–95.

Stabb, E.V. and Ruby, E.G. (2002) RP4-based plasmids for conjugation between *Escherichia coli* and members of the Vibrionaceae. *Methods in Enzymology* 358, 413–426.

Stein, L.D., Bao, Z., Blasiar, D., Blumenthal, T., Brent, M.R., Chen, N., Chinwalla, A., Clarke, L., Clee, C., Coghlan, A., Coulson, A., D'Eustachio, P., Fitch, D.H., Fulton, L.A., Fulton, R.E., Griffiths-Jones, S., Harris, T.W., Hillier, L.W., Kamath, R., Kuwabara, P.E., Mardis, E.R., Marra, M.A., Miner, T.L., Minx, P., Mullikin, J.C., Plumb, R.W., Rogers, J., Schein, J.E., Sohrmann, M., Spieth, J., Stajich, J.E., Wei, C., Willey, D., Wilson, R.K., Durbin, R. and Waterston, R.H. (2003) The genome sequence of *Caenorhabditis briggsae*: a platform for comparative genomics. *PLoS Biology* 1, E45.

Tabara, H., Grishok, A. and Mello, C.C. (1998) RNAi in *C. elegans*: soaking in the genome sequence. *Science* 282, 430–431.

Tavernarakis, N., Wang, S.L., Dorovkov, M., Ryazanov, A. and Driscoll, M. (2000) Heritable and inducible genetic interference by double-stranded RNA encoded by transgenes. *Nature Genetics* 24, 180–183.

Taylor, M.J., Bandi, C. and Hoerauf, A. (2005) *Wolbachia* bacterial endosymbionts of filarial nematodes. *Advances in Parasitology* 60, 245–284.

Timmons, L., Court, D.L. and Fire, A. (2001) Ingestion of bacterially expressed dsRNAs can produce specific and potent genetic interference in *Caenorhabditis elegans. Gene* 263, 103–112.

Vivas, E.I. and Goodrich-Blair, H. (2001) *Xenorhabdus nematophilus* as a model for host-bacterium interactions: *rpoS* is necessary for mutualism with nematodes. *Journal of Bacteriology* 183, 4687–4693.

Waddell, C.S. and Craig, N.L. (1989) Tn*7* transposition: recognition of the attTn*7* target sequence. *Proceedings of the National Academy of Science of the USA* 86, 3958–3962.

Webster, J., Chen, G., Hu, K. and Li, J. (2002) Bacterial metabolites. In: Gaugler, R. (ed.) *Entomopathogenic Nematology*. CAB International, Wallingford, UK, pp. 79–98.

Williams, J.S., Thomas, M. and Clarke, D.J. (2005) The gene *stlA* encodes a phenylalanine ammonia-lyase that is involved in the production of a stilbene antibiotic in *Photorhabdus luminescens* TT01. *Microbiology* 151, 2543–2550.

Winston, W.M., Molodowitch, C. and Hunter, C.P. (2002) Systemic RNAi in *C. elegans* requires the putative transmembrane protein SID-1. *Science* 295, 2456–2459.

Xu, J. and Hurlbert, R.E. (1990) Toxicity of irradiated media for *Xenorhabdus* spp. *Applied and Environmental Microbiology* 56, 815–818.

Zhu, Y.Y., Machleder, E.M., Chenchik, A., Li, R. and Siebert, P.D. (2001) Reverse transcriptase template switching: a SMART approach for full-length cDNA library construction. *Biotechniques* 30, 892–897.

Zioni, S., Glazer, I. and Segal, D. (1992) Phenotypic and genetic analysis of a mutant of *Heterorhabditis bacteriophora* strain HP88. *Journal of Nematology* 24, 359–364.

# IV Genomics and Genetic Engineering

# 12 Genetic Engineering of Bacteria to Improve Efficacy Using the Insecticidal Proteins of *Bacillus* Species

H.-W. Park[1] and B.A. Federici[2]

[1]*John A. Mulrennan, Sr, Public Health Entomology Research and Education Center, College of Engineering Sciences, Technology and Agriculture, Florida A & M University, Panama City, USA;* [2]*Department of Entomology and Interdepartmental Graduate Programmes in Genetics, Genomics and Bioinformatics and Cell, Molecular and Developmental Biology, University of California, Riverside, USA*



## 12.1. Introduction

Entomopathogenic bacteria are characterized by having low invasive power through the insect's midgut, but once within the haemocoel, they are highly pathogenic. *Pseudomonas aeruginosa* and *Serratia marcescens* are well-known examples of the families Pseudomonadaceae and Enterobacteriaceae, respectively (Bucher and Stephens, 1957; Bucher, 1960). Members of the family Vibrionaceae are primarily inhabitants of sea and freshwater, and *Aeromonas punctata* is known to cause black lesions in larvae of *Anopheles annulipes* (Kalucy and Daniel, 1972). Although these Gram-negative organisms have potential for control of disease vectors and agricultural insect pests, the Gram-positive bacteria have proven to be the most useful pathogens for insect control (Priest, 2000). The families Bacillaceae and Streptococcaceae belong to the Gram-positive bacteria, and members of the former family, particularly *Bacillus thuringiensis*, which is used as a bacterial insecticide, have received the most attention because, like chemical insecticides, they are fast-acting. Furthermore, they have high insect specificity, are easy to mass produce and formulate, and can be stored, often for years, with little or no loss of activity (Federici, 1991).

Extensive research on the molecular biology of *B. thuringiensis* beginning early in the 1980s made it possible to construct recombinant new strains with improved efficacy by manipulating the insecticidal endotoxin protein genes and the genetic elements that control their synthesis (Schnepf *et al.*, 1998; Federici *et al.*, 2003, 2006). Furthermore, other bacterial species have also been used to synthesize insecticidal proteins of *B. thuringiensis* to, for example, improve environmental persistence. In this chapter, therefore, we describe key aspects of the basic biology of insecticidal proteins produced by *B. thuringiensis*, including the genetic factors that regulate their synthesis, and, then show how these have been manipulated in this species and others to improve efficacy. We also provide summaries of methods for engineering *B. thuringiensis*, and a protocol for genetic transformation of this species that has worked well in our laboratories.

## 12.2. Basic Biology of *Bacillus thuringiensis*

As a spore-forming bacterium, *B. thuringiensis* is widely distributed in the environment and can be readily isolated on simple media such as nutrient agar from soil, water, plants, insect faeces and grain dust (Schnepf *et al.*, 1998; Federici *et al.*, 2006). When nutrients are sufficient for growth, the spore germinates producing a vegetative cell that grows and reproduces by binary fission. The bacterium continues to multiply until nutrients become insufficient for continued vegetative growth. Under these conditions, the bacterium sporulates producing a spore and a parasporal body, the latter composed primarily of insecticidal protein toxins (Fig. 12.1).

The parasporal body is the principal characteristic used to differentiate this species from closely related species, *Bacillus cereus*, and other bacilli. This parasporal body contains one or more proteins, typically as crystalline inclusions, and most of these are highly toxic to one or more species of insects. The toxins are known as endotoxins and occur in the parasporal body as protoxins, which dissolve after ingestion and are converted to active toxins through cleavage by proteolytic enzymes in the insect gut. The activated toxins bind to the midgut microvillar membrane in sensitive insects, lyse the cells and destroy the most sensitive portions of the midgut epithelium causing death. Thus, *B. thuringiensis* endotoxins are stomach poisons selective for insects that belong to the order of Lepidoptera, Coleoptera and Diptera, and certain other invertebrates such as nematodes.

**Fig. 12.1.** Scanning electron micrograph of purified crystal inclusions of the HD-1 isolate of *Bacillus thuringiensis* ssp. *kurstaki*. The bipyramidal crystals contain Cry1Aa, Cry1Ab, Cry1Ac, whereas the smaller cuboidal crystals contain Cry2Ab. (Modified from Moar *et al.*, 1989.)

Although constituting a single species, *B. thuringiensis* is actually a complex of subspecies consisting of more than 70 subspecies distinguished from one another on the basis of immunological differences in flagellar (H antigen) serotype (Lecadet *et al.*, 1999). Each subspecies name corresponds with a specific H antigen number. For example, *B. thuringiensis* ssp. *kurstaki* is H3a3b3c, whereas *B. thuringiensis* ssp. *israelensis* is H14. Because of the H antigen serotype, subspecies name often does not correlate with insecticidal properties; letters and numbers are used to designate specific isolates. For example, HD-1 is used to designate a specific isolate of *B. thuringiensis* ssp. *kurstaki* that produces four major endotoxin proteins – Cry1Aa, Cry1Ab, Cry1Ac and Cry2A – and has a broad spectrum of activity against lepidopteran pests. Another isolate of *B. thuringiensis* ssp. *kurstaki* is HD-73. This produces only a single endotoxin protein of Cry1Ac and as a result has a much narrower spectrum of activity against insects than does HD-1. *B. thuringiensis* ssp. *kurstaki* isolate HD-1 (Dulmage, 1970), *B. thuringiensis* ssp. *morrisoni* isolate DSM2803 (Krieg *et al.*, 1983) and *B. thuringiensis* ssp. *israelensis* isolate ONR60A (Goldberg and Margalit, 1977) are the most well-known examples of *B. thuringiensis* toxic to lepidopterous, coleopterous and dipterous insects, respectively, and have been developed into commercial bacterial insecticides. More recently, transgenic plants using some of *cry* genes from these isolates have been successfully introduced into the market.

## 12.3. Insecticidal Proteins of *Bacillus thuringiensis*

For most insect pests, the parasporal body which consists of one or more insecticidal toxin proteins accounts for most of this bacterium's activity. These proteins are generally referred to as δ-endotoxins because they assemble into parasporal inclusions within the bacterial cell after synthesis. In the early 1980s, shortly after the development of recombinant DNA techniques, it was discovered that *B. thuringiensis* δ-endotoxins were encoded by genes carried on plasmids. This discovery led quickly to a major research effort in many laboratories to understand the genetics and molecular biology of these toxins. These efforts resulted in cloning and sequencing of numerous δ-endotoxin genes, along with characterization of the toxicity and target spectrum of the protein encoded by each gene. As a consequence, a wide variety of confusing names were being used to refer to δ-endotoxin genes and proteins until Höfte and Whiteley (1989) proposed a simplified nomenclature for naming all insecticidal *B. thuringiensis* genes and proteins based on the spectrum of activity of the proteins as well as their size and apparent relatedness as deduced from nucleotide and amino acid sequence data. In this nomenclature, the proteins are referred to as Cry (for crystal) and Cyt (for cytolytic) proteins. Since this publication, the number of Cry and Cyt proteins has increased dramatically and as more and more genes were sequenced and analysed, it was decided to name genes based on their relatedness as determined primarily from the degree of their deduced amino acid identity. Therefore, the modified nomenclature system in which names of genes were determined solely by their deduced amino acid sequence identity has been suggested (Crickmore *et al.*, 1998). Although the new designations supposedly carry no specific information concerning insecticidal spectrum, because the numbers have been maintained for many of the genes listed by Höfte and Whiteley (1989), and because a high degree of correlation between relatedness and insecticidal spectrum remains, primary insecticidal activity can often be inferred. For example, Cry1 still refers to lepidopteran toxicity; Cry2, to lepidopteran and dipteran toxicity; Cry3, to coleopteran toxicity; and Cry4, to dipteran toxicity (Table 12.1).

**Table 12.1.** Nomenclature for representative insecticidal proteins of *Bacillus thuringiensis*.

| Old nomenclature[a] | New nomenclature[b] | Insect spectrum |
|---|---|---|
| CryIA(a) | Cry1Aa | Lepidoptera |
| CryIA(b) | Cry1Ab | Lepidoptera |
| CryIA(c) | Cry1Ac | Lepidoptera |
| CryIB | Cry1Ba | Lepidoptera |
| CryIC | Cry1Ca | Lepidoptera |
| CryIIA | Cry2Aa | Lepidoptera/diptera |
| CryIIB | Cry2Ab | Lepidoptera |
| CryIIIA | Cry3Aa | Coleoptera |
| CryIIIB | Cry3Ba | Coleoptera |
| CryIVA | Cry4Aa | Diptera |
| CryIVB | Cry4Ba | Diptera |
| CryIVC | Cry10Aa | Diptera |
| CryIVD | Cry11Aa | Diptera |
| Jeg80 | Cry11Ba | Diptera |
| CytA | Cyt1Aa | Diptera, Coleoptera |
| CytB | Cyt2Aa | Diptera |

[a]Höfte and Whiteley, 1989.
[b]Crickmore *et al.*, 1998.

Cry proteins fall into two groups – one with molecular masses of 130–140 kDa, and the other with masses in the range of 65–70 kDa (Schnepf *et al.*, 1998). The former is represented by Cry1 and Cry4 proteins. Only the N-terminal half of these is toxic and the C-terminal half is known to facilitate crystallization of toxin molecules after synthesis (Aronson, 1993; Thompson *et al.*, 1995; Park *et al.*, 2000). With respect to the latter class, typical examples are Cry2A, Cry3A and Cry11A, which lack the C-terminal half characteristic of Cry1 proteins. Therefore, the 65–70 kDa proteins are in essence naturally truncated versions of the 130–140 kDa toxins and composed primarily of the toxic portion of the molecule. Despite the absence of a crystallizing domain, these proteins crystallize readily. Among them, it is known that a 29 kDa protein assists formation of Cry2A crystals, and that a 20 kDa chaperone-like protein enhances net synthesis of Cry11A and promotes Cyt1A crystal formation (Crickmore and Ellar, 1992; Wu and Federici, 1993, 1995; Ge *et al.*, 1998; Park *et al.*, 1999).

Four Cry protein structures have been solved – Cry1Ac (Grochulski *et al.*, 1995), Cry2A (Morse *et al.*, 2001), Cry3A (Li *et al.*, 1991) and Cry4B (Boonserm *et al.*, 2005). All consist of three domains. Domain I contains five to seven antiparallel α-helices in which helix 5 is encircled by the other helices. The long hydrophobic and amphipathic helices of Domain I suggest that this domain forms the lytic pores in the insect midgut. Domain II consists of three antiparallel β-sheets, and the loops at the tips of these are thought to bind the toxin to receptors on microvilli. Domain III consists of two antiparallel β-sheets, which form a β-sandwich thought to maintain structural integrity of the molecule. In addition, it has been shown that the β-sheet structure of Domain III can also participate in receptor binding, membrane penetration and ion channel formation, indicating this domain may have functions other than those proposed originally.

Cyt proteins are highly hydrophobic and all have a mass in the range of 24–28 kDa (Crickmore *et al.*, 1998). They share no significant amino acid sequence identity with Cry proteins, and have only been reported from mosquitocidal strains. The first Cyt protein, Cyt1A, was identified as a component of the parasporal body of *B. thuringiensis* ssp. *israelensis* and is cytolytic to a wide range of invertebrate and vertebrate cells *in vitro* (Thomas and Ellar, 1983). Cyt1A is not very toxic by itself, but synergizes the toxicity of Cry proteins (Wu and Chang, 1985; Ibarra and Federici, 1986; Wu *et al.*, 1994; Crickmore *et al.*, 1995). This synergism accounts for most of high toxicity of *B. thuringiensis* ssp. *israelensis*. It has also been shown that Cry proteins of *B. thuringiensis* ssp. *israelensis* synergize each other, further contributing to its high toxicity (Poncet *et al.*, 1995). In addition to its capacity to synergize Cry proteins of *B. thuringiensis* ssp. *israelensis*, several recent studies show or provide evidence that Cyt1A can overcome resistance to *B. thuringiensis* ssp. *israelensis* Cry proteins and *Bacillus sphaericus* Bin protein, can extend the mosquito target spectrum of *B. sphaericus* and can delay the development of resistance to *B. thuringiensis* ssp. *israelensis* Cry proteins (Georghiou and Wirth, 1997; Wirth *et al.*, 1997, 2000a,b, 2005).

The crystal structure for Cyt2A has been solved, and based on sequence similarities among Cyt proteins, it is assumed all have a similar structure. In contrast to the three-domain structure of activated Cry toxins, the Cyt2A molecule is a single domain consisting of a β-sheet core wrapped in two outer layers of α-helix hairpins (Li *et al.*, 1996).

## 12.4. Genetic Factors Regulating Insecticidal Proteins

The primary genetic factors affecting insecticidal protein synthesis in *B. thuringiensis* are promoters, a 5' mRNA stabilizing sequence and 3' transcriptional termination sequences.

### 12.4.1. Promoters

In *Bacillus* species, the endospore develops in a sporangium consisting of two cellular compartments, the mother cell and the forespore. In *Bacillus subtilis*, the developmental process is temporally regulated at the transcriptional level by the successive activation of six σ factors that by binding to RNA polymerase determine which gene promoters are recognized (Helmann and Moran, 2002). These factors are $\sigma^A$, the primary sigma factor of vegetative cells, and five factors that are activated during sporulation, $\sigma^E$, $\sigma^F$, $\sigma^G$, $\sigma^H$ and $\sigma^K$, in order of their occurrence during sporulation. The $\sigma^A$ and $\sigma^H$ factors are active in the pre-divisional cell, $\sigma^E$ and $\sigma^K$ are active in the mother cell and $\sigma^F$ and $\sigma^G$ are active in the forespore. In *B. thuringiensis*, two genes encoding sigma factors, $\sigma^{35}$ and $\sigma^{28}$, which show, respectively, 88% and 85% amino acid sequence identity with $\sigma^E$ and $\sigma^K$ of *B. subtilis* have been cloned (Adams *et al.*, 1991). In *B. thuringiensis*, there are two primary sporulation-dependent promoters, BtI and BtII. The BtI promoter is transcribed by $\sigma^{35}$ complexed with the RNA polymerase (Brown and Whiteley, 1988), whereas the BtII promoter is transcribed by the $\sigma^{28}$ complexed with the RNA polymerase (Brown and Whiteley, 1990).

Over the years, several *cry* promoters have been identified and their sequences determined. Consensus sequences for promoters recognized by *B. thuringiensis* RNA polymerase containing $\sigma^E$- or $\sigma^K$-like factors have been deduced from alignment of the promoter regions of these genes (Baum and Malvar, 1995). The results indicate that the transcription of many other *cry* genes is likely to be $\sigma^E$- or $\sigma^K$-dependent. Unlike BtI and BtII, the *cry3A* promoter is similar to promoters recognized by $\sigma^A$. The expression of *cry3A* is not dependent on sporulation-specific σ factors in either *B. subtilis* or *B. thuringiensis* (Agaisse and Lereclus, 1994; Lereclus *et al.*, 1995).

### 12.4.2. 5' mRNA stabilizing sequence

The 5' region of the *cry3A* transcript beginning at nucleotide position 129 contains a region that stabilizes this mRNA (Agaisse and Lereclus, 1994). Fusion of this region to the 5' region of the *lacZ* gene transcribed from a promoter inducible in *B. subtilis* increased the stability of the *lacZ* fusion mRNA and resulted in a tenfold increase of both steady-state mRNA and β-galactosidase synthesis (Agaisse and Lereclus, 1996).

The determinant of stability appears to be a consensus Shine-Dalgarno (SD) sequence, designated STAB-SD, close to the 5' end of the *cry3A* mRNA (Agaisse and Lereclus, 1996). Mutations introduced into this region suggest that this sequence provides stability through interaction with the 3' end of the 16S rRNA. Therefore, the binding of a 30S ribosomal subunit to the SD sequence located in the 5' untranslated region of *cry3A* apparently stabilizes the corresponding transcript by protecting it against 5'-3' ribosomal activity. Such SD sequences are also present in similar positions in at least two other members of *cry3* gene family, *cry3Ba1* and *cry3Ba2* (Agaisse and Lereclus, 1996; Crickmore *et al.*, 1998).

### 12.4.3. 3' transcriptional termination sequence

Wong and Chang (1986) showed that a non-coding region near the 3' terminus of *cry1Aa* from *B. thuringiensis* ssp. *kurstaki* HD-1 acts as a positive retroregulator, i.e. serves as a *cis*-acting element that regulates a target gene from a distance. The fusion of this fragment with the 3' end of heterologous genes increased transcript half-life and consequently the amount of Cry protein synthesized.

The activity of 3'-5' exonucleases is affected by RNA secondary structure. In particular, their rate of mRNA degradation is impeded by 3' stem-loop structures. Therefore, it is likely that *cry* and *cyt* gene terminators are involved in mRNA stability by protecting the mRNA from exonucleotic degradation from the 3' end. The putative terminator sequences downstream from various *cry* genes are widely conserved. Recently, it has been shown that the orientation of the *cry3A* transcription terminators was important to enhance truncated *cry1C* transcript stability and protein synthesis (Park *et al.*, 2000).

## 12.5. Construction of Recombinant Bacteria

Due to their high toxicity and specificity, *cry* and *cyt* protein genes of *B. thuringiensis* have been introduced into *B. thuringiensis* and several other bacterial species to improve efficacy using either plasmids that can replicate in the host or by integrating genes into host chromosomal DNA. Although *B. thuringiensis* is still the most successful organism used as a host to synthesize these proteins, other bacterial species discussed below have also been used. Beginning with the use of *B. thuringiensis* as the host cell, we provide examples of how several bacterial species were transformed and genetically engineered to improve efficacy.

### 12.5.1. *Bacillus thuringiensis*

Transfer of plasmids into *B. thuringiensis* was first reported via cell mating, also known as conjugation (González *et al.*, 1982; González and Carlton, 1984). Using this method, transformation efficiency was low, and as these plasmids lacked a selectable marker, screening cells for transformants was slow and cumbersome (Aronson *et al.*, 1986). Several years later, improved protocols for transformation of *B. thuringiensis* using electroporation were published independently, and these new methods accelerated research on the construction of recombinant strains of *B. thuringiensis* (Bone and Ellar, 1989; Lereclus *et al.*, 1989; Masson *et al.*, 1989). These protocols provided high transformation efficiency and made transformants easy to recognize and recover by using antibiotics as selectable markers; their development greatly facilitated basic research and engineering of *B. thuringiensis*. An example of a protocol based on these principles is described below.

*Bacillus thuringiensis* transformation procedure using electroporation:

1. Inoculate 3 ml of Luria-Bertani broth (LB) medium using a *B. thuringiensis* strain freshly grown on a nutrient agar plate at 30°C with appropriate antibiotics where applicable.
2. Incubate it overnight using shaking incubator (30°C, 250 rpm).
3. Take 1 ml of overnight culture and inoculate the fresh 100 ml LB medium (1:100).
4. Let the culture grow at 30°C, 250 rpm.
5. Harvest cells at 5000 g for 10 min at 4°C when $OD_{600}$ reaches 0.6–0.8 (it usually takes about 3 h under this condition).
6. Discard supernatant.
7. Wash pellet three times at 5000 g for 10 min at 4°C using ice-cold sterilized double-distilled water.
8. Resuspend pellet in 2 ml ice-cold 10% glycerol.
9. Make 200 µl aliquots using sterilized 1.5 ml tubes.
10. Add 1–5 µg of plasmid DNA to 200 µl of cell aliquot and mix by pipetting.
11. Transfer the mixture into a 0.2 cm cuvette. Be cautious not to create any bubble while transfer.
12. Apply a pulse. The default setting would be 1.5 kV, 400 Ohm, 25 µF.
13. Immediately add 3 ml of pre-warmed Brain Heart Infusion (BHI) to the mixture.
14. Transfer the whole content into a culture tube.
15. Incubate the tube at 37°C for 1 h at 150 rpm.
16. Plate the culture on to BHI plates containing appropriate antibiotics and incubate plates overnight at 30°C.
17. Check plates for transformants. Colonies normally appear within 24 h.

The most common strategy for constructing recombinant *B. thuringiensis* strains is using a shuttle expression vector, such as pHT3101 (Lereclus *et al.*, 1989) that contains replication origins for both *B. thuringiensis* and *Escherichia coli*, genes for resistance, for example to ampicillin and erythromycin, for easy selection of transformants and a multi-cloning site. A shuttle vector containing the gene of interest is amplified in *E. coli*, isolated, and subsequently introduced into the desirable *B. thuringiensis* strain by electroporation.

In many cases, *cry* and *cyt* genes of *B. thuringiensis* inserted into shuttle vectors were expressed under the control of their own promoters, which typically resulted in a high yield of the encoded protein. In terms of promoter strength, *cyt1A* promoters are the strongest known among *cry* and *cyt* genes (Crickmore and Ellar, 1992; Baum and Malvar, 1995; Federici *et al.*, 2006). In addition, as mentioned above, the *cry3A* upstream 5' mRNA stabilizing sequence (STAB-SD) improves stability of *cry3A* transcripts and concomitantly the yield of certain Cry 3A (Agaisse and Lereclus, 1996). Therefore, to optimize Cry and Cyt protein yields in *B. thuringiensis*, a recombinant expression vector pSTAB was developed (Park *et al.*, 1998, 1999). This vector was constructed by inserting the 660 bp DNA fragment containing *cyt1A* promoters combined with the STAB-SD sequence into the multi-cloning site of pHT3101 (Fig. 12.2).

Using the pSTAB expression vector, which combined these different genetic elements, we were able to significantly increase yields of several Cry proteins. For example, by expressing the *cry3A* gene using this vector, we were able to obtain yields 12-fold greater than those obtained with the wild-type strain of *B. thuringiensis* ssp.

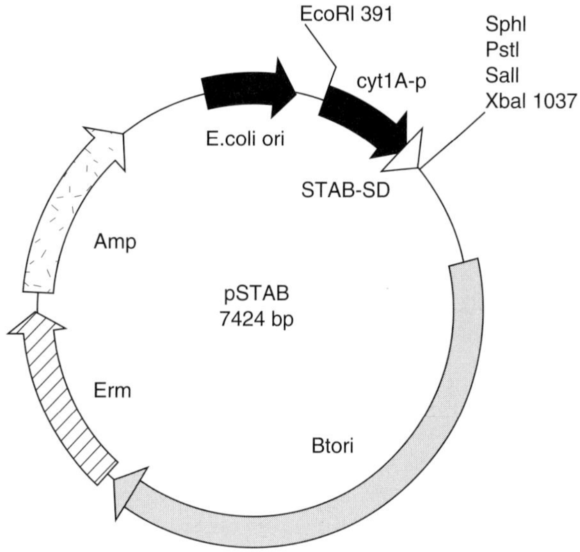


(A)

```
EcoRI
GAATTCTATT  TTCGATTTCA  AATTTTCCAA  ACTTAAATAT  GATTGAATGC   50
CTGAGAAAGG  TAATAGAGAT  GTTTTAGTTT  ATTATGAAGT  ATTAGGGGCG  100
TCTTTTAAAT  TCAATCTATC  AATTTGTGAA  ATATATTACT  CAAAACCCAA  150
TACCATTCTA  AAACTTATTC  AAAATATATA  TTGCTTTAAA  AGAGCATACA  200
TACTAAAAAA  ACAGGCATCT  TTCGAACTAT  AGCGCATAGA  ATACTACGGT  250
                 -35                     -10
GAATCAAAAA  CAAATAAAAT  TTAGGAGGTA  TATTCAAGTA  TACAAAAAAA  300
CTTTAGTGTG  AGGGGATTTA  GATAAAAAGT  ATTCGTTATC  CTTATAAATT  350
AATTCTTAAA  CATGCACCAA  TGTATACATT  AAATAATATT  ATGTGAATTA  400
               -35                     -10
AGTCTATCAA  TTTAATTTAT  TATGTTACTT  TATATTTGAT  TAATAATTGC  450
AAGTTTAAAA  TCATAATTTA  ATGTTGAAAG  GCCACTATTC  TAATTAACTT  500
AAGGAGTTGT  TTATTTGAGC  TCGGTACCCG  GGGATAATCT  TGAAAGGAGG  550
                                                STAB-SD
GATGCCTAAA  AACGAAGAAC  ATTAAAAACA  TATATTTGCA  CCGTCTAATG  600
GATTTATGAA  AAATCATTTT  ATCAGTTTGA  AAATTATGTA  TTATGATAAG  650
      XbaI
AAAGTCTAGA                                                  660
```

(B)

**Fig. 12.2.** The *Bacillus thuringiensis* expression plasmid, pSTAB. (A) Physical map of pSTAB. Amp, ampicillin-resistant gene; Erm, erythromycin-resistant gene; *Escherichia coli* ori, *E. coli* replication origin; Bt ori, *B. thuringiensis* replication origin; cyt1A-p, *cyt1A* promoters. (B) Nucleotide sequence of the 660 bp DNA fragment containing *cyt1A* promoters combined with the STAB-SD sequence. The BtI (in boxes) and BtII (underlined) promoters of *cyt1A* are shown, and the STAB-SD sequence is highlighted as a black box. (Modified from Park *et al.*, 1999.)

*morrisoni* (isolate DSM2803) from which this gene was cloned (Park *et al.*, 1998). The yield of Cry3A obtained per unit medium using *cyt1A* promoters alone, i.e. lacking the STAB-SD sequence, was only about twofold higher than that of the wild-type DSM280 strain (Fig. 12.3). This demonstrates that most of the enhancement was due to inclusion of the STAB-SD sequence.

The significant increase in Cry3A yield obtained using *cyt1A* promoters combined with the STAB-SD sequence led us to test this expression vector for enhancing synthesis of other insecticidal proteins in *B. thuringiensis*. Results of these later studies showed that the level of enhancement using this expression system varies depending upon the candidate protein (Park *et al.*, 1999, 2000, 2005). For example, yields of Cry11B and the binary toxin of *B. sphaericus*, as discussed in the following sections, were increased substantially, as much as eightfold, whereas yields of proteins such as Cry11A and Cry2A increased only 1.5–2 fold.

As our research is primarily directed towards improving mosquitocidal bacteria, our best examples of the successful use of pSTAB come from engineering recombinant strains of *B. thuringiensis* ssp. *israelensis*. We have used this vector to produce several different recombinant strains that vary in complexity, ranging from a strain that produces only a single endotoxin to strains that produce as many as five endotoxins. In the simplest case, we used pSTAB to express the binary (Bin) toxin operon of *B. sphaericus* 2362 in the acrystalliferous strain 4Q7 of *B. thuringiensis* ssp. *israelensis* (Park *et al.*, 2005). The Bin toxin of *B. sphaericus* (Baumann *et al.*, 1987) consists of a 51 kDa binding domain (BinA) and a 42 kDa toxin domain (BinB). Using the pSTAB vector to express the *bin* operon alone (under control of *cyt1A* promoters), synthesis of Bin was eightfold higher than that obtained with wild-type *B. sphaericus* 2362 (Fig. 12.4). Whereas wild-type *B. sphaericus* typically

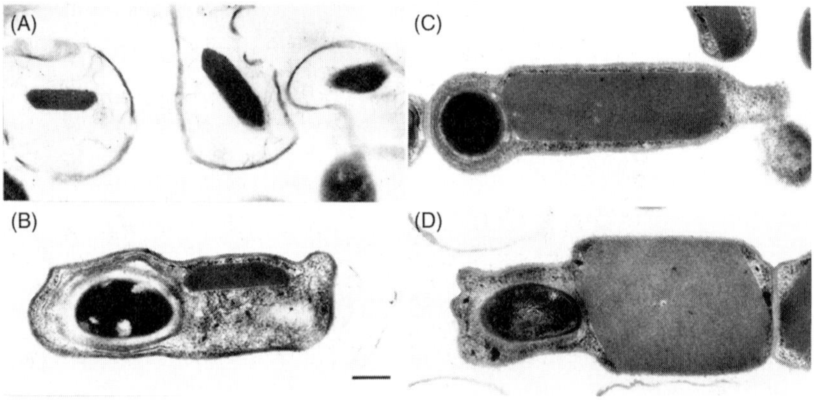


**Fig. 12.3.** Transmission electron micrographs of wild type and recombinant *Bacillus thuringiensis* strains producing Cry3A. (A) Wild-type *B. thuringiensis* ssp. *morrisoni* (strain tenebrionis) DSM 2803. (B) Acrystalliferous strain (4Q7) of *B. thuringiensis* ssp. *israelensis* transformed with pPFT3A (*cry3A* without the STAB-SD sequence under the control of *cyt1A* promoters). (C and D) Cross section (C) and sagittal section (D) through 4Q7 cells transformed with pPFT3As (*cry3A* with the STAB-SD sequence under the control of *cyt1A* promoters). All micrographs are at the same magnification. Bar, 300 nm. (From Park *et al.*, 1998.)

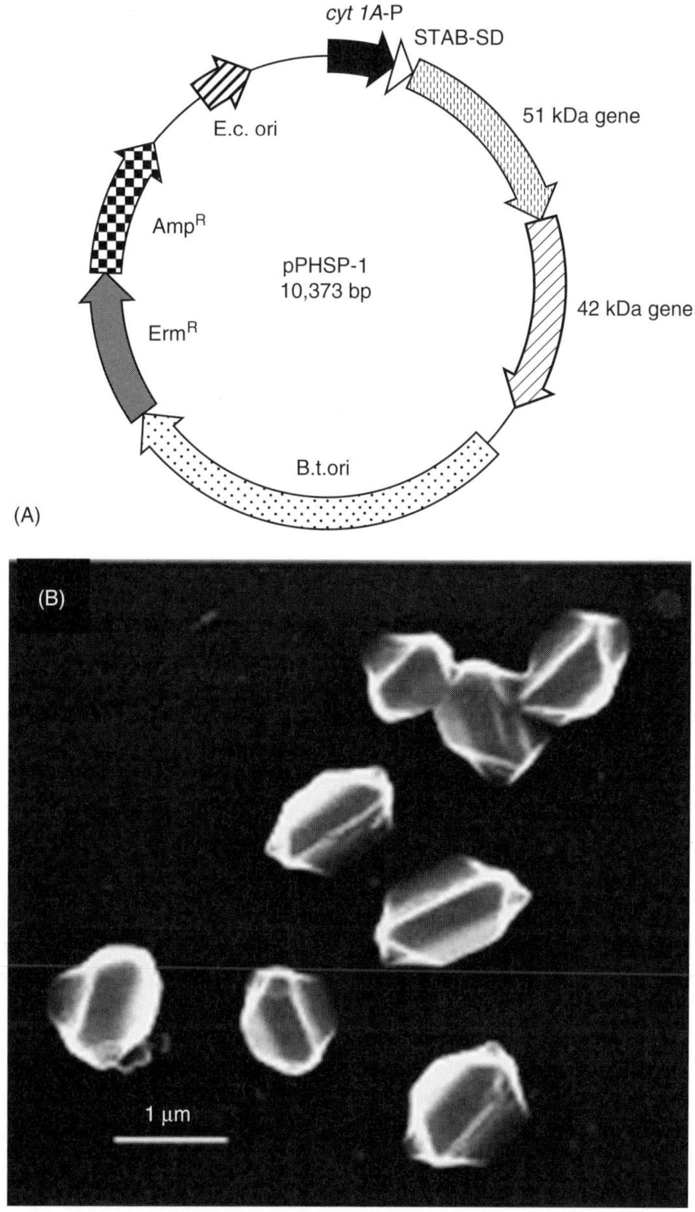


**Fig. 12.4.** Synthesis of the *Bacillus sphaericus* Bin toxin in *Bacillus thuringiensis*. (A) Map of pPHSP-1 that contains the *bin* toxin operon of *B. sphaericus* strain 2362 under control of *cyt1A* promoters combined with the STAB-SD sequence. (B) Scanning electron micrograph of Bin toxin crystals from *B. sphaericus* 2362 synthesized in the 4Q7 strain (acrystalliferous) of *B. thuringiensis* ssp. *israelensis* using the pSTAB expression vector. These crystals are approximately eightfold larger than those produced by wild-type *B. sphaericus* 2362. E.c. ori, *E.coli* origin of replication; $Amp^R$, ampicillin-resistant gene; $Erm^R$, erythromycin-resistant gene; B.t. ori, *B. thuringiensis* replication origin; bp, base pairs. (Modified from Park *et al.*, 2005.)

has an $LC_{50}$ in the range of 8–12 ng/ml against fourth instars of *Culex quinquefasciatus*, the 4Q7 strain that produces the Bin toxin has an $LC_{50}$ of 1.4 ng/ml (Park *et al.*, 2005). However, as this recombinant, like wild-type *B. sphaericus*, only produces a single toxin, it is likely its use would lead the development of resistance in target populations.

To improve toxicity while at the same time preventing or delaying the development of resistance, we made several strains in which we increased toxin complexity and added the Cyt1A protein for resistance management, the efficacy of which we established in several papers (Wirth *et al.*, 1997, 2000a,b, 2005). Previously, Li *et al.* (2000) attempted to make a similar strain. They used a shuttle expression vector pBU-4 to synthesize the Bin toxin of *B. sphaericus* C3–41 along with the Cyt1A protein of *B. thuringiensis* ssp. *israelensis* in an acrystalliferous strain of *B. thuringiensis*. However, the recombinant strain producing the Bin toxin and Cyt1A showed very poor toxicity against both sensitive ($LC_{50}$ = 1.12 μg/ml) and resistant ($LC_{50}$ = 2116.33 μg/ml) colonies of *C. quinquefasciatus*. In our studies, one of the first strains we constructed using this strategy was a recombinant that synthesized the Bin toxin, Cyt1A and Cry11B (Park *et al.*, 2003). In this recombinant, which again used the 4Q7 strain of *B. thuringiensis* ssp. *israelensis* as the host cell, the mosquitocidal proteins were from three different species: (i) Bin from *B. sphaericus* 2362; (ii) Cry11B from *B. thuringiensis* ssp. *Jegathesan*; and (iii) Cyt1A from *B. thuringiensis* ssp. *israelensis*. The Cry11B protein is 58% identical to Cry11A but more toxic than the latter, the most toxic mosquitocidal protein produced by *B. thuringiensis* ssp. *israelensis* (Delécluse *et al.*, 1995). This recombinant was constructed using a dual-plasmid expression system with two different plasmids, each with a different antibiotic resistance gene for selection (Fig. 12.5). The resulting recombinant *B. thuringiensis* produced three distinct crystals (Fig. 12.6), apparently one for each of these proteins, i.e. Cyt1A, Cry11B and the Bin toxin, and was significantly more toxic ($LC_{50}$ = 1.7 ng/ml) to *C. quinquefasciatus* fourth instars than either *B. thuringiensis* ssp. *israelensis* IPS-82 ($LC_{50}$ = 7.9 ng/ml) or *B. sphaericus* 2362 ($LC_{50}$ = 12.6 ng/ml).

To construct a recombinant with an even greater range of endotoxins for both increased toxicity and resistance management, we transformed the IPS-82 strain of *B. thuringiensis* ssp. *israelensis*, which produces the complement of toxins characteristic of this species, with pPHSP-1, the pSTAB plasmid that produces a high level of the *B. sphaericus* Bin toxin (Fig. 12.7). When mortality was obtained after 48 h of exposure, $LC_{50}$s of this recombinant were 0.014 and 3.8 ng/ml, respectively, against *C. quinquefasciatus* and *Culex tarsalis*, whereas those of *B. thuringiensis* ssp. *israelensis* and *B. sphaericus* 2362 were 3.2 and 37.7 ng/ml, and 11.9 and 24.6 ng/ml, respectively (Park *et al.*, 2005). Aside from high efficacy, as noted above, this new bacterium is much less likely to induce resistance in target populations, as it combines Cyt1A with *B. thuringiensis* ssp. *israelensis* Cry toxins and the *B. sphaericus* Bin toxin. The resistance management properties of this bacterium are currently under evaluation. The markedly improved efficacy and resistance-delaying properties of this new bacterium make it an excellent candidate for development and use in vector control programmes, especially to control *Culex* vectors of West Nile and other viruses as well as species of this genus that transmit filarial diseases.

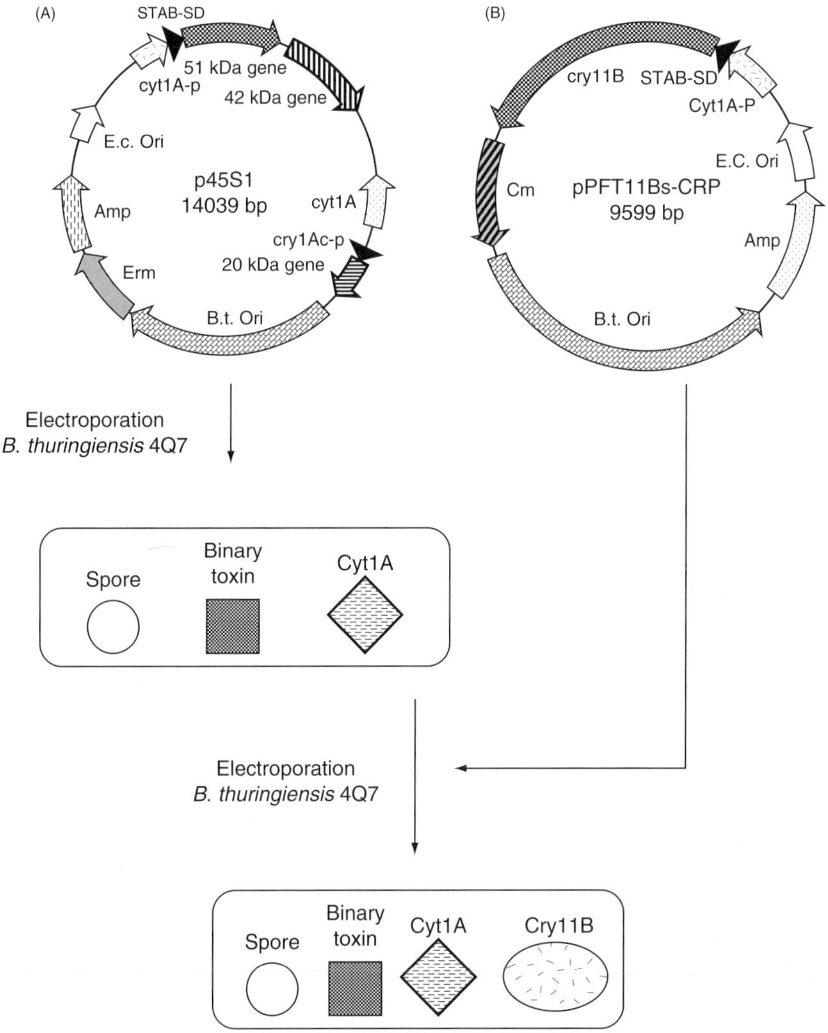


**Fig. 12.5.** Maps of recombinant plasmids and strategy for constructing a strain of Bt that produces Cyt1A, Cry11B and the Bs2362 binary toxin. (A) p45S1 containing *cyt1A* from Bti and a binary toxin gene from Bs2362. (B) pPFT11Bs-CRP containing *cry11B* from Btj. Amp, ampicillin-resistant gene; Erm, erythromycin-resistant gene; Cm, chloramphenicol-resistant gene; cyt1A-p, *cyt1A* promoters; cry1Ac-p, *cry1Ac* promoters; E.c. ori, *Escherichia coli* replication origin; B.t. ori, *Bacillus thuringiensis* replication origin. (From Park *et al.*, 2003.)

Moreover, larvae of certain species of *Anopheles* mosquitoes that are important malaria vectors, such as *Anopheles gambiae* and *Anopheles arabiensis*, should be highly sensitive to this recombinant, as they are not only sensitive to the toxins of *B. thuringiensis* ssp. *israelensis*, but are also highly sensitive to the *B. sphaericus* Bin toxin.

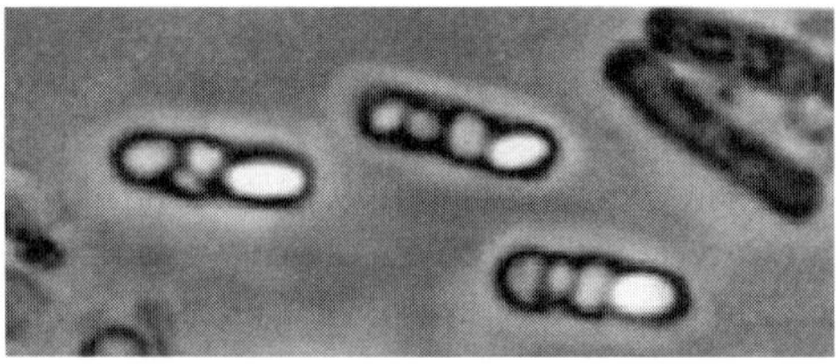

**Fig. 12.6.** Phase-contrast micrograph of *Bacillus thuringiensis* ssp. *israelensis* strain 4Q7/p45S1–11B that produces crystals of Cry11B, Cyt1A and the *Bacillus sphaericus* binary toxin. (From Park *et al.*, 2003.)

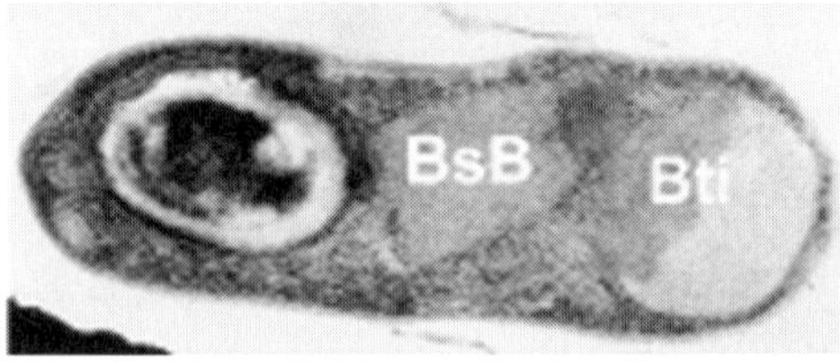


**Fig. 12.7.** Transmission electron micrograph of the recombinant *Bacillus thuringiensis* ssp. *israelensis* strain, BtiIPS-82/BsB, engineered to synthesize the *Bacillus sphaericus* binary toxin. This recombinant strain produces the typical IPS-82 parasporal body (Bti) and Bs2362 binary toxin crystal (BsB). (Modified from Park *et al.*, 2005.)

### 12.5.2. *Bacillus sphaericus*

Mosquitocidal strains of *B. sphaericus* produce several protein toxins. Those referred to as Mtx toxins (of 34–36 or 100 kDa) are produced during vegetative growth, whereas the so-called Bin (for binary) toxin is produced during sporulation (Charles *et al.*, 1996; Federici *et al.*, 2003). The Bin toxin forms a crystal on the inner surface of the exosporium membrane, and this toxin accounts for most of this species' activity, whereas the Mtx toxins are soluble and degrade quickly after synthesis. Highly toxic strains of *B. sphaericus* such as 2362 exhibit activity against *Culex* species equal to, or slightly better than, *B. thuringiensis* ssp. *israelensis*. In addition, *B. sphaericus* has longer residual activity, by at least several days, than *B. thuringiensis* ssp. *israelensis* in various larval habitats, including polluted water. However, the Bin toxin is the only major crystal toxin produced by *B. sphaericus*, and as a result, mosquitoes have developed resistance quickly in the field where this bacterium was used intensively (Sinègre *et al.*, 1994; Rao *et al.*, 1995; Silva-Filha *et al.*, 1995; Yuan *et al.*, 2000; Su and Mulla, 2004).

To improve the efficacy of *B. sphaericus*, there have been several attempts using different transformation strategies to introduce into this species mosquitocidal Cry and Cyt proteins of *B. thuringiensis* ssp. *israelensis* and other subspecies. Trisrisook *et al.* (1990) reported Cry4B production in strains 1593 and 2362 using protoplast transformation. Bar *et al.* (1991) expressed *cry4B* and *cyt1A* genes independently or in combination in strain 2362. Similarly, Poncet *et al.* (1994) synthesized Cry4B and Cry11A independently in strain 2297. More

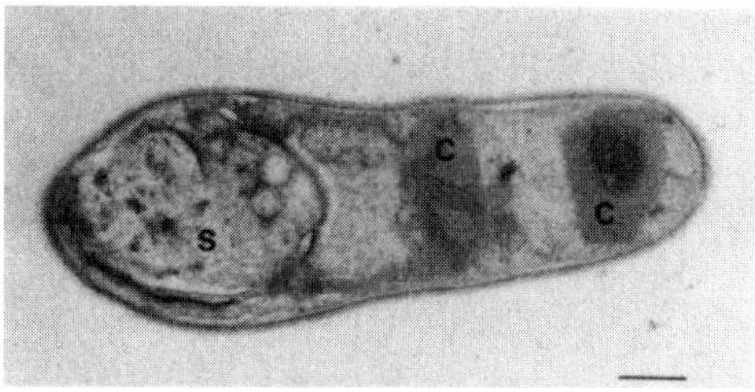


**Fig. 12.8.** Transmission electron micrograph of a recombinant *Bacillus sphaericus* 2297 producing Cry11A of *Bacillus thuringiensis* ssp. *israelensis*. S, spore; C, crystalline inclusion. Bar, 0.2 μm. (Modified from Poncet *et al*., 1997.)

recently, *cyt1Ab* gene from *B. thuringiensis* ssp. *medellin* was introduced into several *B. sphaericus* strains and a reasonable amount of Cyt1Ab was produced only in strain 2297 (Thiéry *et al*., 1998). In all cases, *cry* and *cyt* genes were under the control of their own promoters and the level of synthesis of introduced Cry proteins was very poor due to instability of introduced plasmids.

Later, stable and improved level of synthesis of Cry11A in *B. sphaericus* 2297 was obtained using a new approach (Poncet *et al*., 1997), *in vivo* homologous recombination (Fig. 12.8). In this method, the gene of interest is inserted into the target sequence located on the chromosome without including any other unnecessary sequences such as antibiotic-resistant genes and replication origins. Toxicity of the recombinant strain against *Anopheles stephensi* was enhanced, although against *C. quinquefasciatus*, the toxicity was similar to the wild type. Same group used the same protocol to produce both Cry11A and Cry11B in *B. sphaericus* 2297 (Servant *et al*., 1999). Although Cry11A and Cry11B production was poor in the recombinant strain for unknown reasons, it was toxic to *Aedes aegypti* to which the wild type does not show activity. However, it did not increase the toxicity to *Culex pipiens*.

More recently, an erythromycin resistance-marked pBtoxis (Berry *et al*., 2002), a toxin-coding plasmid of *B. thuringiensis* ssp. *israelensis* was transferred to the restriction-negative strains of *B. sphaericus* 1593 and 2362 by conjugation (Gammon *et al*., 2006). To construct the recombinant *B. sphaericus*, triparental mating was performed using the wild-type VectoBac strain of *B. thuringiensis* ssp. *israelensis* (Valent BioSciences, Long Grove, IL) that contains a natural conjugative plasmid, pXO16 to mobilize the pBtoxis::erm plasmid from strain 4Q5::erm (Delécluse *et al*., 1991). The resulting recombinant *B. sphaericus* strains, which produced Cry11A of *B. thuringiensis* ssp. *israelensis* (Fig. 12.9), were significantly more toxic to *A. aegypti* and were able to overcome resistance to *B. sphaericus* in a resistant colony of *C. quinquefasciatus*. However, the introduced pBtoxis::erm plasmid in both recombinants was lost after serial culturing in the absence of selective antibiotics.

Despite the numerous attempts, researchers have not been able to identify the molecular factors that prevent a high level of foreign gene expression in *B. sphaericus*. Determination of these factors could lead to improved mosquitocidal strains of *B. sphaericus*. Whether these would be more toxic and more persistent than existing recombinant strains of *B. thuringiensis* ssp. *israelensis* awaits future development of improved *B. sphaericus* strains.

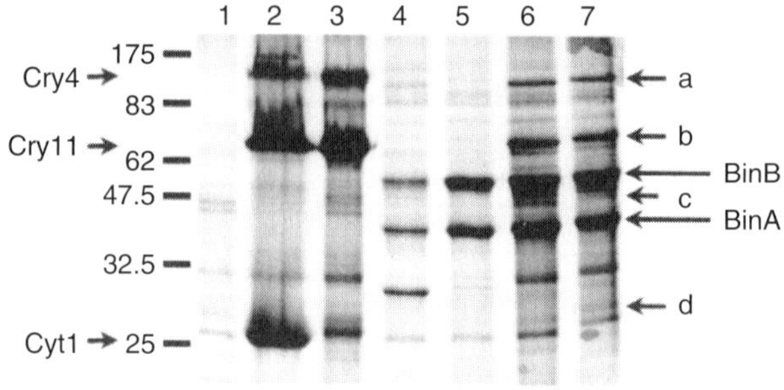


**Fig. 12.9.** Protein profiles of lysed sporulated cultures of *Bacillus thuringiensis* and *Bacillus sphaericus* strains. Proteins from *B. thuringiensis* ssp. *israelensis* strains 4Q7 (lane 1), 4Q5 (lane 2) and 4Q5::erm (lane 3), *B. sphaericus* strains 1593R$^-$ (lane 4) and 2362R$^-$ (lane 5), and recombinant strains 1593 (lane 6) and 2362 (lane 7) harbouring pBtoxis::erm were examined. Bands a, b, c and d in the recombinant strain 1593 (lane 6) were chosen for N-terminal sequencing. (From Gammon *et al.*, 2006.)

### 12.5.3. *Escherichia coli*

Other species of bacteria, such as *E. coli*, have advantages for use as hosts for heterogeneous protein synthesis. Various strains of *E. coli* with different genotypes have been developed so that researchers can choose the appropriate strain depending on their needs. These strains have the advantages of being easy to transform, include a variety of expression systems and are commercially available. Therefore, it is not surprising that *E. coli* was the first bacterial host used to synthesize insecticidal proteins of *B. thuringiensis*, and were the most extensively used until efficient electroporation protocols became available for transformation of *B. thuringiensis*. However, *E. coli* has been used, even from the beginning, as an expression host only to study the properties of endotoxin genes and the proteins they encode, i.e. not for use as a commercial bacterial insecticide.

For Cry proteins, the first *cry* gene *cry1Aa* was cloned from a plasmid of *B. thuringiensis* ssp. *kurstaki* HD-1 into pBR322 and expressed in *E. coli* strain HB101 (Schnepf and Whiteley, 1981). This recombinant *E. coli* strain showed activity similar to Cry1Aa synthesized in *B. thuringiensis*. Later, Oeda *et al.* (1987) cloned the *cry1Ab* gene from *B. thuringiensis* ssp. *aizawai* IPL7 into pUC18 and successfully expressed this gene in *E. coli* strain JM103 using the *tac* promoter and *rrnB* transcription terminator (Brosius *et al.*, 1981). In this study, *E. coli* strain JM103 was incubated at 37°C for 12 h under both IPTG-induced and non-induced conditions to determine whether this induction can enhance protein synthesis. Although there was no difference in protein production between IPTG-induced and non-induced conditions, the recombinant *E. coli* strains did produce visible amounts of insecticidal proteins when analysed by sodium dodecyl sulfate polyacrylamide gel electrophoresis (SDS-PAGE). Similarly, the 3.8 kb *cry4B* gene from *B. thuringiensis* ssp. *israelensis* was cloned into pUC12 and expressed in *E. coli* strain JM107 as well as the minicell-producing strain P678–54 using the *lacZ* promoter (Angsuthanasombat *et al.*, 1987). Cry4B protein produced in minicells was visible by protein gel analysis. All the earlier studies mentioned above showed only low to moderate levels of protein synthesis. In the latter studies, however, Cry4B synthesis in *E. coli* was improved considerably. For example, Boonserm *et al.* (2004) achieved a high level of Cry4A production by using a synthetic promoter in which the *tac* promoter was fused with the 162bp *cry4B* regulatory region including SD sequence (Figs 12.10 and 12.11).

For Cyt Proteins, *cyt1A* was first cloned from *B. thuringiensis* ssp. *israelensis* and expressed in *E. coli* strain JM101 by Ward *et al.* (1984). Subsequently, it was shown that the 20kDa protein encoded by the third open reading frame (*orf*) of the *cry11A* operon of *B. thuringiensis* ssp. *israelensis* is required for efficient synthesis of Cyt1A in *E. coli* (McLean and Whiteley, 1987; Adams *et al.*, 1989).

In addition to Cry and Cyt proteins of *B. thuringiensis*, other insecticidal proteins have also been synthesized in *E. coli*. The vegetative insecticidal proteins (Vips)

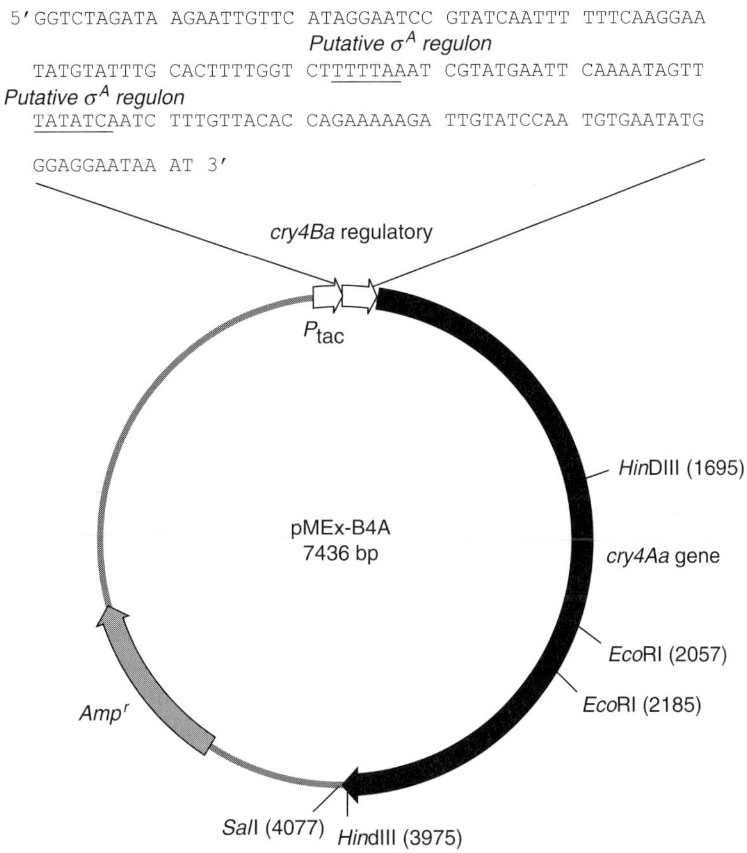


**Fig. 12.10.** Physical map of the pMEx-B4A plasmid. The *cry4A* gene is under control of the *tac* promoter ($P_{tac}$) and the *cry4B* regulatory region. The nucleotide sequences of the 162bp *cry4B* regulatory region are shown with the Shine-Dalgarno sequence (in bold type) and the putative $\sigma^A$ regulon (underlined) found to be effective for expression of *cry4B*..$Amp^r$ indicates the ampicillin-resistance gene. (From Boonserm *et al.*, 2004.)

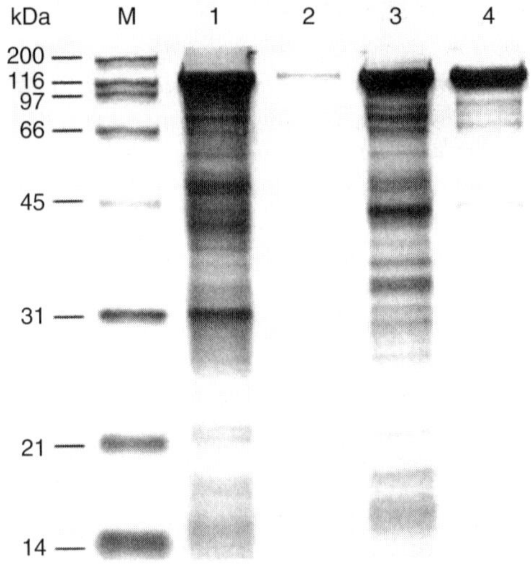


**Fig. 12.11.** Coomassie Blue-stained SDS-PAGE (12% gel) showing the solubility of partially purified Cry4A inclusions extracted from *Escherichia coli* cultures grown at 37°C (lanes 1 and 2) and at 30°C (lanes 3 and 4). Solubilization was performed in carbonate buffer, pH 10.0 for 1 h. Lanes 1 and 3, the total inclusion fractions; lanes 2 and 4, an equivalent amount of the supernatant containing the Cry4A solubilized protoxins after centrifugation. M indicates molecular mass standards. (From Boonserm *et al.*, 2004.)

of *B. thuringiensis* provide an example. Vips are secreted during vegetative growth and do not exhibit any similarity to Cry or Cyt toxins (Estruch *et al.*, 1996). High yields of all three types of Vips – Vip1, Vip2 and Vip3 – have been achieved in *E. coli* using strong T7 promoters alone or in combination with *vip* promoters (Estruch *et al.*, 1996; Doss *et al.*, 2002; Shi *et al.*, 2004; Rang *et al.*, 2005). The expression of genes encoding chitinase from *B. thuringiensis* in *E. coli* provides another example of heterologous expression in the later species. Chitinase has the potential to increase the efficacy of *B. thuringiensis*, apparently because it aids Cry protein action by making the peritropic membrane more porous (Sampson and Gooday, 1998). Thus, chitinase genes from *B. thuringiensis* strains have been cloned into several vectors and expressed in *E. coli*. For example, Arora *et al.* (2003) cloned and expressed the 1.1 kb gene encoding the 36 kDa exochitinase from *B. thuringiensis* ssp. *kurstaki* HD-1. The 36 kDa chitinase synthesized in *E. coli* synergized Vip activity against *Spodoptera litura* larvae. Barboza-Corona *et al.* (2003) cloned and expressed the endochitinase gene *chiA74* from *B. thuringiensis* ssp. *kenyae* strain LBIT-82 in *E. coli* DH5αF' using its own promoter. The *E. coli* strain producing ChiA74 showed lower levels of chitinolytic activity compared with the parental *B. thuringiensis* strain LBIP-82, mainly due to low levels of ChiA74 production. More recently, Driss *et al.* (2005) cloned a new chitinase gene, *chi255*, from a local isolate of *B. thuringiensis* ssp. *kurstaki*, and expressed it in *E. coli* using the expression vector pBAD33-GFPuv. The recombinant *E. coli* strain showed chitinase activity on a colloidal chitin plate.

### 12.5.4. *Pseudomonas fluorescens*

Cry proteins of *B. thuringiensis* are susceptible to rapid degradation in the environment, especially under the wide variety of conditions experienced in agriculture such as ultraviolet light and heat. Frequently the degradation process is so rapid that Cry proteins, which must be ingested by the insect pest, do not last long enough in the field to be effective. To overcome the inherent instability of unprotected insecticidal Cry protein from *B. thuringiensis*, a Gram-negative, plant-colonizing bacterium, *Pseudomonas fluorescens* has been used to synthesize Cry proteins.

The first attempt was reported in 1986 using maize root colonizer strains of *P. fluorescens* (Obukowicz *et al.*, 1986). A 4.6kb *Bam*HI fragment containing *cry1Aa* gene from *B. thuringiensis* ssp. *kurstaki* HD-1 was integrated into the chromosome of *P. fluorescens* strains mediated by the transposon Tn5. Bacterial conjugation was the mechanism of transformation. Strain *E. coli* S17–1, the donor strain that contained *cry1Aa* on pSUP1021, which cannot replicate in non-enteric Gram-negative bacteria and contains Tn5, and tetracycline- and chloramphenicol-resistant genes, was used as the donor strain to introduce this gene into *P. fluorescens* strains 112–12 and Ps3732–7. In transformed *P. fluorescens*, much of the Cry1A was degraded, as detected by Western blotting, indicating this protein was not nearly as stable compared to synthesis in *B. thuringiensis*. It was not determined whether Cry1A synthesized in the recombinant *P. fluorescens* strains formed crystals. To determine the toxicity of the recombinant strains, they were grown to the stationary phase in LB, applied on the surface of an artificial diet, air-dried and neonate larvae of the tobacco hornworm, *Manduca sexta*, were placed on the diet. Mortality was recorded after 4 days. Both recombinant strains producing Cry1Aa gave 100% mortality against these larvae.

Another group in South Africa used *P. fluorescens* strains isolated from sugarcane to produce Cry1Ac (Herrera *et al.*, 1994; Downing *et al.*, 2000). The techniques they used were similar to those described in the study noted above. The same *E. coli* strain was used for conjugation, and *cry1Ac*, under the control of either its own or the *tac* promoter, was integrated into the chromosome of *P. fluorescens* strains using Omegon-Km (Fellay *et al.*, 1989), a transposable element specifically designed for *in vivo* chromosomal insertion in Gram-negative bacteria (Fig. 12.12). Recombinant *P. fluorescens* strains, however, showed only moderate mortality against sugarcane borer, *Eldana saccharina*.

### 12.5.5. *Clavibacter xyli* ssp. *cynodontis*

*Clavibacter xyli* ssp. *cynodontis* is a fastidious Gram-positive bacterium that naturally inhabits the xylem of Bermuda grass and colonizes the vascular system of maize when artificially inoculated (Davis *et al.*, 1984). Homologous recombination was used to introduce *cry1Ac* from *B. thuringiensis* ssp. *kurstaki* HD-73 into the chromosome of this bacterium (Turner *et al.*, 1991; Lampel *et al.*, 1994). The initial attempt to express *cry1Ac* in this species failed. In this failed attempt, the integration plasmid, pCG610, contained three chromosomal DNA fragments from *C. xyli* ssp. *cynodontis* flanked by two copies of the *cry1Ac* gene. One copy of *cry1Ac* was under control of its endogenous *cry1Ac* promoter and the $\lambda_{tR}1$ transcription terminator,

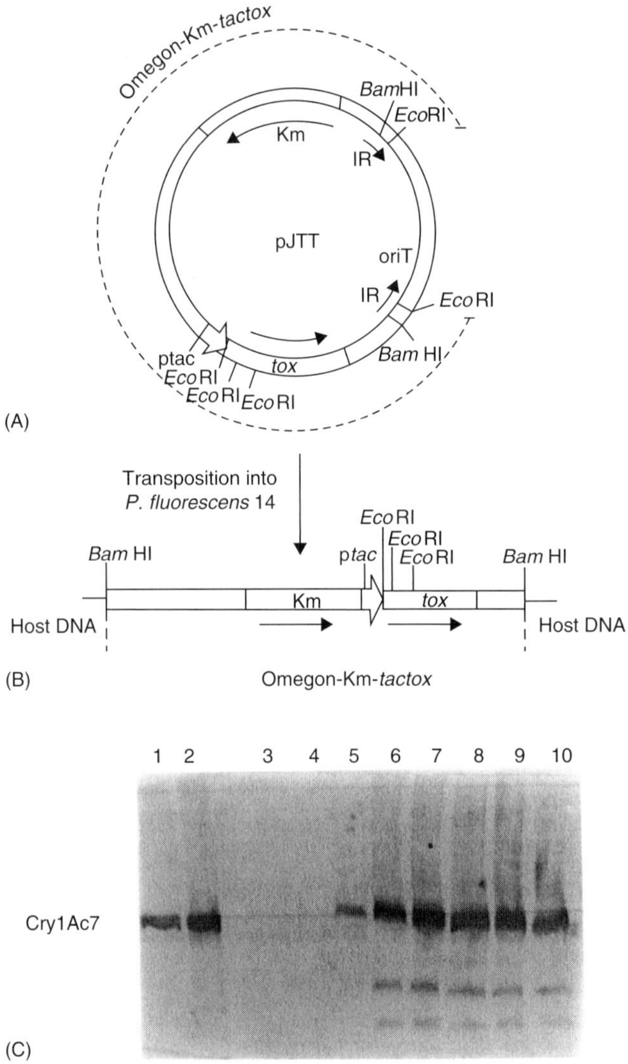


**Fig. 12.12.** Integration of the Omegon-Km-p*tac-cry1Ac7* fragment into the chromosome of *Pseudomonas fluorescens* 14 followed by synthesis of Cry1Ac7 in *P. fluorescens*. (A) The p*tac-cry1Ac7* cassette containing the *cry1Ac7* gene under control of the *tac* promoter from p*tac-tox* was cloned into the *Nde*I site of pJFF350. The latter was mobilized by conjugation into a broad range of Gram-negative bacteria, resulting in pJTT. (B) The plasmid was electroporated into *P. fluorescens* 14 with the transposition of the Omegon-Km-p*tac-cry1Ac7* cassette into the chromosome. IR, inverted repeat; oriT, origin of transfer; Km, kanamycin-resistant gene; ptac, tac promoter; *tox*, *Bacillus thuringiensis cry1Ac7* gene. (C) Western blot analysis of Cry1Ac7 synthesis in recombinant clones. Lanes 1 and 2, *Escherichia coli* carrying *cry1Ac7* in the broad-host-range plasmid, pKTT, and in the integration plasmid, pJTT; lane 3, *P. fluorescens* 14; lane 4, *P. fluorescens* 14 carrying *cry1Ac7* in pKTT; lane 5, *P. fluorescens* 14 in which the *cry1Ac7* gene under control of its own promoter was integrated into the chromosome; lanes 6–10, *P. fluorescens* 14 in which the *cry1Ac7* gene under control of the *tac* promoter was integrated into the chromosome. (Modified from Downing *et al.*, 2000.)

and the other lacked these genetic elements (Turner *et al.*, 1991). After transformation of *C. xyli* ssp. *cynodontis*, however, the introduced *cry1Ac* gene was lost due to recombination between the repeated DNA sequences flanking the integration vector that resulted from the integration into the chromosome. In their later study, integration plasmids were improved by using only a single copy of *cry1Ac*. With this construct, the recombinant *C. xyli* ssp. *cynodontis* produced Cry1Ac (Lampel *et al.*, 1994). Western blot results showed many degradation products, indicating that the Cry1Ac synthesized was not stable. Despite the breakdown of Cry1Ac in the transformed *C. xyli* ssp. *cynodontis*, inoculation of maize plants with this recombinant resulted in significantly lower survival of *Ostrinia nubilalis* (European corn borer), and less feeding damage compared with plants inoculated with wild type. Although using an endophytic bacterial species held some promise, the efficacy obtained by direct transformation of maize (and other crops), i.e. integration of *cry* genes into the maize genome, quickly made the endophytic strategy obsolete.

### 12.5.6. *Cyanobacteria*

Use of formulations of *B. thuringiensis* ssp. *israelensis* for mosquito control requires frequent application because in most habitats these remain near the water surface where larvae feed for only a day or so, or they are inactivated by sunlight. A potential approach to circumvent this problem is to genetically engineer microorganisms living in the upper layers of the water to synthesize Cry endotoxin proteins of *B. thuringiensis* ssp. *israelensis*. Cyanobacteria are strong candidates for this type of genetic engineering owing to their photosynthetic capability and resultant simple nutritional requirements, and because they are widely distributed in the upper layers of water.

Towards this goal, the cyanobacterium, *Agmenellum quadruplicatum* strain PR-6, was engineered in separate studies to synthesize either Cry4B (Angsuthanasombat and Panyim, 1989) or Cry11A (Murphy and Stevens, 1992) of *B. thuringiensis* ssp. *israelensis*. This species was selected because it has a natural mechanism for uptake and integration of exogenous DNA, and therefore an efficient transformation procedure was developed more than 25 years ago (Stevens and Porter, 1980). In both engineering studies, the phycocyanin operon promoter was used to express *cry* genes. In the first study (Angsuthanasombat and Panyim, 1989), approximately $1.5 \times 10^2$ *A. quadruplicatum* transformants per microgram of plasmid DNA were obtained. However, the level of Cry4B synthesized by the recombinant was extremely low. Concomitantly, the recombinant *A. quadruplicatum* showed only 45% mortality against the second instars of *A. aegypti* after 48 h of incubation using $10\,\mathrm{mg\ ml^{-1}}$ of total protein concentration. A few years later, improved Cry protein synthesis was obtained using *cry11A* and a translational gene fusion technique (Murphy and Stevens, 1992). Though 100% mortality with 3–5 µl of recombinant cells against neonates of *C. pipiens* after 6 days of treatment was reported, the amount of toxin per unit volume was not quantified, making it impossible to assess the efficacy per unit volume, and thus compare this recombinant to others.

Subsequently, species of cyanobacteria belonging to the genus *Synechococcus* strains PCC 6803 (Chungjatupornchai, 1990) and PCC 7942 (Soltes-Rak *et al.*, 1993, 1995) were used to produce Cry4B. To enhance Cry4B yield, the *cry4B* gene was placed under control of either the tobacco *psbA* promoter (Chungjatupornchai, 1990), the *lacZ* promoter combined with the endogenous *cry4B* promoter or the ferredoxin (*petF1*) promoter (Soltes-Rak *et al.*, 1993, 1995). Of these expression systems, the *lacZ* promoter combined with the *cry4B* promoter resulted in the highest Cry4B yield in the *Synechococcus* strain. However, even with this best recombinant, larval mortality using neonates of *Culex restuans* was only approximately 70% after 3 days of incubation when a mid- to late-log phase of culture was used (Soltes-Rak *et al.*, 1993).

More recently, in two different studies *Anabaena* sp. strain PCC 7120 was used to express either *cry4A*, *cry11A* and the 20 kDa protein gene (Wu *et al.*, 1997) or *cry4A*, *cry11A*, *cyt1A* and the 20 kDa protein gene (Khasdan *et al.*, 2003). Results from the latter study are shown in (Fig. 12.13). In both cases, greater insecticidal protein synthesis was achieved using a dual promoter system – a cyanobacterial *psbA* promoter and an *E. coli* T7 promoter, and pRL488p, in an *E. coli* – *Anabaena* shuttle vector (Elhai and Wolk, 1990). In a former study (Wu *et al.*, 1997), the recombinant *Anabaena* strain producing Cry4A, Cry11A and the 20 kDa protein was approximately 60-fold more toxic to third instars of *A. aegypti* compared with that producing only Cry4A ($LC_{50}$ ($10^5$ cells $ml^{-1}$) = 53 versus 0.9). The recombinant strain harbouring a plasmid that contained *cry4A* under the control of the *psbA* promoter alone did not show any toxicity against the same mosquito species. In the latter study (Khasdan *et al.*, 2003), the recombinant *Anabaena* strain producing Cry4A, Cry11A, Cyt1A and the 20 kDa protein showed approximately 2.4-fold more toxicity to fourth instars of *A. aegypti* compared with the strain producing Cry4A, Cry11A and the 20 kDa protein ($LC_{50}$ ($10^5$ cells $ml^{-1}$) = 0.83 versus 0.35).

### 12.5.7. *Caulobacter crescentus*

The Gram-negative bacterium, *Caulobacter crescentus*, another species found commonly near the water surface (Poidexter, 1981), has also been used as a host for producing the Cry protein of *B. thuringiensis* ssp. *israelensis*. This species exhibits two distinct cell cycles, a non-motile-stalked cell phase and a monoflagellated swarmer cell phase. The flagellated swarmer stage, this bacterium is motile, and thus distributed throughout the habitat. Therefore, it could be an ideal carrier for biological toxins aimed at the surface-feeding larvae of mosquitoes.

To test this possibility, the *cry4B* gene of *B. thuringiensis* ssp. *israelensis* was placed under the control of *tac* promoter in the presence of the lactose repressor gene and transformed into *C. crescentus* by electroporation (Thanabalu *et al.*, 1992). Recombinant *C. crescentus* cells producing Cry4B were tested against *A. aegypti* larvae using a concentration of $3.2 \times 10^8$ cells $ml^{-1}$. Only 32.5% mortality was obtained after 48 h of incubation. To improve Cry synthesis in *C. crescentus*, two recombinant regulatory sequences that affect transcription were investigated to determine their effect on Cry4B synthesis in *C. crescentus* strain CB15 (Yap *et al.*, 1994). The *cry4B* gene was placed under control of either the: (i) *tac* promoter and the putative ribosome binding sequence (RBS) of the *C. crescentus* 130 kDa surface layer protein gene; or the (ii) *bin* toxin promoter of *B. sphaericus* 2297 and its putative RBS. The *lacZ* gene was placed under control of both expression systems to determine the transcriptional efficiency in *C. crescentus*. The former resulted in approximately 1.3-fold

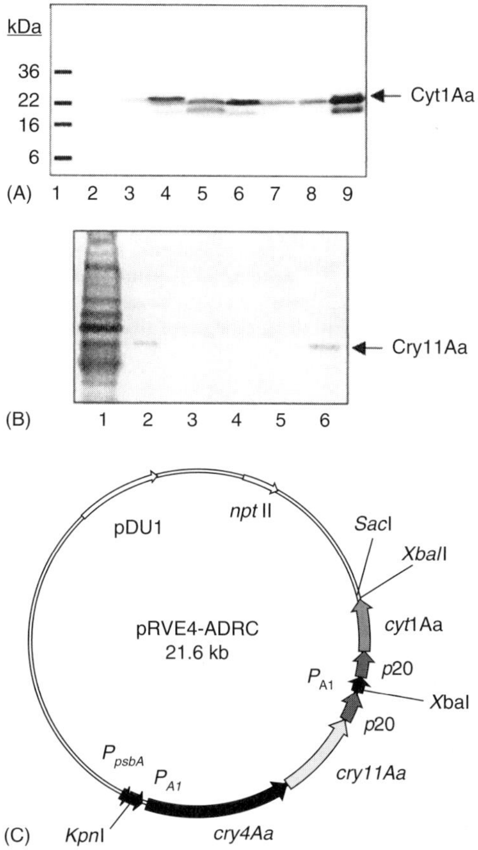


**Fig. 12.13.** Western blot analysis of recombinant *Anabaena* and *Escherichia coli* strains that synthesize Cyt1A (A) and Cry11A (B). Anti-Cyt1A and antiserum against whole *Bacilus thuringiensis* ssp. *israelensis* crystals were used, respectively, in (A) and (B). (A) Lane 1, molecular size marker; lane 2, *Anabaena* PCC 7120; lane 3, *Anabaena* PCC 7120 containing *cyt1A* under control of the *psbA* and T7 promoters; lane 4, *E. coli* XL-Blue MRF' containing *cyt1A* under control of the *psbA* and T7 promoters; lane 5, *Anabaena* PCC 7120 containing *cyt1A* and the 20-kDa protein gene under control of the *psbA* and T7 promoters; lane 6, *E. coli* XL-Blue MRF' containing *cyt1A* and the 20kDa protein gene under control of the *psbA* and T7 promoters; lane 7, *Anabaena* PCC 7120 containing *cry4A* and *cry11A* under control of the *psbA* and T7 promoters, and *cyt1A* and the 20kDa protein gene under control of the T7 promoter; lane 8, *E. coli* XL-Blue MRF' containing *cry4A* and *cry11A* under control of the *psbA* and T7 promoters, and *cyt1A* and the 20kDa protein gene under control of the T7 promoter; lane 9, *B. thuringiensis* ssp. *israelensis*. (B) Lane 1, *B. thuringiensis* ssp. *israelensis*; lane 2, *Anabaena* PCC 7120 containing *cry4A*, *cry11A* and the 20-kDa protein gene under control of the T7 promoter; lane 3, *Anabaena* PCC 7120; lane 4, *Anabaena* PCC 7120 containing *cyt1A* under control of the *psbA* and T7 promoters; lane 5, *Anabaena* PCC 7120 containing *cyt1A* and the 20kDa protein gene under control of the *psbA* and T7 promoters; lane 6, *Anabaena* PCC 7120 containing *cry4A* and *cry11A* under control of the *psbA* and T7 promoters, and *cyt1A* and the 20kDa protein gene under control of the T7 promoter. (C) Physical map of the pRVE4-ADRC used to synthesize *cry4A*, *cry11A* and *cyt1A* of *B. thuringiensis* ssp. *israelensis* in *Anabaena*. $P_{psbA}$, cyanobacterial *psbA* promoter; $P_{A1}$, *E. coli* T7 promoter; *p20*, *B. thuringiensis* ssp. *israelensis* 20kDa protein gene. (Modified from Khasdan *et al.*, 2003.)

higher β-galactosidase activity than the latter (2199 versus 1711 Miller units). When the *C. crescentus* recombinants producing Cry4B were tested against second instars of *A. aegypti*, the former was 18-fold more toxic than the latter ($LC_{50} = 4.0 \times 10^7$ versus $2.2 \times 10^6$ cells $ml^{-1}$). As the two studies mentioned above used different mosquito bioassay procedures, direct comparison of bioassay data to determine the level of improvement obtained with the latter recombinants was not possible.

## 12.6. Conclusions and Future Prospects

As shown above, several different types of bacterial species have been used to construct recombinant bacteria for producing insecticidal proteins of *B. thuringiensis* and *B. sphaericus* depending on the purpose of application. Although *B. thuringiensis* remains the best host to synthesize endotoxin proteins, other bacteria also hold some potential. Major disadvantages of most of the other bacterial species we discussed as hosts were the low level of toxin protein yields and/or instability of the toxin gene(s) after introduction to these species. As molecular biology and genetic engineering techniques advance, we expect that researchers will overcome these barriers and develop much better recombinant bacteria with improved efficacy for insect pest control. Ideally, the design of recombinant bacteria should take into consideration the key principles of resistance management, namely, mixtures of toxins are better than single toxins, especially where the toxins have different modes of action, and where specific proteins are known that delay resistance, such as Cyt proteins in the case of mosquitocidal bacteria, these should be included in the constructs.

The application of recombinant DNA techniques to improving insecticidal bacteria, which began more than two decades ago, was initially met with a high degree of enthusiasm, followed by the establishment of many small biotechnology companies. At the same time, techniques were developed for generating transgenic crops resistant to insects based on the Cry proteins of *B. thuringiensis*. These crops, such as Bt cotton and Bt maize, have been an enormous success, and currently constitute a multibillion dollar industry. Many of the recombinant bacterial insecticides under development in the 1980s and 1990s targeted the same pests on the same crops. In addition, new insecticides, such as imidocloprid and the spinosids came to market. Due to a combination of these events, most of the new biotechnology companies focusing on recombinant bacteria failed. Nevertheless, the extension of the use of *B. thuringiensis* endotoxins in crop plants must be considered one of the key advances, if not the key advance, in pest control technology of the last half of the 20th century. While clearly this success has dimmed interest in recombinant bacterial insecticides (and many other microbial pesticides), there remain an enormous number of crops and markets where these may be useful, and thus justify continued research and development. For example, the market for bacterial insecticides to control nuisance and vector mosquitoes continues to expand, and, as we have shown, recombinants based on *B. thuringiensis* and *B. sphaericus* are much more efficacious than the wild-type species used in current commercial products. With respect to crop pests, there are many crops that have not been transformed with endotoxin genes, including lettuce and cabbage, tomatoes, celery, fruit crops and grapes where lepidopterous insects continue to be major pests. Thus, though the economic prospects may not be as large as they were 20 years ago, many opportunities remain for the development of new and more efficacious recombinant bacterial insecticides. The higher specificity and environmental safety of the recombinants compared to synthetic chemical insecticides, along with increases in efficacy that reduce the cost of production, provide reasons for optimism that these bacteria will play a significant role in future pest and vector control programmes.

## References


Adams, L.F., Visick, J.E. and Whiteley, H.R. (1989) A 20-kilodalton protein is required for efficient production of the *Bacillus thuringiensis* subsp. *israelensis* 27-kilodalton crystal protein in *Escherichia coli*. *Journal of Bacteriology* 171, 521–530.

Adams, L.F., Brown, K.L. and Whiteley, H.R. (1991) Molecular cloning and characterization of two genes encoding sigma factors that direct transcription from a *Bacillus thuringiensis* crystal protein gene promoter. *Journal of Bacteriology* 173, 3846–3854.

Agaisse, H. and Lereclus, D. (1994) Structural and functional analysis of the promoter region involved in full expression of the *cryIIIA* toxin gene of *Bacillus thuringiensis*. *Molecular Microbiology* 13, 94–107.

Agaisse, H. and Lereclus, D. (1996) STAB-SD: a Shine-Dalgarno sequence in the 5' untranslated region is a determinant of mRNA stability. *Molecular Microbiology* 20, 633–643.

Angsuthanasombat, C. and Panyim, S. (1989) Biosynthesis of 130-kilodalton mosquito larvicide in the cyanobacterium *Agmenellum quadruplicatum*. *Applied and Environmental Microbiology* 55, 2428–2430.

Angsuthanasombat, C., Chungjatupornchai, W., Kertbundit, S., Luxananil, P., Settasatian, C., Wilairat, P. and Panyim, S. (1987) Cloning and expression of 130-kd mosquito-larvicidal δ-endotoxin gene of *Bacillus thuringiensis* var. *israelensis* in *Escherichia coli*. *Molecular and General Genetics* 208, 384–389.

Aronson, A.I. (1993) The two faces of *Bacillus thuringiensis*: insecticidal proteins and post-exponential survival. *Molecular Microbiology* 7, 489–496.

Aronson, A.I., Beckman, W. and Dunn, P. (1986) *Bacillus thuringiensis* and related insect pathogens. *Microbiological Review* 50, 1–24.

Arora, N., Ahmad, T., Rojagopal, R. and Bhatnagar, R.K. (2003) A constitutively expressed 36 kDa exochitinase from *Bacillus thuringiensis* HD-1. *Biochemical and Biophysical Research Communications* 307, 620–625.

Bar, E., Lieman-Hurwitz, J., Rahamin, E., Keynan, A. and Sandler, N. (1991) Cloning and expression of *Bacillus thuringiensis israelensis* δ-endotoxin DNA in *B. sphaericus*. *Journal of Invertebrate Pathology* 57, 149–158.

Barboza-Corona, J.E., Nieto-Mazzocco, E., Velázquez-Robledo, R., Salcedo-Hernandez, R., Bautista, M., Jiménez, B. and Ibarra, J.E. (2003) Cloning, sequencing, and expression of the chitinase gene *chiA74* from *Bacillus thuringiensis*. *Applied and Environmental Microbiology* 69, 1023–1029.

Baum, J.A. and Malvar, T. (1995) Regulation of insecticidal crystal protein production in *Bacillus thuringiensis*. *Molecular Microbiology* 18, 1–12.

Baumann, P., Baumann, L., Bowditch, D. and Broadwell, A.H. (1987) Cloning of the gene for the larvicidal toxin of *Bacillus sphaericus* 2362: evidence for a family of related sequences. *Journal of Bacteriology* 169, 4061–4067.

Berry, C., O'Neil, S., Ben-Dov, E., Jones, A.F., Murphy, L., Quail, M.A., Holden, M.T., Harris, D., Zaritsky, A. and Parkhill, J. (2002) Complete sequence and organization of pBtoxis, the toxin-coding plasmid of *Bacillus thuringiensis* subsp. *israelensis. Applied and Environmental Microbiology* 68, 5082–5095.

Boonserm, P., Pornwiroon, W., Katzenmeier, G., Panyim, S. and Angsuthanasombat, C. (2004) Optimized expression in *Escherichia coli* and purification of the functional form of the *Bacillus thuringiensis* Cry4Aa δ-endotoxin. *Protein Expression and Purification* 35, 397–403.

Boonserm, P., Davis, P., Ellar, D.J. and Li, J. (2005) Crystal structure of the mosquito-larvicidal toxin Cry4Ba and its biological implications. *Journal of Molecular Biology* 348, 363–382.

Bone, E.J. and Ellar, D.J. (1989) Transformation of *Bacillus thuringiensis* by electroporation. *FEMS Microbiology Letters* 58, 171–178.

Brosius, J., Dull, T.J., Sleeter, D.D. and Noller, H.F. (1981) Gene organization and primary structure of a ribosomal RNA operon from *E. coli. Journal of Molecular Biology* 148, 107–127.

Brown, K.L. and Whiteley, H.R. (1988) Isolation of a *Bacillus thuringiensis* RNA polymerase capable of transcribing crystal protein genes. *The Proceedings of the National Academy of Sciences of the USA* 85, 4166–4170.

Brown, K.L. and Whiteley, H.R. (1990) Isolation of the second *Bacillus thuringiensis* RNA polymerase that transcribes from a crystal protein gene promoter. *Journal of Bacteriology* 172, 6682–6688.

Bucher, G.E. (1960) Potential bacterial pathogens of insects and their characteristics. *Journal of Insect Pathology* 2, 172–195.

Bucher, G.E. and Stephens, J.M. (1957) A disease of grasshoppers caused by the bacterium *Pseudomonas aeruginosa* (Schroeter) Migula. *Canadian Journal of Microbiology* 3, 611–625.

Charles, J.-F., Nielsen-LeRoux, C. and Delécluse, A. (1996) *Bacillus sphaericus* toxins: molecular biology and mode of action. *Annual Review of Entomology* 41, 451–472.

Chungjatupornchai, W. (1990) Expression of the mosquitocidal protein genes of *Bacillus thuringiensis* subsp. *israelensis* and the herbicide-resistance gene *bar* in *Synechocystis* PCC 6803. *Current Microbiology* 21, 283–288.

Crickmore, N. and Ellar, D.J. (1992) Involvement of a possible chaperonin in the efficient expression of a cloned CryIIA δ-endotoxin gene in *Bacillus thuringiensis. Molecular Microbiology* 6, 1533–1537.

Crickmore, N., Bone, E.J., Williams, J.A. and Ellar, D.J. (1995) Contribution of the individual components of the δ-endotoxin crystal to the mosquitocidal activity of *Bacillus thuringiensis* subsp. *israelensis. FEMS Microbiology Letters* 131, 249–254.

Crickmore, N., Zeigler, D.R., Feitelson, J., Schnepf, E., Van Rie, J., Lereclus, D., Baum, J. and Dean, D.H. (1998) Revision of the nomenclature for the *Bacillus thuringiensis* pesticidal crystal proteins. *Microbiology and Molecular Biology Reviews* 62, 807–813.

Davis, M.J., Gillaspie, A.J., Vidaver, A.K. and Harris, R.W. (1984) *Clavibacter*: a new genus containing some phytopathogenic coryneform bacteria including *Clavibacter xyli* subsp. *xyli* sp. nov., subsp. nov. and *Clavibacter xyli* subsp. *cynodontis* subsp. nov., pathogens that cause ratoon stunting disease of sugarcane and Bermudagrass stunting disease. *International Journal of Systematic Bacteriology* 34, 107–117.

Delécluse, A., Charles, J.-F., Klier, A. and Rapoport, G. (1991) Deletion by *in vivo* recombination shows that the 28-kilodalton polypeptide from *Bacillus thuringiensis* subsp. *israelensis* is not essential for mosquitocidal activity. *Journal of Bacteriology* 173, 3374–3381.

Delécluse, A., Rosso, M.-L. and Ragni, A. (1995) Cloning and expression of a novel toxin gene from *Bacillus thuringiensis* subsp. *jegathesan* encoding a highly mosquitocidal protein. *Applied and Environmental Microbiology* 61, 4230–4235.

Doss, V.A., Kumar, K.A., Jayakumar, R. and Sekar, V. (2002) Cloning and expression of the vegetative insecticidal protein (*vip3V*) gene of *Bacillus thuringiensis* in *Escherichia coli. Protein Expression and Purification* 26, 82–88.

Downing, K.J., Leslie, G. and Thomson, J.A. (2000) Biocontrol of the sugarcane borer *Eldana saccharina* by expression of the *Bacillus thuringiensis cry1Ac7* and *Serratia marcescens chiA* genes in sugarcane-associated bacteria. *Applied and Environmental Microbiology* 66, 2804–2810.

Driss, F., Kallassy-Awad, M., Zouari, N. and Joaua, S. (2005) Molecular characterization of a novel chitinase from *Bacillus thuringiensis* subsp. *kurstaki*. *Journal of Applied Microbiology* 99, 945–953.

Dulmage, H.T. (1970) Insecticidal activity of HD-1, a new isolate of *Bacillus thuringiensis* var. *alesti*. *Journal of Invertebrate Pathology* 15, 232–239.

Elhai, J. and Wolk, C.P. (1990) Developmental regulation and spatial pattern of expression of the structural genes for nitrogenase in the cyanobacterium *Anabaena*. *EMBO Journal* 9, 3379–3388.

Estruch, J.J., Warren, G.W., Mullins, M.A., Nye, G.J., Craig, J.A. and Koziel, M.G. (1996) Vip3A, a novel *Bacillus thuringiensis* vegetative insecticidal protein with a wide spectrum of activities against lepidopteran insects. *The Proceedings of the National Academy of Sciences of the USA* 93, 5389–5394.

Federici, B.A. (1991) Microbial insecticides. *Pesticide Outlook* 2, 22–28.

Federici, B.A., Park, H.-W., Bideshi, D.K., Wirth, M.C. and Johnson, J.J. (2003) Recombinant bacteria for mosquito control. *Journal of Experimental Biology* 206, 3877–3885.

Federici, B.A., Park, H.-W. and Sakano, Y. (2006) Insecticidal protein crystals of *Bacillus thuringiensis*. In: Shively, J.M. (ed.) *Microbiology Monographs, Vol. 1, Inclusions in Prokaryotes*. Springer, Berlin-Heidelberg, pp. 195–236.

Fellay, R., Krisch, H.M., Prentki, P. and Frey, J. (1989) Omegon-Km: a transposable element designed for *in vivo* insertional mutagenesis and cloning of genes in Gram-negative bacteria. *Gene* 76, 215–226.

Gammon, K., Jones, G.W., Hope, S.J., de Oliveira, C.M.F., Regis, L., Silva Filha, M.H.N.L., Dancer, B.N. and Berry, C. (2006) Conjugal transfer of a toxin-coding megaplasmid from *Bacillus thuringiensis* subsp. *israelensis* to mosquitocidal strains of *Bacillus sphaericus*. *Applied and Environmental Microbiology* 72, 1766–1770.

Ge, B., Bideshi, D., Moar, W.J. and Federici, B.A. (1998) Differential effects of helper proteins encoded by the *cry2A* and *cry11A* operons on the formation of Cry2A inclusions in *Bacillus thuringiensis*. *FEMS Microbiology Letters* 165, 35–41.

Georghiou, G. and Wirth, M.C. (1997) Influence of exposure to single versus multiple toxins of *Bacillus thuringiensis* subsp. *israelensis* on development of resistance in the mosquito *Culex quinquefasciatus* (Diptera: Culicidae). *Applied and Environmental Microbiology* 63, 1095–1101.

Goldberg, L.J. and Margalit, J. (1977) A bacterial spore demonstrating rapid larvicidal activity against *Anopheles sergentii*, *Uranotaenia unguiculata*, *Culex univitattus*, *Aedes aegypti*, and *Culex pipiens*. *Mosquito News* 37, 355–358.

González, J.M. Jr and Carlton, B.C. (1984) A large transmissible plasmid is required for crystal toxin production in *Bacillus thuringiensis* variety *israelensis*. *Plasmid* 11, 28–38.

González, J.M. Jr, Brown, B.J. and Carlton, B.C. (1982) Transfer of *Bacillus thuringiensis* plasmids coding for δ-endotoxin among strains of *B. thuringiensis* and *B. cereus*. *The Proceedings of the National Academy of Sciences of the USA* 79, 6951–6955.

Grochulski, P., Masson, L., Borisova, S., Pusztai-Carey, M., Schwartz, J.-L., Brousseau, R. and Cygler, M. (1995) *Bacillus thuringiensis* CryIA(a) insecticidal toxin: crystal structure and channel formation. *Journal of Molecular Biology* 254, 447–464.

Helmann, J.D. and Moran, C.P. Jr (2002) RNA polymerase and sigma factors. In: Sonenshein, A.L., Hoch, J.A. and Losick, R. (eds) Bacillus subtilis *and its Closest Relatives: From Genes to Cells*. ASM Press, Washington, DC, pp. 289–312.

Herrera, G., Snyman, S.J. and Thomson, J.A. (1994) Construction of a bioinsecticidal strain of *Pseudomonas fluorescens* active against the sugarcane borer, *Eldana saccharina*. *Applied and Environmental Microbiology* 60, 682–690.

Höfte, H. and Whiteley, H.R. (1989) Insecticidal crystal proteins of *Bacillus thuringiensis*. *Microbiological Reviews* 53, 242–255.

Ibarra, J.E. and Federici, B.A. (1986) Isolation of a relatively non-toxic 65-kilodalton protein inclusion from the parasporal body of *Bacillus thuringiensis* subsp. *israelensis*. *Journal of Bacteriology* 165, 527–533.

Kalucy, E.C. and Daniel, A. (1972) The reaction of *Anopheles annulipes* larvae to infection by *Aeromonas punctata*. *Journal of Invertebrate Pathology* 19, 189–197.

Khasdan, V., Ben-Dov, E., Manasherob, R., Boussiba, S. and Zaritsky, A. (2003) Mosquito larvicidal activity of transgenic *Anabaena* PCC 7120 expressing toxin genes from *Bacillus thuringiensis* subsp. *israelensis*. *FEMS Microbiology Letters* 227, 189–195.

Krieg, A., Huger, A., Langenbruch, G. and Schnetter, W. (1983) *Bacillus thuringiensis* var. *tenebrionis*: a new pathotype effective against larvae of Coleoptera. *Journal of Applied Entomology* 96, 500–508.

Lampel, J.S., Canter, G.L., Dimock, M.B., Kelly, J.L., Anderson, J.J., Uratani, B.B., Foulke, J.S. Jr. and Turner, J.T. (1994) Integrative cloning, expression, and stability of the *cryIA(c)* gene from *Bacillus thuringiensis* subsp. *kurstaki* in a recombinant strain of *Clavibacter xyli* subsp. *cynodontis*. *Applied and Environmental Microbiology* 60, 501–508.

Lecadet, M.-M., Frachon, E., Dumanoir, V.C., Ripouteau, H., Hamon, S., Laurent, P. and Thiery, I. (1999) Updating the H-antigen classification of *Bacillus thuringiensis*. *Journal of Applied Microbiology* 66, 660–672.

Lereclus, D., Arantes, O., Chaufaux, J. and Lecadet, M.-M. (1989) Transformation and expression of a cloned δ-endotoxin gene in *Bacillus thuringiensis*. *FEMS Microbiology Letters* 60, 211–218.

Lereclus, D., Agaisse, H., Gominet, M. and Chaufaux, J. (1995) Overproduction of encapsulated insecticidal crystal proteins in a *Bacillus thuringiensis spoA* mutant. *Bio/Technology* 13, 67–71.

Li, J., Caroll, J. and Ellar, D.J. (1991) Crystal structure of insecticidal δ-endotoxin from *Bacillus thuringiensis* at 2.5 Å resolution. *Nature* 353, 815–821.

Li, J., Koni, P.A. and Ellar, D.J. (1996) Structure of the mosquitocidal δ-endotoxin CytB from *Bacillus thuringiensis* sp. *kyushuensis* and implications for membrane pore formation. *Journal of Molecular Biology* 257, 129–152.

Li, T., Sun, F., Yuan, Z., Zhang, Y., Yu, J. and Pang, Y. (2000) Coexpression of *cyt1Aa* of *Bacillus thuringiensis* subsp. *israelensis* with *Bacillus sphaericus* binary toxin gene in acrystalliferous strain of *B. thuringiensis*. *Current Microbiology* 40, 322–326.

Masson, L., Préfontaine, G. and Brousseau, R. (1989) Transformation of *Bacillus thuringiensis* vegetative cells by electroporation. *FEMS Microbiology Letters* 60, 273–278.

McLean, K.M. and Whiteley, H.R. (1987) Expression in *Escherichia coli* of a cloned crystal protein gene of *Bacillus thuringiensis* subsp. *israelensis*. *Journal of Bacteriology* 169, 1017–1023.

Moar, W.J., Trumble, J.T. and Federici, B.A. (1989) Comparative toxicity of spores and crystals from the NRD-12 and HD-1 strains of *Bacillus thuringiensis* subsp. *kurstaki* to neonate beet armyworm. *Journal of Economic Entomology* 82, 1593–1603.

Morse, R.J., Yamamoto, T. and Stroud, R.M. (2001) Structure of Cry2Aa suggests an unexpected receptor binding epitope. *Structure* 9, 409–417.

Murphy, R.C. and Stevens, S.E., Jr. (1992) Cloning and expression of the *cryIVD* gene of *Bacillus thuringiensis* subsp. *israelensis* in the cyanobacterium *Agmenellum quadruplicatum* PR-6 and its resulting larvicidal activity. *Applied and Environmental Microbiology* 58, 1650–1655.

Obukowicz, M.G., Perlak, F.J., Kusano-Kretzmer, K., Mayer, E.J., Bolten, S.L. and Watrud, L.S. (1986) Tn-5-mediated integration of the delta-endotoxin gene from *Bacillus thuringiensis* into the chromosome of root-colonizing Pseudomonads. *Journal of Bacteriology* 168, 982–989.

Oeda, K., Oshie, K., Shimizu, M., Nakamura, K., Yamamoto, H., Nakayama, I. and Ohkawa, H. (1987) Nucleotide sequence of the insecticidal protein gene of *Bacillus thuringiensis* strain *aizawai* IPL7 and its high-level expression in *Escherichia coli*. *Gene* 53, 113–119.

Park, H.-W., Ge, B., Bauer, L.S. and Federici, B.A. (1998) Optimization of Cry3A yields in *Bacillus thuringiensis* by use of sporulation-dependent promoters in combination with the STAB-SD mRNA sequence. *Applied and Environmental Microbiology* 64, 3932–3938.

Park, H.-W., Bideshi, D.K., Johnson, J.J. and Federici, B.A. (1999) Differential enhancement of Cry2A versus Cry11A yields in *Bacillus thuringiensis* by use of the *cry3A* STAB mRNA sequence. *FEMS Microbiology Letters* 181, 319–327.

Park, H.-W., Bideshi, D.K. and Federici, B.A. (2000) Molecular genetic manipulation of truncated Cry1C protein synthesis in *Bacillus thuringiensis* to improve stability and yield. *Applied and Environmental Microbiology* 66, 4449–4455.

Park, H.-W., Bideshi, D.K. and Federici, B.A. (2003) Recombinant strain of *Bacillus thuringiensis* producing Cyt1A, Cry11B, and the *Bacillus sphaericus* binary toxin. *Applied and Environmental Microbiology* 69, 1331–1334.

Park, H.-W., Bideshi, D.K., Wirth, M.C., Johnson, J.J., Walton, W.E. and Federici, B.A. (2005) Recombinant larvicidal bacteria with markedly improved efficacy against *Culex* vectors of West Nile virus. *The American Journal of Tropical Medicine and Hygiene* 72, 732–738.

Poidexter, J.S. (1981) The caulobacters: ubiquitous unusual bacteria. *Microbiological Review* 45, 123–179.

Poncet, S., Delécluse, A., Anello, G., Klier, A. and Rapoport, G. (1994) Transfer and expression of the *cryIVB* and *cryIVD* genes of *Bacillus thuringiensis* subsp. *israelensis* in *Bacillus sphaericus* 2297. *FEMS Microbiology Letters* 117, 91–96.

Poncet, S., Delécluse, A., Klier, A. and Rapoport, G. (1995) Evaluation of synergistic interactions among CryIVA, CryIVB and CryIVD toxic components of *Bacillus thuringiensis* subsp. *israelensis* crystals. *Journal of Invertebrate Pathology* 66, 131–133.

Poncet, S., Bernard, C., Dervyn, E., Cayley, J., Klier, A. and Rapoport, G. (1997) Improvement of *Bacillus sphaericus* toxicity against dipteran larvae by integration, via homologous recombination, of the Cry11A toxin gene from *Bacillus thuringiensis* subsp. *israelensis. Applied and Environmental Microbiology* 63, 4413–4420.

Priest, F.G. (2000) Chapter 1.1 Biodiversity of the entomopathogenic, endospore-forming bacteria. In: Charles, J.-F., Delécluse, A. and Nielsen-LeRoux, C. (eds) *Entomopathogenic Bacteria: From Laboratory to Field Application*. Kluwer Academic Publishers, Dordrecht, The Netherlands, pp. 1–22.

Rao, D.R., Mani, T.R., Rajendran, R., Joseph, A.S., Gajanana, A. and Reuben, R. (1995) Development of a high level of resistance to *Bacillus sphaericus* in a field population of *Culex quinquefasciatus* from Kochi, India. *Journal of the American Mosquito Control Association* 11, 1–5.

Rang, C., Gil, P., Neisner, N., Van Rie, J. and Frutos, R. (2005) Novel Vip3-related protein from *Bacillus thuringiensis. Applied and Environmental Microbiology* 71, 6276–6281.

Sampson, M.N. and Gooday, G.W. (1998) Involvement of chitinases of *Bacillus thuringiensis* during pathogenesis in insects. *Microbiology* 144, 2189–2194.

Schnepf, H.E. and Whiteley, H.R. (1981) Cloning and expression of the *Bacillus thuringiensis* crystal protein gene in *Escherichia coli. The Proceedings of the National Academy of Sciences of the USA* 78, 2893–2897.

Schnepf, E., Crickmore, N., Van Rie, J., Lereclus, D., Baum, J., Feitelson, J., Zeigler, D.R. and Dean, D.H. (1998) *Bacillus thuringiensis* and its pesticidal proteins. *Microbiology and Molecular Biology Reviews* 62, 775–806.

Servant, P., Rosso, M.-L., Hamon, S., Poncet, S., Delécluse, A. and Rapoport, G. (1999) Production of Cry11A and Cry11Ba toxins in *Bacillus sphaericus* confers toxicity towards *Aedes aegypti* and resistant *Culex* populations. *Applied and Environmental Microbiology* 65, 3021–3026.

Shi, Y., Xu, W., Yuan, M., Tang, M., Chen, J. and Pang, Y. (2004) Expression of *vip1/vip2* genes in *Escherichia coli* and *Bacillus thuringiensis* and the analysis of their signal peptides. *Journal of Applied Microbiology* 97, 757–765.

Silva-Filha, M.-H., Regis, L., Nielsen-LeRoux, C. and Charles, J.-F. (1995) Low-level resistance to *Bacillus sphaericus* in a field-treated population of *Culex quinquefasciatus* (Diptera: Culicidae). *Journal of Economic Entomology* 88, 525–530.

Sinègre, G., Babinot, M., Quermal, J.M. and Gaven, B. (1994) First field occurrence of *Culex pipiens* resistance to *Bacillus sphaericus* in southern France. Abstracts of the VII European Meeting, Society for Vector Ecology, Barcelona, Spain.

Soltes-Rak, E., Kushner, D.J., Williams, D.D. and Coleman, J.R. (1993) Effect of promoter modification on mosquitocidal *cryIVB* gene expression in *Synechococcus* sp. strain PCC 7942. *Applied and Environmental Microbiology* 59, 2404–2410.

Soltes-Rak, E., Kushner, D.J., Williams, D.D. and Coleman, J.R. (1995) Factors regulating *cryIVB* expression in the cyanobacterium *Synechococcus* PCC 7942. *Molecular and General Genetics* 246, 301–308.

Stevens, S.E. Jr and Porter, R.D. (1980) Transformation in *Agmenellum quadruplicatum. The Proceedings of the National Academy of Sciences of the USA* 77, 6052–6056.

Su, T. and Mulla, M.S. (2004) Documentation of high-level *Bacillus sphaericus* 2362 resistance in field populations of *Culex quinquefasciatus* breeding in polluted water in Thailand. *Journal of the American Mosquito Control Association* 20, 405–411.

Thanabalu, T., Hindley, J., Brenner, S., Oei, C. and Berry, C. (1992) Expression of the mosquitocidal toxins of *Bacillus sphaericus* and *Bacillus thuringiensis* subsp. *israelensis* by recombinant *Caulobacter crescentus*, a vehicle for biological control of aquatic insect larvae. *Applied and Environmental Microbiology* 58, 905–910.

Thiéry, I., Hamon, S., Delécluse, A. and Orduz, S. (1998) The introduction into *Bacillus sphaericus* of the *Bacillus thuringiensis* subsp. *medellin* Cyt1Ab1 gene results in higher susceptibility of resistant mosquito larva populations to *B. sphaericus. Applied and Environmental Microbiology* 64, 3910–3916.

Thomas, W.E. and Ellar, D.J. (1983) Mechanism of action of *Bacillus thuringiensis* var. *israelensis* δ-endotoxin. *FEBS Letters* 154, 362–368.

Thompson, M.A., Schnepf, H.E. and Feitelson, J.S. (1995) Structure, function, and engineering of *Bacillus thuringiensis* toxins. In: Setlow, J.K. (ed.) *Genetic Engineering: Principles and Methods.* Plenum Press, New York, pp. 99–117.

Trisrisook, M., Pantuwatana, S., Bhumiratana, A. and Panbangred, W. (1990) Molecular cloning of the 130-kilodalton mosquitocidal δ-endotoxin gene of *Bacillus thuringiensis* subsp. *israelensis* in *Bacillus sphaericus. Applied and Environmental Microbiology* 56, 1710–1716.

Turner, J.T., Lampel, J.S., Stearman, R.S., Sundin, G.W., Gunyuzlu, P. and Anderson, J.J. (1991) Stability of the δ-endotoxin gene from *Bacillus thuringiensis* subsp. *kurstaki* in a recombinant strain of *Clavibacter xyli* subsp. *cynodontis. Applied and Environmental Microbiology* 57, 3522–3528.

Ward, E.S., Ellar, D.J. and Todd, J.A. (1984) Cloning and expression in *Escherichia coli* of the insecticidal δ-endotoxin gene of *Bacillus thuringiensis* var. *israelensis. FEBS Letters* 175, 377–382.

Wirth, M.C., Georghiou, G.P. and Federici, B.A. (1997) CytA enables CryIV endotoxins of *Bacillus thuringiensis* to overcome high levels of CryIV resistance in the mosquito, *Culex quinquefasciatus. The Proceedings of the National Academy of Sciences of the USA* 94, 10536–10540.

Wirth, M.C., Federici, B.A. and Walton, W.E. (2000a) Cyt1A from *Bacillus thuringiensis* synergizes activity of *Bacillus sphaericus* against *Aedes aegypti* (Diptera: Culicidae). *Applied and Environmental Microbiology* 66, 1093–1097.

Wirth, M.C., Walton, W.E. and Federici, B.A. (2000b) Cyt1A from *Bacillus thuringiensis* restores toxicity of *Bacillus sphaericus* against resistant *Culex quinquefasciatus* (Diptera: Culicidae). *Journal of Medical Entomology* 37, 401–407.

Wirth, M.C., Park, H.-W., Walton, W.E. and Federici, B.A. (2005) Cyt1A of *Bacillus thuringiensis* delays evolution of resistance to Cry11A in the mosquito, *Culex quinquefasciatus. Applied and Environmental Microbiology* 71, 185–189.

Wong, H.C. and Chang, S. (1986) Identification of a positive retroregulator that stabilizes mRNAs in bacteria. *The Proceedings of the National Academy of Sciences of the USA* 83, 3233–3237.

Wu, D. and Chang, F.N. (1985) Synergism in mosquitocidal activity of 26 and 65 kDa protein from *Bacillus thuringiensis* subsp. *israelensis* crystal. *FEBS Letters* 190, 232–236.

Wu, D. and Federici, B.A. (1993) A 20-kilodalton protein preserves cell viability and promotes CytA crystal formation during sporulation in *Bacillus thuringiensis. Journal of Bacteriology* 175, 5276–5280.

Wu, D. and Federici, B.A. (1995) Improved production of the insecticidal CryIVD protein in *Bacillus thuringiensis* using *cryIA(c)* promoter to express the gene for an associated 20-kDa protein. *Applied Microbiology and Biotechnology* 42, 697–702.

Wu, D., Johnson, J.J. and Federici, B.A. (1994) Synergism of mosquitocidal toxicity between CytA and CryIVD proteins using inclusions produced from cloned genes of *Bacillus thuringiensis. Molecular Microbiology* 13, 965–972.

Wu, X., Vennison, S.J., Liu, H., Ben-Dov, E., Zaritsky, A. and Boussiba, S. (1997) Mosquito larvicidal activity of transgenic *Anabaena* strain PCC 7120 expressing combinations of genes from *Bacillus thuringiensis* subsp. *israelensis. Applied and Environmental Microbiology* 63, 4971–4975.

Yap, W.H., Thanabalu, T. and Porter, A.G. (1994) Influence of transcriptional and translational control sequences on the expression of foreign genes in *Caulobacter crescentus. Journal of Bacteriology* 176, 2603–2610.

Yuan, Z., Zhang, Y., Cia, Q. and Liu, E.-Y. (2000) High-level field resistance to *Bacillus sphaericus* C3–41 in *Culex quinquefasciatus* from southern China. *Biocontrol Science and Technology* 10, 41–49.

# 13 Genomic Analysis of the Symbiotic and Entomopathogenic *Photorhabdus* Bacteria

S. GAUDRIAULT[1,2] AND E. DUCHAUD[3]

[1]*INRA, UMR1133 Laboratoire EMIP, F-34095 Montpellier, France;* [2]*Université Montpellier II, UMR1133 Laboratoire EMIP, F-34095 Montpellier, France;* [3]*INRA, UR892, Unité Virologie et Immunologie Moléculaires, F-78350 Jouy-en-Josas, France*

## 13.1. Introduction

There is a growing number of sequenced bacterial genomes. From a functional point of view, projects involving genome-sequencing models for which substantial preliminary data are not available allow researchers to access a wealth of data by comparing sequences obtained with genes present in the databases. Furthermore, whatever the knowledge base for the organism being studied, genome projects allow identification of genes that were not identified using classical genetic strategies. This would include many genes that do not share similarities with known genes in database, paralogous genes leading to redundancy of function and inactive genes resulting from reductive evolution.

Whole-genome-sequencing projects also give information on organization of bacterial genomes. First, genome sequencing reveals the number of chromosomes, which can be linear or circular, the presence of plasmids and the genome size variation within a species (Casjens, 1998). It also provides information on new operons, origin of replication and genome polarity. This increases with the growing number of sequenced genomes. Finally, genome sequencing allows the characterization of guanine-cytosine content (GC) content and its variation, genomic islands and synteny with others organisms. All these data give information on DNA regions acquired by horizontal genetic transfer and therefore on the evolutive history of the bacterium.

*Photorhabdus* species are Gram-negative bacteria belonging to the family of Enterobacteriaceae. *Photorhabdus luminescens* and *Photorhabdus temperata* both interact with invertebrate hosts (Boemare, 2002). They form symbiotic associations with entomopathogenic nematodes in the genus *Heterorhabditis*. Bacterial and nematode taxonomic data reveal a highly specific association between bacterial strain and nematode species. Bacterium–nematode complexes are entomopathogenic. Bacteria alone are also entomopathogenic when injected in the haemolymph of an insect. Recently, a new *Photorhabdus* species, *Photorhabdus asymbiotica*, was described that causes infection in humans (Akhurst *et al.*, 2004).

Genetic analyses of potential virulence factors or symbiosis determinants were rare before 2000. Of 30 original papers dealing with *Photorhabdus*, only four were focused on genetic analyses (Clarke and Dowds, 1994; Zenno and Saigo, 1994; Bintrim and Ensign 1998; Bowen *et al.*, 1998). The interest in this bacterium grew with the discovery of the Tc insecticidal toxins (Bowen *et al.*, 1998). These may be useful in place of, or in association with, the Bt toxins used for biological control, either directly in the field or in transgenic plants.

To better understand the complex lifestyle of *Photorhabdus* and as a prelude to genetic analysis of interactions with invertebrates, a British group in 2000 and a French group in 2003 sequenced *Photorhabdus* genomes. The different approaches they used will be presented in the first part of this chapter. The utilization of the genomic sequence information in functional genomics and comparative genomics will be illustrated in the second and third parts.

## 13.2. Sequencing and Annotation of *Photorhabdus* Genomes

Two genome-sequencing strategies have frequently been used. In the ordered clone approach, a large insert library is constructed in order to obtain a map of overlapping clones covering the whole genome. Selected clones are then sequenced to obtain the whole-genome sequence. In the direct random strategy, preliminary data such as a map are not necessary. A direct shotgun sequencing is performed (see Frangeul *et al.*, 1999, for a review). Due to the small size of microbial genomes, this latter approach is the most widely used for sequencing these genomes. In this section, two *Photorhabdus* genome-sequencing projects will be detailed and the main features of *Photorhabdus* genome will be presented.

### 13.2.1. Partial sequencing of *P. luminescens* W14

The first sequencing performed on a *P. luminescens* strain was done by ffrench-Constant *et al.* (2000) on strain W14. They performed a genomic sample sequence. Strain W14 belongs to the subspecies *akhurstii*. Its nematode vector is unknown. The goal of this sequencing was not to have a total sequence but only a sample of genes encoded by *P. luminescens*.

#### *13.2.1.1. Library*

A unique genomic DNA library with fragment sizes around 1–2 kb was produced in the M13 Janus system (Burland *et al.*, 1993). A total of 2122 random single-sequencing reads (about 400 bp length) were run. There were no assembly steps since these 2122 sequences represented less than 0.2-fold coverage of the genome.

#### *13.2.1.2. Annotation*

Only automatic function prediction was performed. The sequences were submitted to the BLASTX servers at the National Center for Biotechnology Information (NCBI, Bethesda, MD, USA). When a read showed sequence similarities with a gene present in the NCBI database, the read was annotated according to the best hit. When compared to the *Escherichia coli* strain K12 genome, 989 reads showed a significant conservation of sequence between the two genomes. This is consistent with the relatively close phylogenetic relationship between the two organisms. A total of 1133 reads (53%) matched with nothing. Despite this high percentage of sequences that did not match with *E. coli* genes, the authors described numerous genes that potentially encode toxins and virulence factors.

### 13.2.2. Exhaustive sequencing of *P. luminescens* TT01

By contrast, Duchaud and collaborators choose to perform the whole sequencing of one *P. luminescens* strain (Duchaud *et al.*, 2003; Fig. 13.1). Strain TT01 was chosen

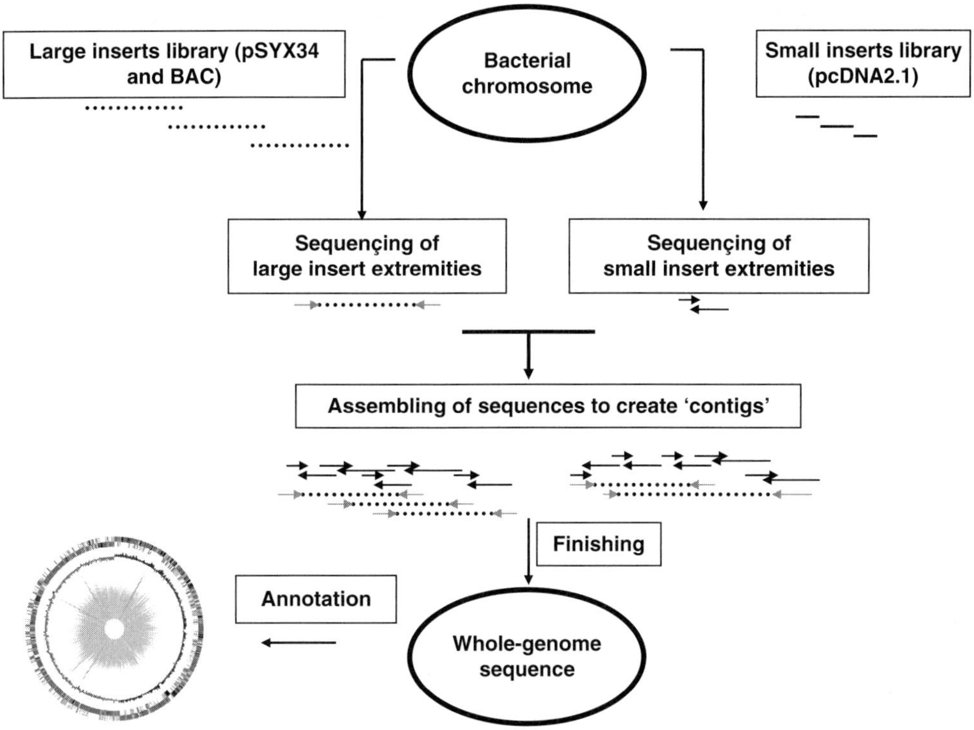


**Fig. 13.1.** Whole-genome shotgun strategy for the *Photorhabdus luminescens* TT01 genome.

because it is a type-strain and because of the ubiquity throughout the world of its nematode species, *Heterorhabditis bacteriophora* (initially isolated from Trinidad).

#### *13.2.2.1. Libraries*

For this sequencing, the complete whole-genome shotgun strategy was used. In order to have the optimal coverage of the genome, three libraries were generated: a small-fragment library (1–3 kb) using pcDNA-2.1 (Invitrogen); a medium-size insert library (5–10 kb) using pSYX34 (Xu and Fomenkov, 1994); a BAC library (30–80 kb) using pBeloBAC11 (Kim *et al.*, 1996) to obtain a 'scaffold' of the genome, which is used during the closing phase.

#### *13.2.2.2. Shotgun*

In an initial step, 63,475 sequencing reads from the three libraries were run. The whole sequencing matches with a sevenfold coverage of the genome.

#### *13.2.2.3. Assembling*

Using the Phred/Phrap/Consed software (Ewing and Green, 1998; Gordon *et al.*, 1998), high-quality sequences were selected and assembled into 472 contigs (307 > 2 kb), i.e. sequence files formed by contiguous sequencing reads. For the closure phase, the CAAT-BOX software was used to predict linkages between contigs (Frangeul *et al.*, 2004).

#### 13.2.2.4. Finishing phase

Walks on individual clones and polymerase chain reaction products amplified from TT01 genomic DNA were used to fill gaps, resolve ambiguities and re-sequence low-quality regions.

#### 13.2.2.5. Annotation

Automatic annotation began during the assembling phase. The coding sequences (CDS) were first defined by combining Genemark predictions with visual inspection of the open reading frames (ORF). An individual protein file (IPF) was created for each predicted CDS (Frangeul *et al.*, 2004). Function predictions were then based on BLASTP similarity searches and on the analysis of motifs using the PFAM databases. IPF could evolve during the finishing phase and when a new assembly was performed. After the finishing phase, functional annotation was manually inspected. The genome sequence and the annotation are now accessible on the PhotoList database (http://genolist.pasteur.fr/PhotoList/), a database constructed according to the SubtiList model (Moszer *et al.*, 2002).

## 13.3. Main Features of the *P. luminescens* Genome

The whole-genome TT01 sequencing project described a unique replicon of 5.7 Mb with an average GC content of 42.8% and 4839 predicted CDS including 157 pseudogenes. Seven complete sets of ribosomal RNA (rRNA) operons and 85 transfer RNA (tRNA) were predicted.

### 13.3.1. Putative proteins playing a role in the *Photorhabdus* life cycle

Strikingly, information from the W14 and TT01 genomes show that the *P. luminescens* genomes encode a large number of proteins potentially playing a role in the elimination of competitors, in host colonization, invasion and bioconversion of the insect cadaver. Surprisingly, the small coverage of the W14 strain partial sequencing project allowed identification of many interesting genes. Descriptions of a sample of such genes follows.

#### 13.3.1.1. Toxins against insects

More toxin genes were predicted in the *P. luminescens* genome than in any other bacterial genome already sequenced. The Tc toxins, toxic by ingestion or injection in insects, were first purified from *P. luminescens* strain W14 (Bowen *et al.*, 1998). The four loci that encode each of these complexes are termed tca, tcb, tcc and tcd. Strain TT01 contains the tcc and tcd loci, an incomplete tca locus and five newly identified loci (Duchaud *et al.*, 2003; Fig. 13.2). Despite the apparent complexity

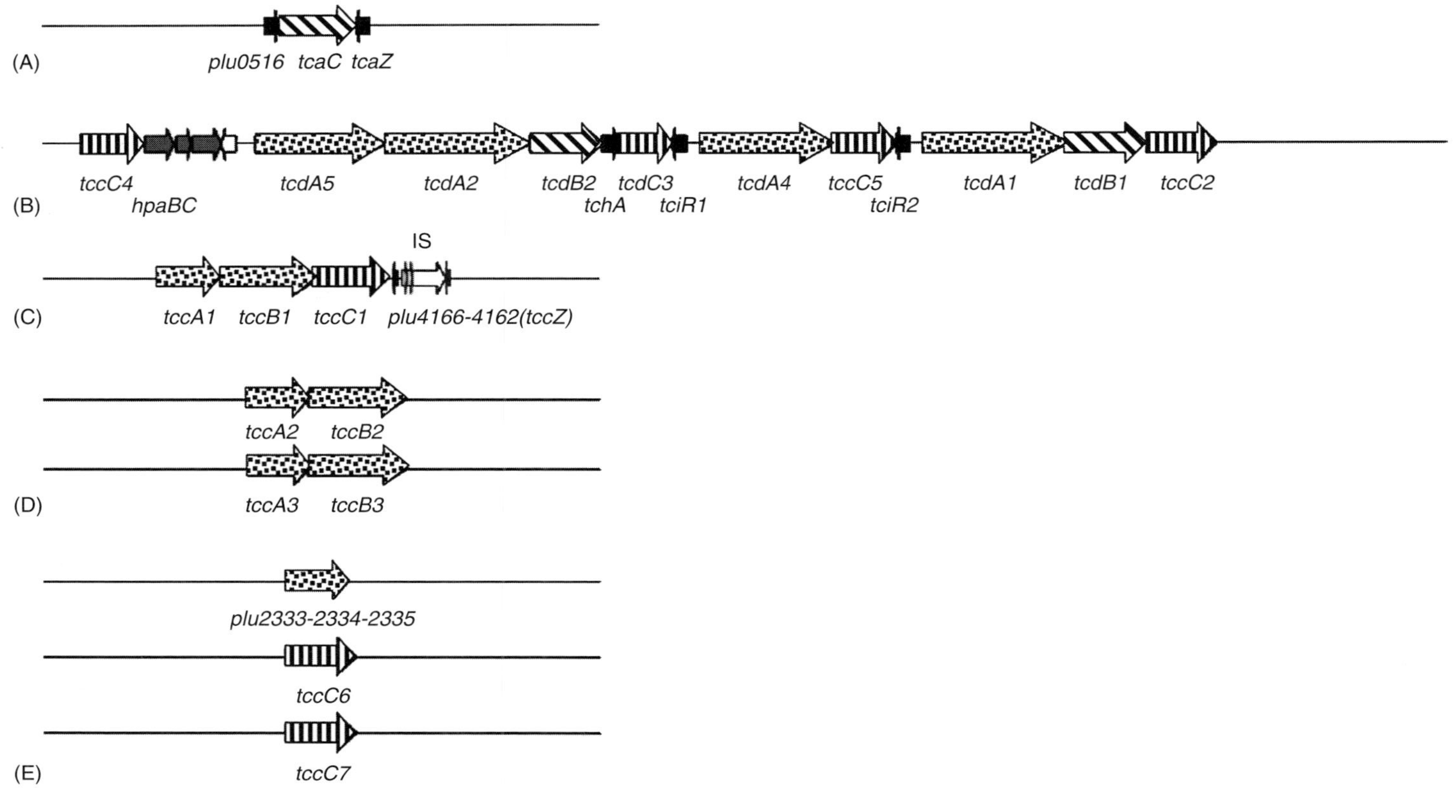


**Fig. 13.2.** Toxin complex loci identified in strain TT01. Nomenclature is according to Waterfield *et al.* (2002). The genes encoding the three conserved elements are motif-coded: tcaAB-like or tcb/tcdA-like, plotted horizontal arrows; tcaC-like, diagonally hatched arrows; tccC-like, vertically hatched arrows. (A) Locus similar to the toxin complex a (*tca*) locus from strain W14. (B) Locus similar to the toxin complex d (*tcd*) island from strain W14. (C) Locus similar to the toxin complex c (*tcc*) locus from strain W14. (D) Locus weakly similar to the toxin complex c (*tcc*) locus from strain W14. Other loci *plu2333*, *plu2334*, *plu2335* correspond to a single pseudogene.

of the loci, only three basic types of genetic elements, defined on the basis of sequence similarity, are apparent: (i) tcaAB or tcdA-like genes; (ii) tcaC or tcdB-like genes; and (iii) tccC-like genes. Homologues of these same three genetic elements have been found in other entomopathogenic bacteria (*Xenorhabdus nematophila*, *Serratia entomophila*, *Pseudomonas entomophila*) and insect-associated bacteria (*Yersinia* spp., *Pseudomonas syringae*). These findings suggest a role of Tc toxins in insect interaction and a horizontal genetic transfer of toxin-complex gene homologues in insect-interacting soil bacteria belonging to different genera.

Another large class of database matches comprises sequences similar to Rtx toxins (Repeats in toxin). The Rtx toxins are cytolytic toxins, metalloproteases and lipases that are virulence factors in many pathogenic Gram-negative bacteria. This large set of putative Rtx toxins and related proteins probably contribute to the insect pathogenicity of *P. luminescens*.

#### *13.3.1.2. Antibiotics and antibiotic resistance*

One of the most important questions about the ecology of *P. luminescens* is how it defends the insect cadaver against different microbial competitors. Analyses of TT01 and W14 genomes showed that they encode numerous proteins similar to polyketide and non-ribosomal peptide synthases that may be part of the biosynthetic pathway of antibiotics, known to be produced by *P. luminescens* (Webster *et al.*, 2002). In addition, TT01 and W14 contain sequence similar to colicin activity, colicin transport and pyocin immunity proteins. This complex pattern indicates a toxin–antitoxin system which may provide a selective advantage to *P. luminescens* against related bacteria producing similar toxins. This system likely contributes to mechanisms leading to the specific bacterium–nematode association.

#### *13.3.1.3. Bioconversion*

*P. luminescens* secretes many enzymes that contribute to insect death and result in bioconversion of the insect cadaver (Schmidt *et al.*, 1988; Wang and Dowds,1993; Clarke and Dowds, 1994, 1995; Bowen *et al.*, 2000; Daborn *et al.*, 2001; Brillard *et al.*, 2002; Valens *et al.*, 2002; Marokhazi *et al.*, 2004). Numerous genes that potentially encode proteases, lipases, haemaglutinins, chitinases, non-RTX haemolysins and ADP-ribosyltransferases were identified in the W14 and TT01 genomes. Some of these results confirm previous biochemical or molecular studies; others help to open new areas for research.

#### *13.3.1.4. Invasion and colonization*

To adapt to invertebrate environments such as the nematode gut, the insect haemolymph and the insect cadaver, *P. luminescens* must be able to sense changes in nutrient levels, cation availability, osmolarity, bacterial density and other cues. Strain TT01 encodes homologues of the five *E. coli* sigma factors, five ECF factors, 32 LuxR family regulators, 37 phage gene repressors, 15 Ner-like regulators, 19 two-component regulators and 20 Lys-R type regulators.

During its complex life cycle, a *Photorhabdus* strain not only needs to detect and adapt to the qualities of its environment, but it must also perform locomotion, attachment and host tissue invasion activities. *P. luminescens* encodes sequences similar to intimins adhesions. The TT01 strain has eleven clusters of fimbrial genes, flagellar operons. Both W14 and TT01 strains have a type three secretion system locus (see 13.4.2).

As iron is often a rate-limiting growth factor in the host, many pathogenic bacteria have high-affinity iron-binding systems that are able to capture iron from host iron chelators. *P. luminescens* W14 and TT01 have numerous sequences that encode proteins similar to those involved in biosynthesis, transport and receptor of siderophore, hemin and ferric iron.

### 13.3.2. Redundant and mobile genetic elements

The impressive number of mobile genetic elements or their remnants suggests that the *Photorhabdus* strain TT01 genome is subject to continuously ongoing gene transfer (Duchaud *et al.*, 2003). Phage remnants represent 4% of the genome. One hundred and ninety-five insertion sequences (IS), IS fragments or transposons were noticed. Seven hundred and eleven ERIC elements were identified in contrast to only 21 ERIC sequences in the *E. coli* K12 chromosome. Thirty-two genomic islands were predicted on the basis of *in silico* features. Moreover six copies of a defective prophage were described. Variable cassettes encoding potential toxin flank the different copies (Hurst *et al.*, 2004). In the W14 strain, four potential pathogenicity islands (PAI) were also described and analysed (ffrench-Constant *et al.*, 2003). Several gene classes were overrepresented and inversion events undergoes inside a P2-related prophage remnant locus (Gaudriault *et al.*, 2004), suggesting frequent rearrangements and a high degree of plasticity.

All these features may help explain the impressive arsenal of toxins allowing the adaptation of *P. luminescens* to two classes of invertebrate hosts (nematode and insect) and its ability to kill a wide variety of insects.

## 13.4. Analogical Post-genomic Analysis

### 13.4.1. Definition

An analogical approach is based on the hypothesis that when two proteins have similar sequences, they have similar function. For this purpose, careful annotation is important for a good analogical approach.

### 13.4.2. Application to a single genomic region, the type three secretion system of *Photorhabdus*

Type three secretion systems have been discovered in Gram-negative bacteria having interactions with mammals, plants and insects (Cornelis and Van Gijsegem, 2000). This system often plays a crucial role in the interaction process. The central function of the TTSS is the delivery of bacterial proteins into eukaryotic cells (Hueck, 1998).

The annotation of strains W14 and TT01 led to the identification of a 25 kb locus encoding proteins similar to the components of the plasmid-encoded type three secretion system (TTSS) of *Yersinia pestis* and the chromosome-encoded TTSS of *Pseudomonas aeruginosa* (Waterfield *et al.*, 2002; Duchaud *et al.*, 2003; Brugirard-Ricaud *et al.*, 2004). A type three secretion system-encoding locus was also identified in the unfinished sequence of *P. asymbiotica* (Brugirard-Ricaud *et al.*, 2004). ffrench-Constant and collaborators, in association with the Sanger Institute, initiated the whole-genome sequencing of one strain of *P. asymbiotica* ssp. *asymbiotica* US 3105–77. The project is in progress and not yet published. Nevertheless, information on raw sequences is accessible at the Sanger Institute site (http://www.sanger.ac.uk/Projects/P_asymbiotica/).

*In silico* analysis revealed in the three strains identical TTSS backbones encoding potential components of the secretion/translocation apparatus and gene expression regulators (Brugirard-Ricaud *et al.*, 2004). The location of the three TTSS is identical suggesting an ancestral origin for the TTSS in the *Photorhabdus* genus. Despite the highly conserved organization and protein sequences of the core components of the secretion machinery, *P. luminescens* TT01 and *P. asymbiotica* TTSS loci encode different potential effectors (Fig. 13.3). *P. luminescens* TT01 encodes a product – LopT – similar to the *Yersinia* cystein protease cytotoxin YopT, that causes cytoskeletal disruption and contributes to the antiphagocytic effect of *Yersinia*. *P. asymbiotica* harbours a gene encoding a protein homologous to the *P. aeruginosa* ExoU effector that displays potent phospholipase activity inducing disruption of epithelial and macrophage cell lines. Analysis of the diversity of the TTSS and the two identified effectors showed that, whereas the TTSS backbone is well conserved in all the *Photorhabdus* strains, the effectors seem to belong to the flexible gene pool as they differ among the different species (Brugirard-Ricaud *et al.*, 2004).

Heterologous expression of LopT in *Yersinia* demonstrated that when it was produced and translocated in Hela cells by TTSS, LopT induced the same modification of the RhoA target as YopT of *Yersinia* (Brugirard-Ricaud *et al.*, 2005). Thus, the *in silico* predicted function of LopT was confirmed.

*In vivo* assays of infection of the cutworm *Spodoptera littoralis* and the locust *Locusta migratoria* showed that a TT01 strain carrying a translational fusion of the *lopT* gene was only detected at sites of cellular defence reactions, such as nodulation, and that TTSS-mutants did not induce nodule formation and underwent phagocytosis by insect macrophage (Brugirard-Ricaud *et al.*, 2005). Thus, *Photorhabdus* sequencing project allowed better understanding of the depression of the insect innate immune system (see Goodrich-Blair *et al.*, Chapter 11, this volume).

### 13.4.3. Application to a whole-genome comparison with genomes of other pathogenic bacteria

Closed genomes are syntenic, i.e. they share the same whole-genome organization and, in particular, the same gene order on the chromosome. This gene order is

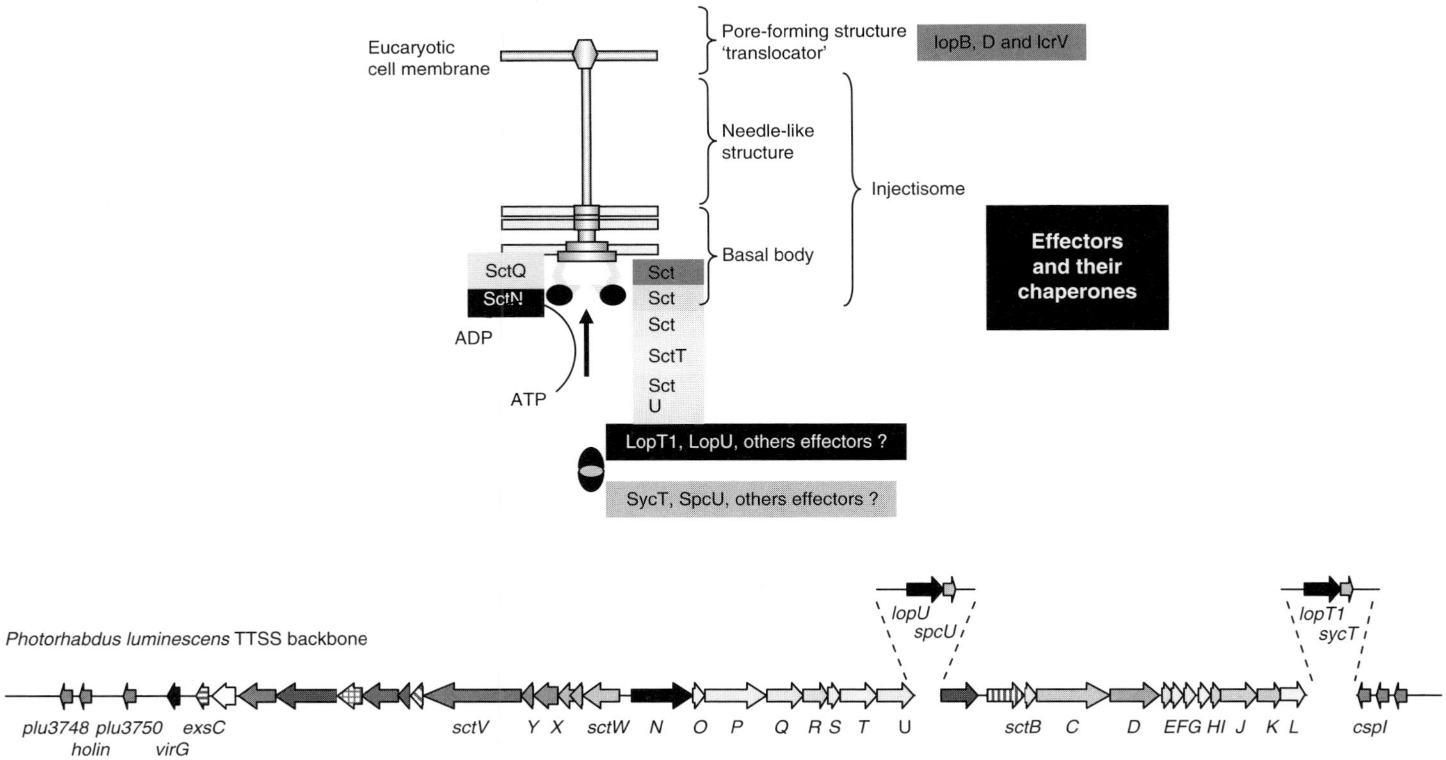


**Fig. 13.3.** Comparative genomics of TTSS organization in *Photorhabdus* spp.

often altered by large-scale genomic rearrangements and duplications that often occur symmetrically around the terminus or the origin of replication. Therefore, alignments of closely related bacterial genomes show an X-shaped pattern (X-alignment; Eisen *et al.*, 2000). The *P. luminescens* life cycle and a number of its phenotypic traits resemble those of *Y. pestis*. Both transit by and colonize insects and both are pathogenic. In order to test if these several common phenotypic characteristics are reflected in the genome, X-alignment was performed between *P. luminescens* and *Y. pestis* CO92.

*P. luminescens* TT01 proteome BLASTP comparisons were undertaken. The threshold was set to a minimum of 50% sequence similarity and a ratio of 0.8–1.2 of the protein length. Genes showing bidirectional best hit were defined as orthologues. A scatter plot of chromosome positions of all orthologous genes between *P. luminescens* and the test genome was built (Duchaud *et al.*, 2003).

A total of 2017 genes in *P. luminescens* have an orthologue in *Y. pestis* (Duchaud *et al.*, 2003). In addition, 77% of the orthologous genes of TT01 and *Y. pestis* CO92 are syntenic (Fig. 13.4A), confirming the close relationship between these two Enterobacteriaceae species. In contrast, orthologues are in similar quantity in *P. aeruginosa*, but X-alignment is not observable (Fig. 13.4D). Interestingly, *Y. pestis* and *P. luminescens* share not only the chromosomal backbone of Enterobacteriaceae, which both share also with *E. coli* (Fig. 13.4C), but they also share many putative mobile regions that do not belong to the enterobacterial core genome. These encode potential toxins, virulence factors and proteins of unknown function (Fig. 13.4B). Genes located in these putative mobile regions and shared by *P. luminescens* and *Y. pestis* are highly interesting potential virulence factors that presumably would have been exchanged by horizontal genetic transfers between the two species.

## 13.5. Post-genomic Analysis by a 'Blind' Approach

### 13.5.1. Definition

By contrast with the analogical approach, a 'blind' approach focuses on genes that give phenotypes, pattern of distribution, etc., of special interest without selecting the gene on the basis of its annotation. This approach will be illustrated by a whole-genome comparison between *Photorhabdus* strains by using a DNA microarray.

### 13.5.2. Whole-genome comparison by DNA microarray analysis

#### *13.5.2.1. Principle*

Bacterial genome structures are usually described as composed of a conserved 'core' genome, which contains the genetic information for basic cellular functions, and a 'flexible' gene pool, which contains mobile and accessory genetic

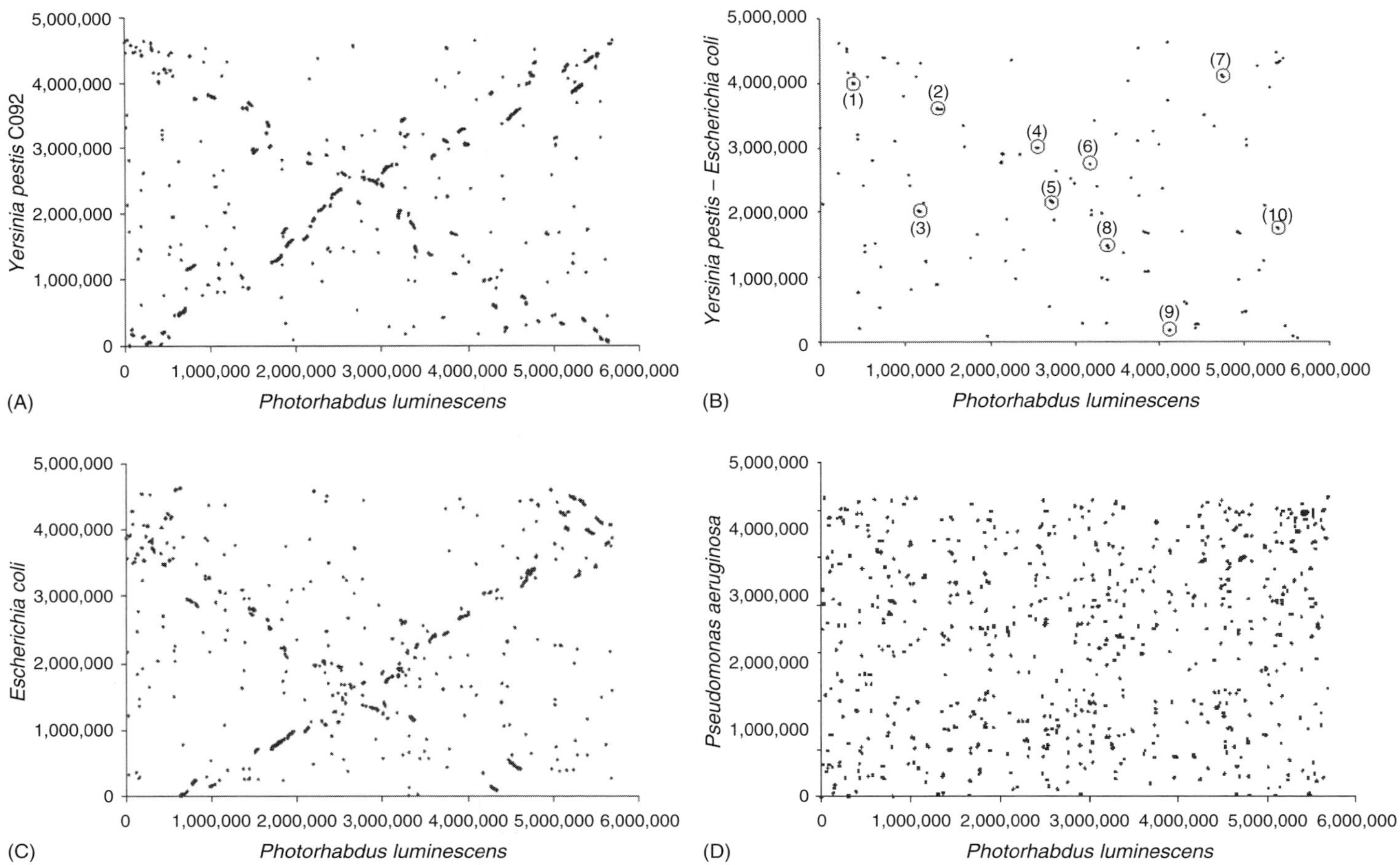


**Fig. 13.4.** Synteny (X-alignment) between *Photorhabdus luminescens* TT01 and *Yersinia pestis* CO92 (A), *Escherichia coli* K12 (C), and *Pseudomonas aeruginosa* (D) and present in *Y. pestis* CO92 but absent from *E. coli* K12 (B). (1) Island 4 of W13 (macrophage-like toxin and phlB/A); (2) NADH-quinone reductase (*nqr* locus); (3) 4-hydroxyphenylacetate catabolism (*hcp* locus); (4) Urease (*ure* locus); (5) Yersiniabactin (HPI locus); (6) iron uptake (*yfe* locus); (7) ribonuclease; (8) haemin/siderophore uptake; (9) haemin uptake (*hmu* uptake); (10) enterobactin.

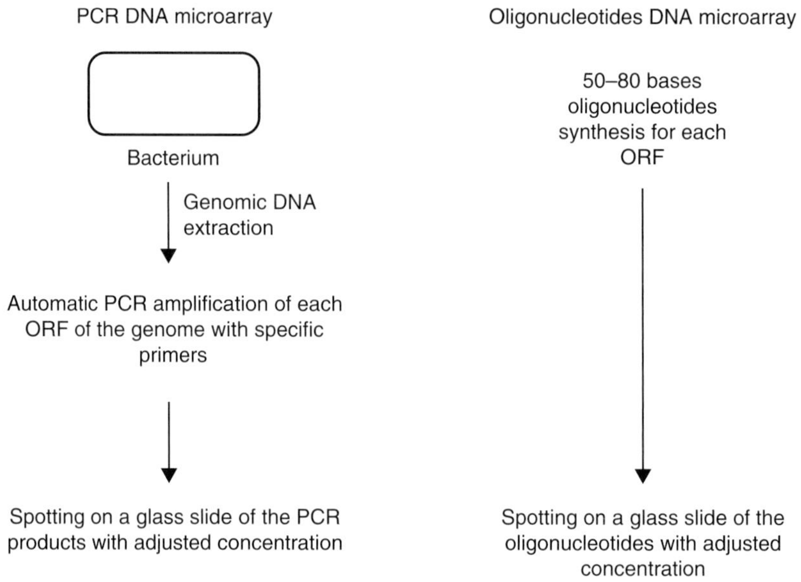


**Fig. 13.5.** Schematic representation of the steps for construction of a bacterial whole-genome DNA microarray.

elements. DNA microarrays can be used to investigate genome fluidity in order to identify conserved versus variable regions in a genome. In some cases, recently transferred blocks of genes act as pathogenicity, symbiosis or adaptation islands. Such methodology has proven to be useful in comparative genomics between species or genera to identify genetic variation that may promote and cause infection and disease (Joyce *et al.*, 2002).

In the whole-genome comparison analysis, a genome reference is needed to construct the DNA microarray. This technology based on the hybridization allows having information on gene or DNA content of closed unsequenced bacteria. A DNA microarray can be made with either oligonucleotides or PCR products representing short fragments of the chromosome or extrachromosomal elements (Fig. 13.5). To compare strains, total DNA of the genome reference and the tested genome is prepared. The DNA samples are labelled with different fluorophores, mixed before hybridization and co-hybridized on the array (Fig. 13.6). Such studies can only detect deletions or duplications in a strain relative to the reference strain. They are unable to detect rearrangements that change genome structure but do not result in alterations in copy number.

### *13.5.2.2. Construction of a TT01 whole-genome PCR DNA microarray*

A PCR whole-genome DNA microarray was constructed for representing genome of the TT01 strain (Gaudriault *et al.*, 2006). Primers were designed for each CDS by use of a modified version of Primer 3 Software (Frangeul *et al.*, 2004) to amplify specific fragments from 300 to 600 bp. Paralogous genes

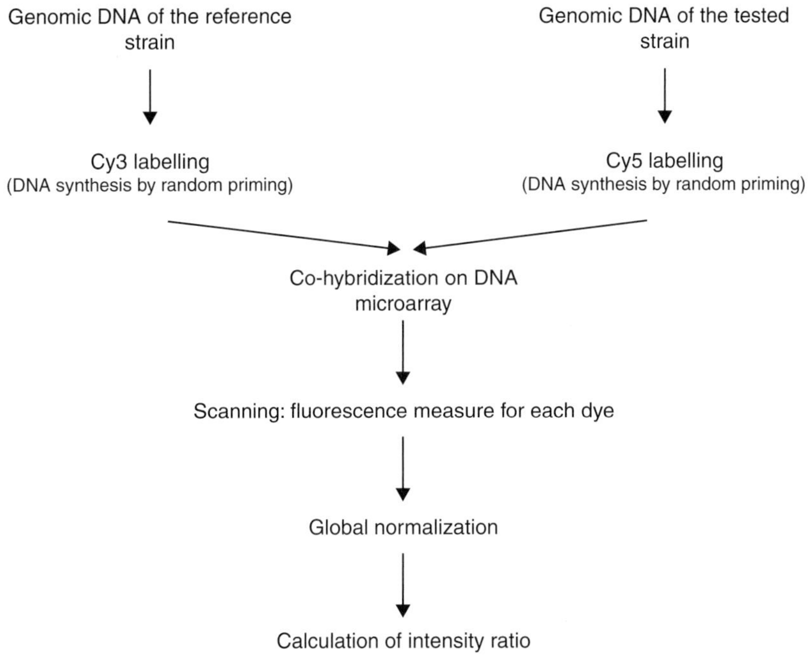


**Fig. 13.6.** Genomic comparison using a DNA microarray: hybridization, normalization and scanning.

(mainly IS and putative prophages) were excluded. Therefore, this *Photorhabdus* DNA microarray was representative of 4134 genes out of the 4839 predicted CDS of TT01 strain. Probes were amplified from *P. luminescens* TT01 genomic DNA, purified and adjusted in concentrations to 30 ng/μl in 50% dimethyl sulfoxide. Quality and quantity of the final matrix were checked by gel electrophoresis of the amplified probes and sequencing of 96 randomly chosen amplified probes. Two replicates of each probe were spotted using the GenIII Amersham spotter.

### *13.5.2.3. Hybridization and data analysis*

Genomic DNA (1 μg) was labelled with Cy3 or Cy5 by random priming and purified in order to remove PCR primers and nucleotides. Cy5 and Cy3 labelled genomic DNA were mixed and hybridized on DNA microarray at 42°C for 12 h. After washes, the DNA microarray was scanned for fluorescence intensity by using scanner. One microarray comparison (TT01, the reference genome versus the test genome) included four slides with two dye-flipped replicates. The signal intensity of each spot in the microarray was quantified by using the ARRAYVISION software (AMERSHAM).

Normalization aims to remove systematic errors by balancing the fluorescence intensities of the two labelling dyes. The dye bias can come from various sources including difference in dye labelling efficiencies or heat and dye sensitivities as well as scanner settings for scanning two channels. Global normalization (using the global median) was applied on the data without background correction (Park *et al.*, 2003;

Fang *et al.*, 2003). This step was done using Microsoft EXCEL software. As the different controls used on each slide demonstrate a good quality for spotting and hybridization, statistical analysis was not used. For each ORF tested, the median from the eight normalized values was calculated and used for determining test genome/TT01 ratios. In order to determine the ratio threshold that indicates the TT01 gene is missing in the test genome, regions of the test strain were randomly selected, amplified and sequenced. Then, 24 genes whose ratios ranged from 0.4 to 1.5 were chosen. When test genome/TT01 ratios are equal to, or less than, 0.6, genes have less than 20% identity with the probe spotted on the microarray. When ratios are equal to, or greater than, 0.98, genes have more than 70% identity with the probe spotted on the microarray. For ratios between 0.7 and 0.97, the percentage identity is variable. Therefore, we fixed the ratio threshold for missing genes to 0.6.

### 13.5.3. Application to the identification of the genomic flexible pool of *Photorhabdus*

To describe further the flexible gene pool of *Photorhabdus* spp. associated with nematodes, a comparison was made between the genomic content of *P. luminescens* TT01 and *P. temperata* ssp. *temperata* XlNach, the type-strain for this species. In strain XlNach, no large regions, such as canonic genomic islands, were absent relative to the reference strain TT01. Then, regions containing at least three contiguous genes missing from XlNach genome that represent at least 50% of the TT01 genomic region were searched. Thirty-one XlNach-missing regions were identified (Table 13.1; Gaudriault *et al.*, 2006). Genes present in these regions are phage remnants or belong mainly to putative functional classes that are likely involved in the *Photorhabdus* life cycle: pilus biosynthesis, antibiotic biosynthesis, insecticidal toxins, iron uptake or amino acid metabolism. Furthermore, this DNA microarray analysis led to the identification of a part of the flexible gene pool of *Photorhabdus* strains (Hacker and Carniel 2001). Indeed, 29 of the XlNach-missing regions fit within *in silico* predicted mobile regions. Thirteen regions belonged to previously described genomic islands (GIs; Duchaud *et al.*, 2003). Furthermore, using the Microbial Genome Annotation System (http://www.genoscope.cns.fr/agc/mage/wwwpkgdb/), the authors identified 16 regions that matched with enterobacterial variable regions (EVRs). The EVRs were gene blocks that were inserted at the location of a synteny rupture in the enterobacterial core genome. Their size (3–62 kb) and their rich content in mobile elements evoked the *Yersinia* 'difference regions' (DFRs), which belong to the intraspecific and interspecific *Yersinia* flexible gene pool (Radnedge *et al.*, 2002; Hinchliffe *et al.*, 2003).

### 13.5.4. Application to the identification of potential symbiosis specificity determinants

Although a few studies have identified *Photorhabdus* genes required for normal growth and development of the nematode (Bintrim and Ensign, 1998; Ciche *et al.*, 2001; Joyce and Clarke, 2003; Bennet and Clarke, 2005; Watson *et al.*, 2005),

**Table 13.1.** XlNach-missing regions described by whole-genome comparison using a DNA microarray.

| Locus | Plu | Size of the region in TT01 (kb) | Products of interest (similarity or function) | Matching genomic island (GI)[a] | Matching Enterobacteriaceae variable region (EVR)[b], other features |
|---|---|---|---|---|---|
| 1 | plu0125 – plu0132 | 10 | Unknown, Sai integrase, etc. | Part of GI plu0125 – plu0169 | |
| 2 | plu0136 – plu0156 | 18 | Unknown, transcriptional regulator, CoA metabolism, helicase, etc. | Part of GI plu0125 – plu0169 | |
| 3 | plu0263 – plu0269 | 8 | Pili cluster VI[b] (Fim-like, type 1 pili) | | Part of EVR plu0260 – plu0271 (11.7 kb, recombinase) |
| 4 | plu0280 – plu0282 | 3 | Phage remnant | | Part of EVR plu0275 – plu0285 (10.4 kb, truncated transposase, DNA ligase, phage protein) |
| 5 | plu0406 – plu0418 | 12 | Phage remnant and Pili Cluster V[b] (mrf-like) | Part of GI plu0404 – plu0419 | |
| 6 | plu0567 – plu0577 | 13 | Sugar transport and metabolism, amino acids synthesis | | Part of EVR plu0570 – plu0574 (5.3 kb, ERIC sequences at 5' extremity, IS) |
| 7 | plu0597 – plu0600 | 5 | Unknown, DNA-methyltransferase | | EVR plu0597 – plu600 (4.7 kb, proximity of a truncated phage gene and a truncated transposase) |
| 8 | plu0752 – plu0764 | 17 | Peptide synthesis and transport, CoA metabolism | Part of GI plu0751– plu0798 | |
| 9 | plu0895 – plu0899 | 16 | Cro/CI transcriptional regulator, antibiotic synthesis | Part of GI plu0884 – plu0901 | |
| 10 | plu0960 – plu0965 | 27 | Insecticidal toxins (loci *tcd* and *tcc*) | Part of GI plu0958 – plu1166 | |

*Continued*

**Table 13.1.** Continued

| Locus | Plu | Size of the region in TT01 (kb) | Products of interest (similarity or function) | Matching genomic island (GI)[a] | Matching Enterobacteriaceae variable region (EVR)[b], other features |
|---|---|---|---|---|---|
| 11 | plu1002 – plu1005 | 4 | Deshydratase, dioxygenase, cyanate and benzoate transport | Part of GI plu0958 – plu1166 | |
| 12 | plu1207 – plu1213 | 13 | Antibiotic synthesis | Part of GI plu1203 – plu1238 | |
| 13 | plu1336 – plu1343 | 11 | Antibiotic synthesis | | Part of EVR plu 1334 – plu1348 (15.4 kb) |
| 14 | plu2727 – plu2729 | 3 | Enterobactin synthetase (entABE) | | EVR plu2727 – plu2729 (3.1 kb, low GC%, flanked by repeats) |
| 15 | plu2792 – plu2799 | 10 | Antibiotic synthesis | | Part of EVR plu 2787 – plu2800 (18.3 kb, repeat containing proteins) |
| 16 | plu3135 – plu3139 | 7 | Citrate synthase, efflux transporter and unknown | Part of GI plu3111– plu3130 | |
| 17 | plu3134 – plu3136 | 5 | *lsr* (luxS synthesis regulated) operon, AI-2 import | | EVR poximity of transposases, ERIC sequence, flanks the GI plu3111–plu3130 |
| 18 | plu3398 – plu3405 | 6 | Phage remnant, unknown proteins | Part of GI plu3379 – plu3538 | |
| 19 | plu3537 – plu3539 | 5 | Aminotransferase, propionate metabolism | Overlaps the right border of GI plu3379 – plu3538 | |
| 20 | plu3724 – plu3726 | 4 | Aminobenzoyl-glutamate uptake and utilization | | Flanks the GI plu3685 – plu3723 |
| 21 | plu4077 – plu4081 | 5 | Truncated aldolase, deshydrogenase, transferase, unknown proteins | | Part of EVR plu4075 – plu4084 (12.3 kb, transposases) |
| 22 | plu4133 – plu4160 | 19 | ABC tranporter, amino acids metabolism, unknown | Part of GI plu4131– plu4246 | |

| | | | | | |
|---|---|---|---|---|---|
| 23 | plu4205 – plu4219 | 16 | Transposase, unknown proteins | Part of GI plu4131– plu4246 | |
| 24 | plu4266 – plu4269 | 5 | Amino acid metabolism, ABC transporter | | Part of EVR plu4254 – plu4310 (61.3 kb, transposase, Rhs family protein, low GC%) |
| 25 | plu4324 – plu4328 | 7 | Unknown proteins | | Part of a EVR plu4318 – plu4331 (16.8 kb, phage proteins, truncated integrase) |
| 26 | plu4336 – plu4348 | 13 | Carotenoid biosynthesis, unknown proteins | | Part of EVR plu4334 – plu4348 (16.4 kb, transposase, NTPase, C-terminal region of group II intron-associated maturase) |
| 27 | plu4589 – plu4591 | 3 | Unknown, transcription regulator Lys-R | | Part of a EVR plu4587 – plu4594 (5.7 kb, tRNA-Gly site insertion at 3' extremity) |
| 28 | plu4621 – plu4630 | 15 | Ferric enterobactin biosynthesis and uptake | | EVR plu4621– plu4630 (13.7 kb, ATP-dependent DNA helicase RecQ at the 5' border) |
| 29 | plu4810 – plu4823 | 15 | Lipopolysaccharide biosynthesis | | Part of EVR plu4796 – plu4833 (38.6 kb, transposase, low GC% by place) |
| 30 | plu4873 – plu4889 | 16 | Formate metabolism, *O*-methyltransferase, reverse transcriptase, macrolide-efflux protein, sugar kinase | | Overlaps a part of EVR plu4872 – plu4884 (11.1 kb, transposase, phage EVRoteins, low GC% by place) |
| 31 | plu4892 – plu4895 | 6 | *O*-methyltransferase, transposase | | Part of EVR plu4890 – plu4895 (9.0 kb, transposase, low GC%) |

[a]Genomic islands described in Duchaud *et al.*, 2003.
[b]Enterobacteriaceae variable regions (EVR) described in Gaudriault *et al.*, 2006.
Grey lines show loci present in *P. temperata* C1 and that consequently may be involved in the specific interaction with *H. bacteriophora*.

little molecular and functional data are available concerning the first step of nematode colonization and nematode specificity. In order to identify regions that are possibly involved in nematode specificity, the genome of *P. temperata* C1 associated with *H. bacteriophora* was added to the previous comparison. (Gaudriault *et al.*, 2006). The eight genomic regions present in both TT01 and C1 strains but missing in the XlNach strain were considered as potentially specific to strains associated with *H. bacteriophora* (grey lines in Table 13.1). To further test the correlation between the TT01- and C1-specific regions and the interaction with *H. bacteriophora*, the distribution of these regions were studied in 13 *Photorhabdus* strains representative of the genus by PCR amplification. Only one locus, the locus *lsr*, had an amplification size clearly correlated with the nematode host species. These data were checked by sequencing of some PCR products of the *lsr* locus. The *lsr*-like loci of *X. nematophila* ATCC 19061 and *Xenorhabdus bovienni* (http://www.xenorhabdus.org/), a genus closely related to *Photorhabdus* (Boemare, 2002), were also added for comparison. It was striking that various *lsrA*, *lsrB* and *lsrR* remnants were observed, showing that the *lsr* locus underwent independent deletions in the matching strains (Fig. 13.7). Therefore, the *lsr* locus in an ancestral locus in *Photorhabdus* and *Xenorhabdus* and the bacterial association with *H. bacteriophora* possibly resulted from selective pressure for the

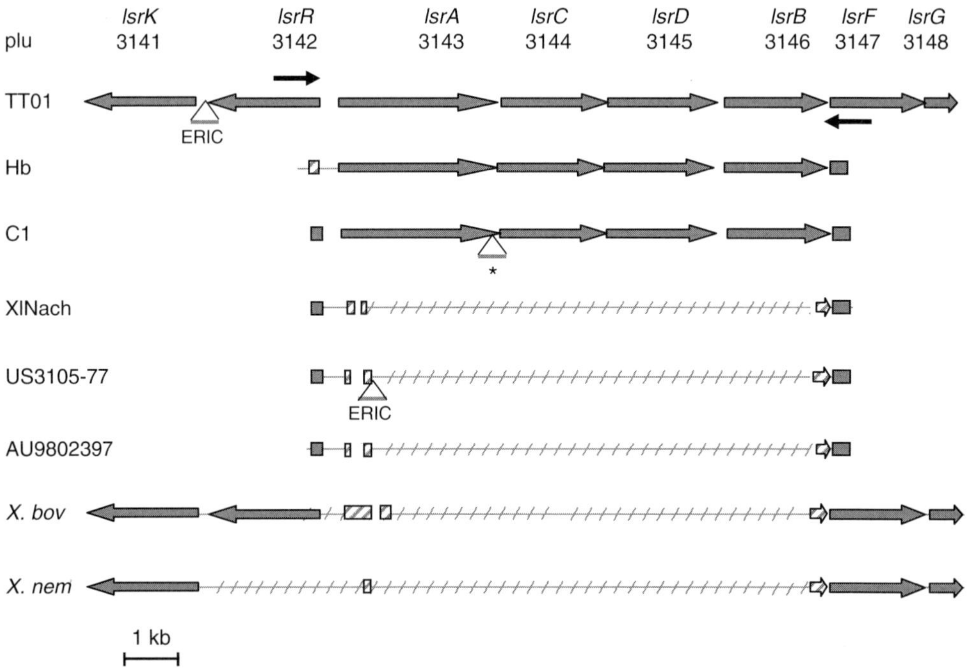


**Fig. 13.7.** Schematic representation of the deletions in the *lsr* region (genes are listed across the top) of several *Photorhabdus* and *Xenorhabdus* strains (listed at left). Horizontal arrows represent primers designed for the long-range PCR analysis of the locus and for sequencing. Grey and hatched arrows or boxes symbolize open reading frames and their remnants, respectively. Insertion of ERIC elements and a 27-nucleotide region(*) are represented.

conservation of the *lsr* locus. In other nematode hosts, the *lsr* locus appears lost by reason of genomic decay.

The *lsr* locus is similar to the *lsr* regions of *Salmonella enterica* serovar *typhimurium* and *E. coli* that encode an inner ABC transporter and a cytoplasmic phosphorylation-processing system of the auto-inducer AI-2, involved in quorum sensing (Taga *et al.*, 2001, 2003; Xavier and Bassler, 2005). In *S. enterica* serovar *typhimurium* and *E. coli*, it was suggested that the Lsr transporter has a role in removing the AI-2 signal from the external environment in order to terminate cell–cell signalling (Taga *et al.*, 2001, 2003; Xavier and Bassler, 2005). In a bacterium–nematode interaction, the termination of cell–cell signalling could be important in allowing a bacterial physiological shift. For example, this may occur in the insect cadaver at the time bacteria re-colonize the nematode intestinal tract of *H. bacteriophora*.

## 13.6. Conclusion and Future Perspectives

The availability of a sequenced genome for *P. luminescens* and its functional analysis should lead to several useful applications. These may include the development of new entomotoxins for crop protection and the genetic engineering of the bacterium–nematode pair for use as biological control agents. The knowledge of additional entomopathogenic and/or symbiotic bacterial genomes, such as the recently published *P. entomophila* genome (Vodovar *et al.*, 2006) and two *Xenorhabdus* genomes (http://www.xenorhabdus.org/), will provide new insights into pathogenic and symbiotic relationships. Nevertheless, knowledge and understanding of phylogenetic relationships among nematode- and insect-associated bacteria and related human-pathogenic bacteria (e.g. *Y. pestis*) must be carefully studied to avoid risks to human health. The anticipated availability of the *P. asymbiotica* genome (http://www.sanger.ac.uk/Projects/P_asymbiotica/) should improve our understanding of the evolutionary history of *Photorhabdus*. The combination of these findings should shed light on the processes by which environmental bacteria may be transformed into emerging human pathogens.

## References


Akhurst, R.J., Boemare, N.E., Janssen, P.H., Peel, M.M., Alfredson, D.A. and Beard, C.E. (2004) Taxonomy of Australian clinical isolates of the genus *Photorhabdus* and proposal of *Photorhabdus asymbiotica* subsp. *asymbiotica* subsp. nov. and *P. asymbiotica* subsp. *australis* subsp. nov. *International Journal of Systematic and Evolutionary Microbiology* 54, 1301–1310.

Bennett, H.P. and Clarke, D.J. (2005) The *pbgPE* operon in *Photorhabdus luminescens* is required for pathogenicity and symbiosis. *Journal of Bacteriology* 187, 77–84.

Bintrim, S.B. and Ensign, J.C. (1998) Insertional inactivation of genes encoding the crystalline inclusion proteins of *Photorhabdus luminescens* results in mutants with pleiotropic phenotypes. *Journal of Bacteriology* 180, 1261–1269.

Boemare, N. (2002) Biology, taxonomy and systematics of *Photorhabdus* and *Xenorhabdus*. In: Gaugler, R. (ed.) *Entomopathogenic Nematology*. CAB International, Wallingford, UK, pp. 35–56.

Bowen, D., Rocheleau, T.A., Blackburn, M., Andreev, O., Golubeva, E., Bhartia, R. and ffrench-Constant, R.H. (1998) Insecticidal toxins from the bacterium *Photorhabdus luminescens*. *Science* 280, 2129–2132.

Bowen, D., Blackburn, M., Rocheleau, T., Grutzmacher, C. and ffrench-Constant, R.H. (2000) Secreted proteases from *Photorhabdus luminescens*: separation of the extracellular proteases from the insecticidal Tc toxin complexes. *Insect Biochemistry and Molecular Biology* 30, 69–74.

Brillard, J., Duchaud, E., Boemare, N., Kunst, F. and Givaudan, A. (2002) The PhlA hemolysin from the entomopathogenic bacterium *Photorhabdus luminescens* belongs to the two-partner secretion family of hemolysins. *Journal of Bacteriology* 184, 3871–3878.

Brugirard-Ricaud, K., Givaudan, A., Parkhill, J., Boemare, N., Kunst, F., Zumbihl, R. and Duchaud, E. (2004) Variation in the effectors of the type III secretion system among *Photorhabdus* species as revealed by genomic analysis. *Journal of Bacteriology* 186, 4376–4381.

Brugirard-Ricaud, K., Duchaud, E., Givaudan, A., Girard, P.A., Kunst, F., Boemare, N., Brehelin, M. and Zumbihl, R. (2005) Site-specific antiphagocytic function of the *Photorhabdus luminescens* type III secretion system during insect colonization. *Cellular Microbiology* 7, 363–371.

Burland, V., Daniels, D.L., Plunkett, G., 3rd and Blattner, F.R. (1993) Genome sequencing on both strands: the Janus strategy. *Nucleic Acids Research* 21, 3385–3390.

Casjens, S. (1998) The diverse and dynamic structure of bacterial genomes. *Annual Review of Genetics* 32, 339–377.

Ciche, T.A., Bintrim, S.B., Horswill, A.R. and Ensign, J.C. (2001) A Phosphopantetheinyl transferase homolog is essential for *Photorhabdus luminescens* to support growth and reproduction of the entomopathogenic nematode *Heterorhabditis bacteriophora*. *Journal of Bacteriology* 183, 3117–3126.

Clarke, D.J. and Dowds, B.C. (1994) The gene coding for polynucleotide phosphorylase in *Photorhabdus* sp. strain K122 is induced at low temperatures. *Journal of Bacteriology* 176, 3775–3784.

Clarke, D.J. and Dowds, B.C. (1995) Virulence mechanisms of *Photorhabdus* sp. strain K122 toward wax moth larvae. *Journal of Invertebrate Pathology* 66, 139–155.

Cornelis, G.R. and Van Gijsegem, F. (2000) Assembly and function of type III secretory systems. *Annual Review of Microbiology* 54, 735–774.

Daborn, P.J., Waterfield, N., Blight, M.A. and ffrench-Constant, R.H. (2001) Measuring virulence factor expression by the pathogenic bacterium *Photorhabdus luminescens* in culture and during insect infection. *Journal of Bacteriology* 183, 5834–5839.

Duchaud, E., Rusniok, C., Frangeul, L., Buchrieser, C., Givaudan, A., Taourit, S., Bocs, S., Boursaux-Eude, C., Chandler, M., Charles, J.F., Dassa, E., Derose, R., Derzelle, S., Freyssinet, G., Gaudriault, S., Medigue, C., Lanois, A., Powell, K., Siguier, P., Vincent, R., Wingate, V., Zouine, M., Glaser, P., Boemare, N., Danchin, A. and Kunst, F. (2003) The genome sequence of the entomopathogenic bacterium *Photorhabdus luminescens*. *Nature Biotechnology* 21, 1307–1313.

Eisen, J.A., Heidelberg, J.F., White, O. and Salzberg, S.L. (2000) Evidence for symmetric chromosomal inversions around the replication origin in bacteria. *Genome Biology* 1(6), 11.1–11.9.

Ewing, B. and Green, P. (1998) Base-calling of automated sequencer traces using phred. II. Error probabilities. *Genome Research* 8, 186–194.

Fang, Y., Brass, A., Hoyle, D.C., Hayes, A., Bashein, A., Oliver, S.G., Waddington, D. and Rattray, M. (2003) A model-based analysis of microarray experimental error and normalisation. *Nucleic Acids Research* 31(16), e96.

ffrench-Constant, R.H., Waterfield, N., Burland, V., Perna, N.T., Daborn, P.J., Bowen, D. and Blattner, F.R. (2000) A genomic sample sequence of the entomopathogenic bacterium *Photorhabdus luminescens* W14: potential implications for virulence. *Applied and Environmental Microbiology* 66, 3310–3329.

ffrench-Constant, R., Waterfield, N., Daborn, P., Joyce, S., Bennett, H., Au, C., Dowling, A., Boundy, S., Reynolds, S. and Clarke, D. (2003) *Photorhabdus*: towards a functional genomic analysis of a symbiont and pathogen. *FEMS Microbiology Review* 26, 433–456.

Frangeul, L., Nelson, K.E., Buchrieser, C., Danchin, A., Glaser, P. and Kunst, F. (1999) Cloning and assembly strategies in microbial genome projects. *Microbiology* 135(Part 10), 2625–2634.

Frangeul, L., Glaser, P., Rusniok, C., Buchrieser, C., Duchaud, E., Dehoux, P. and Kunst, F. (2004) CAAT-Box, Contigs-Assembly and Annotation Tool-Box for genome sequencing projects. *Bioinformatics* 20, 790–797.

Gaudriault, S., Thaler, J.O., Duchaud, E., Kunst, F., Boemare, N. and Givaudan, A. (2004) Identification of a P2-related prophage remnant locus of *Photorhabdus luminescens* encoding an R-type phage tail-like particle. *FEMS Microbiology Letters* 233, 223–231.

Gaudriault, S., Duchaud, E., Lanois, A., Canoy, A.S., Bourot, S., Derose, R., Kunst, F., Boemare, N. and Givaudan, A. (2006) Whole-genome comparison between *Photorhabdus* strains to identify genomic regions involved in the specificity of nematode interaction. *Journal of Bacteriology* 188, 809–813.

Gordon, D., Abajian, C. and Green, P. (1998) Consed: a graphical tool for sequence finishing. *Genome Research* 8, 195–202.

Hacker, J. and Carniel, E. (2001) Ecological fitness, genomic islands and bacterial pathogenicity. A Darwinian view of the evolution of microbes. *EMBO Reports* 2, 376–381.

Hinchliffe, S.J., Isherwood, K.E., Stabler, R.A., Prentice, M.B., Rakin, A., Nichols, R.A., Oyston, P.C., Hinds, J., Titball, R.W. and Wren, B.W. (2003) Application of DNA microarrays to study the evolutionary genomics of *Yersinia pestis* and *Yersinia pseudotuberculosis*. *Genome Research* 13, 2018–2029.

Hueck, C.J. (1998) Type III protein secretion systems in bacterial pathogens of animals and plants. *Microbiology and Molecular Biology Reviews* 62, 379–433.

Hurst, M.R., Glare, T.R. and Jackson, T.A. (2004) Cloning *Serratia entomophila* antifeeding genes – a putative defective prophage active against the grass grub *Costelytra zealandica*. *Journal of Bacteriology* 186, 5116–5128.

Joyce, E.A., Chan, K., Salama, N.R. and Falkow, S. (2002) Redefining bacterial populations: a post-genomic reformation. *Nature Reviews Genetics* 3, 462–473.

Joyce, S.A. and Clarke, D.J. (2003) A *hexA* homologue from *Photorhabdus* regulates pathogenicity, symbiosis and phenotypic variation. *Molecular Microbiology* 47, 1345–1357.

Kim, U.J., Birren, B.W., Slepak, T., Mancino, V., Boysen, C., Kang, H.L., Simon, M.I. and Shizuya, H. (1996) Construction and characterization of a human bacterial artificial chromosome library. *Genomics* 34, 213–218.

Marokhazi, J., Koczan, G., Hudecz, F., Graf, L., Fodor, A. and Venekei, I. (2004) Enzymic characterization with progress curve analysis of a collagen peptidase from an enthomopathogenic bacterium, *Photorhabdus luminescens*. *The Biochemical Journal* 379, 633–640.

Moszer, I., Jones, L.M., Moreira, S., Fabry, C. and Danchin, A. (2002) SubtiList: the reference database for the *Bacillus subtilis* genome. *Nucleic Acids Research* 30, 62–65.

Park, T., Yi, S.G., Kang, S.H., Lee, S., Lee, Y.S. and Simon, R. (2003) Evaluation of normalization methods for microarray data. *BMC Bioinformatics* 4, 33.

Radnedge, L., Agron, P.G., Worsham, P.L. and Andersen, G.L. (2002) Genome plasticity in *Yersinia pestis*. *Microbiology* 138, 1687–1698.

Schmidt, T.M., Bleakley, B. and Nealson, K.H. (1988) Characterization of an extracellular Protease from the insect pathogen *Xenorhabdus luminescens*. *Applied and Environmental Microbiology* 54, 2793–2797.

Taga, M.E., Semmelhack, J.L. and Bassler, B.L. (2001) The LuxS-dependent autoinducer AI-2 controls the expression of an ABC transporter that functions in AI-2 uptake in *Salmonella typhimurium*. *Molecular Microbiology* 42, 777–793.

Taga, M.E., Miller, S.T. and Bassler, B.L. (2003) Lsr-mediated transport and processing of AI-2 in *Salmonella typhimurium*. *Molecular Microbiology* 50, 1311–1327.

Valens, M., Broutelle, A.C., Lefebvre, M. and Blight, M.A. (2002) A zinc metalloprotease inhibitor, Inh, from the insect pathogen *Photorhabdus luminescens*. *Microbiology* 138, 2427–2437.
Vodovar, N., Vallenet, D., Cruveiller, S., Rouy, Z., Barbe, V., Acosta, C., Cattolico, L., Jubin, C., Lajus, A., Segurens, B., Vacherie, B., Wincker, P., Weissenbach, J., Lemaitre, B., Médigue, C. and Boccard, F. (2006) Complete genome sequence of the entomopathogenic and metabolically versatile soil bacterium *Pseudomonas entomophila*. *Nature Biotechnology* 24, 673–679.
Wang, H. and Dowds, B.C. (1993) Phase variation in *Xenorhabdus luminescens*: cloning and sequencing of the lipase gene and analysis of its expression in primary and secondary phases of the bacterium. *Journal of Bacteriology* 175, 1665–1673.
Waterfield, N.R., Daborn, P.J. and ffrench-Constant, R.H. (2002) Genomic islands in *Photorhabdus*. *Trends in Microbiology* 10, 541–545.
Watson, R.J., Joyce, S.A., Spencer, G.V. and Clarke, D.J. (2005) The *exbD* gene of *Photorhabdus temperata* is required for full virulence in insects and symbiosis with the nematode *Heterorhabditis*. *Molecular Microbiology* 56, 763–773.
Webster, J.M., Chen, G., Hu, K. and Li, J. (2002) Bacterial metabolites. In: Gaugler, R. (ed.) *Entomopathogenic Nematology*. CAB International, Wallingford, UK, pp. 99–113.
Xavier, K.B. and Bassler, B.L. (2005) Regulation of uptake and processing of the quorum-sensing autoinducer AI-2 in *Escherichia coli*. *Journal of Bacteriology* 187, 238–248.
Xu, S.Y. and Fomenkov, A. (1994) Construction of pSC101 derivatives with Camr and Tetr for selection or LacZ' for blue/white screening. *Biotechniques* 17, 57.
Zenno, S. and Saigo, K. (1994) Identification of the genes encoding NAD(P)H-flavin oxidoreductases that are similar in sequence to *Escherichia coli* Fre in four species of luminous bacteria: *Photorhabdus luminescens*, *Vibrio fischeri*, *Vibrio harveyi*, and *Vibrio orientalis*. *Journal of Bacteriology* 176, 3544–3551.

# 14 Genomics of Entomopathogenic Viruses

J. Slack, Z. Li, S. Escasa, D. Doucet, T. Ladd, G. Quan and B. Arif

*Great Lakes Forestry Centre, Sault Ste Marie, Canada*



## 14.1. Introduction

Over the last decade, there has been an explosion of genomics information on insect viruses primarily because sequencing technologies have improved drastically in accuracy and speed. Moreover, with enhancement of polymerase chain reaction (PCR) methods, whole-genome sequencing has become an affordable procedure. As a result, knowledge on the genomics of insect viruses has expanded significantly and we are now able to draw more accurate conclusions on the co-evolution of viruses and their natural hosts (Herniou *et al.*, 2004). By far, the most studied group of viruses at the genomics level is the Baculoviruses where 48 genomes belonging to viruses from Lepidoptera, Hymenoptera and Diptera have been totally sequenced. Table 14.1 outlines the groups of insect viruses whose genomes have been sequenced and deposited in GenBank.

Clearly, it will not be possible to cover even part of the list outlined in Table 14.1 and, therefore, this chapter will only focus on baculoviruses. Readers should also refer to Chapter 1 (Erlandson and Theilmann, this volume) for further information on virus classification and molecular methods considered for virus taxonomy.

## 14.2. General Concepts

The *Baculoviridae* is a family of double-stranded DNA (dsDNA) viruses with circular DNA genomes within rod-shaped capsids enclosed by a lipid envelope and are arthropod-specific. The nucleocapsids are approximately 30–60 nm in diameter and 250–300 nm in length (Miller, 1997). Previously, baculoviruses were divided into two genera: Granuloviruses (GVs) and Nucleopolyhedroviruses (NPVs) based on the shape and size of their occlusion bodies (OBs). However, following the total genomic sequencing of one dipteran and three hymenopteran baculoviruses, it became clear that the present classification system of two genera was inadequate and could not accommodate these last four viruses. Comparisons of 29 baculovirus genomes demonstrated that viral phylogeny was more congruent with the classification of the hosts than viral morphological characters that have previously been used in the classification system. Based on this evidence, the current classification system separates dipteran- and hymenopteran-specific NPVs into genera distinct from that of lepidopteran-specific NPVs.

A new classification system had to be adopted. A new system was suggested by Lauzon *et al.* (2004) that was similar to the classification of *Herpesviridae*. Following that, a new classification system was proposed by Jehle *et al.* (2006). The system contains four genera and has the flexibility to accommodate more genera in the future. They are:

1. Alphabaculovirus – lepidopteran-specific NPVs.
2. Betabaculovirus – lepidopteran-specific GVs.
3. Gammabaculovirus – hymenopteran-specific NPVs.
4. Deltabaculovirus – dipteran-specific NPVs.

It is expected that this system will be adopted at the next meeting of the International Committee on Taxonomy of Viruses.

NPVs contain either single or multiple nucleocapsids per envelope and the enveloped virions are further occluded in a granulin matrix in GVs and a polyhedrin for NPVs (Miller, 1997). GVs have been mainly identified in Lepidoptera while NPVs have been found in Lepidoptera, Diptera, Hymenoptera, Coleoptera and other groups. Baculoviruses generally have very species-specific tropisms among the invertebrates with over 600 host species having been described. Since many insect pests have a corresponding baculovirus pathogen, baculoviruses have been studied because of their potential as biological control agents and are currently being used for the control of a number of agricultural and forest pests (Hayakawa *et al.*, 2000).

The replication cycle of lepidopteran baculoviruses is biphasic and distinguished by the production of two different phenotypes: the budded virus (BV) and occlusion-derived virus (ODV). BV, produced early in infection, disseminates infection

**Table 14.1.** Summary of nearly complete or completely sequenced insect virus genomes.

| Virus family or subtaxon | Type | Number of sequenced genomes | Genome size range (kb) | Number of ORFs | Reference |
|---|---|---|---|---|---|
| *Baculoviridae* | Double-stranded DNA, circular | 43 | 82–179 | 89–181 | Wolff *et al*., 2008 |
| Nudivirus | Double-stranded DNA, circular | 3 | 97–228 | 98–154 | Wang *et al*., 2007 |
| *Iridoviridae* (Chloriridovirus and Iridovirus) | Double-stranded DNA, circular | 2 | 190–212 | 126–468 | Delhon *et al*., 2006 |
| *Parvoviridae* (*Densovirinae*) | Single-stranded DNA, linear | 14 | 4–6 | 3–7 | Mukha *et al*., 2006 |
| *Poxviridae* (*Entomopoxvirinae*) | Double-stranded DNA, circular | 2 | 232–236 | 267–292 | Gubser *et al*., 2004 |
| *Polydnaviridae* | Double-stranded DNA, polydisperse circular | 5 | 187–567[a] | 61–156 | Tanaka *et al*., 2007 |
| *Ascoviridae* | Double-stranded DNA, circular | 3 | 156–186 | 123–180 | Asgari *et al*., 2007 |
| Unclassified DNA viruses | Double-stranded DNA, circular | 2 | 124–190 | 108–160 | Abd-Alla *et al*., 2008; Garcia-Maruniak *et al*., 2008 |
| *Reoviridae* (Cypovirus and Dinovernavirus) | Double-stranded RNA, linear | 5 | 23–25[a] | 9–11 | Attoui *et al*., 2005; Li *et al*., 2007; Tan *et al*., 2008 |
| *Birnaviridae* (Entomobirnavirus) | Double-stranded RNA, linear | 1 | 6.6[a] | 2 | Shwed *et al*., 2002 |
| *Nodaviridae* | Single-stranded RNA, linear | 4 | 4.3–4.5[a] | 3–5 | Dasmahapatra *et al*., 1985 |
| *Dicistroviridae* | Single-stranded RNA, linear | 14 | 8–10 | 2 | De Miranda *et al*., 2004; Maori *et al*., 2007 |
| Iflavirus | Single-stranded RNA, linear | 8 | 8.8–10.1 | 1 | Ryabov, 2007 |
| *Tetraviridae* | Single-stranded RNA, linear | 6 | 5.6–7.9[a] | 2–3 | Yi *et al*., 2005 |
| Unclassified RNA viruses | Single-stranded RNA, linear | 4 | 10–11 | 1–4 | van der Wilk *et al*., 1997; Hartley *et al*., 2005; Habayeb *et al*., 2006; Valles *et al*., 2007 |

[a]Size of aggregate genome. Number of segments in *Polydnaviridae*: 15–105; Entomobirnavirus: 2; *Reoviridae*: 9–12; *Nodaviridae*: 2; *Tetraviridae*: 1–2.

throughout susceptible larval tissues and cells while the ODV, produced later in infection, disseminates the viruses from larva to larva (Blissard *et al.*, 2000).

### 14.2.1. Gene content

The study of gene content has the potential to show the extent of variation between baculovirus genomes and the comparison of genomes may provide valuable insights into baculovirus evolution and biology (Herniou *et al.*, 2003). Baculovirus genomes range in size from approximately 80–180 kb (Theilmann *et al.*, 2005). Currently, the baculovirus with the largest genome is *Xestia c-nigrum* NPV (XecnGV) with 178,733 bp (Hayakawa *et al.*, 1999), while the smallest belongs to *Neodiprion lecontei* NPV (NeleNPV) with 81,755 bp (Lauzon *et al.*, 2004). To date, 43 completely sequenced baculovirus genomes are listed in GenBank with 38 NPVs and ten GVs. The number of predicted open reading frames (ORFs) found in sequenced baculoviruses encoding 50 or more amino acids range from approximately 89–181 ORFs. The average G + C content is quite variable in baculoviruses, ranging from low 30% to low 60%. A distinctive feature of most sequenced baculoviruses is the presence of repeat regions or homologous regions (*hrs*) dispersed throughout the genome ranging from approximately three *hr*/repeat regions to 17. In NPVs, most *hrs* contain 30 bp palindromes within direct repeats and are similar to other NPV *hrs*, whereas GV repeat regions are more variable and often lack palindromes (Wormleaton *et al.*, 2003). Hymenopteran baculovirus genomes contain repeated regions that do not conform to the typical structure of *hrs* in lepidopteran NPVs.

### 14.2.2. Genomes and phylogenetics

Comparative analysis of baculoviruses provided insight into their evolutionary history in that differences and similarities in amino acid sequences and in gene order aided in the division of baculoviruses into groups sharing gene characteristics and overall genome relatedness. More closely related viruses share a higher degree of gene co-linearity; Hu *et al.* (1998) developed a method called gene parity plots that compared the positions of homologous genes in different genomes and is used to show conservation between baculovirus genomes. Another method for comparing and classifying genomes is the use of phylogenetic trees (Herniou *et al.*, 2003). Traditionally, phylogenies were done by analysing individual genes within genomes such as *polyhedrin* or *dna polymerase*. However, the use of a concatenation of shared proteins has been shown to produce more reliable trees (Herniou *et al.*, 2003). This is done by producing concatamers of the core proteins found in all baculoviruses, aligning their amino acid sequences and generating trees that reflect different taxonomic divisions. Gene order and phylogeny provide essential information on the evolution and relatedness of baculoviruses.

Comparative analysis of all completely sequenced baculoviruses revealed a set of core genes conserved in all genomes that have essential roles in the baculovirus life cycle (Garcia-Maruniak *et al.*, 2004). There were 29 conserved genes found in all baculoviruses, until recently when McCarthy *et al.* (2008) identified *ac143/odv-e18* as a core gene with a homologue found in *Culex nigripalpus* NPV (CuniNPV; Afonso *et al.*, 2000). This brings up the core baculovirus conserved genes to 30. A homologue to *ac143* was previously not found in this genome; however, a gene that clustered with the *ac142* homologue was found in the same orientation, and was similar in size to *ac143* and was identified as CuniNPV ORF 31. The predicted protein, cuni31, had homology and similar transmembrane domain structures to all *ac143* homologues (McCarthy *et al.*, 2008). Most of the 30 core genes have a known function within the genome, either required for RNA transcription, DNA replication or as structural and auxiliary proteins. Currently, the functions of seven of the conserved genes are unknown: *38K, p33, ac68, ac96, ac109, ac81* and *ac115* (Herniou *et al.*, 2003).

Identification of genes that are essential or that stimulate DNA replication in baculoviruses has provided a basis for elucidating the process by which they replicate their genomes (Kool *et al.*, 1995). Baculovirus DNA replicates in the nucleus and they carry their own complement of genes encoding DNA replication proteins. Four of these genes are found in all sequenced baculoviruses to date: *DNA polymerase, DNA helicase* (*p143*), *lef-1* and *lef-2* (Herniou *et al.*, 2003). A list of additional genes involved in DNA replication found in lepidopteran baculoviruses also includes: *lef-3, ie-1* and *me53* (Herniou *et al.*, 2003; Lange and Jehle, 2003) and an additional gene is found in lepidopteran and dipteran baculoviruses: *dbp* (DNA binding protein; Lauzon *et al.*, 2004).

### 14.2.3. Gene expression

Baculovirus gene expression occurs in a regulated manner, primarily at the level of transcription. Baculovirus transcription is categorized into three classes based on when in the cycle a gene is expressed: early, late and very late. DNA replication marks the demarcation of early and late gene expression. Control of the expression is mainly done by the promoter region of the gene. Host RNA polymerase II transcribes early genes (such as *ie-1* found in lepidopteran baculoviruses but not in those of Hymenoptera), while viral RNA polymerase transcribes late and very late genes (Miller, 1997). Transcription of early genes is initiated upstream of the ATG start codon from highly conserved promoter elements that imitate eukaryotic TATA elements (TATA+) or arthropod-like initiator element(INR+) ‘CA(G/T)T’ or a combination of both (Hefferon and Miller, 2002). Early gene products are detectable at approximately 1 h post infection of cells and are required for DNA replication. Late RNA polymerase is composed of four viral proteins: *lef-4, lef-8, lef-9 and p47*, which are present in all sequenced baculoviruses. Two other genes in the core set are *lef-5* (unknown function) and *vlf-1* (required for very late gene transcription; Herniou *et al.*, 2003). Additional transcription genes present in some but not all baculoviruses include: *39K, lef-6* and *lef-11* (Herniou *et al.*, 2003; Lange and Jehle, 2003; Lauzon *et al.*, 2004). Late genes are expressed following the onset of DNA replication. Transcription is initiated from a late promoter element with a consensus element TAAG.

Baculoviruses share a number of structural proteins as the virions of NPVs and GVs have similar appearance. Conserved baculovirus structural protein genes include: *odv-e56* (ODV envelope protein), *p6.9* (DNA binding protein), *gp41* (integument protein also associated with BV production), *p74* (associated with occluded virions and is required for oral infectivity), *odv-ec27*, *vp91*, *vp1054* (capsid-associated protein), *odv-e18* (ODV envelope protein) and *vp39*. Genes found in some but not all baculoviruses include *polyhedrin*, *F protein*, *gp64* (membrane fusion proteins mediate the fusion of BV to cell membranes and the release of nucleocapsids), *odv-e66* (ODV envelope protein), *fp25K*, *odv-e25* (ODV envelope protein) and *pk-1* (Lange and Jehle, 2003; Lauzon *et al.*, 2004).

The genomes of lepidopteran baculoviruses contain auxiliary genes that have been defined as not essential for viral replication but may provide the virus a selective advantage in nature (O'Reilly, 1997). There is one known auxiliary gene present in all baculoviruses: *alk-exo* (alkaline exonuclease gene). The lack of auxiliary genes distinguishes the genomes of hymenopteran baculoviruses from those of Lepidoptera (Lauzon *et al.*, 2004). Auxiliary genes found in some but not all baculoviruses include: *fgf* genes (fibroblast growth factor genes), which play important roles in cell proliferation, differentiation and tissue repair (Ornitz and Nobuyuki, 2001); *ubiquitin* (implicated in protein degradation; Doherty and Mayer, 1992); *sod* (superoxide dismutase, which catalyses the dismutation of the superoxide radical $O_2$-into $H_2O$ and $O_2$; Pardini, 1995); *chitinase* (involved in degradation of insect cuticle during moulting); *cathepsin* (involved with insect liquefaction; Slack *et al.*, 1995); *ecdysteroid UDP-glucosyltransferae* (*egt*, conjugates ecdysone with sugar molecules to inhibit larval moulting: O'Reilley, 1997). The broad conservation of these genes may suggest that they provide an essential function for virus replication (Herniou *et al.*, 2003).

One of the main uses of baculoviruses is in the control of forest and agricultural insect pests. Pests have natural enemies such as predators and contain a variety of parasites such as bacteria, fungi and viruses and represent attractive alternatives as control agents. Using viruses as control agents against insect pests are environmentally and ecologically safe methods of insect pest control (Feng *et al.*, 2001). From an environmental- and health-safety standpoint, baculovirus pesticides are outstanding alternatives to synthetic chemical pesticides (Wood and Granados, 1991). Many insect viruses are attractive as biological control agents because their host ranges are generally narrow, infecting only a few species or genera. However, often wild-type viruses are not virulent enough to kill insects quickly and they act very slowly. Both viruses and their larval hosts have co-evolved over millions of years and during this they have in many cases developed mechanisms to accommodate each other. Because of the time required to kill the insect pests, extensive defoliation or crop damage often occurs before death of the insects. By understanding the underlying mechanisms of the virus replication strategy including detailed studies on genome structure, evolution, gene expression in the natural host, protein–protein interactions during the infection process, effects of the host's developmental process on virus replication, larval immunity and how it is circumvented by the virus, virulence factors, etc. will aid in the development of baculoviruses as successful pest control agents (Feng *et al.*, 2001).

With the advent of gene manipulation technologies, it is possible to generate a baculovirus with enhanced effect on the pest but which is also environmentally benign. Recombinant DNA technology has made it possible to introduce exogenous genes into, or delete genes from, the viral genome so the resulting virus is more effective against the pest than the original wild-type virus. Any gene coding for a protein that disrupts normal larval development or behaviour and reduces feeding damage caused by the insect is a potential candidate for expression by a recombinant baculovirus for insect control (Bonning and Hammock, 1996). A number of exogenous genes from different origins and having different functions have been successfully engineered into baculoviruses to improve its control potential. They include toxins, peptide hormones, enzymes and transcription factors. Introduced toxins include: mite neurotoxins, wasp toxins and scorpion toxins, which result in recombinant viruses that show increased insecticidal activity and/or feeding inhibition (Feng *et al.*, 2001). Peptide hormones such as diuretic, prothoracicotropic and eclosion hormones from insects, as well as the enzyme, juvenile hormone esterase, result in recombinant viruses that show adverse larval development in the insects, resulting in reduced lethal times and feeding damage (Feng *et al.*, 2001). Developmentally regulated insect transcription factors (which control gene expression) are controlled by various insect hormones such as ecdysone and juvenile hormone. They can be used to produce recombinant viruses with increased insecticidal activity (due to over-expression of the inserted transcriptional factor gene; Feng *et al.*, 2001; Inceoglu *et al.*, 2001). In addition to the enhancement of virulence, one of the objectives for constructing a recombinant virus is to broaden the host specificity of the wild type to infect more or fewer insect species. This is more of a commercially attractive path than an environmental one. Understanding baculovirus host ranges, gene content, genome organization and evolutionary analysis of baculoviruses are necessary when moving forward into genome technology and the manipulation of baculoviruses.

Of the 43 genomes that have been fully sequenced, only three were viruses infecting Hymenoptera and only one infecting Diptera. The rest were all from viruses infecting Lepidoptera. A number of generalized conclusions drawn from the genomics of lepidopteran baculoviruses had to be revised once some of the genomes of viruses infecting more ancient orders of insects were fully sequenced. For example, IE-1 was thought to be an essential protein for baculovirus replication without which, the replication cycle would come to a quick halt. However, the genomes of the three sequenced hymenopteran viruses did not contain the gene encoding IE-1 (Garcia-Maruniak *et al.*, 2004; Lauzon *et al.*, 2004; Duffy *et al.*, 2006). Similarly, lepidopteran baculoviruses have a biphasic replication cycle producing both BVs and ODVs. Genomics analyses of hymenopteran baculoviruses revealed that their genomes do not encode proteins essential for BV function nor do they encode proteins that were shown to be necessary for the productions of BVs (Garcia-Maruniak *et al.*, 2004; Lauzon *et al.*, 2004). Hence, studies on the genomics of viruses from different orders of insects have broadened our views on the properties and replication strategies of baculoviruses. Essential to the analyses of viral genomes were the incredible advancements in the field of bioinformatics without which, analyses of the sequenced genomes would have revealed much less knowledge.

## 14.3. Analyses of DNA and Protein Sequences

A very large number of programs have been developed and aided significantly in the analyses of sequenced genomes. Below is list of some of the major programs that served as essential tools in the study of genomics and proteomics.

### 14.3.1. Web-based software

GenBank, DNA Database of Japan (DDBJ), the European Bioinformatics Institute (EBI) – these are the three depository banks of nucleic acid and protein sequences that are publicly accessible. They share and update their database daily as part of the International Nucleotide Database Collaboration. Apart of the database, GenBank contains bibliographic and biological annotations. The advantage of GenBank is that it provides free programs for retrieving and analysis of data such as PUBMED, ORF FINDER, BLAST and ENTREZ.

#### *14.3.1.1. BLAST*

It is probably the most widely used program on bioinformatics. It compares and calculates sequence similarity and identity. Query sequences are usually compared to GenBank database and the BLAST search will identify sequences deposited in the database to the query. The BLAST program has a variety of subprograms to accommodate the query sequence including:

- BLASTN – for nucleotide comparisons;
- BLASTP – for protein sequences;
- BLASTX – for a six-frame nucleotide translation to protein;
- TBLASTX – nucleotide six-frame translation-nucleotide six-frame translation;
- TBLASTN – for protein to nucleotide six-frame translation.

Other useful BLAST programs for specific purpose exist such as BL2SEQ that aligns two sequences using the BLAST engine. RPS-BLAST performs comparisons against sequences to search and identify conserved domains.

#### *14.3.1.2. EXPASY*

(Expert Protein Analysis System, http://www.expasy.org/tools/). It is used to acquire protein sequences and related information as well as in proteins and proteomics analyses. This system contains a number of tools used in proteomics including:

1. UniProt Knowledgebase (Swiss-Prot and TrEMBL) – protein knowledgebase.
2. PROSITE – to identify protein families and domains.
3. SWISS-2D PAGE – data on 2-D PAGE.
4. World-2DPAGE Repository – it is a repository of standard-based data for PAGE-based proteomics data in the literature.
5. SWISS-MODEL Repository – for protein modelling.
6. ENZYME – enzyme nomenclature.

All the analyses tools are listed on the web site of EXPASY as well as links to other programs for the analysis of proteins sequences and proteomics. The databases are linked to other international databases and updated regularly.

#### *14.3.1.3. ORF FINDER*

As the name ORF FINDER indicates, this program, which is rooted in the GenBank database, does exactly that on a query sequence or another one from the database.

#### *14.3.1.4. SuperPose*

SuperPose is a web server for protein superposition and maintained by Canadian Bioinformatics Help Desk. SuperPose employs a simple interface that requires only program database (PDB) files or accession numbers as input. All other superposition decisions are made by the program. SuperPose calculates protein superpositions using a modified quaternion approach. From a superposition of two or more structures, SuperPose generates sequence alignments, structure alignments, PDB coordinates, root mean square deviation (RMSD) statistics, Difference Distance Plots and interactive images of the superimposed structures. SuperPose superimposes structures that differ substantially in sequence, size or shape. It is also capable of much larger range of superposition queries and situations than many stand-alone programs.

### 14.3.2. Commercially available and free softwares

#### *14.3.2.1. DNASTAR Lasergene*

This software package is composed of seven interactive modules based on functional units:

- SeqBuilder – editing sequences;
- SeqMan Pro – to assemble sequences and to identify single nucleotide polymorphism;
- MegAlign – used extensively to align sequences;
- PrimerSelect – designing primers for PCR and primer walking;
- Protean – prediction of proteins;
- GeneQuest – to identify potential genes;
- EditSeq – utility program used for importing certain files.

It integrates data between the different modules so that all edits are brought in line with the other modules. The latter is a very useful feature of DNASTAR.

#### *14.3.2.2. MacVector*

This is an excellent software for Apple Macintosh computers and version 10 has just been released. The software provides all the usual applications and also includes phylogenetic constructions. It is a highly intuitive program and relatively easy to use.

#### *14.3.2.3. ANTHEPROT (ANalyse THE PROTeins)*

This is a package of tools integrated into a graphical user interface for the analyses of protein sequences and structures. It predicts function, delivers physico/chemical profiles and three-dimensional display. It also connects to web servers for large-scale sequence comparisons and data handling.

#### *14.3.2.4. CLUSTAL*

This is a widely used software for multiple alignments of sets of sequences by a modification of the Multiple Alignment method of Feng and Doolittle. It can accommodate a number of formats including FASTA, NBRF/PIR, GCG RSF, GDE, CLUSTAL, GCG/MSF and EMBL/Swissprot. Two main variations exist, CLUSTALW (command line interface) and CLUSTALX (graphic user interface).

#### 14.3.2.5. T-COFFEE

(Tree-based Consistency Objective Function For AlignmEnt Evaluation). This is another alternative package for multiple sequence alignment packages (for DNA and proteins). It is reputed to be slightly slower than other software but the designers claim improvement in accuracy. Given a set of sequences (proteins or DNA), T-COFFEE generates a multiple sequence alignment which combines sequences and structures. T-COFFEE makes it possible to combine a collection of multiple/pairwise, global/local alignments into a single one. T-COFFEE also makes it possible to estimate the level of consistency of each position within the new alignment with the rest of the alignments. This consistency is usually an indicator of alignment accuracy.

T-COFFEE. Comes in different flavours. 3D-COFFEE combines the alignment of sequences and structures. M-COFFEE combines the output of several multiple sequence alignment packages. R-COFFEE aligns multiple RNA sequences using their predicted secondary structures. All these various modes are available on the web server and are included in the main distribution.

#### 14.3.2.6. GeneDoc

This is a multifaceted software that provides means to visualize, analyse and edit multiple DNA and protein sequence alignments. GeneDoc provides tools for visualizing, editing and analysing multiple sequence alignments of protein and nucleic acid sequences. As such, the multiple sequence alignments are excellent starting points for any mutagenesis experiments dealing with the structure and function of a macromolecule. The builders of the software explain that in the evolutionary context, the software can divide the sequences of superfamilies into clearly defined families. The analyses functions permit the investigator to determine with reasonable confidence residues that are important in structure, hence functions. It has other functions that have been utilized by molecular biologists to further analyse their data.

#### 14.3.2.7. PHYLIP

This is a package that contains various programs for phylogenetic analysis and construction of evolutionary trees that one sees so often in the literature associated with genomic sequences. There are a number of methods used to construct phylogenetic trees and the package includes those of likelihood, distance matrix and maximum parsimony methods and the software is capable of handling several data types including distance matrices, distinct characters, gene frequencies, etc.

#### 14.3.2.8. PAUP (phylogenetic analysis using parsimony)

It is probably the most widely used software package for the construction of phylogenetic trees. Improvement on recent versions of the software were made by including maximum likelihood and distance methods. The recent version 4.0 uses NEXUS common data format similar to MacClade 3 thus permitting interchange of data between the softwares. PAUP has included many options in its program such as distance matrix, maximum likelihood and parsimony plus statistical tests.

#### *14.3.2.9. TREEVIEW*

Basically it is a program for displaying the contents of PHYLIP, NEXUS, etc. It generates graphic files for editing by other programs. The editor package in the software allows for editing a tree in a variety of ways and produces publication quality trees. The present program reads trees with as many as 1000 taxa.

## 14.4. Genetic Modification of Baculoviruses

Genetic characteristics of baculoviruses make them amenable to manipulation of the virus to aid in basic investigations and in practical applications as pest control agents and as vectors for the expression of transgenes. Typically, exogenous genes encoding for foreign proteins have been cloned into non-essential loci under the control of viral promoters. Loci such as those of polyhedrin (*polh*) and ecdysteroid UDP-glucosyltransferase (*egt*) have been used extensively for the insertion of foreign genes. Non-essential genes can be deleted from the genome without adversely affecting the virus replication cycle. Cloning a gene of interest in a manner that could affect the expression of an essential gene would produce an undesirable affect on the virus replication cycle or the viral structure. Strong promoters, such as those of *polh* or *p10*, have been used for the expression of foreign genes in medical, pharmaceutical or industrial applications. A range of viral promoters have been used depending on desired time of expression (early versus late promoters), the strength of the promoter and the desired outcome. Promoters such as those of *polh*, *p10*, *ie-1* or of an *iap* have been used extensively.

The existence of powerful promoters has been exploited to abundantly express a protein of commercial interest. In this manner, recombinant proteins are modified post translation and could be produced in relatively large quantities. One advantage in some of the insect cell culture system is that the foreign proteins can be produced in a serum-free medium thus alleviating the undesirable downstream effects of serum. Several commercial ventures around the world have been started and are presently taking advantage of the technologies of baculoviruses in insect cells to produce large amounts of proteins for commercial and research applications in a variety of sectors. Recently, a new cell line, commercially known as Hi Five, was derived from the *Trichoplusia ni* cell line. This line (BTI 5B1–4) produces more proteins than any of the previously derived insect cell lines and is being used more and more these days for the abundant expression of exogenous proteins (Granados *et al.*, 2007).

### 14.4.1. Homologous recombination

Traditionally, recombinant NPVs have been generated in cell lines using homologous recombination. Basically, a plasmid transfer vector is constructed that carries the gene of interest under the control of a viral promoter. It is usually in the form of a cloning cassette with sequences homologous to the locus where the gene is to be located. The gene of interest is followed by a polyadenylation signal to allow the recombinant messenger RNA (mRNA) to be stabilized with a poly A + tail. A selectable marker gene such as *lacZ* or a gene expressing the green fluorescent protein, under the control of another viral promoter, is also contained within the transfer vector. Transfer vectors can be designed to transfer several recombinant genes into the same virus to affect expression of several recombinant proteins (Sridhar *et al.*, 1994). The transfer vector contains arms on either side of the transgene that have homology to the non-essential loci where the transfer vector cassette is desired to incorporate. Susceptible cell lines is transfected with both the viral DNA and the plasmid vector and during replication, the flanking arms align by base pairing with the genomic locus where the transgene is to be inserted and by homologous recombination, the transgene is transferred to the desired locus. The marker gene signals the establishment of infection and the medium containing the recombinant virus is then harvested. Normally, plaque purification assays are employed to purify the recombinant virus from the parental strain. Usually, PCR and restriction endonuclease (REN) analysis are employed to ensure that no parental strain remains in the purified recombinant stock. REN is time-consuming as several rounds of plaque purification must be done to produce a pure clone. Frequently, single crossovers occur, where only one of the arms within the vector is transferred to the viral genome. Therefore, many clones are purified and double crossover is verified by PCR and sequencing.

### 14.4.2. Direct cloning

In this method, a unique restriction endonuclease site or homing endonuclease site is introduced into a non-essential region of a baculovirus genome (Lihoradova *et al.*, 2007). Foreign insert DNA fragments are generated by PCR using primers that incorporate the same unique endonuclease cutting site onto the ends of the PCR product. Viral DNA and foreign insert DNA are then purified and enzymatically digested with the same unique site cutting endonuclease. The resulting linearized viral DNA and cut insert DNA are then ligated together using T4 ligase enzyme. The insert DNA is added in excess to ensure that recircularized viral DNA will contain the insert. The ligation is then transfected onto insect cells and only viral DNA that has recircularized will be viable to produce virus. Transfections can be serially diluted in microtitre plates such that only single viral clones are produced. There are no bacterial intermediate steps in direct cloning. With proper design of parent virus vectors and optimization of protocols, direct cloning could be used for high through put baculovirus cloning.

### 14.4.3. Recombinant proteins

Several methods are available to the genetic engineer in the production of baculoviruses expressing recombinant proteins. Recombinant viruses can also be used to produce antisense RNA and act to cause the down regulation of critical endogenous insect genes. Perhaps as the use of chemical insecticides decreases, the potential of these environmentally benign control methods will become more pronounced. The safety of these viruses could be further improved by developing suicidal viruses that can only live a few generations under field conditions.

### 14.4.4. Bacmid systems

Site-specific transposition systems utilizing the bacmids have become a powerful tool in the mutagenesis of baculoviruses and have been used with a number of viruses to generate recombinants to study gene function or to produce a desired product (Invitrogen, 2004).

The bacmid method of creating and selecting recombinant baculoviruses was published by Luckow *et al.* (1993). Bacmids have revolutionized baculovirus expression vector cloning technology and have been marketed in kit form by Invitrogen Corporation (Bac-to-Bac™). Past baculovirus cloning systems which rely on homologous recombination involve a small part of the baculovirus genome being propagated in *Escherichia coli* in a shuttle vector plasmid. In contrast, bacmid systems propagate the whole baculovirus genome in *E. coli* as a bacmid. Bacmid propagation in *E. coli* was made possible by engineering into the baculovirus genome a mini-F replicon (Lovett and Helinski, 1976) and a kanamycin-resistant ($Kan^r$) gene. The mini-F replicon is a low copy number DNA replication origin and the $Kan^r$ gene permits antibiotic selective retention of the bacmid in *E. coli*. The mini-F replicon, $Kan^r$ gene and other non-viral elements of the bacmid system replace the non-essential baculovirus *polh* gene and bacmid DNA will replicate as baculoviruses upon transfection into insect cells.

In past baculovirus cloning systems, foreign genes were introduced into baculovirus genomes by homologous recombination in insect cells. Foreign genes are introduced into bacmid baculovirus genomes by transposition in *E. coli* using components of the Tn7 transposon (Craig, 1991). The Tn7 transposon is a mobile DNA element that will transpose itself into a 68 bp target DNA sequence called *att*Tn7 (Lichtenstein and Brenner, 1982; McKown *et al.*, 1988). Tn7 encodes transposition genes that are collectively called TnsABCDE (Orle and Craig, 1991). The TnsABCDE gene products catalyse cleavage of the Tn7 transposon at Tn7L and Tn7R DNA sites (Arciszewska *et al.*, 1989) and insertion of the resulting excised Tn7 transposon into an *att*Tn7 DNA site.

Bacmid genomes have been engineered to contain an *att*Tn7 site into which a foreign gene cassette can be transposed from a mini-Tn7 donor plasmid. The mini-Tn7 donor plasmid has Tn7L and Tn7R DNA elements flanking an insertion cassette which includes a baculovirus promoter, downstream multiple cloning site and gentamycin-resistance ($Gen^r$) gene. To make a recombinant bacmid, foreign genes must first be cloned into the multiple cloning site of the mini-Tn7 donor plasmid which is then transformed into bacmid containing *E. coli*. Transposition of the foreign gene cassette from the mini-Tn7 donor plasmid into the bacmid's *att*Tn7 site is catalysed by presence of a helper plasmid expressing TnsABCDE genes.

The bacmid system design is such that the *att*Tn7 site is in frame with a *LacZα* gene and transposition of the donor plasmid cassette into the *att*Tn7 site disrupts the reading frame of *LacZα*. In the presence of β-galactosidase colourimetric substrates, such as X-gal (5-bromo-4-chloro-3-indolyl-b-D-galactopyranoside), *E. coli* colonies harbouring the parent bacmid are blue and recombinant bacmid colonies with *att*Tn7 site insertions are white. Thus, there is a simple visual screening system for recombinant bacmids.

The antibiotics tetracycline and gentamycin are needed for creation of recombinant bacmids. The helper plasmid has a tetracycline-resistance gene (*Tet$^r$*) and the recombinant bacmid receives a *Gen$^r$* gene from the mini-Tn7 donor plasmid after transposition. Helper plasmid-associated transposase activities on the donor plasmid ensure that gentamycin resistance is only associated with recombinant bacmids. The ability to positively select recombinant bacmids in *E. coli* with antibiotics is a powerful feature of the bacmid system.

Once recombinant bacmid clones are selected, they can be amplified in *E. coli* and their purified DNA can be transfected on to insect cells. It is assumed that the baculoviruses produced from these transfections are pure clones because they have been already selected in bacteria. Bacmid technology shifts much baculovirus cloning away from insect cell culture and towards *E. coli* for which there is broader availability of equipment and trained personnel. In addition blue/white bacterial colony selection is an easily automated process. Problems may occur when baculovirologists are not present to verify the final recombinant baculoviruses in the context of insect cells. In the original bacmid paper, Luckow *et al.* (1993) recommended plaque purification of bacmid viral stocks intended for continuous or large-scale production. There has also been at least one report of genetic instability of bacmids (Pijlman *et al.*, 2003, 2004).

Bacmids have been enthusiastically adopted by the baculovirus researchers investigating baculovirus genetics because the system permits the targeted introduction of lethal mutations into essential baculovirus genes (Lin and Blissard, 2002; Okano *et al.*, 2004; McCarthy and Theilmann, 2008).

## 14.5. Conclusions

The versatility of baculoviruses makes them a most interesting group of viruses with applications in a variety of sectors such as pest control, expression system, vaccine production and gene therapy. The amenability to genetic modification opens even other avenues to explore. The availability of excellent cell lines that can be maintained in a well-defined and serum-free medium allows for further use in the medical and pharmaceutical fields. Future exploitation of baculoviruses should be very exciting.

## References

Abd-Alla, A.M., Cousserans, F, Parker, A.G., Jehle, J.A., Parker, N.J., Vlak, J.M., Robinson, A.S. and Bergoin, M. (2008) Genome analysis of a *Glossina pallidipes* salivary gland hypertrophy virus reveals a novel, large, double-stranded circular DNA virus. *Journal of Virology* 82, 4595–4611.

Afonso, C.L., Tulman, E.R., Lu, Z, Balinsky, C.A., Moser, B.A., Becnel, J.J., Rock, D.L. and Kutish, G.F. (2000) Genome sequence of a baculovirus pathogenic for *Culex nigripalpus*. *Journal of Virology* 75, 11157–11165.

Arciszewska, L.K., Drake, D. and Craig, N.L. (1989) Transposon Tn7. cis-Acting sequences in transposition and transposition immunity. *Journal of Molecular Biology* 207, 35–52.

Asgari, S., Davis, J., Wood, D., Wilson, P. and McGrath, A. (2007) Sequence and organization of the *Heliothis virescens* ascovirus genome. *Journal of General Virology* 88, 1120–1132.

Attoui, H., Mohd Jaafar, F., Belhouchet, M., Biagini, P., Cantaloube, J.F., de Micco, P. and de Lamballerie, X. (2005) Expansion of family *Reoviridae* to include nine-segmented dsRNA viruses: isolation and characterization of a new virus designated Aedes pseudoscutellaris reovirus assigned to a proposed genus (Dinovernavirus). *Virology* 20, 212–223

Blissard, G., Black, B., Crook, N., Keddie, B.A., Possee, R., Rohrmann, G., Theilmann, D. and Volkmann, L. (2000) *Baculoviridae*. In: van Regenmortel, M.H.V., Fauquet, C.M., Bishop, D.H.L., Carstens, E.B., Estes, M.K., Lemon, S.M., Maniloff, J., Mayo, M.A., McGeotch, D.J., Pringle, C.R. and Wickner, R.B. (eds) *Virus Taxonomy. Seventh Report of the International Committee on Taxonomy of Viruses*. Academic Press, San Diego, California, pp. 195–202.

Bonning, B.C. and Hammock, B.D. (1996) Development of recombinant baculoviruses for insect control. *Annual Review of Entomology* 41, 191–210.

Craig, N.L. (1991) Tn7: a target site-specific transposon. *Molecular Microbiology* 5, 2569–2573.

Dasmahapatra, B., Dasgupta, R., Ghosh, A. and Kaesberg, P. (1985) Structure of the black beetle virus genome and its functional implications. *Journal of Molecular Biology* 182, 183–189.

Delhon, G., Tulman, E.R., Afonso, C.L., Lu, Z., Becnel, J.J., Moser, B.A., Kutish, G.F. and Rock, D.L. (2006) Genome of invertebrate iridescent virus type 3 (mosquito iridescent virus). *Journal of Virology* 80, 39–49.

de Miranda, J.R., Drebot, M., Tyler, S., Shen, M., Cameron, C.E., Stoltz, D.B. and Camazine, S.M. (2004) Complete nucleotide sequence of Kashmir bee virus and comparison with acute bee paralysis virus. *Journal of General Virology* 85, 2263–2270.

Doherty, F.J. and Mayer, R.J. (1992) *Intracellular Proteins*. IRL Press, Oxford.

Duffy, S.P., Young, A.M., Morin, B., Lucarotti, C.J., Koop, B.F. and Levin, D.B. (2006) Sequence analysis and organization of the *Neodiprion abietis* nucleopolyhedrovirus genome. *Journal of Virology* 80, 6952–6963.

Feng, Q., Arif, B.M., Palli, S.R., Sohi, S.S. and Retnakaran, A. (2001) Molecular modifications of baculoviruses for the control of forest insect pests. *Advances in Virus Research* 87, 263–291.

Garcia-Maruniak, A., Maruniak, J.E., Zanotto, P.M., Doumbouya, A.E., Liu, J.C., Merritt, T.M. and Lanoie, J.S. (2004) Sequence analysis of the genome of the *Neodiprion sertifer* nucleopolyhedrovirus. *Journal of Virology* 78, 7036–7051.

Garcia-Maruniak, A., Maruniak, J.E., Farmerie, W. and Boucias, D.G. (2008) Sequence analysis of a non-classified, non-occluded DNA virus that causes salivary gland hypertrophy of *Musca domestica*, MdSGHV. *Virology*, 377(1), 184–196.

Granados, R.R., Li, G. and Blissard, G.W. (2007) Insect cell culture and biotechnology. *Virologica Sinica* 22, 83–93.

Gubser, C., Hué, S., Kellam, P. and Smith, G.L. (2004) Poxvirus genomes: a phylogenetic analysis. *Journal of General Virology* 85, 105–117.

Habayeb, M.S., Ekengren, S.K. and Hultmark, D. (2006) Nora virus, a persistent virus in Drosophila, defines a new picorna-like virus family. *Journal of General Virology* 87, 3045–3051.

Hartley, C.J., Greenwood, D.R., Gilbert, R.J., Masoumi, A., Gordon, K.H., Hanzlik, T.N., Fry, E.E., Stuart, D.I. and Scotti, P.D. (2005) Kelp fly virus: a novel group of insect picorna-like viruses as defined by genome sequence analysis and a distinctive virion structure. *Journal of Virology* 79, 13385–13398.

Hayakawa, T., Ko, R., Okano, K., Seong, S., Goto, C. and Maeda, S. (1999) Sequence analysis of the *Xestia c-nigrum* granulovirus genome. *Virology* 262, 277–297.

Hayakawa, T., Rohrmann, G.F. and Hashimoto, Y. (2000) Patterns of genome organization and content in lepidopteran baculoviruses. *Virology* 278, 1–12.

Hefferon, K.L. and Miller, L.K. (2002) Reconstructing the replication complex of AcMNPV. *European Journal of Biochemistry* 269, 6233–6240.
Herniou, E.A., Olszewski, J.A., Cory, J.S. and O'Reilly, D.R. (2003) The genome sequence and evolution of baculoviruses. *Annual Review of Entomology* 48, 211–234.
Herniou, E.A., Olszewski, J.A., O'Reilly, D.R. and Cory, J.S. (2004) Ancient coevolution of baculoviruses and their insect hosts. *Journal of Virology* 78, 3244–3251.
Hu, Z.H., Arif, B.M., Jin, F., Martens, J.W.M., Chen, X.W., Sun, J.S., Zuidema, D.G., Goldbach, R.W. and Vlak, J.M. (1998) Distinct gene arrangement in *Buzura suppressaria* single-nucleocapsid nucleopolyhedrovirus genome. *Journal of General Virology* 79, 2841–2851.
Inceoglu, A.B., Kamita, S.G., Hinton, A.C., Huang, Q., Severson, T.F., Kang, K. and Hammock, B.D. (2001) Recombinant baculoviruses for insect control. *Pest Management Science* 57, 981–987.
Jehle, J.A., Blissard, G.W., Bonning, B.C., Cory, J.S., Herniou, E.A., Rohrmann, G.F., Theilmann, D.A., Thiem, S.M. and Vlak, J.M. (2006) On the classification and nomenclature of baculoviruses: a proposal for revision. *Archives in Virology* 151, 1257–1266.
Kool, M., Ahrens, C.H., Vlak, J.M. and Rohrmann, G.F. (1995) Replication of baculovirus DNA. *Journal of General Virology* 76, 2103–2118.
Lange, M. and Jehle, J.A. (2003) The genome of *Cryptophlebia leucotreta* granulovirus. *Virology* 317, 220–236.
Lauzon, H.A.M., Lucarotti, C.J., Krell, P.J., Feng, Q., Retnakaran, A. and Arif, B.M. (2004) Sequence and organization of *Neodiprion lecontei* nucleopolyhedrovirus genome. *Journal of Virology* 78, 7023–7035.
Li, Y., Zhang, J., Li, Y., Tan, L., Chen, W., Luo, H. and Hu, Y. (2007) Phylogenetic analysis of *Heliothis armigera* cytoplasmic polyhedrosis virus type 14 and a series of dwarf segments found in the genome. *Journal of General Virology* 88, 991–997.
Lichtenstein, C. and Brenner, S. (1982) Unique insertion site of Tn7 in the E. coli chromosome. *Nature* 297, 601–603.
Lihoradova, O., Ogay, I., Abdukarimov, Azimova Sh S., Lynn, D.E. and Slack, J.M. (2007) The homingbac baculovirus cloning system: an alternative way to introduce foreign DNA into baculovirus genomes. *Journal of Virological Methods*, 140: 59–65.
Lin, G. and Blissard, G.W. (2002) Analysis of an Autographa californica nucleopolyhedrovirus lef-11 knockout: LEF-11 is essential for viral DNA replication. *Journal of Virology* 76, 2770–2779.
Lovett, M.A. and Helinski, D.R. (1976) Method for the isolation of the replication region of a bacterial replicon: construction of a mini-F'kn plasmid. *Journal of Bacteriology* 127, 982–987.
Luckow, V.A., Lee, S.C., Barry, G.F. and Olins, P.O. (1993) Efficient generation of infectious recombinant baculoviruses by site-specific transposon-mediated insertion of foreign genes into a baculovirus genome propagated in Escherichia coli. *Journal of Virology* 67, 4566–4579.
Maori, E., Lavi, S., Mozes-Koch, R., Gantman, Y., Peretz, Y., Edelbaum, O., Tanne, E. and Sela, I. (2007) Isolation and characterization of Israeli acute paralysis virus, a dicistrovirus affecting honeybees in Israel: evidence for diversity due to intra- and inter-species recombination. *Journal of General Virology* 88, 3428–3438.
McCarthy, C.B. and Theilmann, D.A. (2008) AcMNPV ac143 (*odv-e18*) is essential for mediating budded virus production and is the 30th baculovirus core gene. *Virology* 375, 277–291.
McKown, R.L., Orle, K.A., Chen, T. and Craig, N.L. (1988) Sequence requirements of Escherichia coli attTn7, a specific site of transposon Tn7 insertion. *Journal of Bacteriology* 170, 352–358.
Miller, L.K. (1997). *The Baculoviruses*. Plenum Press. New York.
Mukha, D.V., Chumachenko, A.G., Dykstra, M.J., Kurtti, T.J. and Schal, C. (2006) Characterization of a new densovirus infecting the German cockroach, *Blattella germanica*. *Journal of General Virology* 87, 1567–1575.
Okano, K., Vanarsdall, A.L. and Rohrmann, G.F. (2004) Characterization of a baculovirus lacking the alkaline nuclease gene. *Journal of Virology* 78, 10650–10656.

O'Reilly, D.R. (1997) Auxiliary genes of baculoviruses. In: Miller, L.K. (ed.) *The Baculoviruses*, Plenum, New York, pp. 267–300.

Orle, K.A. and Craig, N.L. (1991) Identification of transposition proteins encoded by the bacterial transposon Tn7. *Gene* 104, 125–131.

Ornitz, D.M. and Nobuyuki, I. (2001) Protein family review: fibroblast growth factors. *Genome Biology* 2, 3005–3012.

Pardini, R.S.T. (1995) oxicity of oxygen from naturally occurring redox-active pro-oxidants. *Archives in Insect Biochemistry and Physiology* 29, 101–118.

Pijlman, G.P., van Schijndel, J.E. and Vlak, J.M. (2003) Spontaneous excision of BAC vector sequences from bacmid-derived baculovirus expression vectors upon passage in insect cells. *Journal of General Virology* 84, 2669–2678.

Pijlman, G.P., de Vrij, J., van den End, F.J., Vlak, J.M. and Martens, D.E. (2004) Evaluation of baculovirus expression vectors with enhanced stability in continuous cascaded insect-cell bioreactors. *Biotechnology and Bioengineering* 87, 743–753.

Ryabov, E.V. (2007) A novel virus isolated from the aphid *Brevicoryne brassicae* with similarity to Hymenoptera picorna-like viruses. *Journal of General Virology* 88, 2590–2595.

Shwed, P.S., Dobos, P., Cameron, L.A., Vakharia, V.N. and Duncan, R. (2002) Birnavirus VP1 proteins form a distinct subgroup of RNA-dependent RNA polymerases lacking a GDD motif. *Virology* 10, 241–250.

Slack, J.M., Kuzio, J. and Faulkner, P. (1995) Characterization of *v-cath*, a cathepsin L-like proteinase expressed by the baculovirus *Autographa californica* multiple nuclear polyhedrosis virus. *Journal of General Virology* 76, 1091–1098.

Sridhar, P., Awasthi, A.K., Azim, B., Habib, S., Jain, S., Makherjee, B., Rajan, A. and Hasnain, S.E. (1994) Baculovirus vector-mediated expression of heterologous genes in insect cells. *Journal of Bioscience* 19, 603–614.

Tan, L., Zhang, J., Li, Y., Li, Y., Jiang, H., Cao, X. and Hu, Y. (2008) The complete nucleotide sequence of the type 5 *Helicoverpa armigera* cytoplasmic polyhedrosis virus genome. *Virus Genes* 36, 587–593.

Tanaka, K., Lapointe, R., Barney, W.E., Makkay, A.M., Stoltz, D., Cusson, M. and Webb, B.A. (2007) Shared and species-specific features among ichnovirus genomes. *Virology* 20, 26–35.

Theilmann, D.A., Blissard, G.W., Bonning, B., Jehle, J., O'Reilly, D.R., Rohrmann, G.F., Thiem, S. and Vlak, J.M. (2005) Family *Baculoviridae*. In: Fauquet, C.M., Mayo, M.A., Maniloff, J., Desselberger, U. and Ball, L.A. (eds) *Virus Taxonomy Eighth Report of the International Committee of Taxonomy of Viruses*. Elsevier Academic Press, San Diego, California, pp. 177–185.

Valles, S.M., Strong, C.A. and Hashimoto, Y. (2007) A new positive-strand RNA virus with unique genome characteristics from the red imported fire ant, *Solenopsis invicta*. *Virology* 365, 457–463.

van der Wilk, F., Dullemans, A.M., Verbeek, M. and Van den Heuvel, J.F. (1997) Nucleotide sequence and genomic organization of *Acyrthosiphon pisum* virus. *Virology* 238, 353–362.

Wang, Y., Kleespies, R.G., Huger, A.M. and Jehle, J.A. (2007) The genome of Gryllus bimaculatus nudivirus indicates an ancient diversification of baculovirus–related nonoccluded nucliviruses of insects. *Journal of Virology* 81(10), 5395–5406.

Wolff, J.L., Valicente, F.H., Martins, R., Oliveira, J.V. and Zanotto, P.M. (2008) Analysis of the genome of *Spodoptera frugiperda* nucleopolyhedrovirus (SfMNPV-19) and of the high genomic heterogeneity in group II nucleopolyhedroviruses. *Journal of General Virology* 89, 1202–1211.

Wood, H.A. and Granados, R.R. (1991) Genetically engineered baculoviruses as agents for pest control. *Annual Review of Microbiology* 45, 69–87.

Wormleaton, S., Kuzio, J. and Winstanley, D. (2003) The complete sequence of the *Adoxophyes orana* granulovirus genome. *Virology* 311, 350–365.

Yi, F., Zhang, J., Yu, H., Liu, C., Wang, J. and Hu, Y. (2005) Isolation and identification of a new tetravirus from *Dendrolimus punctatus* larvae collected from Ynnan Province, China. *Journal of General Virology* 86, 789–796.

# 15 Genomics and Genetic Improvement of Entomopathogenic Nematodes

H. Koltai

*Department of Ornamental Horticulture, ARO, Volcani Center, Bet Dagan, Israel*



## 15.1. Introduction

Recent advances in the decoding of the sequence of genomes and progress in the ability to look at many genes, proteins and genetic pathways have led to a great revolution in biological sciences during the last few years. Initially, sequencing of genomes and gene expression products (expressed sequence tags (ESTs)) of human and model organisms was accomplished. The need to characterize the huge volume of resulting sequences, to identify genes and to establish their biological function, led to the development of functional genomics; i.e. the set of tools that allow elucidation of gene expression and function. Researchers endeavour to investigate multiple genes, rather than one gene at a time, while working towards the ultimate understanding of genetic networks. Subsequent efforts are turning now to sequencing of various other organisms, many of medical or agricultural importance. As a result, the complete genomes of over 165,000 organisms can be found on the World Wide Web at sites such as the Entrez Genomes site, the European Molecular Biology Laboratory and the DNA databank of Japan, exceeding 100 gigabases and representing all the main domains of life (http://www.ncbi.nlm.nih.gov/Genbank/index.html).

The sequencing of organisms of agricultural importance, along with the use of the functional genomics tools developed for model organisms, has led to the emergence of agricultural genomics, whose goals included both basic understanding and applied outcomes. In this chapter, I will present several issues of agricultural genomics relating to entomopathogenic nematodes. These will include agriculture database resources, bioinformatics and functional genomics tools and the available and developing resources and tools that could be applicable for entomopathogenic nematodes. I will present some considerations regarding the use of a model nematode for the study of entomopathogenic nematodes, and will discuss the application of agricultural genomics to the potential use of entomopathogenic nematodes to manage agricultural pests.

## 15.2. Genomic Sequencing and Bioinformatics

### 15.2.1. Sequencing projects: genome sequence and databases

Since the completion of the human genome sequencing in 2001 (Venter *et al.*, 2001), genome sequencing for many other organisms is now in progress or has already been completed. This effort has led to a proliferation of consortiums, centres and companies that participate in sequencing projects. A list of these may be found at http://www.ncbi.nlm.nih.gov/genomes/static/lcenters.html.

Initially, following the human genome, genomes of model organisms were sequenced. These include the yeast *Saccharomyces cerevisiae* (reviewed by Goffeau *et al.*, 1996), the fruit fly *Drosophila melanogaster* (Adams *et al.*, 2000), the roundworm *Caenorhabditis elegans* (The *C. elegans* Sequencing Consortium, 1998) and the plant *Arabidopsis thaliana* (Arabidopsis Genome Initiative, 2000).

Several major genomics initiatives of agriculturally important organisms are in progress or were completed. Among plants, this includes the completed rice genome (Goff *et al.*, 2002; Yu *et al.*, 2002) and in progress are genomes of the model legumes *Medicago truncatula* (barrel medic), *Lotus corniculatus* var. *japonicus* (lotus), *Lycopersicon esculentum* (tomato) and *Populus trichocarpa* (black cottonwood). Among agriculturally important insects are the pea aphid *Acyrthosiphon pisum*, the honeybee *Apis mellifera*, the tropical butterfly *Bicyclus anynana* and the silkworm *Bombyx mori*. Several nematode genomes are currently being sequenced.

These include the root-knot nematodes *Meloidogyne arenaria*, *Meloidogyne artiellia*, *Meloidogyne chitwoodi* and *Meloidogyne hapla* (http://www.ncbi.nlm.nih.gov/sites/entrez/). Most significantly, sequencing of the genome of the entomopathogenic nematode *Heterorhabditis bacteriophora* is under way. The results promise to provide enormous amount of data on insect-parasitic nematodes, which will lead to development of knowledge regarding their biology and life history (discussed below).

### 15.2.2. Sequencing projects: ESTs sequence and databases

For larger genomes, or for lower-budget sequencing projects, a more reasonable approach for gene discovery is through the development of databases of ESTs, which provide a wealth of information in a relatively short time. An EST is a set of single-pass sequenced complementary DNAs (cDNAs) from a messenger RNA (mRNA) population derived from a specified cell population (representing, for example, a specific tissue, organ, developmental state or environmental condition).

Many major EST sequencing programmes of agriculturally relevant organisms have been initiated recently; the number of EST sequences totalled more than 50 million as of February 2008 and it continues to grow (dbEST, GenBank, accessible at http://www.ncbi.nlm.nih.gov/dbEST/dbEST_summary.html).

Among the agriculturally relevant nematodes, *M. hapla* has 24,452 sequenced ESTs, *Heterodera glycines* (soybean cyst nematode) 24,444, *Meloidogyne incognita* (southern root-knot nematode) 20,334 ESTs, *Meloidogyne javanica* (root-knot nematode) 7587, *Globodera rostochiensis* (cyst nematode) 11,851 and *Steinernema feltiae* has 83.

However, the majority of ESTs are short, are of relatively low quality and represent regions that are not necessarily translated to proteins, e.g. the 3' untranslated region (UTR). This makes their functional annotation difficult. Nevertheless, they provide a profile of the mRNA population and offer a quick method for cloning a large number of genes that are expressed in a cell population.

In particular, useful information can be found by clustering ESTs based on sequence overlaps, to form unigenes. Unigenes are sequences that are longer and more accurate, and which, therefore, better represent the underlying genes. Databases of unigenes may be found in several sites, including in the web site of The J. Craig Venter Institute (http://www.tigr.org/tdb/tgi/index.shtml) or in that of National Centre for Biotechnology Information (NCBI) (http://www.ncbi.nlm.nih.gov/entrez/query.fcgi?db=unigene).

It should be noted, however, that until genomic or EST databases are completed, the 'unigene' designation may change based on emergence of newly found EST (and genomic) sequences. Moreover, comparison of multiple ESTs for a given gene may reflect genes that have several splice variants (Wang *et al.*, 2003; Mecham *et al.*, 2004) or reveal polymorphisms in gene sequences, embedded between individuals or populations.

### 15.2.3. New and emerging technologies of sequencing

A new generation of sequence technologies is emerging. These non-Sanger-based technologies (e.g. Mardis, 2008; Schuster, 2008) are enabling sequencing DNA at an unprecedented pace. This high speed of sequencing, allows, on the one hand, whole-genome scale of sequence analysis for multiple species, such that in a few years genome sequence of all species with interest will be available. On the other, these technologies promote utilization of multiple usages for sequence analysis. Examples are mutation discovery, gene expression profiles (discussed below), discovering non-coding RNAs, exploring chromatin packaging and discovering of clinical isolates and genetic differences between populations and individuals (Mardis, 2008). It should be noted, however, that these new sequencing technologies are still developing. For example, sequence reads are still short, in comparison to the Sanger-based sequencing technology, and sequence fidelity should be improved (Schuster, 2008). In the next few years, it is expected that the new sequencing technologies will revolutionize the genomics field, to allow new scientific achievements and novel biological applications.

### 15.2.4. Integration of databases

One of the major recent advances in genomics is the integration of databases. Each and every database contains information on a certain aspect of genome biology. The integration of these, including a user interface that allows a specific search on each of the data sets' components, may revolutionize the ability to understand biology, allowing characterization of a component, or a pathway, from several different perspectives. An example of a retrieval system for an integrated database is the NCBI text-based search. This is used to search the major databases, including the literature (PubMed), Nucleotide and Protein Sequences, Protein Structures, Complete Genomes, Taxonomy and others (http://www.ncbi.nih.gov/Database/datamodel/index.html; Fig. 15.1). For example, UniGene cluster contains information related to such information as the tissue types in which the gene has been expressed, its map location and the related literature.

Other integrated retrieval systems may be useful specifically for nematodes. These include, for example, WormBase, which provides information concerning the genetics, genomics and biology of *C. elegans* and related nematodes (http://www.wormbase.org/).

### 15.2.5. Entomopathogenic-related gene sequencing and bioinformatics tools

Most of the highly curated, connected databases are built for model organisms only (e.g. *C. elegans* from the Nematoda). Until these tools become available for non-model organisms, and until sequences of *H. bacteriophora* genome are available, isolation of entomopathogenic genes may rely on a semi-high-throughput approach. For example, subtraction libraries may be generated: ESTs from two physiological conditions are cloned and subtracted by subtractive hybridization

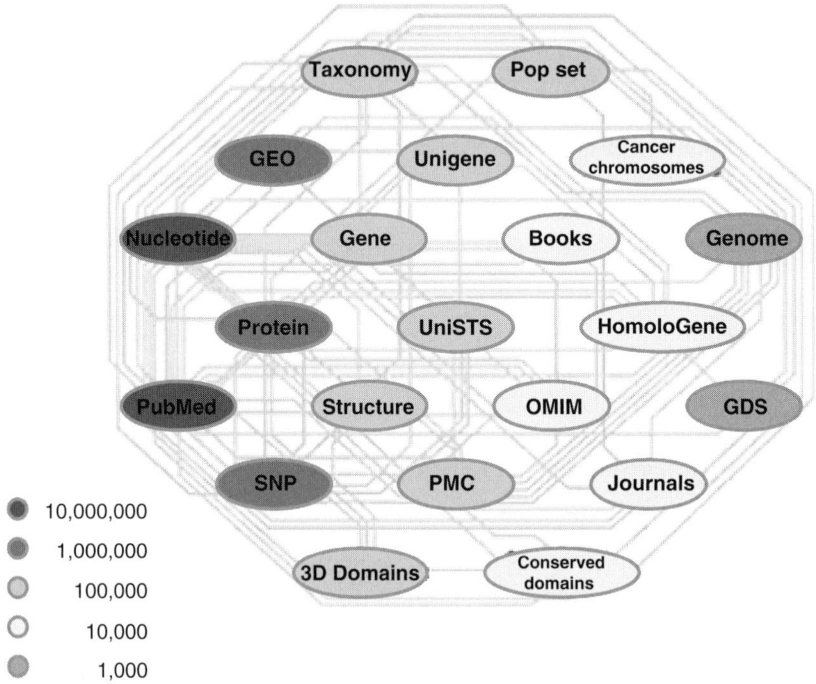


**Fig. 15.1.** Entrez is the integrated, text-based search and retrieval system used at NCBI for the major databases, including PubMed, Nucleotide and Protein Sequences, Protein Structures, Complete Genomes, Taxonomy and others. The colour indicates the approximate number of records in the database. (Taken with permission, NCBI web site-Entrez data model.)

to form an EST population which is specific to the physiological condition under study. Then detection of differentially expressed RNAs is possible. This approach for gene isolation was done for entomopathogenic nematodes to isolate genes specifically expressed in infective juveniles (IJs) during their desiccation stress response (Gal *et al.*, 2003).

Once genes are isolated, their function may be predicted. Their sequences may be blasted via the Basic Local Alignment Search Tool (BLAST) of NCBI (http://www.ncbi.nlm.nih.gov/BLAST/). BLAST search can help the researcher infer gene function by identifying areas of established homology between sequences, based on their evolutionary relationships, and help identify members of gene families. BLAST finds regions of local similarity between sequences by comparing nucleotide or protein sequences to sequence databases and calculating the statistical significance of matches. Once matched sequences are found, an access to a wealth of biological information may be available; the BLAST search is linked to other databases, such as literature citations and related sequences of proteins.

Several types of BLAST searches exist. Of these, BLASTN can be used to see if the sequence of interest already exists in a public database by comparing nucleotide sequences. The BLASTX compares translational products of the nucleotide query sequence to a protein database. Because BLASTX translates the query sequence in all six reading frames, it is particularly useful when the reading frame of the sequence of interest is unknown or it contains errors that may lead to frame shifts or other coding errors. Thus, BLASTX is often the first analysis performed with a newly determined nucleotide sequence and is used extensively in analysing EST sequences. This tool enabled us to identify several classes of genes during our annotation of genes whose expression was induced during *S. feltiae* IJ desiccation (Gal *et al.*, 2003). These classes included stress-related genes, genes that are homologous to hypothetical *C. elegans* proteins and novel genes that may be involved in traits specific to *S. feltiae* (Gal *et al.*, 2003).

Other types of BLAST search include the protein query versus translated database (TBLASTN) and translated query versus translated database (TBLASTX). These are useful tools for finding protein homologous or novel genes in unannotated, error-prone nucleotide databases. BLAST tools may also be used for blasting multiple sequences; one of the options is using *Standalone* BLAST *executables*. These are command line programs which run BLAST searches against local downloaded copies of the NCBI BLAST databases. The Standalone executables are available at the anonymous FTP location: ftp://ftp.ncbi.nih.gov/blast/executables/.

Once homologous sequence of interest are found, an annotation (i.e. the biological information) may be attached to the isolated gene. It should be noted that a wide variation in annotation terminology is being used. This inhibits effective searching by computers, as well as people, and makes more difficult the effective characterization of genes and genetic pathways. Recently, a collaborative effort to address the need for consistent descriptions of gene products in different databases emerged, designated The Gene Ontology Project (GO). The GO collaborators are developing vocabularies (ontologies) that are controlled at three structural levels, describing gene products in terms of their associated biological processes, cellular components and molecular functions in a species-independent manner (http://www.geneontology.org/index.shtml). The vocabularies may be queried at different levels. For example, GO can be used to find all the gene products in the *C. elegans* genome that are involved in signal transduction, or it could be queried at a different level of resolution to find all the receptor tyrosine kinases. This structure also allows annotators to assign properties to gene products at different levels, depending on how much is known about it.

As before, this database is built especially for model organisms or organisms that have advanced functional genomics tools (as Affymetrix oligo arrays, see below). Thus, for a proper annotation of an entomopathogenic gene, via the GO project, its *C. elegans* homologue should be identified and examined. This inhibits GO-assistant research of unique features of entomopathogenic nematodes, which may not be represented by *C. elegans* genes (discussed below).

### 15.2.6. Initiatives for genomics studies of entomopathogenic nematodes

Only recently, efforts are turning to sequencing the genome of the insect-parasitic nematode *H. bacteriophora* TTO1; *Photorhabdus luminescens* ssp. *luminescens*, the bacterial symbiont of *H. bacteriophora* TTO1, has been fully sequenced. The 5.7 Mb bacterial genome contains 4839 predicted open reading frames, unveiling genes and pathways that may account for biological traits – such as symbiotic and parasitic associations of the bacteria with the nematode and insects, respectively (see St Leger and Wang, Chapter 16, this volume).

The recently NSF-financed *H. bacteriophora* genome sequencing should lead to a dramatic increase in our knowledge of the underlying mechanisms controlling important biological traits of *H. bacteriophora*. This will complement the completion of the *P. luminescens* genome and help develop knowledge unique to this complex, symbiotic association. The *H. bacteriophora* genome sequencing was performed by The Washington University Genome Sequencing Center (WUGSC; Fig. 15.2; P. Grewal and The Entomopathogenic Nematode Genome Team and Consortium, USA, 2007, personal communication). Following sequencing of other nematodes' genomes (*C. elegans* and *Caenorhabditis briggsae*), it was revealed that these are highly repetitive. Thus, coverage of at least eight times was used for proper assembly of the genome. The approach for *H. bacteriophora* genome sequencing was based on whole-genome shotgun (WGS) sequencing of the genome (Fig. 15.2). Shotgun sequencing is a method used for sequencing long DNA strands. Since the chain termination sequencing method can only be used for fairly short strands, it is necessary to divide longer sequences up and then assemble the results to give the overall sequence. Shotgun sequencing uses a faster but more complex process to assemble random pieces of the sequence (Weber and Myers, 1997).

For the WGS approach, a *H. bacteriophora* genomic DNA library were constructed from high molecular weight genomic DNA obtained from axenic eggs from an inbred strain. Then, a randomly sheared shotgun library was constructed with fragments sheared to the desired size (4 and 6 kb) and the quality of the library was determined.

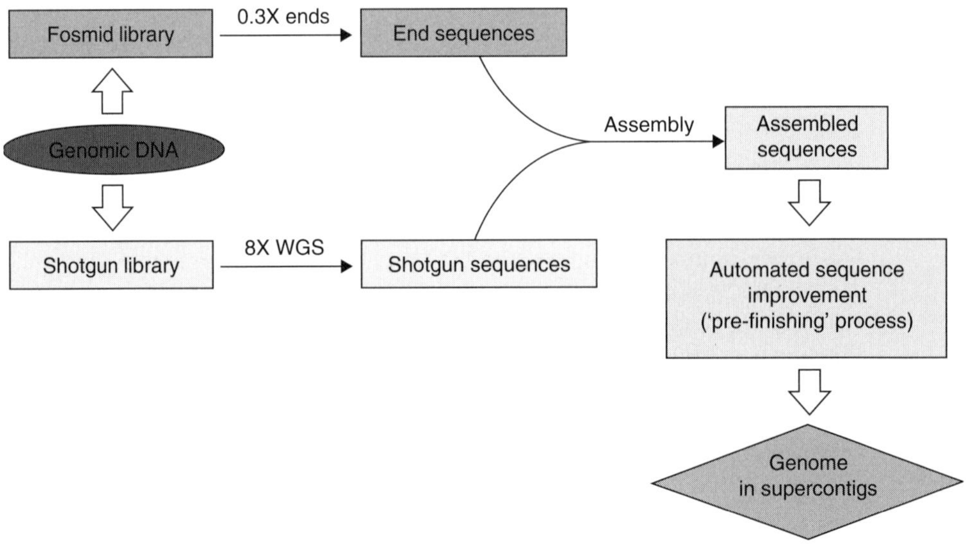


**Fig. 15.2.** The approach taken for sequencing of *Heterorhabditis bacteriophora* genome. (Taken with permission, Parwinder Grewal.)

In addition to the WGS approach, fosmids (e.g. a low-copy-number cosmid vector in which clone sequences are relatively stable) were constructed and sequenced (Fig. 15.2). The fosmids contain intermediate size inserts (~40 kb) yielding read pair information over a larger area. These were sequenced to add 0.3 times coverage of the genome. Notably, the fosmid library is critical to the assembly process, as it adds end pair sequence information from intermediate size clones to complement the data from the WGS. Once all the WGS and fosmid end sequence data were collected and trimmed for vector sequence, assembly was performed.

The sequencing start date was 15 January 2006 and the anticipated completion date is 14 January 2009; sequencing data are released every 3 months to a web site (www.oardc.ohio-state.edu/nematodes; P. Grewal and The Entomopathogenic Nematode Genome Team and Consortium, USA, 2007, personal communication). Clearly, the *H. bacteriophora* genomics project opens new avenues in functional genomics of entomopathogenic nematode; these will be discussed below.

## 15.3. Functional Genomics: Towards Deciphering of Genomics and ESTs Sequences

### 15.3.1. Transcriptomics

Transcriptomics is a new, still-evolving powerful tool which enables quantitative, real-time study of gene expression on a genome scale. Products and functions at the transcript level may be assigned to a majority of the genes that comprise an organism.

Some of the transcriptomics studies use gene microarrays. These are mainly applied by hybridization of labelled gene transcription products with solid surfaces. On these, either short (50–70 mer) synthesized oligonucleotides that represent a gene, an intron or an exon sequence ('oligonucleotide microarrays'; e.g. those of Affymetrix) or cDNA clones that are reverse transcribed from gene transcription products ('cDNA microarrays') are spotted, such that numerous genes may be represented on an array (reviewed by Aharoni and Vorst, 2002).

Following hybridization with fluorescent labelled RNA, image analysis is performed. This important aspect of microarray experiments can have a potentially large impact on subsequent analyses (Yang *et al.*, 2002a). Then, interpretation of the huge amounts of data extracted from microarray experiments is performed. This comprises not only a biological challenge but also a major computational challenge, and success depends on computational tools for the efficient integration and interpretation of large data sets.

The first stage of data analysis is normalization; this may be performed by one or a combination of multiple approaches developed during the last few years (e.g. Yang *et al.*, 2002b). Second, the normalized data set is analysed by, for example, grouping or classifying samples and genes according to similar expression patterns by unsupervised methods such as hierarchical and K-means clustering (Eisen *et al.*, 1998). Reduction of dimensionality can be performed by mapping the high dimensional expression data sets into lower-dimensional space using principal components analysis (Raychaudhuri, *et al.*, 2000) or self-organizing maps

(Toronen *et al.*, 1999), and significantly, differentially regulated genes may be identified by filtering the data according to *P* values and fold change.

Notably, these methods are useful for describing changes in gene expression, but are of limited use to describe cellular responses in the context of available knowledge. A more advanced tool, biologically speaking, is pathway analysis. Pathway analysis relates the microarray data directly to biology, by incorporating differentially expressed genes to known biological pathways (e.g. Curtis *et al.*, 2005).

The number of publications associated with microarray studies is dramatically increasing, demonstrating the usefulness of DNA microarrays as a tool for conducting quantitative, large-scale experiments on gene expression. These studies have led to the elucidation of mechanisms and the prediction of biological processes, the assignment of functions to previously unannotated genes, the grouping of genes into functional pathways and the prediction of the activities of new compounds (reviewed by Stoughton, 2005; Plomin and Schalkwyk, 2007). Specifically for nematodes, such functional genomics tools were established mainly for the free-living, model nematode *C. elegans*. These tools were utilized to annotate the nematode genome sequence. The annotated genome will serve as a solid platform on which may be built a better understanding of nematode biology and its relevance to other organisms (e.g. Hillier *et al.*, 2005; Reisner *et al.*, 2005).

Currently, gene microarrays are not available for entomopathogenic nematodes. An approach to microarray-based genomics research is the use of cross-species hybridization (CSH), by hybridizing RNA of the studied organism to a microarray chip which contains transcripts of genes of a closely related species (e.g. Rifkin *et al.*, 2003; Held *et al.*, 2004; Renn *et al.*, 2004; Rise *et al.*, 2004; Snape *et al.*, 2004; Nowrousian *et al.*, 2005; Bar-Or *et al.*, 2006, 2007a,b; Koltai and Weingarten-Baror, 2008). Previously it was suggested that CSH might result in biologically meaningful expression profiling, even for distantly related organisms, as long as sequence divergence is limited for a given gene (Renn *et al.*, 2004). Because of the molecular and genomics tools available for *C. elegans*, a possible program might be the use of *C. elegans* gene-microarray to study expression profiles of entomopathogenic nematodes.

Nevertheless, many genes, and especially those that correlate to traits controlling response to a changing environment (including the host), may evolve rapidly. For example, a comparative study of gene expression in *C. elegans* suggested that genes involved in specialized stages of development (e.g. dauer larvae) may rapidly evolve, and are therefore less conserved between species (Mitreva *et al.*, 2004). Since there may be only limited homology between rapidly evolving genes from different species, CSH may have limited capability in accurate reflection of processes involving non-conserved genes of entomopathogenic nematodes.

Notably, the newly emerging technologies of sequencing (discussed above, in 19.2.3) may provide a plausible and attractive solution for transcription profiling of entomopathogenic nematodes. Neither previous sequence information nor homology to other species is needed. Rather, high-speed sequencing of multiple transcripts is feasible, for any of the species of interest, to allow quantitative identification of gene expression, spatially and temporally (Torres *et al.*, 2008).

The *H. bacteriophora* genome project opens new avenues for transcriptomics studies. Based on the retrieved sequences, microarray platforms may be constructed which represent *H. bacteriophora* genes. These may allow examination, at a genome scale, of changes in gene expression profiles during different physiological conditions, such as stress response, insect pathogenicity, symbiosis with the bacteria and enhancement of reproduction and shelf life.

Taking into consideration the high cost of microarray experiments, the lack of specific platforms for each of the entomopathogenic nematodes and the inherent problematic nature of CSH, a valid approach for quantification of gene expression may be the sensitive method of quantitative PCR. Although this approach may not support mass analysis of multiple genes, it provides a sensitive and accurate method for determination of the relative amount of gene transcription products. With entomopathogenic nematodes it was applied in the study of stress-induced genes (Gal *et al.*, 2003).

### 15.3.2. Proteomics

Proteomics studies allow quantitative study of proteins on a large scale. The sum of all proteins in a cell (i.e. the proteome) may differ from cell to cell and may constantly change during development or response to the environment. A single organism may have different composition of proteins in different parts of its body, in different stages of its life cycle and under different environmental conditions. Thus, although proteomics may be considered as the next step in studies that tend to analyse biological systems, its complexity is still posing a challenge for comprehensive analysis.

Several current tools are in use for proteomics studies. High-resolution two-dimensional gel electrophoresis (2DE) may be used for protein separation according to their isoelectric point (pI) and their molecular weight (MW) (reviewed by Kersten *et al.*, 2002; Patterson and Aebersold, 2003). Mass spectrometers can determine the mass of a protein or peptide with a high degree of accuracy, and thus can be used to distinguish protein products from closely related species. Tandem mass spectrometry or 'MS/MS' can provide structural information on molecular ions that can be isolated and fragmented within the instrument, whereas ionization of proteins and peptides, at high sensitivity and without excessive fragmentation, may be accomplished by matrix-assisted laser desorption ionization ('MALDI'). MALDI was most commonly coupled with time-of-flight (TOF) 'mass analysers', and this led to the development of commercial 'MALDI-TOF' mass spectrometers (reviewed by Patterson and Aebersold, 2003).

The analyses of spots of 2DE separated proteins with mass spectrometry (MS), led to an unprecedented progress in proteomics studies, enabling identification of a collection of proteins found in a particular cell type under a particular set of environmental conditions. Notably, the mass of a eukaryotic protein is not a uniquely identifying feature. Nevertheless, the masses of the various peptides generated by fragmentation of an isolated protein with an enzyme of known cleavage specificity could uniquely identify a protein (Patterson and Aebersold, 2003).

As a result, many of the advances in proteomics technology have been driven by MS hardware and methodology developments. To support high-throughput protein identification by MS, several computer programs were developed. SEQUEST (MacCoss *et al.*, 2002) and MASCOT (Perkins *et al.*, 1999) were developed for searching uninterpreted tandem mass spectra of peptides against either a protein or nucleotide sequence database; other computer programs such as DTASELECT (Tabb *et al.*, 2002) were developed for assembling of peptide identifications to proteins.

Using the 2DE and MS technology, several proteins were isolated from entomopathogenic nematodes. Proteins expressed during induction of the dormant state of *S. feltiae* IJs following exposure to osmotic stress included actin, ATP synthase, Proteasome regulatory particle (ATPase-like), GroEL chaperonin, GroES co-chaperonin (both are members of the Hsp60 family of chaperons), a transposase family member and a stress protein (46.3 kDa) (Chen and Glazer, 2005; Chen *et al.*, 2005); a LEA-like protein was expressed during the process of inducing the *S. feltiae* IJs into a quiescent anhydrobiotic state (Solomon *et al.*, 2000). Another study examined the activity of trehalose-6-phosphate synthase (T6PS), an enzyme involved in the synthesis of trehalose, demonstrating its increasing titer during both heat- and cold-shocked nematodes during the first 3 h of exposure (Jagdale *et al.*, 2005).

The combination of MS technology, the use of the above-mentioned computer programs, and the significant increase in available genomic sequence information have allowed the methods and technology to evolve away from 'one protein, one analysis' and towards the analysis of thousands of proteins in a single experiment (MacCoss, 2005). With entomopathogenic nematodes, a high-throughput proteomics approach has yet to be applied. Significantly, the *H. bacteriophora* genome project may allow deduction of nucleotide sequences from amino acid sequences, promoting identification of multiple, entomopathogenic proteins.

### 15.3.3. Gene silencing

One other functional genomics tool, RNAi ('i' for 'interference'), deserves special consideration. This technique involves sequence-specific gene silencing induced by double-stranded RNA. Double-stranded RNAs are cleaved by the RNAse II-like enzyme Dicer to form 21–23 nucleotide small interfering RNAs (siRNAs). The resulting siRNAs are incorporated into an RNA-induced silencing complex (RISC). The RISC targets and cleaves mRNA complementary to the siRNAs (reviewed by Campbell and Choy, 2005). This natural mechanism has been demonstrated in organisms ranging from trypanosomes to nematodes to vertebrates; it was first exploited experimentally in *C. elegans* (Timmons and Fire, 1998; Fire, 1999; Kamath and Ahringer, 2003; Kamath *et al.*, 2003).

The RNAi technique has been applied to some additional nematode species, including plant parasites (*Globodera pallida* and *H. glycines*) and human parasites (e.g. *Brugia malayi*) (Aboobaker and Blaxter, 2004). This approach may provide a plausible (and perhaps the only) way to elucidate gene function on the genomics-transcriptomics scale (Aboobaker and Blaxter, 2004).

RNAi-mediated gene silencing by soaking, feeding or microinjection methods (Maeda *et al.*, 2001; Timmons *et al.*, 2001; Kamath *et al.*, 2003) was unsuccessful in the insect-parasitic nematodes *Steinernema* spp. (Gal, Glazer and Koltai, unpublished data; Dix, Tyson and Burnell, personal communication). However, only recently, by soaking L1 stage of *H. bacteriophora* with double-stranded RNA of *H. bacteriophora* genes, Ciche and Sternberg (2007) demonstrated an ability to reduce transcripts level of these genes, in a gene-specific manner. These results suggest that post-embryonic RNAi in *H. bacteriophora* may be a potent approach for manipulation of gene expression in this entomopathogenic nematode. This approach may dramatically enhance the ability to analyse function of multiple genes and genetic pathways involved in biological processes of interest.

## 15.4. Genetic Improvement

The generation of transgenic entomopathogenic nematodes may provide significant benefit to agriculture. To create a genetically modified organism (GMO), a gene of interest (homologous or heterologous) may be introduced into an organism by transformation in a way that will make it heritable. The first successful transformation of an entomopathogenic nematode was reported by Hashmi *et al.* (1995). Foreign genes were introduced into *H. bacteriophora* HP88 by microinjection using vectors carrying the *C. elegans* genes coding for the roller phenotype and 16 kDa heat-shock protein (hsp16) gene. Expression of a translational fusion of hsp16/lacZ was detected in the body musculature, hypodermis and pharyngeal muscles.

Genetic enhancement for thermotolerance in *H. bacteriophora* was achieved by transforming it with the gene encoding a 70 kDa heat-shock protein (hsp70) from *C. elegans* (Hashmi *et al.*, 1998). Successful transformation of the gene was confirmed, transcripts were increased several-fold in transgenic nematodes upon heat shock, and survival of IJs dramatically increased in the transgenic nematodes (Hashmi *et al.*, 1998; Wilson *et al.*, 1999). This strain was released in turfgrass microplots in the field in the spring, summer and fall of 1996. Although transgenic and wild-type strains did not differ in their ability to persist (Gaugler *et al.*, 1997; Wilson *et al.*, 1999), the risk assessment study supported the regulatory view that the transgenic nematode strain is an unlikely environmental threat; it was suggested that subsequent regulatory reviews in the USA will probably continue to be decided on a case-by-case basis according to organism phenotype rather than the techniques used to generate them (Gaugler *et al.*, 1997; Wilson *et al.*, 1999).

Another study generated *S. feltiae* transgenic nematodes carrying a cloned stress-resistance gene *trehalose-phosphate synthase 1* from yeast (Vellai *et al.*, 1999). A dramatic increase in the tolerance of transgenic IJs was demonstrated, further demonstrating the potential of the transgenic approach in genetic improvement of entomopathogenic nematodes. Considerably, the vast amount of information regarding *H. bacteriophora* genes and their expression, expected from the *H. bacteriophora* genome sequencing and the anticipated functional genomics studies, may further enhance this approach by providing numerous genes for transformation of entomopathogenic nematodes.

## 15.5. Conclusion and Future Prospects

Genomics studies of entomopathogenic nematodes, and especially the initiative of the genome sequencing project, are expected to revolutionize research on these organisms in over 100 academic, government and industrial laboratories worldwide. Undoubtedly, sequencing of the *H. bacteriophora* genome will enable construction of functional genomics tools (e.g. entomopathogenic nematode microarray platforms and databases of predicted genes and proteins). These, in turn, should lead to a proliferation of studies within other research fields (e.g. metabolomics, aiming to examine the set of metabolites synthesized in a biological system, reviewed by Fiehn, 2002).

More specifically, the genomics data and tools developed for entomopathogenic nematodes may provide a glimpse into the complex phenomena involved in pathogenesis in insects, symbiosis with bacteria, nematode stress response and reproduction. Due to the important role *H. bacteriophora* can play in biological control, the application of these new findings should lead to improving all aspects of producing, packaging and stabilizing nematode products for use in agriculture.

*H. bacteriophora* is serving as a model organism for other entomopathogenic species used as biological control agents for insect pests worldwide. However, entomopathogenic nematodes are phylogenetically diverse; insect parasitism has emerged several times (Dorris *et al.*, 1999). As a result, several difficulties may be foreseen. A technical difficulty may arise in the usage of *H. bacteriophora* platforms (such as gene microarrays) for CSH of other entomopathogenic nematodes. This may be especially difficult with rapidly evolving genes since these are often correlated with specialized development and adaptation to the environment. Nevertheless, studies are under way in our laboratory and those of others to evaluate the relative performance of CSH compared to species-specific microarray hybridizations (SSH), to ensure that CSH closely reflects the biological process analysed by SSH (e.g. Gilad *et al.*, 2005; Bar-Or *et al.*, 2006, 2007a,b).

However, *H. bacteriophora* platforms, may not be applicable to other entomopathogenic species, For example, the molecular mechanisms by which *H. bacteriophora* is associated with bacteria may be different from those of *S. feltiae* (given their probably independent origins; Dorris *et al.*, 1999).

Genomic analyses have a number of inherent difficulties that must be addressed or overcome. First, databases used for analysis are variable in quality; some EST libraries are redundant, some are not, and their assigned unigenes may split or merge due to appearance of new or better annotated ESTs. Second, an immaturity of functional genomics platforms used for analysis (such as microarray platforms) and of computational tools for the conversion of the data to usable knowledge may lead to reduction in the trustfulness of the results and to inaccuracy in drawing biological conclusions. Evidently, measurements of gene expression generated from identical RNA preparations that were obtained using three commercially available microarray platforms resulted in only poor correlations in gene expression levels suggesting the need for establishing industrial manufacturing standards and further validation of the technology by independent and thorough means (Tan *et al.*, 2003).

Nevertheless, genomics, as a field of science, is rapidly evolving. The interdisciplinary research of computational biology is constantly progressing towards the improvement of high-throughput laboratory techniques and computational tools for accurate data production and analysis. An example may be the evolution of tools for microarray analysis: it was demonstrated by Mecham *et al.* (2004) that data derived from sequence-verified probes show increased precision in technical replicates, increased accuracy translating data from one generation microarray to another and increased accuracy translating data from oligonucleotide to cDNA microarrays. In more general terms, increasing of microarray hybridization specificity should be exercised: in addition to sequence verification of probes, multiple effectors of specificity should be examined, quantified and accordingly manipulated, to achieve a higher level of hybridization specificity. Enhancement of hybridization specificity, and in addition, increased accuracy, reproducibility and sensitivity of microarray hybridization should lead to data with higher validity (Koltai and Weingarten-Baror, 2008).

The improvement in genomic databases and related analytical tools may lead to knowledge obtained from correct interpretation of the data. Information accumulated in this way and from various biological experiments will lead to a more comprehensive understanding of biological systems and mapping of all genetic pathways, their interactions, and their divergences and convergences. This will help lay the genetic bases for interpreting or managing complex biological components and interactions between organisms.

Despite the challenges and difficulties faced, the *H. bacteriophora* genomic project irrefutably provides an important base for comparative analysis, on a genomic scale, of genes and pathways that affect traits of interest in a system of agricultural relevance. From an evolutionary point of view, the *H. bacteriophora* genome may serve as a bridge between *C. elegans* and more distantly related nematode parasites. The availability of the genome sequences of two closely related free-living nematodes, *C. elegans* and *C. briggsae*, together with a full complement of genetic and molecular tools, offers new means for studies of nematode evolution and adaptation to parasitic life habit.

It should be noted that a genomics approach for the study of entomopathogenic nematodes cannot be used as a stand-alone tool. Rather, it should be integrated with other research approaches, including those which involve the study of individual genes in the course of in-depth studies of a biological system. This integration can ultimately lead to development of genetically improved strains of entomopathogenic nematodes for use in agriculture. This new, budding field of the genomics of entomopathogenic nematodes will pave the way towards a global understanding of nematode biology.

## References

Aboobaker, A.A. and Blaxter, M.L. (2004) Functional genomics for parasitic nematodes and platyhelminths. *Trends in Parasitology* 20, 178–184.

Adams, M.D., Celniker, S.E., Holt, R.A., Evans, C.A., Gocayne, J.D., Amanatides, P.G., Scherer, S.E., Li, P.W., Hoskins, R.A., Galle, R.F., George, R.A., Lewis, S.E., Richards, S., Ashburner, M., Henderson,

S.N., Sutton, G.G., Wortman, J.R., Yandell, M.D., Zhang, Q., Chen, L.X., Brandon, R.C., Rogers, Y.H., Blazej, R.G., Champe, M., Pfeiffer, B.D., Wan, K.H., Doyle, C., Baxter, E.G., Helt, G., Nelson, C.R., Gabor, G.L., Abril, J.F., Agbayani, A., An, H.J., Andrews-Pfannkoch, C., Baldwin, D., Ballew, R.M., Basu, A., Baxendale, J., Bayraktaroglu, L., Beasley, E.M., Beeson, K.Y., Benos, P.V., Berman, B.P., Bhandari, D., Bolshakov, S., Borkova, D., Botchan, M.R., Bouck, J., Brokstein, P., Brottier, P., Burtis, K.C., Busam, D.A., Butler, H., Cadieu, E., Center, A., Chandra, I., Cherry, J.M., Cawley, S., Dahlke, C., Davenport, L.B., Davies, P., de Pablos, B., Delcher, A., Deng, Z., Mays, A.D., Dew, I., Dietz, S.M., Dodson, K., Doup, L.E., Downes, M., Dugan-Rocha, S., Dunkov, B.C., Dunn, P., Durbin, K.J., Evangelista, C.C., Ferraz, C., Ferriera, S., Fleischmann, W., Fosler, C., Gabrielian, A.E., Garg, N.S., Gelbart, W.M., Glasser, K., Glodek, A., Gong, F., Gorrell, J.H., Gu, Z., Guan, P., Harris, M., Harris, N.L., Harvey, D., Heiman, T.J., Hernandez, J.R., Houck, J., Hostin, D., Houston, K.A., Howland, T.J., Wei, M.H., Ibegwam, C., Jalali, M., Kalush, F., Karpen, G.H., Ke, Z., Kennison, J.A., Ketchum, K.A., Kimmel, B.E., Kodira, C.D., Kraft, C., Kravitz, S., Kulp, D., Lai, Z., Lasko, P., Lei, Y., Levitsky, A.A., Li, J., Li, Z., Liang, Y., Lin, X., Liu, X., Mattei, B., McIntosh, T.C., McLeod, M.P., McPherson, D., Merkulov, G., Milshina, N.V., Mobarry, C., Morris, J., Moshrefi, A., Mount, S.M., Moy, M., Murphy, B., Murphy, L., Muzny, D.M., Nelson, D.L., Nelson, D.R., Nelson, K.A., Nixon, K., Nusskern, D.R., Pacleb, J.M., Palazzolo, M., Pittman, G.S., Pan, S., Pollard, J., Puri, V., Reese, M.G., Reinert, K., Remington, K., Saunders, R.D., Scheeler, F., Shen, H., Shue, B.C., Siden-Kiamos, I., Simpson, M., Skupski, M.P., Smith, T., Spier, E., Spradling, A.C., Stapleton, M., Strong, R., Sun, E., Svirskas, R., Tector, C., Turner, R., Venter, E., Wang, A.H., Wang, X., Wang, Z.Y., Wassarman, D.A., Weinstock, G.M., Weissenbach, J., Williams, S.M., Woodage, T., Worley, K.C., Wu, D., Yang, S., Yao, Q.A., Ye, J., Yeh, R.F., Zaveri, J.S., Zhan, M., Zhang, G., Zhao, Q., Zheng, L., Zheng, X.H., Zhong, F.N., Zhong, W., Zhou, X., Zhu, S., Zhu, X., Smith, H.O., Gibbs, R.A., Myers, E.W., Rubin, G.M. and Venter, J.C. (2000) The genome sequence of *Drosophila melanogaster. Science* 287, 2185–2195.

Aharoni, A. and Vorst, O. (2002) DNA microarrays for functional plant genomics. *Plant Molecular Biology* 48, 99–118.

Arabidopsis Genome Initiative (2000) Analysis of the genome sequence of the flowering plant *Arabidopsis thaliana. Nature* 408, 796–815.

Bar-Or, C., Bar-Eyal, M., Gal, T.Z., Kapulnik, Y., Czosnek, H. and Koltai, H. (2006) Derivation of species-specific hybridization-like knowledge out of cross-species hybridization results. *BMC Genomics* 7, 110.

Bar-Or, C., Czosnek, H. and Koltai, H. (2007a) Cross-species microarray hybridizations: a developing tool for studying diversity. *Trends in Genetics* 23, 200–207.

Bar-Or, C., Novikov, E., Reiner, A., Czosnek, H. and Koltai, H. (2007b) Utilizing microarray spot characteristics to improve cross-species hybridization results. *Genomics* 90, 636–645.

Campbell, T.N. and Choy, F.Y. (2005) RNA interference: past, present and future. *Current Issues in Molecular Biology* 7, 1–6.

Chen, S. and Glazer, I. (2005) A novel method for long-term storage of the entomopathogenic nematode *Steinernema feltiae* at room temperature. *Biological Control* 32, 104.

Chen, S., Gollop, N. and Glazer, I. (2005) Cross-stress tolerance and expression of stress-related proteins in osmotically desiccated entomopathogenic *Steinernema feltiae* IS-6. *Parasitology* 131, 695–703.

Ciche, T.A. and Sternberg, P.W. (2007) Postembryonic RNAi in *Heterorhabditis bacteriophora*: a nematode insect parasite and host for insect pathogenic symbionts. *BMC Developmental Biology* 7, 101.

Curtis, R.K., Oresic, M. and Vidal-Puig, A. (2005) Pathways to the analysis of microarray data. *Trends in Biotechnology* 23, 429–435.

Dorris, M., De Ley, P. and Blaxter, M.L. (1999) Molecular analysis of nematode diversity and the evolution of parasitism. *Parasitology Today* 15, 188–193.

Eisen, M.B., Spellman, P.T., Brown, P.O. and Botstein, D. (1998) Cluster analysis and display of genome-wide expression patterns. *Proceedings of the National Academy of Sciences of the USA* 95, 14863–14868.

Fiehn, O. (2002) Metabolomics – the link between genotypes and phenotypes. *Plant Molecular Biology* 48, 155–171.

Fire, A. (1999) RNA-triggered gene silencing. *Trends in Genetics* 15, 358–363.

Gal, T.Z., Glazer, I. and Koltai, H. (2003) Differential gene expression during desiccation stress in the insect-killing nematode *Steinernema feltiae* IS-6. *Journal of Parasitology* 89, 761–766.

Gaugler, R., Wilson, M. and Shearer, P. (1997) Field release and environmental fate of a transgenic entomopathogenic nematode. *Biological Control* 9, 75–80.

Gilad, Y., Rifkin, S.A., Bertone, P., Gerstein, M. and White, K.P. (2005) Multi-species microarrays reveal the effect of sequence divergence on gene expression profiles. *Genome Research* 15, 674–680.

Goffeau, A., Barrell, B.G., Bussey, H., Davis, R.W., Dujon, B., Feldmann, H., Galibert, F., Hoheisel, J.D., Jacq, C., Johnston, M., Louis, E.J., Mewes, H.W., Murakami, Y., Philippsen, P., Tettelin, H. and Oliver, S.G. (1996) Life with 6000 genes. *Science* 274, 546, 563–567.

Hashmi, S., Hashmi, G. and Gaugler, R. (1995) Genetic transformation of an entomopathogenic nematode by microinjection. *Journal of Invertebrate Pathology* 66, 293–296.

Hashmi, S., Hashmi, G., Glazer, I. and Gaugler, R. (1998) Thermal response of *Heterorhabditis bacteriophora* transformed with the *Caenorhabditis elegans hsp70* encoding gene. *The Journal of Experimental Zoology* 281, 164–170.

Held, M., Gase, K. and Baldwin, I.T. (2004) Microarrays in ecological research: a case study of a cDNA microarray for plant-herbivore interactions. *BMC Ecology* 4, 13.

Hillier, L.W., Coulson, A., Murray, J.I., Bao, Z., Sulston, J.E. and Waterston, R.H. (2005) Genomics in *C. elegans*: So many genes, such a little worm. *Genome Research* 15, 1651–1660.

Goff, S.A., Ricke, D., Lan, T.H., Presting, G., Wang, R., Dunn, M., Glazebrook, J., Sessions, A., Oeller, P., Varma, H., Hadley, D., Hutchison, D., Martin, C., Katagiri, F., Lange, B.M., Moughamer, T., Xia, Y., Budworth, P., Zhong, J., Miguel, T., Paszkowski, U., Zhang, S., Colbert, M., Sun, W.L., Chen, L., Cooper, B., Park, S., Wood, T.C., Mao, L., Quail, P., Wing, R., Dean, R., Yu, Y., Zharkikh, A., Shen, R., Sahasrabudhe, S., Thomas, A., Cannings, R., Gutin, A., Pruss, D., Reid, J., Tavtigian, S., Mitchell, J., Eldredge, G., Scholl, T., Miller, R.M., Bhatnagar, S., Adey, N., Rubano, T., Tusneem, N., Robinson, R., Feldhaus, J., Macalma, T., Oliphant, A. and Briggs, S. (2002) A draft sequence of the rice genome (*Oryza sativa* L. ssp. *japonica*). *Science* 296, 92–100.

Jagdale, G.B., Grewal, P.S. and Salminen, S.O. (2005) Both heat-shock and cold-shock influence trehalose metabolism in an entomopathogenic nematode. *Journal of Parasitology* 91, 988–994.

Kamath, R.S. and Ahringer, J. (2003) Genome-wide RNAi screening in *Caenorhabditis elegans*. *Methods* 30, 313–321.

Kamath, R.S., Fraser, A.G., Dong, Y., Poulin, G., Durbin, R., Gotta, M., Kanapin, A., Le Bot, N., Moreno, S., Sohrmann, M., Welchman, D.P., Zipperlen, P. and Ahringer, J. (2003) Systematic functional analysis of the *Caenorhabditis elegans* genome using RNAi. *Nature* 421, 231–237.

Kersten, B., Burkle, L., Kuhn, E.J., Giavalisco, P., Konthur, Z., Lueking, A., Walter, G., Eickhoff, H. and Schneider, U. (2002) Large-scale plant proteomics. *Plant Molecular Biology* 48, 133–141.

Koltai, H. and Weingarten-Baror C. (2008) Specificity of DNA microarray hybridization: characterization, effectors and approaches for data correction. *Nucleic Acids Research*, 36, 2395–2405.

MacCoss, M.J. (2005) Computational analysis of shotgun proteomics data. *Current Opinion in Chemical Biology* 9, 88–94.

MacCoss, M.J., Wu, C.C. and Yates, J.R. 3rd (2002) Probability-based validation of protein identifications using a modified SEQUEST algorithm. *Analytical Chemistry* 74, 5593–5599.

Maeda, I., Kohara, Y., Yamamoto, M. and Sugimoto, A. (2001) Large-scale analysis of gene function in *Caenorhabditis elegans* by high-throughput RNAi. *Current Biology* 11, 171–176.

Mardis, E.R. (2008) The impact of next-generation sequencing technology on genetics. *Trends in Genetics* 24, 133–141.

Mecham, B.H., Wetmore, D.Z., Szallasi, Z., Sadovsky, Y., Kohane, I. and Mariani, T.J. (2004) Increased measurement accuracy for sequence-verified microarray probes. *Physiological Genomics* 18, 308–315.

Mitreva, M., McCarter, J.P., Martin, J., Dante, M., Wylie, T., Chiapelli, B., Pape, D., Clifton, S.W., Nutman, T.B. and Waterston, R.H. (2004) Comparative genomics of gene expression in the parasitic and free-living nematodes *Strongyloides stercoralis* and *Caenorhabditis elegans*. *Genome Research* 14, 209–220.

Nowrousian, M., Ringelberg, C., Dunlap, J.C., Loros, J.J. and Kuck, U. (2005) Cross-species microarray hybridization to identify developmentally regulated genes in the filamentous fungus *Sordaria macrospora*. *Molecular Genetics and Genomics* 273, 137–149.

Patterson, S.D. and Aebersold, R.H. (2003) Proteomics: the first decade and beyond. *Nature Genetics* 33, 311–323.

Perkins, D.N., Pappin, D.J., Creasy, D.M. and Cottrell, J.S. (1999) Probability-based protein identification by searching sequence databases using mass spectrometry data. *Electrophoresis* 20, 3551–3567.

Plomin, R. and Schalkwyk, L.C. (2007) Microarrays. *Developmental Science* 10, 19–23.

Raychaudhuri, S., Stuart, J.M. and Altman, R.B. (2000) Principal components analysis to summarize microarray experiments: application to sporulation time series. *Pacific Symposium on Biocomputing*, 455–466.

Reisner, K., Asikainen, S., Vartiainen, S. and Wong, G. (2005) Developmental and biological insights obtained from gene expression profiling of the nematode *Caenorhabditis elegans*. *Current Genomics* 6, 97–107.

Renn, S.C., Aubin-Horth, N. and Hofmann, H.A. (2004) Biologically meaningful expression profiling across species using heterologous hybridization to a cDNA microarray. *BMC Genomics* 5, 42.

Rifkin, S.A., Kim, J. and White, K.P. (2003) Evolution of gene expression in the *Drosophila melanogaster* subgroup. *Nature Genetics* 33, 138–144.

Rise, M.L., von Schalburg, K.R., Brown, G.D., Mawer, M.A., Devlin, R.H., Kuipers, N., Busby, M., Beetz-Sargent, M., Alberto, R., Gibbs, A.R., Hunt, P., Shukin, R., Zeznik, J.A., Nelson, C., Jones, S.R.M., Smailus, D.E., Jones, S.J.M., Schelin, J.A., Marra, M.A., Buterfield, Y.S.N., Stott, J.M., Ng, S.H.S., Davidson, W.S. and Koop, B.F. (2004) Development and application of a salmonid EST database and cDNA microarray: data mining and inter specific hybridization characteristics. *Genome Research* 14, 478–490.

Schuster, S.C. (2008) Next-generation sequencing transforms today's biology. *Nature Methods* 5, 16–18.

Snape, J.R., Maund, S.J., Pickford, D.B. and Hutchinson, T.H. (2004) Ecotoxicogenomics: the challenge of integrating genomics into aquatic and terrestrial ecotoxicology. *Aquatic Toxicology* 67, 143–154.

Solomon, A., Salomon, R., Paperna, I. and Glazer, I. (2000) Desiccation stress of entomopathogenic nematodes induces the accumulation of a novel heat-stable protein. *Parasitology* 121, 409–415.

Stoughton, R.B. (2005) Applications of DNA microarrays in biology. *Annual Review of Biochemistry* 74, 53–82.

Tabb, D.L., McDonald, W.H. and Yates, J.R. 3rd (2002) DTASelect and contrast: tools for assembling and comparing protein identifications from shotgun proteomics. *Journal of Proteome Research* 1, 21–26.

Tan, P.K., Downey, T.J., Spitznagel, E.L. Jr, Xu, P., Fu, D., Dimitrov, D.S., Lempicki, R.A., Raaka, B.M. and Cam, M.C. (2003) Evaluation of gene expression measurements from commercial microarray platforms. *Nucleic Acids Research* 31, 5676–5684.

The *C. elegans* Sequencing Consortium (1998) Genome sequence of the Nematode *C. elegans*: platform for investigating biology. *Science* 282, 2012–2018.

Timmons, L. and Fire, A. (1998) Specific interference by ingested dsRNA. *Nature* 395, 854.

Timmons, L., Court, D.L. and Fire, A. (2001) Ingestion of bacterially expressed dsRNAs can produce specific and potent genetic interference in *Caenorhabditis elegans*. *Gene* 263, 103–112.

Toronen, P., Kolehmainen, M., Wong, G. and Castren, E. (1999) Analysis of gene expression data using self-organizing maps. *FEBS Letters* 451, 142–146.

Torres, T.T., Metta, M., Ottenwalder, B. and Schlotterer, C. (2008) Gene expression profiling by massively parallel sequencing. *Genome Research* 18, 172–177.

Vellai, T., Molnár, A., Lakatos, L., Ba'nfalvi, T., Fodor, A. and Sáringer, G. (1999) Transgenic nematodes carrying a cloned stress resistance gene from yeast. In: Glazer, I., Richardson, P., Boemare, N. and Coudert, F. (eds) *COST 819 Entomopathogenic Nematodes – Survival of Entomopathogenic Nematodes*. Office for Official Publications of the EC, Luxembourg EUR 18855 EN, pp. 105–119.

Venter, J.C., Adams, M.D., Myers, E.W., Li, P.W., Mural, R.J., Sutton, G.G., Smith, H.O., Yandell, M., Evans, C.A., Holt, R.A., Gocayne, J.D., Amanatides, P., Ballew, R.M., Huson, D.H., Wortman, J.R., Zhang, Q., Kodira, C.D., Zheng, X.H., Chen, L., Skupski, M., Subramanian, G., Thomas, P.D., Zhang, J., Gabor Miklos, G.L., Nelson, C., Broder, S., Clark, A.G., Nadeau, J., McKusick, V.A., Zinder, N., Levine, A.J., Roberts, R.J., Simon, M., Slayman, C., Hunkapiller, M., Bolanos, R., Delcher, A., Dew, I., Fasulo, D., Flanigan, M., Florea, L., Halpern, A., Hannenhalli, S., Kravitz, S., Levy, S., Mobarry, C., Reinert, K., Remington, K., Abu-Threideh, J., Beasley, E., Biddick, K., Bonazzi, V., Brandon, R., Cargill, M., Chandramouliswaran, I., Charlab, R., Chaturvedi, K., Deng, Z., Di Francesco, V., Dunn, P., Eilbeck, K., Evangelista, C., Gabrielian, A.E., Gan, W., Ge, W., Gong, F., Gu, Z., Guan, P., Heiman, T.J., Higgins, M.E., Ji, R.R., Ke, Z., Ketchum, K.A., Lai, Z., Lei, Y., Li, Z., Li, J., Liang, Y., Lin, X., Lu, F., Merkulov, G.V., Milshina, N., Moore, H.M., Naik, A.K., Narayan, V.A., Neelam, B., Nusskern, D., Rusch, D.B., Salzberg, S., Shao, W., Shue, B., Sun, J., Wang, Z., Wang, A., Wang, X., Wang, J., Wei, M., Wides, R., Xiao, C., Yan, C., Yao, A., Ye, J., Zhan, M., Zhang, W., Zhang, H., Zhao, Q., Zheng, L., Zhong, F., Zhong, W., Zhu, S., Zhao, S., Gilbert, D., Baumhueter, S., Spier, G., Carter, C., Cravchik, A., Woodage, T., Ali, F., An, H., Awe, A., Baldwin, D., Baden, H., Barnstead, M., Barrow, I., Beeson, K., Busam, D., Carver, A., Center, A., Cheng, M.L., Curry, L., Danaher, S., Davenport, L., Desilets, R., Dietz, S., Dodson, K., Doup, L., Ferriera, S., Garg, N., Gluecksmann, A., Hart, B., Haynes, J., Haynes, C., Heiner, C., Hladun, S., Hostin, D., Houck, J., Howland, T., Ibegwam, C., Johnson, J., Kalush, F., Kline, L., Koduru, S., Love, A., Mann, F., May, D., McCawley, S., McIntosh, T., McMullen, I., Moy, M., Moy, L., Murphy, B., Nelson, K., Pfannkoch, C., Pratts, E., Puri, V., Qureshi, H., Reardon, M., Rodriguez, R., Rogers, Y.H., Romblad, D., Ruhfel, B., Scott, R., Sitter, C., Smallwood, M., Stewart, E., Strong, R., Suh, E., Thomas, R., Tint, N.N., Tse, S., Vech, C., Wang, G., Wetter, J., Williams, S., Williams, M., Windsor, S., Winn-Deen, E., Wolfe, K., Zaveri, J., Zaveri, K., Abril, J.F., Guigo, R., Campbell, M.J., Sjolander, K.V., Karlak, B., Kejariwal, A., Mi, H., Lazareva, B., Hatton, T., Narechania, A., Diemer, K., Muruganujan, A., Guo, N., Sato, S., Bafna, V., Istrail, S., Lippert, R., Schwartz, R., Walenz, B., Yooseph, S., Allen, D., Basu, A., Baxendale, J., Blick, L., Caminha, M., Carnes-Stine, J., Caulk, P., Chiang, Y.H., Coyne, M., Dahlke, C., Mays, A., Dombroski, M., Donnelly, M., Ely, D., Esparham, S., Fosler, C., Gire, H., Glanowski, S., Glasser, K., Glodek, A., Gorokhov, M., Graham, K., Gropman, B., Harris, M., Heil, J., Henderson, S., Hoover, J., Jennings, D., Jordan, C., Jordan, J., Kasha, J., Kagan, L., Kraft, C., Levitsky, A., Lewis, M., Liu, X., Lopez, J., Ma, D., Majoros, W., McDaniel, J., Murphy, S., Newman, M., Nguyen, T., Nguyen, N., Nodell, M., Pan, S., Peck, J., Peterson, M., Rowe, W., Sanders, R., Scott, J., Simpson, M., Smith, T., Sprague, A., Stockwell, T., Turner, R., Venter, E., Wang, M., Wen, M., Wu, D., Wu, M., Xia, A., Zandieh, A. and Zhu, X. (2001) The sequence of the human genome. *Science* 291, 1304–1351.

Wang, H., Hubbell, E., Hu, J.S., Mei, G., Cline, M., Lu, G., Clark, T., Siani-Rose, M.A., Ares, M., Kulp, D.C. and Haussler, D. (2003) Gene structure-based splice variant deconvolution using a microarray platform. *Bioinformatics* 19(Suppl 1), 315–322.

Weber, J.L. and Myers, E.W. (1997) Human whole-genome shotgun sequencing. *Genome Research* 7, 401–409.

Wilson, M., Xin, W., Hashmi, S. and Gaugler, R. (1999) Risk assessment and fitness of a transgenic entomopathogenic nematode. *Biological Control* 15, 81–87.

Yang, Y.H., Dudoit, S., Luu, P., Lin, D.M., Peng, V., Ngai, J. and Speed, T.P. (2002a) Normalization for cDNA microarray data: a robust composite method addressing single and multiple slide systematic variation. *Nucleic Acids Research* 30, e15.

Yang, Y.H., Buckley, M.J., Dudoit, S. and Speed, T.P. (2002b) Comparison of methods for image analysis on cDNA microarray data. *Journal of Computational and Graphical Statistics* 11, 108–136.

Yu, J., Hu, S., Wang, J., Wong, G.K., Li, S., Liu, B., Deng, Y., Dai, L., Zhou, Y., Zhang, X., Cao, M., Liu, J., Sun, J., Tang, J., Chen, Y., Huang, X., Lin, W., Ye, C., Tong, W., Cong, L., Geng, J., Han, Y., Li, L., Li, W., Hu, G., Huang, X., Li, W., Li, J., Liu, Z., Li, L., Liu, J., Qi, Q., Liu, J., Li, L., Li, T., Wang, X., Lu, H., Wu, T., Zhu, M., Ni, P., Han, H., Dong, W., Ren, X., Feng, X., Cui, P., Li, X., Wang, H., Xu, X., Zhai, W., Xu, Z., Zhang, J., He, S., Zhang, J., Xu, J., Zhang, K., Zheng, X., Dong, J., Zeng, W., Tao, L., Ye, J., Tan, J., Ren, X., Chen, X., He, J., Liu, D., Tian, W., Tian, C., Xia, H., Bao, Q., Li, G., Gao, H., Cao, T., Wang, J., Zhao, W., Li, P., Chen, W., Wang, X., Zhang, Y., Hu, J., Wang, J., Liu, S., Yang, J., Zhang, G., Xiong, Y., Li, Z., Mao, L., Zhou, C., Zhu, Z., Chen, R., Hao, B., Zheng, W., Chen, S., Guo, W., Li, G., Liu, S., Tao, M., Wang, J., Zhu, L., Yuan, L. and Yang, H. (2002) A draft sequence of the rice genome (*Oryza sativa* L. ssp. *indica*). *Science* 296, 79–92.

# 16 Entomopathonic Fungi and the Genomics Era

R.J. St Leger[1] and C. Wang[2]

[1]*Department of Entomology, University of Maryland, College Park, USA;*
[2]*Institute of Plant Physiology and Ecology, Shanghai Institute for Biological Sciences, Chinese Academy of Sciences, Shanghai, China*

## 16.1. Introduction

Insect-pathogenic fungi are key regulatory factors in insect pest populations. Unlike bacteria and viruses that have to be ingested to cause disease, fungi infect insects by direct penetration of the cuticle. They therefore allow microbial control of insects, which feed by sucking plant or animal juices, as well as for the many coleopteran pests that have no known virulent viral or bacterial diseases. Notwithstanding the potential of many fungi as insect control agents, only a handful have been commercialized and most attention has focused on the ascomycetes *Metarhizium anisopliae* and *Beauveria bassiana*. The devastating plagues of locusts in the mid-1980s provided a sense of urgency to insect fungus research in general, and use of *M. anisopliae* in particular, and a requirement for standardized products that advanced our knowledge in a range of areas such as formulation, quality control and storage (St Leger, 2007; Thomas and Read, 2007). Industrial production of *M. anisopliae* is now highly automated allowing *Metarhizium* products to be competitively priced compared with established insecticides (Langewald and Kooyman, 2007; Thomas and Read, 2007).

However, the slow speed of kill and inconsistent results of biologicals in general compared with chemicals has deterred development. An example is the use of *M. anisopliae* to kill adult mosquitoes inside Tanzanian houses as the current protocol only reduces the number of bites fourfold (Scholte *et al.*, 2005). Scholte *et al.* (2004, 2005) do not believe that this would provide adequate protection against disease transmission, but argue that improving infection rates using a more aggressive fungal strain will contribute in a significant and sustainable manner to the control of vector-borne diseases such as malaria, dengue and filariasis. More virulent mosquitocidal strains of *M. anisopliae* might be found by screening wild strains. However, intensive searches in the 1960s and 1970s failed to identify strains active at low doses. This is consistent with intensive efforts to control other pests; historically, fungal pathogens of plant and insect pests have not met expectations because of low virulence (slow kill and high inoculum loads). Gressel *et al.* (2007) suggested that this was because an evolutionary balance has developed between microorganisms and their hosts, even when the biocontrol agent is used at very high levels. Thus, sufficient virulence for cost-effective biocontrol may require transferring genes to the microorganism (Wraight *et al.*, 2001; Gressel *et al.*, 2007). Ultimately, various traits of fungal pathogens, including host range, production capacity, stability and virulence and saprophytic competence may be enhanced through genetic manipulations (Wraight *et al.*, 2001).

Over the past decade, significant progress has been made in uncovering the genes and core signalling pathways regulating infection processes. Earlier, pre-functional genomics work uncovering the genes and core signalling pathways regulating infection processes in *M. anisopliae* is reviewed in St Leger and Screen (2001). The addition and expression of pesticidal genes in *M. anisopliae* is straightforward and can improve pathogen performance (St Leger, 2001). In the first genetically improved entomopathogenic fungus, additional copies of the gene encoding the regulated cuticle degrading protease (*Pr1*) were inserted into the genome of *M. anisopliae* such that the gene was constitutively overexpressed (St Leger *et al.*, 1996a). The toxicity of *Pr1*, expressed in the haemolymph, caused a significant reduction in the time of death of infected lepidopterous larvae and reduced food consumption compared to the wild-type fungus. The development of a native strain of *M. anisopliae* that constitutively expresses a homologous gene should not change host range and is unlikely to raise public concern. However, recent developments include engineering a second generation of transgenic hypervirulent *M. anisopliae* that express much more acute toxins than Pr1A, such as the 70 aa AaIT neurotoxin from the scorpion *Androctonus australis* (Wang and St Leger, 2007a). There is an inherent uncertainty because of the paucity of our knowledge concerning the fate of fungal genotypes at the population and ecosystem level. We made a preliminary attempt to remedy this defect with a field trial with a strain carrying the *gfp* gene as a marker (Hu and St Leger, 2002). Recently, expressed sequence tag (EST) approaches have been used to probe the intimate associations between fungi and their hosts (Freimoser *et al.*, 2003, 2005; Wang *et al.*, 2005a; http:\\TEGR.umd.edu).

This chapter has two aims: first, to provide a brief didactic overview of the current usage of molecular techniques in understanding entomopathogenicity; and second, to provide an introduction to the state of knowledge of the molecular biology relevant to entomopathogenic fungi. Using this information, workers may make an initial judgement as to the applicability of these techniques to their research problems, and potential applied scientists can see which avenues of research might fruitfully be followed. Sambrook *et al.* (2001) and Ausubel *et al.* (2002) have written extensive compilations of standard techniques providing much more detail than we are able to in this overview.

### 16.1.1. The potential of biotechnology

The traditional approach to discovering a biopesticide has included searching natural ecosystems to discover an organism that attacks the target pest. The advanced-engineered biopesticide approach would begin by designing the ideal biocontrol organism using genetic engineering or other techniques (Stowell, 1994). Recently, molecular biology methods have been applied to elucidate pathogenic processes in both *B. bassiana* and *M. anisopliae*. Inherent and developed advantages of working with these fungi include ready cultivability, synchronous germination and growth, gene cloning, EST collections, microarray analysis and quantitative reverse transcriptase-polymerase chain reaction (RT-PCR) (Screen and St Leger, 2000; Wang and St Leger, 2005; Cho *et al.*, 2006), gene knockout (Wang and St Leger, 2006), transformation systems and promoters that allow expression of foreign genes (e.g. Screen *et al.*, 2001) and significant ecological and genetic differences between strains to facilitate comparative studies on life strategies (Bidochka *et al.*, 1994).

### 16.1.2. An introduction to genetic engineering

Genetic engineering may be defined as the genetic alternation of cells or organisms by methods that require the *in vitro* modification of DNA. The cloning, analysis and modification of DNA fragments are accomplished by a small but powerful set of techniques. The field of recombinant DNA began its rapid growth after the purification and characterization of a set of DNA cutting and modifying enzymes. The restriction enzymes are used to cut DNA at specific recognition sites, whereas the enzyme ligase is used to rejoin these DNA fragments. Other enzymes are used to label DNA (e.g. T4 polynucleotide kinase for 5'labelling with radioactive nucleotides), modifying the ends of DNA (e.g. Klenow fragment for removal of 3'protruding ends) or DNA sequencing (e.g. Taq polymerase). The application of these enzymes to practical problems requires the use of a variety of gel electrophoresis methods for the resolution and separation of RNA and DNA molecules. Electrophoresis methods are used for the purification of RNA or DNA molecules, for restriction site mapping, DNA sequencing and many other procedures.

## 16.2. Procedures for Isolating Pathogen DNA

The application of molecular techniques to pathogen genomics often depends on the ability to purify high-molecular-weight DNA. There are various methods of doing this and a correspondingly wide range of commercial kits are available. In addition, the *Fungal Genetics Newsletter* has many papers describing improved methods for purifying small or large amounts of DNA.

Two major factors influencing satisfactory purification of high-molecular-weight DNA are shearing and the presence of nucleases. To avoid shear forces, the lysate (containing DNA) should be treated gently and pipetting with narrow pipette tips should be avoided. To prevent nuclease activity, the cells can be frozen (preferably under liquid nitrogen) and should only be thawed in the extraction buffer. The DNeasy Plant Mini kit (Qiagen) allows rapid and efficient isolation of high-quality DNA from fungi, but is expensive. Some specific methods are described below.

### 16.2.1. Preparation of very high-quality DNA from protoplasts (after Bidochka *et al.*, 1994)

1. Prepare protoplasts by shaking mycelia with 0.8% Novozyme 234 (complex enzyme mixture) in 1.2 M sorbitol, 10 mM Tris-HC1 (pH 7.5) for up to 3 h. The protoplasts are filtered through two layers of sterile cheese cloth and centrifuged.
2. Burst protoplasts by resuspending them in 10 mM Tris, 2% sodium dodecyl sulfate (SDS), 1 mM ethylenediaminetetraacetic acid (EDTA) and treat sequentially with RNase A and proteinase K.
3. Extract proteins with phenol/chloroform, precipitate DNA with 1/10 volume 3 M sodium acetate and 1 volume isopropanol. Purified DNA is washed sequentially with ice-cold 100% and 80% ethanol and air-dried or dried under a speed vac.

### 16.2.2. Preparation from mycelia

Genomic DNA of lower quality can be obtained by lyophilizing mycelia and homogenizing under liquid nitrogen (St Leger *et al.*, 1992). The homogenate is further processed as described above. Homogenization of the mycelia eliminates the protoplast formation step, but is still laborious, particularly when DNA is being isolated from many fungal samples. Using a FastPrep FB120 (ThermoSavant) according to the manufacturers instructions provides good results extracting nucleic acids from soil, infected insects or mycelia.

### 16.2.3. High-throughput extraction of DNA from mycelium

A transfer DNA (T-DNA) insertion library containing thousands of mutants has been constructed (Section 16.7.6.1). In order to get the T-DNA flanking sequences from a large number of mutants quickly, we developed a simple method to isolate genomic DNA that avoids using liquid nitrogen and lyophilization. Sufficient DNA can be obtained by this method for regular polymerase chain reaction (PCR) and PCR walking (Section 16.7.5.1), but not for Southern blot analysis.

1. Inoculate conidia into 1 ml of sabouraud dextrose broth (SDB) and incubate at 27°C/250 rpm for 2–3 days.
2. Collect the mycelium by centrifugation and break up the mycelial pellets using a glass rod.
Immediately add 400 µl of fungal DNA extraction buffer (0.2 M Tris-Cl (pH 7.5) + 0.5 M NaCl + 10 mM EDTA (pH 8.0) + 1% (w/v) SDS), vortex and add 400 µl of phenol–chloroform–isoamylalcohol (25:24:1).
3. Vortex for 5 min, centrifuge at RT/10,000 rpm for 8 min and transfer the aqueous top layer to a new microcentrifuge tube, being careful not to transfer the middle pellet layer. For PCR walking proceed to step 4. For regular PCR, go to step 8.
4. Add 1 µl of RNase solution (20 mg/ml, Qiagen) and mix well. Incubate at 37°C for 30 min.
5. Add 500 µl phenol–chloroform–isoamylalcohol and mix well.

6. Centrifuge at RT/10,000 rpm for 8 min.
7. Transfer the aqueous top layer to a new microcentrifuge tube.
8. Add 2.5 volume of 100% ethanol and gently mix by inverting five to six times.
9. Incubate at room temperature (RT) for 2 min.
10. Centrifuge at RT/12,000 rpm for 10 min.
11. Pour of the ethanol. Add 200 µl pre-chilled 70% ethanol.
12. Centrifuge at RT/12,000 rpm for 2 min.
13. Decant the tube and dry pellet (30 min). Add 50 µl of $ddH_2O$ to dissolve the DNA.

### 16.2.4. Extraction of DNA from an infected insect

Identifying the presence of a particular pathogen genotype using specific DNA probes is greatly facilitated by the ability to extract fungal DNA from infected insects or soil. A method to extract DNA from dead grasshoppers (Bidochka *et al.*, 1995) appears to be broadly applicable.

1. Freeze-dry each soil, insect or plant sample, grind under liquid nitrogen (or employ the FastPrep FB120), extract with phenol–chloroform and precipitate DNA with 0.3 M sodium acetate/isopropanol.
2. Remove polysaccharides and polyphenols, which co-precipitate with DNA by selective precipitation of DNA with polyethylene glycol (PEG) (Rowland and Nguyen, 1993).
3. Blot samples on to nitrocellulose following standard protocols (Sambrook *et al.*, 2001).
4. Incubate the blots in laundry detergent (LaFrance, Dial Corporation) to remove contaminating proteins. PCR techniques can be used to improve the sensitivity of DNA blotting as long as the DNA is separated from the polyphenols (Bidochka *et al.*, 1994).

## 16.3. Procedures for Isolating Pathogen RNA

Clean RNA is a fundamental requirement in molecular biology, particularly in the gene cloning processes and in analysing gene expression. It is absolutely crucial that precautions are taken to minimize RNase contamination (e.g. the use of diethyl pyrocarbonate (DEPC)-treated deionized water, baking of glassware at 149°C for 4–6 h). Whenever possible use sterile, disposable plastic ware as this is free of RNases and can be used for the preparation and storage of RNA without any pretreatments. All the work should be done in a flow hood, the surface of which has been treated with RNase-inactivating agents such as RNase Away (Molecular Bio-products, CA). There are many available kits for purifying RNA and they use similar methodologies. The RNeasy Plant Mini Kit (Qiagen) can extract sufficient high-quality RNA from 0.1 to 0.5 g mycelia for microarray hybridization. The mycelia are transferred into a 2 ml screw cap tube charged with *c.*0.5 g 1.0 mm silica beads (BioSpec Products, Inc.). After flash-freezing in liquid nitrogen, the mycelia are broken using a FastPrep

FB120 (ThermoSavant). The RNA is extracted using the buffers and the columns provided in the kit according to the manufacturers instructions.

## 16.4. Pulsed Field Gel Electrophoresis

The utility of conventional agarose gel electrophoresis drops sharply for DNA molecules >25 kb, as electrophoretic mobilities become increasingly independent of molecular size. Pulse field gel electrophoresis (PFGE) has been developed to overcome the constraint of the mobility of larger DNA molecules. In PFGE, molecules are subjected to electric fields applied alternatively in two different directions. It has been suggested that separation of molecules is achieved because in order to migrate, the molecules must first reorient themselves in response to each changing orientation of the electric field; the time taken by the molecules to reorient themselves depends on the molecular size (Bustamante *et al.*, 1993). Most of the PFGE protocols for electrophoretic karyotyping have been developed empirically. The counter-clamped homogeneous electric field (CHEF) PFGE system has been used to determine the size and number of chromosomes of certain entomopathogenic fungi (Shimizu *et al.*, 1993; Wang *et al.*, 2003). This method has also been used to compare the electrophoretic karyotypes of various entomopathogenic fungi. As such, it offers a major advance over the traditional time-consuming strategy of linkage analysis. For example, densitometric analysis of PFGE gels suggested that three Brazilian strains of *M. anisopliae* possess eight chromosomes, with two chromosomes migrating as doublets under the electrophoretic conditions used. The genome size was estimated as varying between 23.39 and 31.88 Mb, not including possible doublet chromosomes (Valadares-Inglis and Peberdy, 1998).

Following electrophoresis, gels can be blotted and hybridized successively to a series of genes used as probes to establish their chromosomal location. These techniques have great potential for genome mapping and elucidating the nature of the variation between the races of a pathogen.

Bio-Rad has new technology – the CHEF Mapper XA system. This is the most flexible of their pulsed field gel electrophoresis units. The special features of this unit should give greater speed of separation, greater accuracy and higher resolution than with other systems. To date researchers on fungi have employed Bio-Rad's CHEF Drive II and III systems, which are still available. The protocol using the Drive-III is given here but would apply equally well with other Bio-Rad machines.

### 16.4.1. Preparation of protoplasts

Protoplasts are prepared from young mycelia growing in liquid shake cultures. The mycelia are centrifuged and washed with osmotic stabilizer containing varying amounts of sorbitol NaCl/$MgSO_4$. Novozyme 234 (a complex enzyme system) is added to the osmoticum and incubated at 30°C with gentle shaking for 60–90 min. This usually yields $10^7$–$10^8$ protoplasts/ml. The protoplasts are filtered through two layers of sterile cheese cloth, centrifuged and then resuspended and washed three times with sorbitol, Tris-HC1 and calcium chloride (STC) solution.

### 16.4.2. Preparation of agar plugs

Each protoplast suspension is equilibrated at 42°C and added to an equal amount of molten 1.4% low-melting agarose (PFGE grade agarose, Bio-Rad) in a protoplast mould. The protoplast plugs are incubated overnight at 50°C with proteinase K solution to dissolve the protoplast membrane and release nuclear contents. The plugs are then washed with EDTA to remove excess proteinase K and can be stored for up to 4 months at 4°C before use.

### 16.4.3. Run conditions of CHEF gel

CHEF analysis is carried out using Bio-Rad's CHEF Drive III. The gel is cast and loaded as described in the manufacturer's manual. Different voltage, switching intervals and total run-time conditions were tried. Consecutive 1800 and 2500 s switching time intervals, 1.8–2.5 V/cm, 105–120° angle and total run-time of 72–96 h gave reasonable separation of chromosomes from different strains of *M. anisopliae*.

## 16.5. Construction of Cloning and Expression Vector Components and Markers

The availability of different gene transfer systems with different characteristics permits a molecular genetic study of many biologically interesting processes by isolation, characterization and functional analysis of the genes and gene products involved. To perform these studies, specific vectors are constructed which facilitate genetic manipulation such as cloning of a gene, gene disruption or gene replacement. Although there already exists a wide selection of filamentous fungi vectors encoding marker proteins for vector studies, it may be that the exact vector for a particular purpose has not yet been made and so a tailor-made vector has to be constructed. Perhaps, for example, this will involve introducing a gene that encodes a toxin or enzyme, or a new marker or promoter. Thus, besides knowing how to employ existing vectors in transformation systems, it is necessary to know how to make new ones when circumstances dictate. Here, we will consider the options available for making such constructions along with the circumstances under which they may be employed. To illustrate the possibilities of genetic manipulation for molecular genetic studies examples will be given of research on *Metarhizium* that is progressing in our laboratory.

### 16.5.1. Vector construction

Basic vector construction starts with a precursor plasmid, the new DNA component to be ligated into it and a procedure to identify the new construction. The major procedures involved in this process are the isolation of specific DNA fragments from gels and the cloning of them into the burgeoning vector construction.

### 16.5.2. Isolation of DNA fragments from gels

For agarose gels, Tris-acetate electrophoresis buffer is used, the gel is stained with ethidium bromide and the DNA bands are viewed under UV illumination in the standard way (Sambrook *et al.*, 2001). Using a scalpel, the appropriate fragment is then excised in a minimum-sized gel slice, and this is then processed to extract the DNA. There are a number of ways to do this. The methods that work well include the QIAquick Gel Extraction Kit from QIAGEN and low-melt agarose gel electrophoresis. We recommend the QIAquick kit because of its ease of use and low cost.

#### *16.5.2.1. Cloning manipulations*

It is important to set up the cloning ligation with the correct concentration, and relative molar ratio of vector and insert fragment. Methods used to calculate these conditions have been described elsewhere (Ochman *et al.*, 1990; Sambrook *et al.*, 2001). After selecting the appropriate concentration conditions for the ligation, it can be set up and undertaken as described in Ausubel *et al.* (2002) using T4 DNA ligase (5–7 Weiss units), ligation buffer and vector and fragment DNA. The Quick ligase from NEB Biolabs gives high efficiency of ligation in just 5 min at RT.

#### *16.5.2.2. Modifying plasmid vectors*

There is a wide choice of possible promoters to drive heterologous gene expression (reviewed by Davies, 1992). Of those commonly used, the phosphoglycerate kinase (PGK), alcohol dehydrogenase 1 (ADH1) and glyceraldehyde 3-phosphate dehydrogenase (GAPDH) promoters generally give a good level of constitutive expression. The available expression vectors, e.g. pBARGPE1 or pBARMTE1 from Fungal Genetics Stock Center, use the *Aspergilluse gpdA* or *Neurospora mtr* promoter. Until recently, the plasmid vectors have not had homologous counterparts in *M. anisopliae* and therefore constituted a target for improvement to optimize DNA delivery systems. Nakazato *et al.* (2006) generated a plasmid pMaTEFGFPBAR from the well-known pAN5.2 that contained the *gfp* reporter gene under control of the promoter from *Metarhizium* translation elongation factor-1α. Of the regulated promoters, the GAL series are very tightly repressed by glucose and induced by galactose. Expression signals of the glucamylase (GLA) gene are induced by starch (Davies, 1992). The inducible *Neurospora crassa* quinic acid (QA) promoter has been used to drive expression of genes of the plant pathogen *Colletotrichum trifolli* and may have broad applicability. We identified a collagen-like protein MCL1 exclusively expressed in insect haemolymph *in vivo* or *in vitro* (Wang and St Leger, 2006). The collagen protein is laborious to assay. To test its promoter in different environmental conditions with a more easily assayed protein, a region of *c.*2 kb upstream from the start codon ATG was amplified and inserted into the plasmid pGPS3Bar to generate pPCBar. The *gfp* gene amplified from pEGFP (Clonetech) was integrated into pPCBar using a PCR-Fusion kit (Clontech), resulting in the plasmid pPCGFPBar (Fig. 16.1A). Following transformation, the fluorescent signal was only detectable from *Metarhizium* hyphal bodies or the fungus grown in haemolymph *in vitro* (Fig. 16.1B). The *Mcl1* promoter is currently being used for targeted expression of toxin genes into the haemolymph (Wang and St Leger, 2007a).

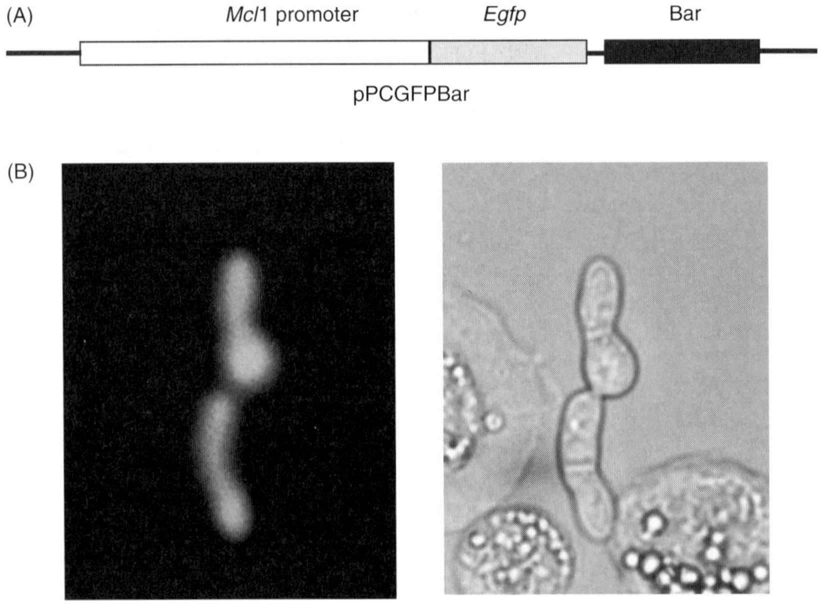


**Fig. 16.1.** Construction of a *gfp* expression vector under the control of *Mcl1* promoter. (A) Schematic map of the vector. (B) GFP signal detected in *Metarhizium* hyphal bodies (left) and the same cells shown with bright-field microscopy (right).

#### *16.5.2.3. Fusion construct*

Reporters such as green fluorescent protein (GFP) can also be used to track the physical location of a protein by fusing the coding regions of the two genes together. Depending on the functional domain of the target protein, the reporter gene can be fused with either the N- or C-terminus of the target protein. There are different commercial vectors available from Clontech or Invitrogen that can be used to express genes from pathogens in model organisms. We have used Invitrogen pYes2 series vectors to perform functional studies of *Metarhizium* genes in yeast, i.e. free of other *Metarhizium* proteins. The vectors can be propagated in *Escherichia coli* and transformed into yeast *Saccharomyces cerevisiae* under the control of *GAL1* promoter. To generate a fusion construct, the stop codon has to be omitted from the target gene for C-terminal tagging or from the reporter gene itself for N-terminal tagging. As an example of this approach, we successfully localized a *Metarhizium* MPL1 protein on lipid droplets by N-terminal fusion with GFP (Fig. 16.2) (Wang and St Leger, 2007c).

### 16.5.3. Selective markers for transformation

#### *16.5.3.1. Nutritional selective markers*

The first group of selective markers, and the most widely used in the transformation of *Aspergillus nidulans*, comprises the nutritional selective markers.

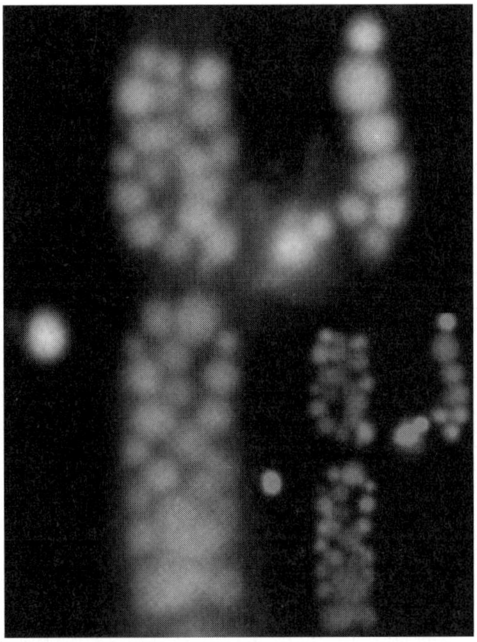

**Fig. 16.2.** GFP-MPL1 fusion to localize MPL1 distribution. Left, GFP labelled MPL1 protein on lipid droplids; Right, nile red staining of neutral lipid.

Transformation with nutritional markers is based on having auxotrophic mutant strains that can be transformed to prototrophy for the selective marker being used. Mutants in general and auxotrophs in particular will be difficult to obtain in many fastidious entomopathogens of economic importance. Although stable auxotrophs are readily obtained by chemical mutagenesis of *M. anisopliae* (Al-Aidroos, 1980), this raises the possibility of additional alterations in the genome which could influence virulence. It is possible to select for spontaneous mutants in the *niaD* and *pyrC* genes thus eliminating this potential problem and providing an attractive system in genetically poorly characterized species. In the case of *pyrC* they can be selected by resistance to 5-fluoro-orotic acid (Benito *et al.*, 1992) and in the case of *niaD* by resistance against chlorate (Unkles *et al.*, 1989). Such an approach was applied to *B. bassiana* using the *niaD* gene for nitrate non-utilization, but the high level of natural resistance to chlorate shown by the wild-type fungus will probably limit the applicability of this technique. This is a pity, since it is possible to select both for and against the mutant and wild-type phenotype. These markers are also particularly useful for genetic manipulation strategies, such as gene disruption studies. In short, if a mutation in the fungus has been identified and a heterologous gene is available from another fungus, then the likelihood is that a transformation system can be developed. If an appropriate mutation is not evident then it is probably not worth the time it takes to isolate one. The broad host range of fungal promoters will probably allow employment of one of the already existing dominant markers for experiments.

#### *16.5.3.2. Dominant selective strategy*

The most straightforward and industrially available approach for transformation is to develop a dominant selective strategy. This involves transforming an existing wild-type cell, such as one sensitive to a drug, with a cloned gene that is selectable in that cell, such as a gene for drug resistance. The only requirement is that the gene should be dominant or semi-dominant. A gene can be used to transform cells even if the gene has no eukaryotic origin of DNA replication. This is possible because the transforming DNA can integrate into the chromosomal DNA of the recipient cell, either by homologous recombination at the locus of the resident gene or by non-homologous recombination elsewhere in the genome (May, 1992). The dominant selective markers used to transform many filamentous fungi are resistance genes for antibiotics (oligomycin, bleomycin, G418, hygromycin and phleomycin) and mutant β-tubulin genes that give resistance to the antimicrotubule compound benomyl. Unfortunately, except for *Verticillium lecanii* which is susceptible to hygromycin, most strains of most entomopathogens are naturally resistant to antibiotics and until recently could only be transformed to benomyl resistance. However, the Ignite/basta-resistance (*bar*) gene originally from *Streptomyces hygroscopicus* provides a novel compact fungal selectable marker (Avalos *et al.*, 1989). It has been successfully used to transform *M. anisopliae* (Screen *et al.*, 2001) and *B. bassiana* (Fang *et al.*, 2004) to phosphinothricin/glufosinate resistance.

## 16.6. Transformation Systems

Techniques developed to transform filamentous fungi have followed the same three conceptual steps applied to all organisms. The first step is the preparation of 'competent' cells (i.e. cells capable of taking up foreign DNA). The second step is inducing the cells to take up the transforming DNA and the third step is applying a selective pressure to the cells so that only those cells which take up and express the DNA (i.e. transformants) are capable of growing. The first transformation systems for *M. anisopliae* (Bernier *et al.*, 1988; Goettel *et al.*, 1990) were based on the original work of Tilburn *et al.* (1983) with *A. nidulans*. Competence is achieved in part by digesting away the cell wall (which is a barrier to DNA entry) from log phase mycelium using a complex enzyme mixture such as Novozyme 234. Transformation is achieved by mixing osmotically stabilized protoplasts (in 0.6 M KCl or 1.2 M sorbitol) with DNA in the presence of buffered 10–50 mM $CaCl_2$ and polyethylene–glycol which induces membrane fusions allowing $CaCl_2$ precipitated DNA to be internalized. The following generalizations can be made about the results. Transforming DNA is integrated into the genome, frequently in multiple copies, sometimes at a homologous locus, and most systems rely on the expression of a foreign gene for the selection of transformants. Filamentous fungi are often very permissive with respect to the expression of foreign genes (Covert and Cullen, 1992) and this is also demonstrated by *M. anisopliae*, for example, the utilization of the benomyl-resistance gene of *N. crassa* to transform *M. anisopliae* to benomyl resistance. Also, promoters which function in one filamentous fungus can often be used to confer expression of a molecular gene in other genera or species. For example, the glyceraldehyde phosphate dehydrogenase promoter of *A. nidulans* can drive expression of the bacterial beta-glucuronidase (GUS) gene or the *Prl* gene in *M. anisopliae* (St Leger *et al.*, 1996a). This has important consequences, not only for the development of transformation systems, but also for our ability to produce improved pathogens secreting foreign proteins and probably reflects the genetic relatedness of Ascomycete fungi.

### 16.6.1. Problems with transformation systems for entomopathogenic fungi

Genetic studies of insect-pathogens employing these conventional protoplast fusion methods were traditionally hampered by low transformation frequencies. Transformation frequencies utilizing tradition $Ca^{2+}$/PEG technology are usually tenfold to 100-fold less than the $10^3$/μg DNA routinely achieved in *A. nidulans*. Furthermore, the efficiencies of protoplast-mediated procedures vary between strains of *M. anisopliae* and *V. lecanii*, indicating that time-consuming optimization of conditions may be required for each strain. Also, the stability of transforming DNA in the genome can vary markedly between strains. For example, the vector, pNOM-102, containing the GUS expression gene stably transforms some, but not all, strains of *M. anisopliae* and many strains exhibit transient expression. Several alternative methods to $Ca^{2+}$/PEG such as electroporation and biolistic transformation have been used to deliver DNA into cells (St Leger *et al.*, 1995). Compared to the $Ca^{2+}$/PEG method, no significant improvements of transformation frequency were observed but these techniques are less laborious. The biolistic process was comparatively straightforward and enabled co-transformations of *M. anisopliae* and *V. lecanii*, albeit at comparatively lower rates compared to the protoplast method. This technique has some promise for routine transformations of entomopathogens, particularly as conditions optimized for *M. anisopliae* worked almost equally well for *V. lecanii*.

At least for yeasts, LiAc/PEG-mediated transformation is highly efficient and simple. A recent study employed this method with blastospores of *B. bassiana* achieving a respectable 24 transformants per microgram DNA (Ying and Feng, 2006). The advantages of using blastospores include easy handling, unicellularity as well as a single nucleus per cell. It would be very useful if this technique was broadly applicable to the many fungi that can produce large amount of unicellular blastospores in liquid cultures. However, our own unpublished trial applying this technique to *M. anisopliae* blastospores resulted in <1 transformant per microgram DNA.

Perhaps, the most promising new method has adapted *Agrobacterium*-mediated transformation (AMT) to *B. bassiana* (Fang *et al.*, 2004; Leclerque *et al.*, 2004) and *M. anisopliae* (Fang *et al.*, 2006). It has also been used to generate insertional mutants of *M. anisopliae* and will be considered further in Section 16.7.6.1.

### 16.6.2. Protocol modifications to electroporation and biolistic transformation procedures for entomopathogenic fungi

For both electroporation and biolistic transformation, regeneration media (RM) consist of 1% potato dextrose broth (Difco), 0.2% yeast extract, 0.5 M $MgSO_4$ (or

1.2 M sorbitol) and 0.3% $NaNO_3$. RM is solidified with 1.2% noble agar (RA) or 1.2% sea plaque low-gelling-temperature agarose (RLGA). For selection of benomyl-resistant (BenR) transformants, RLGA plates are overlaid with molten (42°C) RLGA containing 5 μg/ml benomyl (Sigma) from a stock solution of 5 mg/ml in dimethyl sulfoxide.

#### *16.6.2.1. Transformation by electroporation*

Published protocols were optimized for *M. anisopliae* (St Leger *et al.*, 1995). Using this method with some combinations of fungal strains and vectors gives high levels of transformants, although many of them are unstable.

1. Germinate 10-day-old conidia for up to 9 h in 0.2% yeast extract at 27°C with shaking.
2. When germ-tubes begin to emerge add 1 mg/ml crude cellulase (Sigma, type C-1184) and 1 mg/ml glucuronidase (Sigma, type H1) and resume incubation for up to 2 h.
3. Centrifuge germlings before they produce protoplasts, wash twice in electroporation buffer (1 mM HEPES, 50 mM mannitol, pH 7.5, 1% (wt/vol) PEG 6000 containing 10 μg/ml aurin tricarboxylic acid and resuspend in electroporation buffer at approximately $5 \times 10^7$ spores per millilitre.
4. Gently mix aliquots of 100 μl with 10 μg plasmid DNA and keep on ice for 15 min.
5. Deliver high-voltage 5 ms pulses (25 μFD capacitor and a field strength of 12.5 kV/cm) to samples in disposable cuvettes (Bio-Rad Laboratories: inter-electrode distance of 0.2 cm) by using a gene pulse apparatus (Bio-Rad Laboratories).
6. Immediately following electroporation mix the spore suspension with 1 ml of potato dextrose broth (PDB) plus 0.5 M $MgSO_4$ and incubate at 25°C for 3 h.
7. Plate aliquots containing $2 \times 10^5$ germlings on to 10 ml PDA and incubate for 10 h before selection using an RLGA overlay containing 5 μg/ml benomyl. Transfer colonies arising on transformation plates to PDA containing benomyl and X-gluc (20 μg/ml).

#### *16.6.2.2. Transformation by particle bombardment*

High-velocity ballistic transformation of *M. anisopliae* germlings can be performed with the DuPont particle delivery system PDS 1000/HE by modifying methods used with other filamentous fungi as described (St Leger *et al.*, 1995).

1. Mix 25 μl of a tungsten (MIO particles, mean diameter 1 μm) suspension (60 mg/ml) with 2.5 μg (2 μg/μl) of plasmid DNA, 25 μl of 2.5 M $CaCl_2$ and 10 μl of 0.1 M spermidine.
2. Vortex the particle-DNA mixture at 4°C for 15 min and incubate for 10 min on ice. Collect the particles by centrifugation and wash sequentially with water, 70% ethanol and absolute ethanol.
3. Resuspend the particles in 20 μl of ethanol by sonication in a sonicator water bath (Branson 100) for 5 s.
4. Spread 6 μl of coated particles on a kapton flying disc and use for bombardment.

5. Propel the particles towards the germinating conidia by release of helium at 1200 psi. Before DNA is introduced into conida, they should be germinated for 9 h at 27°C with shaking (100 rpm) in 0.2% yeast extract. For transformation, spread germlings on to RA containing 0.1% Igepal (Alltech Associates, Inc.). Once seeded, place the plates into the chamber of the particle delivery system at a distance of 14 cm from the launch site. Bombard the germlings once or twice with tungsten particles coated with mixes of the plasmid DNAs. Selection for benomyl resistance can be achieved with an overlay applied 10 h after bombardment.

## 16.7. Gene Cloning Strategies

### 16.7.1. Identifying pathogenicity-related genes

The characteristics of a disease may be defined differently at the molecular level, the cellular level, the whole-insect pathogen level and the population level. Thus, the processes and gene products which are involved in the ability of the fungus to cause disease will be many and varied and the insect pathologist could be interested in several broad classes of pathogenicity genes. Some genes will encode receptors that detect either directly or indirectly the presence of the host (e.g. a GTP-regulated adenylate cyclase, tyrosine protein kinases, serine and threonine protein kinases and phosphoprotein phosphatase (St Leger, 1993)). These act to change second messenger levels or are themselves activated by second messengers to trigger differentiation (St Leger, 1993). Activation of such receptors and signal transduction pathways may result in the induction of the generic pathogenicity genes. Another class of pathogenicity genes may inactivate host defences. Other pathogenicity genes may encode toxins that are required for disease symptoms, e.g. destruxins and hydrophobins. A fourth category of pathogenicity genes encode enzymes that allow the fungus to overcome host barriers. Much of our research has concentrated on these genes. A fifth category of pathogenicity genes may be a heterogeneous group. To determine whether such pathogenicity genes exist and what are the characteristics of each class, it will probably be necessary to characterize multiple genes conferring pathogenicity to diverse hosts, preferably from several pathogen species. Several approaches have been used to isolate genes from insect pathogens. One can broadly distinguish between cloning strategies involved when the protein product of the gene to be cloned is known and identified biochemically as playing a role in disease, and isolation of genes with unknown products. The latter approach can identify whole new, previously unsuspected stratagems of pathogen attack. In cases where there is evidence that a messenger RNA (mRNA) of interest is transcribed at a particular stage of development, or in response to a specific induction signal, it is possible to take advantage of differential gene expression to isolate a complementary DNA (cDNA) clone. Three approaches have been successfully applied to *M. anisopliae*, differential hybridization, differential display (DD) and EST screening. The last two are currently employed and will be described here, along with subtractive hybridization.

### 16.7.2. Isolation of genes from libraries

Isolation of single-copy genes from a complex genome and either isolation of cDNA clones or EST analysis of a complex mRNA population has traditionally required the generation of recombinant DNA libraries which contain complete representation of genomic or cDNA sequences. Consequently, the main problem in generating a useful library is the creation of the huge population of clones and the solutions to this problem are essentially similar for genomic and cDNA libraries. The genomic DNA or cDNA is first prepared for insertion into the host vector. The vector and target DNA are then ligated and introduced into *E. coli* by packaging into phage lambda heads *in vitro*. These procedures are described in detail in Sambrook *et al.* (2001) and Ausubel *et al.* (2002) but much the easiest way to get a library is to get one from someone else; several laboratories have genomic or cDNA libraries for important entomopathogens. Alternatively, use one of the comprehensive kits available from Stratagene or Clontech. It takes about 2 weeks to prepare a library using a kit. These companies will also make a custom library from DNA and total or poly(A) RNA provided to them. Stratagene has the greater choice of vectors including the very useful unidirectional ZAP vectors, but Clontech is generally less expensive.

### 16.7.3. Heterologous probing

A simple method to clone an individual gene from a library is to 'clone by phone', that is, to use a gene obtained from one organism as a heterologous probe to pull out the related gene from a DNA library of a second organism. This requires that the nucleotide sequences of the equivalent genes be similar which is more likely if the two organisms are related. Thus, the *Pr1* protease gene from *M. anisopliae* was used to clone the corresponding protease gene from *B. bassiana* using relaxed hybridization and washing conditions (Joshi *et al.*, 1995), but it is less likely that the equivalent bacterial gene could have been employed in the same way. Some genes are much more strongly conserved than others over broad taxonomic distances, e.g. heat-shock genes, ribosomal RNA, histone and tubulin genes and genes involved in signal transduction (e.g. calmodulin and some kinases) being good examples. Probing a library with a heterologous probe is essentially standard (Sambrook *et al.*, 2001). We recommend using a random primer labelling method for the probe. The BRL random primers labelling system provides a very convenient way of labelling nucleic acids to high specific radio activity and can be easily adapted to label DNA in agarose gel.

### 16.7.4. Antibody detection

Several screening methods that rely on the expression of the cloned gene have been devised. Most of these use antibodies against the gene product of expression. Expression is obtained by cloning cDNA (to avoid the problem of introns) in expression vectors so that they are expressed as galactosidase fusion proteins. To screen clones for expression, protein from plaques is fixed to a nitrocellulose filter, and the filter is treated first with antibody, and then with a labelled second antibody to detect the binding of the first antibody. Again, several companies (e.g. Amersham,

Promega) provide kits containing secondary antibodies (usually labelled with alkaline phosphatase or horseradish peroxidase) and full instructions on their use. The most efficient and easily screened vectors are those that allow directional cDNA cloning (e.g. the Uni-ZAP XR or lambda ZAP express unidirectional vectors from Stratagene) as these double the number of clones detected by antibody screening. The primary prerequisite for this technique is a protein of sufficient purity to generate antibodies. The most convenient way to separate a mix of polypeptides is by sodium dodecyl sulfate polyacrylamide gel electrophoresis (SDS-PAGE). The Coomassie blue-stained band of interest can then be cut out and injected directly into a rabbit. It is also possible to electroblot the protein on to immobilon-Pm (Millipore, USA) membranes, quick-stain with Coomassie blue and use the band to obtain sequencing data directly with an automated sequencer (St Leger *et al.*, 1996b).

### 16.7.5. The polymerase chain reaction

Heterologous probing has been largely superseded by PCR-based protocols. PCR is used in gene cloning, identification of pathogens, genetic analysis and genetic engineering. PCR is artificial amplification of a DNA sequence by repeated cycles of replication and strand separation. It requires a DNA thermocycler and two PCR primer oligonucleotides which are short (typically 17–25 nucleotides) pieces of single-stranded DNA that match the sequences at either end of the target sequence in the template DNA. When PCR works correctly, it gives rise to a single DNA band visible on a gel. However, if the primers are not specific enough, several bands may be visible. With luck, these may be eliminated by optimizing the $Mg^{2+}$ concentration (a kit sold by Invitrogen is very useful for this purpose) or increasing the annealing temperature. For the 17-mer universal primers that recognize sequences in the pUC plasmids (common cloning plasmids) on either side of the site where the target DNA will be inserted, typical PCR conditions are: denature at 94°C for 20 s, anneal at 55°C for 20 s and extend at 72°C for 30 s for a total of 25–30 cycles. It is usually best to start with a modest concentration of $MgCl_2$, such as 1.5 mM, and then alter if non-specific primings occur. Optimization of PCR has been reviewed by Sambrook *et al.* (2001). If using a 'Universal' primer, these are available from Pharmacia, Gibco- BRL, New England Biolabs and others.

The major snag with PCR is that you need to know the sequence, or at least the ends of the sequence, in order to make the PCR primers. There are several ways to deal with this but in each case it is necessary to know the sequence of something to get started. One procedure is to use degenerate primers. If the protein product of the gene is known it can be sequenced, and the genetic code can be deduced from the sequence. The sequence generally contains many ambiguities, because of the degeneracy of the genetic code. The least ambiguous region of 11–20 bases is chosen and a synthetic oligonucleotide (or pool of related oligonucleotides to increase the possibility that at least a portion of the probe will match the DNA sequence) is then chemically synthesized (many companies with automated oligonucleotide synthesizers make primers as a service). A similar strategy can be used if DNA sequence data are available from a related organism. The big disadvantage of this technique is that many incorrect clones may hybridize. To some extent this problem may be overcome by the use of carefully controlled hybridization conditions.

Once DNA has been obtained by PCR it can be sequenced or cloned. To clone something into a plasmid requires restriction sites. These can be obtained by adding restriction enzyme cut sites at the far ends of the primers. As long as the primer has enough bases at the 3' end to match the target site, adding a few bases at the 5' end will not affect the reaction. The bases making up the restriction site will be copied and appear at the ends of the newly made DNA. The PCR product and the plasmid can then be treated with the same restriction enzyme to generate sticky ends for cloning.

A large number of modifications have been made to the basic PCR scheme. Regarding the use of PCR in genetic engineering most modifications fall into two main categories: (i) changing one or two bases of a DNA sequence; and (ii) rearranging large stretches of DNA.

Mutating a single base simply involves making a PCR primer with the desired base alteration. As long as the primer is long enough to bind to the correct location on either side of the mutation, the DNA product will incorporate the change made in the primer (Fan *et al.*, 2007). Large rearrangements could involve making a hybrid gene, for example, to join the chitin-binding domain of a chitinase from the silkworm to the front half of the chitinase Bbchit1 from *B. bassiana*. This involved using an overlap primer that matches part of both DNA segments. The PCR reaction was run using a primer for the front end of the chitin-binding domain, a primer for the rear end of Bbchit1, and the overlap primer to produce a hybrid gene. Some variations of this procedure make the two halves separately and ligate them together (Fan *et al.*, 2007).

It is possible by using RT-PCR to obtain PCR products from mRNA. This can be useful, for example, if you wish to obtain coding sequences that can be expressed in bacteria because they are not interrupted by introns. This procedure, commonly referred to as RT-PCR, uses the enzyme RT to convert mRNA to single-strand cDNA and then PCR is performed on the DNA. RT-PCR has also largely replaced Northern blot analysis for semi-quantitative measurement of gene expression. RT-PCR is used on mRNA isolated from a cell using PCR primers specific for a particular gene. If the gene is expressed, then a PCR band will be detected on a gel. RT-PCR can be applied to an organism growing under different conditions or different developmental stages to determine which factors bring about expression of the target gene.

Real-time PCR is a powerful tool to quantify gene expression level or gene copy number (see Wong and Medrano, 2005, for review). This procedure uses real-time PCR machines and fluorescent probes that allow an increase in the amount of DNA during the PCR reaction to be followed by watching an increase in fluorescence. The reactions are run in glass capillary tubes to allow light to excite the dye. The most commonly used dyes are SYBR Green and SYBR red as their fluorescence increase when they bind to DNA. These dyes do not recognize any specific sequence so they just measure total DNA. Fluorescent probes can be made that are specific for a particular sequence of DNA. They consist of short single-stranded DNA probes attached to a fluorescent dye. Only when the DNA probe binds to the correct target DNA does the fluorescence increase. This approach can be used to rapidly identify the pathogen causing a disease. Instruments such as Stratagene MX4000 monitor four different fluorescent dyes at once, which allows four diagnostic reactions to be run in the same tube.

### *16.7.5.1. Cloning of flanking sequences of known DNA sequence*

To generate DNA by PCR, regions of known sequence on either side of the unknown target sequence are needed to allow primer binding. In contrast, if a sequence of a long DNA molecule is already known it is possible to explore either side of this molecule into the unknown flanking regions upstream and downstream by a variety of techniques collectively known as PCR walking. These techniques are important in cloning the full coding sequence of a gene, identifying regulatory sequences and analysing DNA insertion mutants. The three methods currently employed are: (i) inverse PCR; (ii) adaptor ligation PCR; and (iii) random primer PCR. We are currently using an adaptor ligation PCR walking technique called YADE (Y-shaped adaptor dependent extension) (Xiao *et al.*, 2002) (Fig. 16.3). The advantages of YADE over the other techniques are: (i) the Y-shaped adaptor decreases non-specific amplification as compared to other adaptor-based PCR walking strategies; and (ii) it is much cheaper than the commercially available PCR walking kit. This is particularly important when a large amount of unknown sequences need to be cloned and combined with the rapid small-scale DNA extraction method (Section 16.2.3). YADE allows high-throughput cloning of T-DNA insertion mutants. Typically, we go from DNA extraction of 96 mutants to PCR products in a day. The following procedures are used for cloning of T-DNA flanking sequences in insertion mutants from a T-DNA insertion library but they should be suitable for any unknown flanking sequences.

Primers used for preparation of adaptor:

Y-shaped adaptor long chain: 5' > cggtaggatcccgcagaacgacggccag < 3'

Y-shaped adaptor short chain: 5' > ctggccgtccaagacgc < 3'. This short chain is for blunt-end DNA. For sticky ends obtained using restriction enzymes (*Bam*HI, for an example), four nucleotides (-gatc- for *Bam*HI) can be added to the 5' of the short

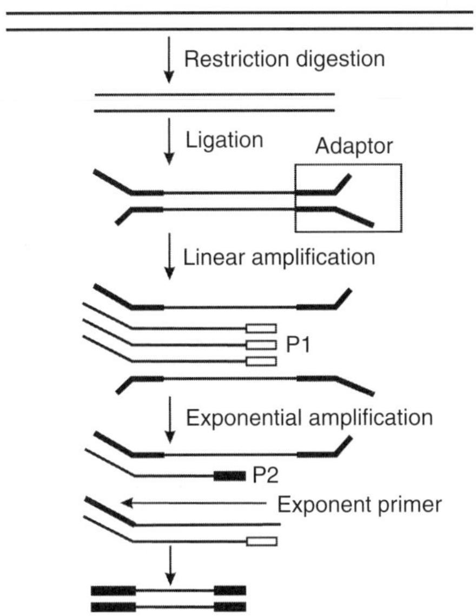


**Fig. 16.3.** Schematic diagram of YADE method.

chain to form 5' > gatcctggccgtccaagacgc < 3' in preparation to using an adaptor for *Bam*HI-digested DNA. It is convenient to choose restriction enzymes with the same sticky end, so a single adaptor can be used for two restriction enzymes. For example, the *Xba*I adaptor works for *Xba*I, *Spe*I and *Nhe*I.

Exponent primer: 5'> cggtaggatcccgcagaa c < 3'. This primer is the same as the first 18 nucleotides in Y-shaped adaptor long chain.

Y-shaped adaptor preparation

| Phosphorylation of short chain | | Annealation | |
|---|---|---|---|
| 10× $T_4$ polynucleotide kinase buffer | 2 μl | 10× annealing buffer* | 4 μl |
| Adaptor short chain (0.68 μg/μl) | 2 μl | Phosphorylated short chain | 20 μl |
| $T_4$ polynucleotide kinase | 2 μl | Adaptor long chain (0.68 μg/μl) | 2 μl |
| ATP (10 mM) | 2 μl | $ddH_2O$ | 14 μl |
| $ddH_20$ | 12 μl | Total volumn | 40 μl |
| Total volume | 20 μl | 1 M NaCl, 0.1 M Tris-Cl (pH 7), 10 mM EDTA (pH 8) | |
| 37°C for 30 min, 72°C for 10 min | | 65°C for 10 min cool to RT in bath | |

Restriction enzyme digestion and ligation to adaptor

| Blunt-ended restriction enzyme digestion | | Ligation of digested DNA + adaptor | |
|---|---|---|---|
| Genomic DNA (from quick isolation method) | 10 μl | 10× ligation buffer | 1 μl |
| 10× enzyme buffer | 2 μl | $T_4$ DNA ligase | 1 μl |
| Restriction enzymes* | 1 μl | Adaptor | 1 μl |
| 10× BSA | 2 μl | Digested DNA | 1 μl |
| $ddH_2O$ | 5 μl | $ddH_2O$ | 6 μl |
| Total volume | 20 μl | Total volume | 10 μl |
| 37°C for 10 h, 72°C for 20 min | | 16°C for 16 h | |

*Based on our experiences, restriction enzymes, *DraI*, *EcoRV*, *ScaI* and *SmaI*, work well for *M. anisopliae* and *B. bassiana*. With these four enzymes, we have cloned more than 30 genes and more than 200 flanking sequences of T-DNA.

Linear PCR amplification

| Linear PCR mixture | | Thermocycle meters | |
|---|---|---|---|
| 10× PCR buffer | 2.5 μl | 94°C | 3 min |
| 25 mM $MgCl_2$ | 1.5 μl | 94°C | 30 s |
| 10 mM dNTPs | 0.2 μl | 60°C | 30 s 40 cycles |
| P1 (5 μM) | 1 μl | 72°C | 2 min 30 s |
| Taq DNA polymerase (5 U/μl) | 0.2 μl | 72°C | 7 min |
| Ligation DNA product | 1 μl | 4°C | Hold |
| $ddH_2O$ | 18.6 μl | Wait until PCR machine gets to 94°C before putting tubes in | |
| Total volume | 25 μl | | |
| Prepare the mixture on ice | | | |

Exponential PCR amplification

| Exponential PCR mixture | | Thermocycle meters | |
|---|---|---|---|
| 10x PCR buffer | 2.5 μl | 94°C | 3 min |
| 25 mM $MgCl_2$ | 1.5 μl | 94°C | 30 s |
| 10 mM dNTPs | 0.2 μl | 60°C | 30 s 35 cycles |
| P2 (5 μM) | 1 μl | 72°C | 2 min 30 s |
| Exponential primer (5 μM) | 1 μl | 72°C | 7 min |
| Taq DNA polymerase (5 U/μl) | 0.2 μl | 4°C | Hold |
| Product from linear PCR amplification | 1 μl | Wait until PCR machine gets to 94°C before putting tubes in | |
| $ddH_2O$ | 16.6 μl | | |
| Total volume | 25 μl | | |
| Prepare the mixture on ice | | | |

### 16.7.6. Transformation as a tool for isolating pathogenicity genes

Many genes of pathological importance cannot be isolated by the foregoing approaches because their protein products are currently unknown, and prospects for finding an abundant mRNA are low. The classic way to isolate such genes is by complementation of a mutant lacking the gene function with a library of DNA fragments prepared from an isolate that has the gene function. The presence of the gene in question is recognized by its ability to cause the mutant to behave like the wild type. This laborious approach has been superseded by insertional mutagenesis so that the mutated gene is tagged by transforming DNA and can subsequently be cloned. The two methods most widely used to tag genes are restriction enzyme-mediated integration (REMI) of transforming plasmid and *Agrobacterium tumefaciens*-mediated transformation (ATMT) (Mullins and Kang, 2001). REMI generates insertions into genomic restriction sites in an apparently random manner, some of which cause mutations (Kahmann and Basse, 1999). This provides a simple method of insertional mutagenesis to tag genes based on their mutant phenotypes. The integrated plasmid (which contains an *E. coli* selectable gene) along with flanking genomic DNA can be excised from some of these mutants (by cutting genomic DNA with a restriction enzyme that does not cut within the plasmid), cloned in *E. coli* cells and the flanking DNA sequenced to identify the disrupted gene. Problems with REMI include multiple and untagged insertion sequences that can be a considerable nuisance when it comes to cloning the gene causing the mutant phenotype. For this and other reasons, ATMT has become the preferred method for generating insertional libraries (Michielse *et al.*, 2005), so REMI will not be considered further. Genome-wide functional screening using *Agrobacterium* insertional mutagenesis is under way in *M. anisopliae* to identify genes involved in particular phenotypes or morphogenetic process, e.g. sporulation, appressoria formation, attenuation of virulence.

However, ATMT is having a broader impact than just generating insertion libraries. Genetic studies of entomopathogenic fungi were traditionally hampered by low transformation frequencies. This has been remedied by adapting a method of ATMT that is relatively straightforward for both *B. bassiana* and *M. anisopliae* (Fang *et al.*, 2004, 2006; Leclerque *et al.*, 2004). In natural *Agrobacterium* infections of plant cells, the bacterium transfers a piece of single-stranded DNA that becomes integrated into the host genome. Termed the T-DNA (transfer DNA), this DNA segment is bound by left and right border sequences each of which is composed of a 25 bp imperfect direct repeat. Conveniently for transformation studies, any DNA, irrespective of its source, engineered to fit between the T-DNA left and right borders can be integrated into the host of choice. Of greatest interest, AMT has been used successfully to transform various fungi including members of the Ascomycetes, Basidiomycetes, Zygomycetes and Oomycetes (Michielse *et al.*, 2005, for review). The ability of *Agrobacterium* to transfer its DNA to fungi belonging to various classes is indicative of the potential of this transformation system for introducing biotechnology to fungi such as *Erynia* spp. and *Lagenedium* spp. that have so far not been transformed. *Agrobacterium* may therefore provide a simple standardized method for transformation of essentially any entomopathogenic species which would obviously be novel and useful.

### *16.7.6.1. Agrobacterium-mediated transformation protocols*

The transformation efficiency is dependent on the *Agrobacterium* strains used. The highest transformation efficiency with either *M. anisopliae* or *B. bassiana* was with *A. tumefaciens* AGL-1. Different transformation efficiencies were observed with various *M. anisopliae* strains, the transformation procedures described below are optimized for the commonly used *M. anisopliae* ARSEF2575. In contrast, no strain preferences were observed with strains of *B. bassiana*. The transformation efficiency was also found to be dependent on selection markers. When the benomyl-resistance gene *BenA3* was used a marker, an average of 17 transformants appeared per 90 mm plate 15 days after selection. Using the herbicide-resistance gene *bar* as a selectable marker, >50 stable transformants could be obtained per plate within 3–9 days. The T-DNA exists as a single copy in 96% of the transformants.

T-DNA inserts both heterologously and homologously into the fungal genome. The homologous insertion rate is about 14% in both *M. anisopliae* and *B. bassiana*, which makes this method reasonably efficient for gene disruption. We developed a master gene disruption vector (Fig. 16.4), where the bar gene cassette can be flanked by the target gene. A GFP gene cassette placed outside of the disruption vehicle provides a marker. When heterologous insertion happens, transformants show herbicide resistance and GFP, while only herbicide resistance exists in transformants with homologous recombination, allowing disruptants to be screened microcopically. In our experience, more than 90% of non-fluorescent transformants are real disruptants.

The high transformation efficiency coupled with mostly single-copy insertion events make this method suitable for construction of large DNA insertion libraries. We constructed a 10,320 transformant T-DNA insertion library of *M. anisopliae* ARSEF2575 in just 2 months using plasimd pPK2-BAR-GFP. All transformants are stable based on GFP fluorescence. So far, 150 conidiation mutants, 30 sectorization mutants and 25 virulence mutants have been obtained. PCR walking based on YADE (Section 16.7.5.1) has obtained sufficient flanking sequence for identification of 56 of the genes involved in conidiation and virulence.

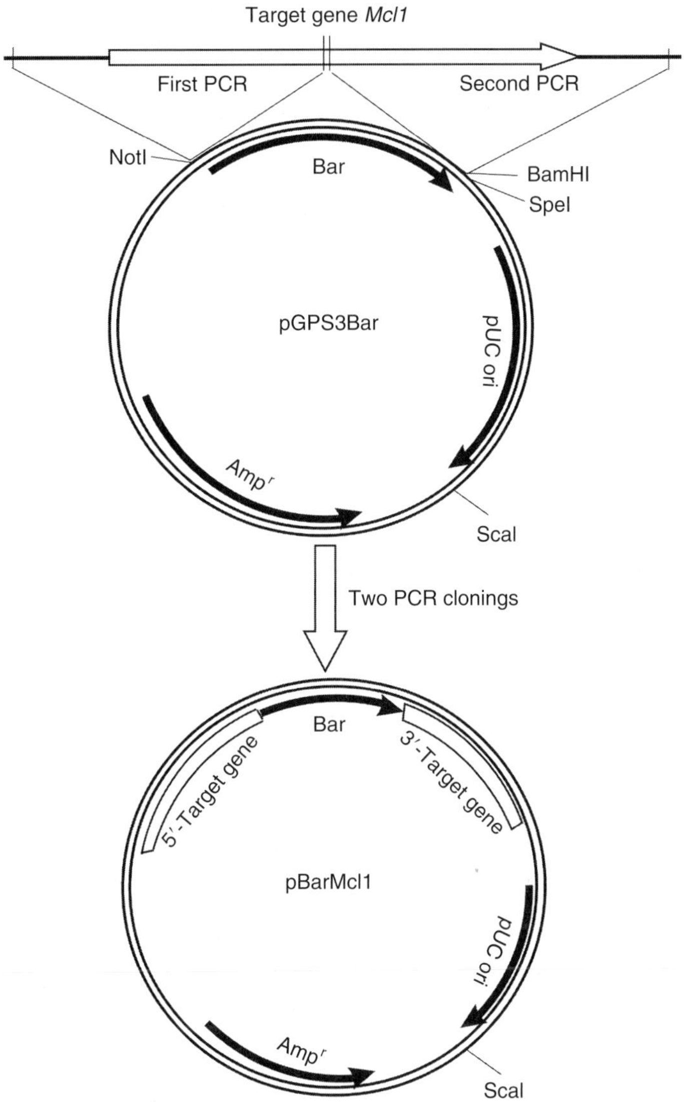


**Fig. 16.4.** Schematic procedures for generating a gene knockout cassette.

The following transformation protocol is optimized for *M. anisopliae* ARSEF2575 using the herbicide-resistance gene *bar* as a selectable marker:

1. A single colony of *A. tumefaciens* AGL-1 containing pPK2-BAR-GFP was transferred to LB broth (10 ml) supplemented with Carbenicillin (50 μg/ml) and Kanamycin (50 μg/ml) and incubated at 250 rpm/29°C until the $OD_{600}$ is 0.6–0.8 (~16 h). The culture was diluted with induction medium (IM; Covert *et al.*, 2001) to an $OD_{600}$ value of 0.15 and incubated at 250 rpm/27°C for 4 h ($OD_{600} \cong$ 0.4–0.8).

**2.** *M. anisopliae* conidia ($10^6$ conidia $ml^{-1}$) were mixed with equal volumes of *A. tumefaciens* cells in IM ($OD_{660nm}$ = 0.4–0.8). The mixture was spread on black filter paper resting on IM agar plates, and incubated at 27°C for 2 days.
**3.** After co-cultivation, the black filter paper was transferred on to M-100 plates containing 300 μg $ml^{-1}$ cefotaxime (to kill *A. tumefaciens* cells) and 200 μg $ml^{-1}$ phosphino thricin (PPT) (to select fungal transformants).
**4.** After 2 days growth, the black filter paper was overlaid with M-100 agar containing 200 μg $ml^{-1}$ PPT. Transformants are visible at approximately 3–4 days.

Stock solution, media and chemicals used in this method are indicated below:

Salt stock solutions

| 2.5× MM salts for IMAS | | M-100 trace element solution | | M-100 salt solution | |
|---|---|---|---|---|---|
| Ingredients | For 1 l | Ingredients | For 500 ml | Ingredients | For 1 l |
| $KH_2PO_4$ | 3.625 g | $H_3BO_3$ | 30 mg | $KH_2PO_4$ | 16 g |
| $K_2HPO_4$ | 5.125 g | $MnCl_2 \cdot 4H_2O$ | 70 mg | $Na_2SO_4$ | 4 g |
| NaCl | 0.375 g | ZnCl2 | 200 mg | KCl | 8 g |
| $MgSO_4$ | 0.610 g | $Na_2MoO_4 \cdot 2H_2O$ | 20 mg | $MgSO_4$ | 0.9757 g |
| $CaCl_2.2H_2O$ | 0.165 g | $FeCl_3$ | 30 mg | $CaCl_2$ | 1 g |
| $FeSO_4.7H_2O$ | 0.0062 g | $CuSO_4 \cdot 5H_2O$ | 200 mg | M-100 | 8 ml |
| $(NH_4)_2SO_4$ | 1.250 g | | | | |

Dissolve each salt once at a time. Do not autoclave. Store at RT. For 2.5× MM salts for IMAS, final solution typically contains a small amount of white precipitate.

IM solution and plates

| IM solution | | IM plates | |
|---|---|---|---|
| Ingredients | For 100 ml | Ingredients | For 100 ml |
| 1× MM salts | 40 ml of 2.5× stock | 1× MM salts | 40 ml of 2.5× stock |
| 10 mM glucose | 0.18 g | 5 mM glucose | 0.09 g |
| 0.5% glycerol | 0.5 ml | 0.5% glycerol | 0.5 ml |
| $dH_2O$ volume | To 94 ml final | $dH_2O$ | To 94 ml final |
| | volume | 1.5% agar | 1.5 g |

Autoclave, cool to 50°C, then add 4 ml of 1M MES (pH 5.3), add 2 ml of 10 mM acetosyringone (AS) just before use.

M-100 agar medium (M-100 plates)

| Ingredients | For 100 ml |
|---|---|
| Glucose | 1 g |
| $KNO_3$ | 0.3 g |
| M-100 salt solution | 6.25 ml |
| $dH_2O$ | To 100 ml final volume |
| 1.5% agar | 1.5 g |

## 16.8. Analysing Differential Gene Expression

Analysis of mRNA and proteins is widely used to compare patterns of gene expression between cells or tissues of different kinds and under different conditions. Classical genetics and conventional gene analysis have been powerful tools for dissecting host–pathogen interactions that are affected by the gain or loss of function of single proteins. Such strategies have been less fruitful for understanding disease processes that are controlled by many genes. So, the analysis of differential gene expression – known as functional genomics – has become one of the most widely used strategies for discovering and understanding the molecular circuitry underlying disease processes. Several of the ingenious techniques available (reviewed by Liang and Pardee, 2003) have been applied to insect pathogens.

### 16.8.1. Subtractive hybridization

Subtractive hybridization (SSH) is a method to identify DNA/RNA present in one sample but not in the others, for example, if it was desired to compare a specific strain of a pathogen with a strain with a broad host range to determine if the specific strain has lost DNA sequence reflecting loss of genes for opportunism. The basic procedure involves digesting the different DNA samples with restriction enzymes and hybridizing them together. This results in hybrid molecules for all regions of the DNA except for regions of gene loss. If a large surplus of the 'tester' DNA from the specific strain is used then all matching (hybridizing) DNA will be subtracted out leaving only the segments of DNA exclusively present in the broad host range strain. Subtractive hybridization can also be used for isolating mRNA exclusively up-regulated in response to a particular environmental stimulus. Messenger RNA from the control culture is used to subtract out mRNA from the induced culture. This will leave behind only the mRNA specific for the environmental stimulus. As RNA molecules are single-stranded and will not hybridize to each other this requires that cDNA be prepared from the control culture and used to hybridize to the mRNA from the induced culture. The most commonly used kit for this procedure is the Clontech PCR-Select Custom cDNA Subtraction. This is also available as a rather expensive service from Clontech. SSH has been largely replaced for most applications by EST analysis (Section 16.9).

### 16.8.2. The use of reverse transcription-differential display-PCR (RT-DD-PCR) technique to identify differentially regulated genes

DD has been developed as a tool to detect and characterize differentially expressed transcripts in eukaryotic cells that have been subjected to different environmental and developmental conditions. We currently use EST and microarray analysis (see below) for this purpose. However, DD is useful for estimating the proportion of genes that are differentially expressed under different circumstances. It can thus be used as an assay to guide library construction for EST analysis. In addition, DD does not have the limitations of other methods such as subtractive and differential hybridization, which require large amounts of RNA, are rather difficult to establish and are usually less reproducible. Subtractive and differential hybridization methods are mainly qualitative and do not detect quantitative changes; DD detects all mRNA species expressed by the cell. Comparing the expressed mRNA patterns from different cells thus makes it possible to detect both qualitative and quantitative changes. The experimental design involves anchored oligo-dT primers which anneal to the beginning of the poly(A) tails of mRNA. These are in conjunction with arbitrarily defined 13-mer oligonucleotides (AP) for subsequent PCR amplification. The products are radioactively or fluorescently labelled during PCR amplifications. The amplified fragments of cDNA are then separated by size on large denaturing polyacrylamide gels. The differentially expressed bands are cut out, amplified and used in Northern blots to confirm their mode of expression. Suitable cDNA fragments are ligated into PCR cloning vector and are either sequenced directly or used to pull out genomic or cDNA clone from a suitable library. Most laboratories use kits from Genhunter (Brookline, MA, USA) kits, when commencing their studies. However, if partial sequences are available specific primers can be ordered and used instead of arbitrary primers. The general protocol for DD is described below.

#### *16.8.2.1. Purification of total RNA and DNase treatment*

The integrity of the RNA preparation is the most important factor in RNA-DD. Although mRNA can be used in DD, total RNA is preferred as the substrate because of its cleaner background signal. Any of the commercial kits will do (Section 16.2) and Genhunter have one of their own (RNApure). DNase treatment of the total RNA is often a prerequisite to DD because any traces of DNA increase the chances of having false positives by this method.

#### *16.8.2.2. Differential display*

Typically, 0.02–2 μg RNA will achieve a clear band pattern with RT (Song *et al.*, 1995). It is best to run duplicates of each RNA sample to minimize errors in the PCR amplification which could lead to spurious results. RNA is incubated with one of the oligo-dT primers, dNTP and RT. The RT reaction is carried out for 1 h at 37°C. The anchored oligo-dT primers consist of 11 or 12 Ts plus one or two additional 3' bases which provide specificity. The use of one-base anchored oligo-dT primers instead of two-base anchored oligo-dT primers reduces the number of RT reactions for each RNA species due to the degeneracy of the primers.

The commonly used cycling parameters are: 94°C for 30 s, 40°C for 2 min, 72°C for 30 s for 40 cycles followed by 72°C for 5 min (RNAimage, GenHunter). Usually $^{33}P$ is used in the amplification but the new RNA spectra kits provide a fluorescent alternative. The amplified PCR products are separated on a 6% denaturing sequencing polyacrylamide gel. The gel is blotted on to a filter paper and dried under vacuum at 80°C for 2 h. The dried gel is marked with photoluminescent markers or radioactive ink and exposed to an X-ray film overnight. False positives can sometimes count for a significant percentage of the total number of bands observed. One solution to this is to run duplicates, or triplicates if possible, and to repeat the experiment for the lanes in which potential candidate cDNA bands are observed. Inosine will pair with similar binding strength to all four bases.

Consequently, the use of inosine at ambiguous positions during the synthesis of primers will increase the rate of reproducibility since many of the false positives are caused by imperfect annealing of primers to sequences within mRNA (Rohrwild *et al.*, 1995).

### *16.8.2.3. Recovery, reamplification and cloning of cDNA fragments*

After developing the X-ray film, cDNA bands of interest are marked and cut out. Each gel slice is soaked in water and boiled to facilitate diffusion of the cDNA from the polyacrylamide gel. The cDNA is purified by precipitating with 3 M sodium acetate and 100% ethanol using glycogen as a carrier. Reamplification is done with the same primer set and PCR conditions as the first PCR reaction except that no radioisotope is added. Part of the reamplified cDNA is run on a 1.5% agarose gel and the rest is stored at −20°C for further use. The gel bands are cut out and used as probes on Northern blots to confirm the differential expression of the transcripts of interest. Smearing or multiple bands often necessitate gel purification which may result in decreased ligation efficiency. For gel purification, all nuclease contamination must be removed. Spin columns for gel purification should be avoided since they can result in loss of ligation efficiency. Electroelution and agarase are satisfactory. The reamplification of cDNA fragments sometimes fails or a smear of bands is produced due to the arbitrary primer not being totally complementary to the cDNA of interest. This situation can be dealt with by altering the annealing and elongation temperatures during reamplification; thus, reducing the annealing temperature has helped in some reamplification reactions. Reamplified cDNA probes are cloned into various vectors but the most commonly used one is the pCRII (TA cloning kit, Invitrogen) vector. Taq polymerase has a non-template-dependent activity which adds a single deoxyadenosine (A) to the 3' ends of PCR products. The linearized TA vectors (such as pCRII) have a single 3' deoxythymidine (T) residue. This allows PCR inserts to ligate efficiently with the vector. It is best to use fresh PCR inserts since the inserts gradually loose the 3' A-overhangs resulting in reduced ligation efficiency. Competent *E. coli* cells are transformed and recombinant colonies are selected. To determine the sequence and orientation of the insert, selected colonies are grown overnight and the plasmid is isolated and sequenced using an automated sequencer.

Since the method of DD was first described, there have been several reports of modifications in the basic protocol with each having certain advantages. Some of these are mentioned here:

1. Specific primers may be used in place of arbitrary primers in the PCR step. We have used the primers from the most conserved regions of subtilisin-like enzymes and certain protein kinases to clone and analyse all the possible genes with sequence homologies.
2. DD has been used to detect transcripts from very few cells using the whole cell lysate, without RNA being purified. The cells are suspended in phosphate buffer saline in the presence of transfer RNA (tRNA) which acts as a substrate for RNase activity limiting the degradation of transcripts of interest. The cells are lysed, incubated with proteinase K and heated at 95°C. The supernatant is used for RT reactions (O'Brien *et al.*, 1994).

3. Various commercial rivals to DD are available, but all are based on similar technologies. Of particular interest is the 'GeneSnareTM differential expression kit', sold by Sigma but using Seegene's gene-fishing technology. Sigma reports that this kit is less complicated than DD and provides substantial improvements in terms of reproducibility and the elimination of false positives.

## 16.9. EST Screening

In the early 1990s it was realized that sequencing expressed genes (cDNA), instead of whole genomes, is a sensible initial approach to gene discovery. The usual strategy is to do a single run of sequencing at the 5'or 3'ends of randomly picked cDNA clones from a cDNA library, generating a comprehensive collection of such ESTs. EST sequencing not only results in the discovery of many novel genes but also provides information on the relative abundance in expression of each gene based on the number of times a corresponding cDNA sequence was represented in a cDNA library from differentiating cells or cells exposed to different environments.

A broad sampling of ESTs from cDNA libraries thus provides the best backdrop against which to test the various populations of differentially expressed genes. SSH, for example, is comparatively less powerful than EST analysis for developing resources useful for functional genomics studies. In part, this is because, while it is theoretically possible to retain representation of quantitatively different but common sequences in two samples, in practice SSH eliminates most mutually expressed genes. Thus, SSH being sequence-dependent, not prevalence-dependent, misses many truly differentially expressed genes. SSH is sometimes justified as providing a targeted approach as opposed to a general approach, thus greatly reducing the overall effort and funding required. In terms of effort and funding this is no longer convincing. Getting good subtraction libraries is time-consuming and technically difficult. In contrast, automated sequencing has become fast, easy and cheap. It can therefore make more sense to put funds into a much more extensive random-sequencing project employing multiple cDNA libraries. Microarrays would then be the method of choice to identify differential gene expression between biotypes. A further advantage of a random approach is that there is currently little sequence data for most insect pathogens and the arrays will contain a larger proportion of the transcriptome increasing the possibility of building a broadly based genomic resource that would be of greater use to the entire community. Where SSH and DD may have an advantage is in obtaining rare or low copy-number gene sequences related to insect virulence. EST analysis is the method of choice for discovering common and moderately expressed genes but it can miss others.

## 16.10. Microarray Analysis

Using traditional methods to assay gene expression, e.g. Northern blot or RT-PCR, researchers are only able to survey a relatively small number of genes at a time. Microarrays allow scientists simultaneously to analyse expression patterns of thousands of genes providing a global overview of cellular responses to environmental changes. Thus, the application of *M. anisopliae* to host tissues *in vitro* and subsequent analysis with microarrays showed that sets of functionally related genes are coordinately induced or repressed by strains ARSEF 2575 and ARSEF 324 in response to host-related stimuli. To date, we have identified more than 700 genes up-regulated by ARSEF 2575 during adaptation to host cuticle or haemolymph and these provide great insight into the very intricate mechanisms by which *M. anisopliae* has adapted to survive in these environments (Freimoser *et al.*, 2005; Wang and St Leger, 2005; Wang *et al.*, 2005a,b).

If similar pathogenicity genes are hypothesized to be involved in related species then obtaining large amounts of sequence data from them both may be redundant unless a rationale is made in terms of the synergistic effects of studying two systems and the ability to investigate the molecular evolution of the two species since they diverged. Otherwise a high degree of relationship between pathogen species may allow microarrays developed from the sequences of one species to be used for the other species to test for commonalities and differences in gene expression. In the absence of DNA sequence data much can be achieved by cross-species hybridizations to gene arrays (Cossins and Crawford, 2005). Heterologous hybridization was used by us to compare *Metarhizium* strains obtaining information on physiological processes in poorly characterized strains that do not usually occur in ARSEF2575 (Wang and St Leger, 2005). It has also provided a fast and powerful tool facilitating the merging of functional genomics with physiology, ecology and evolution (Renn *et al.*, 2004) in species of yeasts (Daran-Lapujade *et al.*, 2003; Moran *et al.*, 2004), fish (Cossins and Crawford, 2005), mammals (Ji *et al.*, 2004) and plants (Alba *et al.*, 2004).

The principal disadvantage of microarray analysis is that the procedures are very expensive and there are no commercial chips available for insect-pathogenic fungi. In most of our studies, we have used subsets of sequences suitable for specific research applications, e.g. genes expressed in response to cuticle as this is obviously cheaper. There are two ways to create your own arrays and they employ cDNAs and oligonucleotides. We used cDNA rather than oligonucleotides as their longer sequences facilitate heterologous gene expression studies. We obtained cDNAs by PCR, amplifying individual genes from EST plasmids using universal primers. Each product is spotted in triplicate on poly-lysine-coated glass slides. Printing, hybridization and scanning were performed with an Affymetrix 417 Arrayer and 418 Scanner by an experienced team of technicians. For oligo arrays, there are companies providing services. The prerequisite is to provide sequences of the different genes you want arrayed. For a 'mini' array of less than a hundred genes, sufficient perhaps to cover a particular functional category such as toxin-producing genes, the companies will design specific primers for each gene (50–70 bp long) and then use the purified products for slide printing.

There are commercial software packages available (http://genome-www5.stanford.edu/resources/restech.shtml) for data analysis. However, the TM4 system developed by TIGR (http://www.tm4.org/) is a free, well-designed and user-friendly open-source software. It includes packages from Spotfinder, Data Analysis, Multiexperiments Viewer and Data Manager. Most journals require the deposit of standardized microarray data in public databases before publication. This can be generated by Data Manager.

## 16.11. Targeted Gene Mutagenesis

A typical research objective of a pathologist is to determine what processes are essential for the target fungus to cause disease. The random mutagenesis procedures described above can accomplish this task by detecting previously unsuspected stratagems of pathogenicity. EST and microarray approaches also allow experimenters to look 'under the boot' of infection processes and let the pathogen inform on what it is doing during infection processes (St Leger, 2007). These techniques allow side effects occurring in constructed strains to be accessed and a greater range of engineering possibilities to be exploited from knowledge of the interrelated regulatory and metabolic processes going on in cells. Although they dramatically accelerate the gain of information, the bottleneck is in translating gene expression profiling on to hypothesis-driven approaches which result in clearly proven function of single genes and new ideas for genetically enhancing virulence. This requires a directed approach to specifically disrupt a predetermined gene that is conceived to be important for pathogenicity. If a gene is altered or eliminated, then pathogenicity will be reduced to an extent dependent on the importance of the gene in pathogenicity. Termed gene replacement, the method comprises the direct substitution of the wild-type gene with its mutant allele, which has been disrupted by insertion of a selectable marker (e.g. antibiotic resistance) within its coding region via homologous recombination mediated by transformation. Construction of deletion strains for highly expressed genes has already led to the identification of an immune evasion protein (Wang and St Leger, 2006), separate adhesins essential for binding to insect cuticle and plant surfaces (Wang and St Leger, 2007b) and a perilipin that regulates lipolysis, osmotic pressure and formation of infection structures (Wang and St Leger, 2007c). These genes all have as yet un-realized potential to engineer changes in host range; either increasing it or diminishing it, and illustrate the power of expression profiling for revealing previously unsuspected stratagems of infection.

Targeted mutagenesis in filamentous fungi usually requires gene disruption cassettes with large homologous DNA-flanking regions (>1000 bp). Such constructs give a homologous recombination frequency of 0.5–30.0%. Several methods have been described to obtain large homologous DNA flanks in gene-disruption cassettes and these are reviewed by Wendland (2003).

The most challenging step for gene disruption is generation of the knockout cassette by flanking a selective marker gene with the targeted gene sequence. Overlap PCR is an efficient way to generate the construct; however, for long-length amplification it is easy to introduce erroneous nucleotide matches. The GPS-Mutagenesis system (New England Biolabs) has been successfully used for generating disruption constructs for fungi (e.g. Zwiers and de Waard, 2001). The kit enables *in vitro* insertion of a transposable element (the marker gene in a donor vector) into target DNA (plasmid) by a transposase. The EZ::TN Transposome kit (Epicentre) works in a similar fashion. A drawback of these kits is that the insertions occur randomly entailing laborious screening to identify a plasmid with an ideal insertion. We chose a more targeted approach with a PCR cloning method to generate the knockout construct. The example we give is for the *Mcl1* gene (Wang and St Leger, 2006). The phosphinothricin acetyltransferase (*bar*) gene was amplified from the plasmid pBARGPE1 (Fungal Genetics Stock Center) inserted into the plasmid pGPS3 (New England Biolabs) and the product labelled as pGPS3Bar. The 5' (*c*.2.5 kb) and 3' (*c*.2 kb) ends of the target gene *Mcl1* were amplified separately and the two PCR products were inserted into pGPS3Bar using two steps cloning by flanking the *bar* gene to generate the disrupting construct pBarMcl1 (Fig. 16.2). The construct pBarMcl1 was linearized before transformation of protoplasts (Wang and St Leger, 2006).

## 16.12. RNA Interference

Post-transcriptional gene silencing using RNA interference (RNAi) has become one of the most important techniques in current life sciences (Cogoni and Macino, 2000; Hammond *et al.*, 2001, for reviews). Comprehensive kits have been developed by various companies; however, almost all of these are for mammalian systems or for model organisms such as the fruit fly *Drosophila melanogaster* and the nematode *Caenorhabditis elegans*. The technique is especially powerful for functional studies of lethal genes as it usually reduces rather than eliminates expression.

RNAi seems to work equally well in fungi. The phenomena of quelling and meiotic silencing have been found in *N. crassa*. Orthologues of RNAi components, e.g. dicer, RNA-dependent RNA polymerase and RISC (RNA-induced silencing complex) are found in most, if not all, fungal genomes (Nakayashiki, 2005, for review). Transformation with inverted repeat transgenes containing sequences of mycotoxin-specific regulatory genes suppressed mycotoxin production by *Aspergillus* spp. and *Fusarium* spp. (McDonald *et al.*, 2005). We generated a construct by inversely inserting a partial sequence (*c*.500 bp) of a high osmolarity sensor gene of *Metarhizium* into the *Bam*HI site of pBARGPE1 (Fungal Genetics Stock Center). The insert is constitutively controlled by the *gpdA* promoter. The transformants demonstrated greater than threefold reduction of RNA intensity as well as the decrease of virulence (Wang *et al.*, 2008).

## 16.13. Conclusions

Fungal biology is entering the era of functional genomics as more sequence data for entire fungal genomes become available. Currently about 200 fungal genomes have been sequenced (http://www.genomesonline.org). To date no insect pathogen has been sequenced but this situation is likely to change in the near future, considering the community wide efforts to promote the genome sequencing and large-scale EST analysis of two major insect-pathogenic fungi. Information and technology resources derived from these sequencing efforts will undoubtedly transform the way we study these pathogens. In the meantime good advantage can be taken of large-scale EST collections for *M. anisopliae* and *B. bassiana*, the ascomycete fungal genome sequences already available and the very extensive work by many groups on deciphering plant pathogen genes involved in development and overcoming host defences. This work complements the studies of insect pathogens. Moreover, by using plant pathogen research to inform their own work, with its similar goals of understanding pathogenicity, researchers can potentially increase progress in studies of those insect pathogens where very little research has focused. Microarray-based assays will be further exploited to monitor genome-wide changes in gene expression in both pathogens and their hosts during the various stages of infection. Since cellular activities are controlled not only at the transcriptional but also at the translational and post-translational levels, it will also be necessary to monitor the pattern of proteins synthesized under a given condition. Most of the necessary tools, including protein profiling using 2D gel analysis and/or mass spectrometry, have already been adapted for filamentous fungi. The judicious application of these techniques will identify a subset of genes from pathogenic fungi which might play a role in pathogenesis and would provide a resource of genes for engineering improved pathogenicity in other pathogens. To systematically characterize the function of these genes, the development of efficient gene knockout and modification tools is critical. To date, ATMT seems to have the most potential as a tool for both targeted and random mutagenesis in fungi. It has also opened the field of molecular genetics to entomopathogenic zygomycete and oomycete fungi that were difficult to transform because of lack of suitable vectors, or poor growth in culture.

## References

Al-Aidroos, K. (1980) Demonstration of a parasexual cycle in the entomopathogenic fungus *Metarhizium anisopliae. Canadian Journal of Genetics and Cytology* 22, 309–314.

Alba, R., Fei, Z., Payton, P., Liu, Y., Moore, S.L., Debbie, P., Cohn, J., D'Ascenzo, M., Gordon, J.S., Rose, J.K., Martin, G., Tanksley, S.D., Bouzayen, M., Jahn, M.M. and Giovannoni, J. (2004) ESTs, cDNA microarrays, and gene expression profiling: tools for dissecting plant physiology and development. *Plant Journal* 39, 697–714.

Ausubel, E.M., Brent, R., Kingston, R.E., Moore, D.D., Seidman, J.G., Smith, J.A. and Struhl, K. (eds) (2002) *Current Protocols in Molecular Biology*. Wiley Interscience, New York.

Avalos, J., Geever, R.F. and Case, M.E. (1989) Bialaphos resistance as a dominant selectable marker in *Neurospora crassa. Current Genetics* 16, 369–372.

Benito, E.P., Diaz-Minguez, J.M., Iturriaga, E.A., Campuzano, V. and Eslava, A.P. (1992) Cloning and sequence analysis of the *Mucor circinelloides* pyr G gene encoding orotidine-5'-monophosphate decarboxylase: use of pyr G for homologous transformation. *Gene* 116, 59–67.

Bernier, L., Cooper, R.M., Charnley, A.K. and Clarkson, J.M. (1988) Transformation of the entomopathogenic fungus *Meturhizium anisopliae* to benomyl resistance. *FEMS Microbiology Letters* 60, 261–266.

Bidochka, M.J., St Leger, R.J. and Roberts, D.W. (1994) Differentiation of species and strains of entomopathogenic fungi by random amplification of polymorphic DNA (RAPD). *Current Genetics* 25, 107–113.

Bidochka, M.J., Walsh, S.R.A., Ramos, M.E., St Leger, R.J., Silver, J.C. and Roberts, D.W. (1995) Pathotypes in the *Entomophaga grylli* species complex of grasshoppers pathogens differentiated by molecular methods. *Applied and Environmental Microbiology* 61, 556–560.

Bustamante, C., Gumeri, S. and Smith, S.B. (1993) Towards molecular description of pulse-field gel electroohoresis. *Trends in Biotechnology* 11, 23–30.

Cho, E.-M., Boucias, D. and Keyhani, N.O. (2006) EST analysis of cDNA libraries from the entomopathogenic fungus *Beauveria (Cordyceps) bassiana* II: fungal cells sporulating on chitin and producing oosporein. *Microbiology* 152, 2855–2864.

Cogoni, C. and Macino, G. (2000) Post-transcriptional gene silencing across kingdoms. *Genes and Development* 10, 638–643.

Cossins, A.R. and Crawford, D.L. (2005) Fish as models for environmental genomics. *Nature Reviews Genetics* 6, 324–333.

Covert, S.F. and Cullen, D.J. (1992) Heterologous protein expression in filamentous fungi. In: Leatham, G.E. (ed.) *Frontiers in Industrial Mycology*. Chapman and Hall, New York, pp. 66–77.

Covert, S.F., Kapoor, P., Lee, M., Briley, A. and Nairn, C.J. (2001) *Agrobacterium tumefaciens*-mediated transformation of *Fusarium circinatum*. *Mycological Research* 105, 259–264.

Daran-Lapujade, P., Daran, J.M., Kotter, P., Petit, T., Piper, M.D. and Pronk, J.T. (2003) Comparative genotyping of the *Saccharomyces cerevisiae* laboratory strains S288C and CEN.PK113-7D using oligonucleotide microarrays. *FEMS Yeast Research* 4, 259–269.

Davies, R.W. (1992) Genetic engineering of filamentous fungi for heterologous gene expression and protein secretion. In: Murray, J.A.H. (ed.) *Transgenesis, Applications of Gene Transfer*. Wiley, Chichester, UK, pp. 81–104.

Fan, Y., Fang, W., Guo, S., Pei, X., Zhang, Y., Xiao, Y., Li, D., Jin, K., Bidochka, M.J. and Pei, Y. (2007) Increased insect virulence in *Beauveria bassiana* strains overexpressing an engineered chitinase. *Applied and Environmental Microbiology* 73, 295–302.

Fang, W., Zhang, Y., Yang, X., Zheng, X., Duan, H., Li, Y. and Pei, Y. (2004) *Agrobacterium tumefaciens*-mediated transformation of *Beauveria bassiana* using an herbicide resistance gene as a selection marker. *Journal of Invertebrate Pathology* 85, 18–24.

Fang, W., Pei, Y. and Bidochka, M.J. (2006) Transformation of *Metarhizium anisopliae* mediated by *Agrobacterium tumefaciens*. *Canadian Journal of Microbiology* 52, 623–626.

Freimoser, F.M., Screen, S., Bagga, S., Hu, G. and St Leger, R.J. (2003) Expressed sequence tag (EST) analysis of two subspecies of *Metarhizium anisopliae* reveals a plethora of secreted proteins with potential activity in insect hosts. *Microbiology* 149, 239–247.

Freimoser, F.M., Hu, G. and St Leger, R.J. (2005) Variation in gene expression patterns as the insect pathogen *Metarhizium anisopliae* adapts to different host cuticles or nutrient deprivation *in vitro*. *Microbiology* 151, 361–371.

Goettel, M.S., St Leger, R.J., Bhairi, S., Jung, M.K., Oakley, B.R., Roberts, D.W. and Staples, R.C. (1990) Pathogenicity and growth of *Metarhizium anisopliae* stably transformed to benomyl resistance. *Current Genetics* 17, 129–132.

Gressel, J., Meir, S., Herschkovitz, Y., Al-Ahmad, H., Greenspoon, I., Babalola, O. and Amsellem, Z. (2007) Approaches to and successes in developing transgenically enhanced mycoherbicides. In: Vurro, M. and Gressel, J. (eds) *Novel Biotechnologies for Biocontrol Agent Enhancement and Management*. Springer, Dordrecht, The Netherlands, pp. 277–296.

Hammond, S.M., Caudy, A.A. and Hannon, G.J. (2001) Post-transcriptional gene silencing by double-stranded RNA. *Nature Reviews Genetics* 2, 110–119.

Hu, G. and St Leger, R.J. (2002) Field studies using a recombinant mycoinsecticide (*Metarhizium anisopliae*) reveal that it is rhizosphere competent. *Applied and Environmental Microbiology* 68, 6383–6387.

Ji, W., Zhou, W., Gregg, K., Yu, N., Davis, S. and Davis, S. (2004) A method for cross-species gene expression analysis with high-density oligonucleotide arrays. *Nucleic Acids Research* 32, e93.

Joshi, L., St Leger, R.J. and Bidochka, M.J. (1995) Cloning of a cuticle-degrading protease from the entomopathogenic fungus, *Beauveria bassiana*. *FEMS Microbiology Letters* 125, 211–218.

Kahmann, R. and Basse, C. (1999) REMI (Restriction Enzyme Mediated Integration) and its impact on the isolation of pathogenicity genes in fungi attacking plants. *European Journal of Plant Pathology* 105, 221–229.

Langewald, J. and Kooyman, C. (2007) Green muscle, a fungal biopesticide for control of grasshoppers and locusts in Africa. In: Vincent, C., Goettel, M.S. and Lazarovitis, G. (eds) *Biological Control, a Global Perspective*. CAB International, Wallingford, UK.

Leclerque, A., Wan, H., Abschutz, A., Chen, S., Mitina, G.V., Zimmermann, G. and Schairer, H.U. (2004) *Agrobacterium*-mediated insertional mutagenesis (AIM) of the entomopathogenic fungus *Beauveria bassiana*. *Current Genetics* 45, 111–119.

Liang, P. and Pardee, A.B. (2003) Analysing differential gene expression in cancer. *Nature Reviews Cancer* 3, 869–876.

May, G. (1992) Fungal technology. In: Kinghom, J.R. and Turner, G. (eds) *Applied Molecular Genetics of Filamentous Fungi*. Blackie, Glasgow, UK, pp. 1–27.

McDonald, T., Brownm, D., Keller, N.P. and Hammond, T.M. (2005) RNA silencing of mycotoxin production in *Aspergillus* and *Fusarium* species. *Molecular Plant–Microbe Interaction* 18, 539–545.

Michielse, C.B., Hooykaas, P.J., van den Hondel, C.A. and Ram, A.F. (2005) *Agrobacterium*-mediated transformation as a tool for functional genomics in fungi. *Current Genetics* 48, 1–16.

Moran, G., Stokes, C., Thewes, S., Hube, B., Coleman, D.C. and Sullivan, D. (2004) Comparative genomics using *Candida albicans* DNA microarrays reveals absence and divergence of virulence-associated genes in *Candida dubliniensis*. *Microbiology* 150, 3363–3382.

Mullins, E.D. and Kang, S. (2001) Transformation: a tool for studying fungal pathogens of plants. *Cellular and Molecular Life Sciences* 58, 2043–2052.

Nakayashiki, H. (2005) RNA silencing in fungi: mechanisms and applications. *FEBS Letters* 579, 5950–5957.

Nakazato, L., Dutra, V., Broetto, L., Staats, C.C., Vainstein, M.H. and Schrank, A. (2006) Development of an expression vector for Metarhizium anisopliae based on the tef-1alpha homologous promoter. *Applied Microbiology and Biotechnology* 73, 521–528.

O'Brien, D.P., Billadeau, D. and Van Ness, B. (1994) RT-PCR assay for detection of transcripts from very few cells using whole cell lysates. *Biotechniques* 16, 586–589.

Ochman, H., Medhora, M.M., Garza, D. and Hartl, D.C. (1990) Amplification of flanking sequences by inverse PCR. In: Innis, M.A., Gelfrand, D.H., Sironsky, J.J. and White, T.K. (eds) *PCR Protocols: a Guide to Methods and Applications*. Academic Press, New York, pp. 219–227.

Rohrwild, M., Alpha, R.S., Liang, P. and Pardee, A.B. (1995) Inosine – containing primers for mRNA differential display. *Trends in Genetics* 11, p. 300.

Renn, S.C., Aubin-Horth, N. and Hofmann, H.A. (2004) Biologically meaningful expression profiling across species using heterologous hybridization to a cDNA microarray. *BMC Genomics* 5, 42.

Rowland, L.J. and Nguyen, B. (1993) Use of polyethylene glycol for purification of DNA from leaf tissue of woody plants. *Biotechniques* 14, 734–736.

Sambrook, J., Fritsch, E.F. and Maniatis, T. (2001) *Molecular Cloning. A Laboratory Manual*, 3rd edn. Cold Spring Harbor Laboratory Press, Cold Spring Harbour, New York, pp. 1–1940.

Scholte, E.J., Knols, B.G., Samson, R.A. and Takken, W. (2004) Entomopathogenic fungi for mosquito control: a review. *Journal of Insect Science* 4, 1–24.

Scholte, E.J., Ng'habi, K., Kihonda, J., Tokken, W., Paaijmans, K., Abdulla, S., Killeen, G.F. and Knols, B.G. (2005) An entomopathogenic fungus for control of adult African malaria mosquitoes. *Science* 308, 1641–1642.

Screen, S.E. and St Leger, R.J. (2000) Cloning, expression, and substrate specificity of a fungal chymotrypsin. Evidence for lateral gene transfer from an actinomycete bacterium. *Journal of Biological Chemistry* 275, 6689–6694.

Screen, S.E., Hu, G. and St Leger, R.J. (2001) Transformants of *Metarhizium anisopliae* sf. *anisopliae* overexpressing chitinase from *Metarhizium anisopliae* sf. acridum show early induction of native chitinase but are not altered in pathogenicity to *Manduca sexta*. *Journal of Invertebrate Pathology* 78, 260–266.

Shimizu, S., Yoshioka, H. and Matsumoto, T. (1993) Electrophoretic karyotyping of the entomogenous fungus *Paecilomyces fumoroseus*. *Letters in Applied Microbiology* 16, 183–186.

Song, P., Yamamoto, E. and Allen, R.D. (1995) Improved procedure for differential display of transcripts from cotton tissues. *Plant Molecular Biology Reporter* 13, 174–181.

St Leger, R.J. (1993) Biology and mechanisms of invasion of deuteromycete fungal pathogens. In: Beckage, N.C., Thompson, S.N. and Federici, B.A. (eds) *Parasites and Pathogens of Insects*, Vol. 2. Academic Press, San Diego, California, pp. 211–229.

St Leger, R.J. (1995) The role of cuticle-degrading proteases in fungal pathogens of insects. *Canadian Journal of Botany* 73(Suppl. 1), s1119–s1125.

St Leger, R.J. (2001) Development and testing of genetically improved mycoinsecticides. In: Gressel, J., Butt, T., Harman, G., Pilgeram, A., St Leger, R. and Nuss, D. (eds) *Enhancing Biocontrol Agents and Handling Risks. Proceedings of a NATO Advanced Research Workshop, 9–15 June, 2001*. IOS Press, Amsterdam, The Netherlands, pp. 229–239.

St Leger, R.J. (2007) *Metarhizium anisopliae* as a model for studying bioinsecticidal host pathogen interactions. In: Vurro, M. and Gressel, J. (eds) *Novel Biotechnologies for Biocontrol Agent Enhancement and Management*. Springer, Dordrecht, The Netherlands, pp. 179–204.

St Leger, R.J. and Screen, S. (2001) Prospects for strain improvement of fungal pathogens of insects and weeds. In: Butt, T.M., Jackson, C. and Morgan, N. (eds) *Fungal Biocontrol Agents: Progress, Problems And Potential*. CAB International, Wallingford, UK, pp. 219–238.

St Leger, R.J., Roberts, D.W. and Staples, R.C. (1992) Molecular cloning and regulatory analysis of the cuticle-degrading protease structural gene from the entomopathogenic fungus Metarhizium anisopliae. *European Journal of Biochemistry* 204, 991–001.

St Leger, R.J., Shimizu, S., Joshi, L., Bidochka, M.J. and Roberts, D.W. (1995) Co-transformation of *Metarhizium anisopliae* by electroporation or using the gene gun to produce stable GUS transformants. *FEMS Microbiology Letters* 131, 289–294.

St Leger, R.J., Joshi, L., Bidochka, M.J. and Roberts, D.W. (1996a) Construction of an improved mycoinsecticide over-expressing a toxic protease. *Proceedings of the National Academy of Sciences of the USA* 93, 6349–6354.

St Leger, R.J., Joshi, L., Bidochka, M.J., Rizzo, N.W. and Roberts, D.W. (1996b) Characterization and ultrastructural localization of chitinases from *Metarhizium misopliae, M. flavoviride*, and *Beauveria bassiana* during fungal invasion of host (*Manduca sexta*) cuticle. *Applied and Environmental Microbiology* 62, 907–912.

Stowell, L.J. (1994) Factors influencing acceptance and development of biopesticides. In: Kim, L. (ed.) *Advances in Engineered Pesticides*. Marcel Dekker, New York, pp. 249–260.

Thomas, M.B. and Read, A.F. (2007) Can fungal biopesticides control malaria? *Nature Review Microbiology* 5, 377–383.

Tilburn, J., Scazzocchio, C., Taylor, G.G., Zabicky-Zissman, J.H., Lockington, R.A. and Davies, R.W. (1983) Transformation by integration in *Aspergillus nidulans*. *Gene* 26, 205–221.

Unkles, S.E., Campbell, E.I., De Ruiter-Jacobs, Y.M.J.T., Broekhuijsen, M., Macro, J.A., Carrez, D., Contreras, R., Van den Hondel, C.A.M.J.J. and Kinghorn, J.R. (1989) The development of a homologous transformation system for *Aspergillus oryzae* based on the nitrate assimilation pathway: a convenient and general selection system for filamentous fungal transformation. *Molecular Genetics and Genomics* 218, 99–104.

Valadares-Inglis, M.C. and Peberdy, J.F. (1998) Variation in the electrophoretic karyotype of Brazilian strains of *Metarhizium anisopliae*. *Genetics and Molecular Biology* 21, 10–13.

Wang, C. and St Leger, R.J. (2005) Developmental and transcriptional responses to host and nonhost cuticles by the specific locust pathogen *Metarhizium anisopliae* var. *acridum*. *Eukaryotic Cell* 4, 937–947.

Wang, C. and St Leger, R.J. (2006) A collagenous protective coat enables *Metarhizium anisopliae* to evade insect immune responses. *Proceedings of the National Academy of Sciences of the USA* 103, 6647–6652.

Wang, C. and St Leger, R.J. (2007a) A scorpion neurotoxin increases the potency of a fungal insecticide. *Nature Biotechnology* 25, 1455–1456.

Wang, C. and St Leger, R.J. (2007b) The MAD1 adhesin of *Metarhizium anisopliae* links adhesion with blastospore production and virulence to insects: the MAD2 adhesion enables attachment to plants. *Eukaryotic Cell* 6, 808–816.

Wang, C. and St Leger, R.J. (2007c) *Metarhizium anisopliae* perilipin homolog MPL1 regulates lipid metabolism, appressorial turgor pressure and virulence. *Journal of Biological Chemistry* 282, 21110–21115.

Wang, C., Duan, Z. and St Leger, R.J. (2008) MOS1 Osmosensor of *Metarhizium anisopliae* is required for adaptation to insect host hemolymph. *Eukaryotic Cell* 7, 302–309. doi:10.1128/EC.00310-07.

Wang, C., Skrobek, A. and Butt, T.M. (2003) Concurrence of losing a chromosome and the ability to produce destruxins in a mutant of *Metarhizium anisopliae. FEMS Microbiology Letters* 226, 373–378.

Wang, C., Hu, G. and St Leger, R.J. (2005a) Differential gene expression by *Metarhizium anisopliae* growing in root exudate and host (*Manduca sexta*) cuticle or hemolymph reveals mechanisms of physiological adaptation. *Fungal Genetics and Biology* 42, 704–718.

Wang, C., Butt, T.M. and St Leger, R.J. (2005b) Colony sectorization of *Metarhizium anisopliae* is a sign of ageing. *Microbiology* 151, 3223–3236.

Wendland, J. (2003) PCR-based methods facilitate targeted gene manipulations and cloning procedures. *Current Genetics* 44, 115–123.

Wong, M.L. and Medrano, J.F. (2005) Real-time PCR for mRNA quantitation. *Biotechniques* 39, 75–85.

Wraight, S.P., Jackson, M.A. and Kock, S.L. de (2001) Production, stabilization and formulation of fungal biocontrol agents. In: Butt, T.M., Jackson, C. and Magan, N. (eds) *Fungi as Biocontrol Agents: Progress, Problems and Potential*. CAB International, Wallingford, UK, pp. 253–287.

Xiao, Y.H., Luo, M., Fang, W.G., Luo, K.M., Hou, L., Li, D.M. and Pei, Y. (2002) PCR walking in cotton genome using YADE method. *Acta Genetica Sinica* 29, 62–66.

Ying, S.H. and Feng, M.G. (2006) Novel blastospore-based transformation system for integration of phosphinothricin resistance and green fluorescence protein genes into *Beauveria bassiana. Applied Microbiology and Biotechnology* 72, 206–210.

Zwiers, L.H. and de Waard, M.A. (2001) Efficient *Agrobacterium tumefaciens*-mediated gene disruption in the phytopathogen *Mycosphaerella graminicola. Current Genetics* 39, 388–393.

# Index

Note: Page numbers in *italics* refer to figures and tables